3 1386 00002 8098

D0080856

INTERNATIONAL ENCYCLOPEDIA OF PHARMACOLOGY AND THERAPEUTICS

Sponsored by the International Union of Pharmacology (IUPHAR)
(Chairman: B. Uvnäs, Stockholm)

Executive Editor: G. Peters, Lausanne

Section 7

BIOSTATISTICS IN PHARMACOLOGY

Section Editor

A. L. DELAUNOIS

J. F. en C. Heymans Instituut
voor Farmakodynamie en Terapie
Rijksuniversiteit
Gent

Volume 1

INTERNATIONAL ENCYCLOPEDIA OF
PHARMACOLOGY AND THERAPEUTICS

BIOSTATISTICS IN PHARMACOLOGY

VOLUME 1

CONTRIBUTORS

A. L. DELAUNOIS
L. J. MARTIN
E. OLBRICH
A. A. WEBER

PERGAMON PRESS
OXFORD · NEW YORK · TORONTO
SYDNEY · BRAUNSCHWEIG

Pergamon Press Ltd., Headington Hill Hall, Oxford
Pergamon Press Inc., Maxwell House, Fairview Park, Elmsford, New York 10523
Pergamon of Canada Ltd., 207 Queen's Quay West, Toronto 1
Pergamon Press (Aust.) Pty. Ltd., 19a Boundary Street, Rushcutters Bay, N.S.W. 2011, Australia
Vieweg & Sohn GmbH, Burgplatz 1, Braunschweig

First edition 1973
Library of Congress Catalog Card No. 72-78883

Printed in Hungary
ISBN 0 08 016556 7

CONTENTS

VOLUME 1

VOLUME 2

LIST OF CONTRIBUTORS

DELAUNOIS, A. L., Professor, J. F. en C. Heymans Instituut voor Farmakodynamie en Terapie, Rijksuniversiteit, De Pintelaan 135, 9000 Gent, Belgium.

MARTIN, L. J., Professor, Faculté de Médecine et de Pharmacie, Statistique Médicale, Université Libre de Bruxelles, 100 rue Belliard, B 1040 Bruxelles, Belgium.

OLBRICH, E., Dr. med., Müllerstrasse 59, Innsbruck, Austria.

WEBER, A. A., Statistician, Institut de Médecine Sociale et Préventive, Université de Lausanne, Hôpital Cantonal, CH-1011, Lausanne, Switzerland.

PREFACE

As MENTIONED in "Short Historical Survey", psychology and pharmacology were the medical disciplines where, in the early stages of scientific endeavour, mathematical methods have been applied. P. Ehrlich and W. Wiechowski were probably the first research workers in biological fields to use statistics in order to acquire more and better information from their experimental data.

During the last thirty years, organic chemistry made available new substances which possess powerful pharmacological properties. The complexity of the action of drugs and the response of organisms to them, the intricate physiological systems, the diversity of individuals and species, etc. and on the other hand the stringent legal regulations which have been imposed before the substances are accepted as therapeutic agents, make it necessary, for scientific and economic reasons, to adopt the statistical method as a fundamental element of medical science and research.

Statistics are nowadays an integral part of medicine, they are not a luxury nor a magic. They elicit information from experimental data through a logical and simple sequence of manipulations: Hypothesis—Experiment—Conclusion and they evaluate the uncertainty of the experimental evidence.

As the purpose of the International Encyclopedia of Pharmacology and Therapeutics is "to collect, analyse, assess, integrate and disseminate the available data important to pharmacology and therapeutics so as to assist research workers in basic and clinical sciences and those responsible for teaching therapeutics to medical students and graduate physicians" it is obvious that a section had to be dedicated to "biostatistics".

Such a section had to have two objectives:

a) to cover the pharmacological and therapeutic areas. As it is rather difficult to precisely define these areas, practically the whole of the medical sciences are dealt with;

b) to give as much information as possible on what is known of the applications of statistics in the medical and biological fields. An encyclopedia has to be complete and the information has to be usable for many years to come.

These considerations precluded a conception of Section 7 in the form of a "livre de cuisine". They explain also why some chapters have been developed beyond the expected level.

The enormous area to be covered in "Biostatistics" made the task of collecting the necessary material extremely difficult. The authors of the different chapters, aware of the complexity of their charge, have done their utmost to be as general, as complete and as up-to-date as possible. They have used examples from all disciplines in the medical sciences.

Section 7 consists of three volumes and is written for research workers, physicians and students. It shows *how* the techniques have to be used and *why* statistics must be compiled as shown.

Without unduly emphasizing the role of pure mathematical statistics it aims at developing intuitive commonsense reasoning.

Volume 1 deals with sampling, data, preparation and reduction of data, graphs and distributions and transformations.

The theory of sampling and collecting data is particularly important for epidemiology, preventive medicine and in evaluating conclusions based on large numbers of experimental data. The knowledge of biological distributions and the transformations necessary to make statistical methods applicable to them is of course very important. The available information on this subject is dispersed over a vast amount of literature and generally hard to find for non-specialized investigators. It is therefore fortunate and convenient that the available material, as well as new aspects, have been brought together in Chapter 6 of this volume.

Volume 2 handles estimation and testing, confidence intervals, analysis of variance, regression, correlation, sequential analysis, non-parametric methods and biological assay.

The handling of results, their estimation and the testing of hypothesis, the analysis of variance, the regression and the correlation are subjects which are difficult and often misunderstood. A large part of Volume 2 has been dedicated to these topics and to their significance.

A chapter on sequential analysis will help the clinical investigator to conduct controlledtrials by sequential methods, eliminating unnecessary use of inferior treatments.

A chapter on non-parametric methods and one on biological assay will be appreciated by all those involved in drug evaluation.

Volume 3 covers intensive design and use of the computer in statistical calculations and contains a comprehensive collection of the tables discussed in the other volumes.

Special techniques of recent date are described in this volume. Especially in an era of computers it would be a shortcoming not to cover the applications of these tremendous techniques in statistical methods. The repetitions of the tables from the different chapters will be of practical help when studying and applying the described methods.

A. L. DELAUNOIS

CHAPTER 1

SHORT HISTORICAL SURVEY

A. L. Delaunois

Gent, Belgium

THERE are many records which show a trend, even before our era, towards the use of probability. The *census* of the Chinese, the Egyptians, the *Politeiai* of Aristotle, the *Census civitatis* of the Romans, the *Brevium Rerum Fiscalium* of Charlemagne, etc., indicate the importance ancient governments attributed to social knowledge in order to be able to organize peace and to prepare war.

Most of those early "statistical" problems have in fact little or nothing in common with statistics in the modern sense of the word. The problems were limited to the "comparative description of states" *(Staatenkunde)* and treated more or less systematically the geography, history, political organization, social institutions, commerce and military strength of the states then existing. This system was cultivated later by the Italians and the Germans and reached its culmination in the German universities of the seventeenth and eighteenth centuries.

From this term *Staatenkunde* originates the name of "statistics", which probably occurs for the first time in the preface of Gottfried Achenwall's (1719–72) *Abriss der Staatswissenschaft der heutigen vornehmsten europäischen Reiche und Republiken* (1749) (Walker, 1929).

The theory of probability was first introduced into a mathematical treatise by Luca Pacioli in his *Suma* (1494) (Smith, 1913) although some years earlier a statement was already made in a commentary on Dante's *Divine Comedy* (Venice, 1477) concerning the different throws that can be made with three dice (Todhunter, 1865). In 1663, a treatise, *De Ludo Aleae*, by Cardan (1501–76) (Todhunter, 1865), who was himself an inveterate gambler, was published containing, as Libri mentions: "résolutions de plusieurs questions d'analyse combinatoire".

From their astronomical studies, Kepler and Galileo developed the mathematical theory of probability and made brief reference to the subject of chance. But the real history of the theory of probability begins in the seventeenth century, when its foundations were well laid in 1654 by Blaise Pascal (1623–62) and Pierre de Fermat (1608–65) (Todhunter, 1865). Although neither of them published anything on the subject, some of the correspondence between them has been preserved. From these letters, which were the result of certain questions put to Pascal by Chevalier de Méré, a gambler with unusual ability "even for mathematics", the course of development of the theory can be followed. Three years later Christiaan Huygens (1629–95) published a small pamphlet entitled *De Ratiociniis in Ludo Aleae* on the mathematical treatment of the chances of winning certain card and dice games. At about the same time, Captain John Graunt (1620–74) published his *Observations* treating *Natural and Political Observations made upon the Bills of Mortality—with reference to the Government, Religion, Trade, Growth, Ayre, Diseases*" (Walker, 1929). It was the work of C. Huygens that stimulated Jacques Bernoulli (1654–1705), Pierre Rémond de Montmort (1678–1719), Abraham De Moivre (1667–1754), John Arbuthnott (± 1700), Francis Roberts (Robartes) (± 1700), and others to study the theories and problems concerning games of chance.

From J. Bernoulli appeared posthumously in 1713 the *Ars Conjectandi* which is in many ways a landmark in the history of probability. Unfortunately Bernoulli left unfinished the last section of his book when he died, and the world had to wait for Adolphe Quetelet (1796–1874) to get an exposition of the application of the theory of probability to civil, moral and economic affairs. Bernoulli attempted in *Pars Quarta* to prove that if the number of observations is made sufficiently large any predetermined degree of accuracy can be achieved. Pierre-Rémond de Monfort published in 1708 his *Essai d'Analyse sur les Jeux de Hazards* and Abraham De Moivre in 1711 the very important *De Mensura Sortis*, later expanded into *The Doctrine of Chances* and *Approximatio*, in which he treated the discovery of the normal curve.

The presentation of data in tables has been introduced by H. P. Anchersen (1700–65) in his descriptive *Statuum cultiorum in tabulis* (1741) (Westergaard, 1932) where he gives a synoptical table concerning fifteen European countries. This *Tabellenstatistik* consisted only of a typographical arrangement of the facts but invited naturally the use of numbers, thus leading to the statistical tables we are accustomed to.

One has to wait until 1812 for the *Théorie Analytique des Probabilités* of Pierre Simon, Marquis de Laplace (1749–1827), to have the greatest single work on the subject of probability. Two years later appeared from the same author *Essai philosophique sur les probabilités* (1814), work that had a powerful effect on the thinking of Europe (Walker, 1929).

Carl Friedrich Gauss (1777–1855) made use of the mathematical investigations of Laplace to reach the principle of least squares already published previously by A. M. Legendre (1805) (Walker, 1929). The most important points of Gauss's outstanding mathematical work for modern statistics are:

(1) Definition of the measure of precision $h = 1/\sigma\sqrt{2}$.
(2) Enunciation of the principle of least squares.
(3) The general idea that among a system of conflicting values obtained from measurements, none of which have perfect accuracy, there may exist a most probable value which can be found by numerical methods.
(4) The standard deviation of a mean being the standard deviation of the distribution divided by the square root of the number of values.
(5) Gauss's approach to the normal curve.

He also introduced, a year after Friedrich Wilhelm Bessel (1784–1846), the term "probable error", suggesting the computation of this probable error by a method that gives what we now call the "median deviation".

A student of Gauss, Johann Franz Encke (1791–1865), applied in his work *Über die Methode der kleinsten Quadrat* (1834) (Encke, 1834) the theory of probability to errors of observation and gave practical direction for computation. This work, which served as the basis for numerous textbooks, is written in a simple, clear and readable style. It contains the formulas for the standard error of a mean, the standard error of a standard deviation, the "gross-score formula" for computing a standard deviation, the formula for deriving the probable error from the standard deviation, the formula for computing a standard deviation from any arbitrary origin, the probable error of a sum and a weighted sum, etc.

Although the word "statistics" is not used, this work might almost be considered as a text in the mathematical theory of statistics.

At about the same period, Adolphe Quetelet (1796–1874) used the same methods of research to fuse them into one powerful tool for investigating social phenomena. He made important contributions to the statistics of population and effected a great improvement in the organization and administration of official statistics and in the technique of statistical methods. He developed a general method for anthropometric measurements which he published in *Sur l'Homme* (1835) and *Anthropométrie* (1871) (Walker, 1929). In 1885 Hermann Ebbinghaus (1850–1909) made one of the earliest attempts to apply the law of error in a purely psychological study (Ebbinghaus, 1885). He applied precise scientific methods of quantitative analysis in the field of higher mental processes and pointed the way for many investigators to follow.

Wilhelm Lexis (1837–1914), one of the many notable German writers of the last quarter of the nineteenth century, applied the theory of probability to the practical affairs of the world. In his work *Zur Theorie der Masserschei-*

nungen in der menschlichen Gesellschaft (1877) (Lexis, 1877) he stated the famous "Lexian ratio" between the observed dispersion in a statistical series and the theoretical value of the standard deviation $\sigma = \sqrt{npq}$.

It would be very difficult to exaggerate the influence of Sir Francis Galton (1822–1911) and his associates of the English School upon modern statistical methods in psychology (Pearson, 1914/1924). It was his discovery and publication of the method of correlation which seems to have attracted Karl Pearson in the field of statistical theory. The many outstanding papers of K. Pearson were devoted to the probable errors of frequency constants and the influence of random selection on variation and correlation. The fundamental work of K. Pearson and L. N. G. Filon established the mathematical theory and the general method of procedure and has been the foundation on which later research has been built. After these fundamental treatises, there appeared a large number of papers dealing with particular aspects of the theory of probable errors. Among the more important authors may be mentioned A. L. Bowley, W. F. Sheppard, W. Gilson, R. Pearl, F. Y. Edgeworth, W. S. Gosset (Student), A. Rhind, D. Heron, M. Greenwood, H. E. Soper, G. Thomson, E. L. Dodd, R. A. Fisher, F. Yates, A. W. Young, L. Isserlis, K. Smith, etc., all of whom made important contributions to the subject (Walker, 1929).

Of particular interest to biostatistics is the biological assay. This assay has to be considered as a rather recent development in scientific methods. Biological assay started about the beginning of the nineteenth century when Ehrlich tried to standardize diphtheria antitoxin. From then on standardization of drugs and materials by means of reactions of living matter has been practised in pharmacology and in other branches of science. Papers of H. Dale, E. Gautier, P. Hartley, A. A. Miles deal with the development of pharmacological standardization, and, in a valuable publication, J. H. Burn emphasized the importance of introducing as a general practice the use of standard preparations of drugs against which others could be assayed (Finney, 1952). Later J. H. Gaddum, J. O. Irwin, C. I. Bliss, McK. Cattell, D. J. Finney, and others described many of the methods and ideas introduced in recent years.

But the use of statistical methods has not been limited to the standardization of drugs. These methods have been shown to be of great importance in medical trials to compare the effectiveness of different therapeutic or prophylactic treatments. Besides the "classical" statistical tests, the sequential method of design and analysis was developed around 1940. A. Wald, A. B. Hill, L. J. Witts, I. Bross, P. Armitage, L. Schmetterer, F. X. Wohlzogen and others have treated the many aspects of the subject and have shown the advantages of the method which permits continuous examination of the results of a trial (Armitage, 1960; Fülgraff, 1965).

This short historical survey does not claim to be exhaustive. Especially during recent years much statistical research has been done in the medical and biological field, work that would call for a much more exhaustive chapter to handle the latest developments of this branch of science.

REFERENCES

ARMITAGE, P. (1960) *Sequential Medical Trials.* Blackwell, Oxford.

BERNOULLI, J. (1713) *Ars Conjectandi, Opus Posthumus Accedit Tractatus de Seriebus Infinitis et Epistola Gallice Scripta de Ludo Pilae Reticularis.* Basel.

EBBINGHAUS, H. (1885) *Über das Gedächtnis.* Leipzig.

ENCKE, J. F. (1834) *Berliner Astron. Jahrb.*

FINNEY, D. J. (1952) *Statistical Method in Biological Assay.* C. Griffin, London.

FÜLGRAFF, G. (1965) Sequentielle statistische Prüfverfahren in der Pharmakologie. *Arzneimittel Forschung,* April.

LEXIS, W. (1877) *Zur Theorie der Masserscheinungen in der menschlichen Gesellschaft.* Freiburg.

PEARSON, K. (1914/1924) *Life, Letters and Labours of Francis Galton.* London.

SMITH, C. (1913) *History,* Vol. II, p. 529, footnote.

TODHUNTER, I. (1865) *A History of the Mathematical Theory of Probability.* Macmillan, Cambridge, London.

WALKER, H. M. (1929) *Studies in the History of Statistical Method.* Williams & Wilkins, Baltimore.

WESTERGAARD, H. (1932) *Contributions to the History of Statistics.* P. S. King & Son, London.

CHAPTER 2

SCOPE AND MEANING OF BIOSTATISTICS

Eugen Olbrich
Innsbruck, Austria

IT IS the aim of every science to generalize, in other words to discover laws. If such laws exist, and are known to us, we can predict and sometimes direct events.

Generalizations in science are reached by way of hypotheses. The raw material of these is current knowledge, from which we form hypotheses by an intellectual process; at this stage they are no more than proposals for a still unknown law. Hypotheses have therefore to be tested by means of experiments. Scientifically valid hypotheses are mainly those which can be tested experimentally.

Since the statement of a hypothesis is an attempt at generalization it is concerned not with isolated cases but with a group or population. *The theory of the quantitative study of populations is statistics.*

A hypothesis can hardly ever be tested on an entire *population* or "universe", if only because this would be too large. Thus we have generally to be content with testing a part of the population, a *sample*, and with assuming that the results obtained from the sample can be applied to the population. *Statistics show why, under what conditions, and to what extent this assumption is justified.*

This means arguing from the particular to the general, or reasoning by induction. *Statistics is thus the science of inductive reasoning.* Knowledge gained by induction is never absolutely certain, but only more or less probable. Inductive reasoning is the only way to gain *new* knowledge, though.

For an inference from the particular to the general to be reliable the sample must be sufficiently representative of the population. The idea is often trivially expressed by saying that inferences from a sample apply only to the population from which it is drawn. Absurd as this statement may sound, the principle is very often violated. For it is by no means easy to define a population, and perhaps harder still to get a clear picture of the population which a given sample, such as a group of hospital patients, represents.

An experiment will clearly give one of two results:

(a) Reasonable agreement with the hypothesis—more precisely, with what the hypothesis leads us to expect.
(b) Sufficient divergence from the hypothesis to make it highly unlikely to be tenable.

It is essential to recognize that result (a) does *not* prove that the hypothesis is correct: the same result might accord equally well with some other hypothesis. But result (b) permits the conclusion, with a degree of probability which can be freely chosen, that the hypothesis is false. Some very important principles follow:

The experiment must be designed to indicate that a hypothesis is false. Hence it is essential that experiments should be *planned. Statistics teaches the planning of experiments.* Certain important aspects of planning have to be considered, only one of which was stated above.

A hypothesis usually predicts the effect of a procedure, a treatment, a process. From what has already been said it will be clear that with a *positive* formulation of the hypothesis—"A causes B"—interpretation of the results will involve difficulties. If the effects are great enough, we can indeed say that the result is not incompatible with the hypothesis, but again, this by no means proves that the hypothesis is correct. If, however, the effects do not reach the level at which they can be statistically evaluated, the result is still not necessarily incompatible with the hypothesis; the effects might simply be small in relation to the experimental error.

It is therefore preferable always to formulate the hypothesis to be tested in the form of a *null hypothesis*, thus: "A does *not* cause B". Now either of the possible results can be given a useful interpretation. If the effect is great enough, then the result is incompatible (e.g. with 99% probability) with the null hypothesis. If the effects are less than the selected degree of probability requires, then the hypothesis is not invalidated.

But results are all too often wrongly interpreted in the light of a positive formulation of the hypothesis, from which unwarranted assumptions are readily drawn, at least by the reader. It should be pointed out that the null hypothesis is not intended to play the part of devil's advocate, always to maintain the opposite to what one would like to prove, although the conditions usually mean that this is the case. But in a trial to confirm that a drug is harmless, the null hypothesis is in conformity with the result for which one hopes.

Even if only null hypotheses are tested, two fallacies are evidently possible:

(a) Rejection of a correct (null) hypothesis (fallacy I).
(b) Non-rejection of a false (null) hypothesis (fallacy II).

These two sources of error are interdependent. If the risk of fallacy I is kept as small as possible, the risk of fallacy II increases, and *vice versa*. *Statistics teaches how best to overcome this dilemma.*

Various peculiarities of biological experiments have become known which are not usually encountered in physical experiments. It is well worth tracing the differences of method between these two branches of natural science. In the first place, the need for experiments was first recognized in physics by Galilei; furthermore, many biologists still believe that they must experiment in the manner of physicists.

Even the physicist tests his hypotheses by means of samples. But classical physics, which is all that is considered here, does not treat these samples as populations, or as discontinua, which in fact they certainly are—vast numbers of molecules; it treats them as if they were continua *(Ersatzprinzip)*. The property to be studied or its molecular equivalent can hardly be determined or measured in the single individuals, namely the molecules. But this need not concern us here. In physics, it is the state of the sample itself which is measured, for instance the temperature of a volume of liquid or the pressure of a volume of gas. For at least 100 years the theory has held good that the temperature of a body is a function of the *average* velocity of its molecules, and that the pressure of a gas depends on the *average* number and intensity of the collisions of its molecules with a given area of its surface. It is thus conceivable that temperature or gas pressure could be expressed as an *average* of the appropriate molecular values—if these were open to direct measurement, which in general they are not.

We may accept, then, (1) that the physical properties of classical physics are only averages of what in reality are highly variable individual values—Maxwell's distribution law, for instance, shows how wide the variation can be; and (2) that these averages can be directly measured.

Biology is another matter. The samples are coarsely divided; the individual, in general terms the element of the population, as it were obtrudes itself. Unlike the physicist, the biologist cannot behave as though the discontinuum were a continuum, for he must calculate the measurements of the population, the average, from the recorded individual values. For him, only the individual values are directly available; for the physicist, only the average, as a rule. These wholly different starting points, which from the outset require different methods, must be borne in mind, as must the fact that the difference arises only from the different extent to which physical and biological samples are divided. One is tempted to draw an analogy with true solutions on the one hand and colloids on the other. Just as macromolecular substances such as proteins can only assume a colloidal distribution in solution, and hence are also called eucolloids, so the coarsely divided populations or "collectives" of the biologist might be called eucollectives. Conversely, in inanimate nature one might speak of masked collectives.

While this first contrast between (classical) physics and biology is easy enough to grasp, the next calls for certain fundamental, that is to say statistical, knowledge.

It is always held to be a characteristic of biological experiments that their results are subject to much greater variation than those of physical experiments, so that their interpretation is more difficult because it is less certain. Thus expressed, this is incorrect. Single molecular measurements also vary considerably—once again, consider Maxwell's distribution law. In statistical language this might be expressed thus: individual values in inanimate nature also show considerable scatter. But as we have just seen, classical physics is not concerned with individual values: it measures averages directly. And this second contrast to be considered signifies no more than that when a physical measurement is repeated the results correspond within very narrow margins of error, as opposed to repeated calculations of averages from comparable biological samples. How can this be explained?

Let us take a simple example. We have two baskets, each containing 100 apples; the apples are of the same kind, but not, of course, all the same size. We place the first basket on the scales: it weighs 20 kg net. (This is the same as the physical method: direct measurement of a summary value.) We now wish to estimate how much *one* apple weighs. The best approximation appears to be the average weight (20 kg/100 = 200 g), because we consider it to be the most likely individual weight—this, incidentally, is justified only if the weights of the apples are distributed symmetrically. But we shall certainly not be surprised if we find that an apple taken at random from the basket weighs only 170 g and another 220 g. We knew from the start that the apples were of different sizes and that our estimate was inevitably uncertain. The measure of uncertainty is the standard deviation, σ, which is the square root of the variance, σ^2. This, too, can be estimated. For this we need a sufficient number of individual weights (N) and a formula.

It is to be observed (1) that to estimate the degree of uncertainty of our assertion (that an apple weighs 200 g) we need a sufficient number of individual measurements. Thus *repetition* is especially important in biological experiments; generally more than one individual should be allotted to each trial batch. And (2) that the standard deviation of the sample, s, which is our estimate for the standard deviation of the whole population, σ, becomes better with increasing number of apples, as the sample becomes larger. Mathematically speaking, as N increases, s tends towards the true value σ.

Here it should be noted that modern statistics, i.e. the statistics of small samples which biology requires, began with the discovery of the exact relationship between the estimated standard deviation, s, and the size of the sample. This was achieved by the chemist Gosset, known by his pseudonym "Student". The solution of the problem is given by the "t-distribution".

To return to our two baskets of apples: if we are asked to estimate from the data of the first basket (100 apples, 20 kg) what is the average weight of the apples in the second basket, we shall reply with considerably more confidence than before: "200 g", although it has involved extrapolation. We are relying on the fact that the apples in both baskets are of the same kind, in other words samples of the same population. We should of course be surprised if our estimate proved correct within milligrams, but we might be equally surprised if the apples in the second basket turned out to weigh 21 kg instead of 20 kg.

The estimated uncertainty of the *average* of a sample can in fact be shown to be $\sqrt{N}$ times smaller than that of an *individual* value. Supposing that in our example we found $s = \pm 30$ g, then the estimated standard deviation of the mean for $N = 100$ would be $\pm 30/10 = \pm 3$ g.

In a *physical* sample the number N of molecules is always enormous, how ever small the volume. Hence the standard deviation of the mean, $\sigma/\sqrt{N}$, is practically zero even if σ is by no means very small. Biological samples, on the other hand, are generally quite small, so that $\sigma/\sqrt{N}$ never vanishes. This is why biological experiments are less exactly reproducible and thus more difficult to interpret.

Bacteriology holds a special place among the biological sciences in this respect. The unit with which the bacteriologist deals is often a colony or suspension of bacteria. This is a similar situation to that of a physical experiment. The third contrast between physics and biology will, however, show a fundamental difference.

But first let us observe the following: (1) To estimate the standard deviation of the mean, we must begin by estimating the standard deviation of the individual values. Thus we cannot avoid having to measure individual values. (2) The estimated standard deviation of the mean becomes smaller as the size of the samples increases; if N is large enough it becomes so small that it can be ignored.

Since the standard deviation of the mean decreases only with the *square root* of N, the percentage gain in certainty as N increases becomes smaller and smaller, until it no longer justifies the extra money, time, laboratory animals, etc., involved. For the sake of economy the technique of *sequential analysis* is often advisable: a limited experiment is first done, and from the results obtained it is possible to calculate how far the experiment has to be extended before the null hypothesis can be rejected.

We now come to the third contrast between physical and biological methods. One principle of classical experimental method is as follows: if you wish to study the effect of a factor, vary it and measure the effects, while keeping other factors constant. In biology this is often only partly possible, because numerous factors are beyond our control.

What is to be done instead? For we must somehow isolate the factor to be studied as far as possible from the interference of other factors, known or unknown. Only one method meets the requirements: in alloting individuals to trial groups chance must be given a free rein. This is the only way in which the interference of other factors can be made, on the average, the same for all groups. With this more or less homogeneous background the particular factor can be observed and measured in sufficient isolation. This random allocation or randomization is the starting point of modern statistics, of biometry. The importance of using some kind of mechanical aid to randomization, such as coins, dice, cards, or tables of "random numbers", cannot be overemphasized.

Even in biological experiments, extraneous factors will of course be kept constant as far as possible. This can be done with factors which are clearly defined or capable of definition. In practice, these are all the factors which are known to affect the homogeneity of a population of live creatures, described as *heteropoietic factors*. It is a basic rule that the groups should be matched for *age* and *sex*, if the problem allows. An important point has to be made here: in assessing therapeutic trials the results obtained are usually compared with those of other workers or with previous trials. Such comparisons, which of course are repeatedly drawn, are to be interpreted with the greatest caution, because for numerous reasons the samples cannot be treated as if they were drawn from the same population. In statistics, a clear distinction is made between *survey* and *experiment*. The distinction does not by any means arise solely from the fact that in an experiment all groups are tested at the same time and in the same place, though it is usual to ensure that these important factors—*time* and *environment*—are kept as constant as possible. The difference lies deeper. When existing, merely assembled data are evaluated (survey), it will often happen that a factor with which we are concerned is associated with one or other of the heteropoietic factors. With surveys, then, there is always the possibility that any information extracted may be false, or rather, inadvertently falsified, distorted. With experiments, one tries to prevent such falsification or *bias* at the planning stage.

In animal experiments care must also be taken to ensure that the animals

are similarly fed, and that they are equally well treated. What appear to be real differences between groups may arise from differences in activity (for instance, because of variously sized containers). With man, it should be obvious that social class, and much besides, has to be considered.

If heteropoietic factors are kept as constant as possible, the precaution can be expected to decrease the scatter or experimental error; as far as is feasible the subjects of the experiment are therefore homogenized. The homogenization is a part of planning, not of evaluating the experiment. It is a thankless task to evaluate when because of defective planning no clear interpretation is possible, or to assess differences against a scale that is too crudely drawn.

But the best possible homogenization is not the be-all of planning experiments. On the contrary, the more we standardize the more sensitive the experiment and also the narrower the basis for induction, i.e. the more limited the applicability of the conclusions drawn from the experimental findings. There is a valuable device for overcoming this difficulty: the construction of *blocks* of as far as possible similar individuals.

If we wish to compare the effects of two treatments, A and B, and also to establish whether the two treatments have any significant effects, we clearly need three groups: a control group 0, and a group each for treatments A and B. The crude design of the experiment is in this form:

	0	A	B
	...	...	...
	...	...	...
	⋮	⋮	⋮
	...	...	...
Total	...	...	...

Each of the 3 groups contains e.g. 10 individuals (animals) allocated at random.

From the 30 individual values given by the experiment, a total variance or total s^2 is calculated. By means of a calculation, analysis of variance or of scatter, this total variance can be broken down, in this simple case into two figures—namely, variance *within* groups, and variance *between* groups (between treatments). If the variance between groups exceeds a certain multiple (given in tables) of the variance within groups, then the null hypothesis—that there is no difference between the groups—can be rejected with a degree of confidence chosen beforehand. If the experimental animals are heterogeneous it may easily happen that the variance within groups, which is the unit of our scale, is so large that a genuine difference between groups passes unrecognized.

If we are fortunate enough to have a supply of animals in accordance with modern experimental design, so that we can always get animals from the same litter, our planning can be better. These litter-mates form blocks, i.e. groups of individuals whose responses are certainly more similar than those of animals merely thrown together. The same experiment, but with blocks, has the following plan:

	Treatment			
Block	0	A	B	Total
1	...	...	...	...
2	...	...	...	...
⋮	⋮	⋮	⋮	⋮
10	...	...	...	...
Total	...	...	...	...

Litter-mates are allotted to the 3 groups at random.

The total variance can now be broken down into *three* figures: (a) between treatments, (b) between blocks, (c) remainder. Structurally, (a) is the same as in the first plan. Taken together, (b) and (c) are comparable with "variance within groups". It need not follow that (c), representing the experimental error in this plan, will inevitably be smaller than the experimental error in the first plan, but in general it can be expected to be so. The point will be discussed in detail in the following chapters. As a rule, smaller differences between treatments will come to light with plan 2 than with plan 1. Since animals from different litters are involved in the experiment, the basis for induction is not narrowed; yet plan 2 is more sensitive than plan 1. Putting it rather differently, to attain the same level of confidence, plan 1 requires more animals than plan 2, and thus more time and money: plan 2 is more economical.

If this comparison is considered further, one does not have to be a mathematician to see that difference in sensitivity between various experimental designs can be directly measured. This should indicate the extent to which we can control our experimental work if we make use of biometry.

CHAPTER 3

SAMPLING METHODS

A. A. Weber

Lausanne, Switzerland

3.1. INTRODUCTION

Sampling methods, the subject treated in the present chapter, are a relatively recent field of study. Most of the papers have been published since 1930, or perhaps 1920. Sampling methods were originally devised for enumerative or descriptive surveys, where the purpose was to count all the individuals displaying a certain property, or to describe the characteristics of some population as a whole. Later others conducted surveys for analytical ends—seeking to determine by means of a sample whether, for instance, the number of sick persons differed from one region to another, or between groups of persons in a particular population.

It is practically impossible to draw a clear distinction between sample surveys of a descriptive type and those of an analytic type. In general the same survey will serve both purposes. For an administrator a survey of morbidity will indicate the number of cases for whom medical care should be provided, while at the same time the epidemiologist will be informed as to the high risk groups of the population, or possibly as to the forces influencing the morbidity. The essential distinction between the two applications of the results of a survey is that in the first case the sampled population, no matter how large it may be, is always considered to be finite, whereas in the second case this finite population is assumed to be a large sample of some infinite population or hypothetical universe.

Sampling techniques can be employed as well within the context of an experiment. For instance, to compare the effectiveness of different insecticides, an orchard can be divided into a certain number of "plots" and each plot treated according to some pre-arranged plan. It would be fastidious to count all the insects still remaining after treatment; usually they are counted on a sample of branches or leaves in each plot. An excellent account of the

use of sampling methods in experimental investigations is given by Yates and Zacopanay (1935).

In this chapter an attempt has been made to produce a concise account of the main problems in sample surveys—their preparation, execution and utilization. There are numerous works of reference where the methods outlined here are presented systematically and where the formulas given here are derived rigourously. In this connexion may be mentioned, *inter alia*, W. G. Cochran's *Sampling Techniques* (1963) and F. Yates' *Sampling Methods for Censuses and Surveys* (1960), which are particularly adapted to the application of sampling techniques in biometry.

3.2. SAMPLING PROCEDURES

3.2.1. General Considerations

The aim of sampling is to produce an estimate of certain characteristics of a population by means of a suitably chosen sample. Whenever the population as a whole is homogeneous, or if it can be made homogeneous without difficulty, it is of no importance how or where the sample is taken. However, when the population is heterogeneous the sample must necessarily be selected according to a rigid scheme, as otherwise it would not be representative of the population and the results obtained would be invalidated by uncertainty. It would be incorrect, for example, to estimate the characteristics of out-patients attending a clinic by a sample made up of the first ten patients arriving at the clinic each day. The early morning visitors would probably be working people, coming for treatment before going to their work and they are not representative of the total population of out-patients. In all such cases the sample must be spread out in an appropriate manner over the total population. In this regard various procedures have been elaborated in relation to the information available concerning the distribution of the variate under consideration within the particular population.

In general there are two main methods of sampling. The selection may be empirical, or reasoned, or alternatively it may be probabilistic or random. In the first method the sample is so selected that one may assume it to be representative of the population. The most common methods of empirical selection are: (1) the representative unit method where experts are entrusted with the task of breaking down the population into homogeneous classes and then, within each class, of selecting a unit that in their judgement is the most representative, and (2) the quota method which consists in selecting a sample

such that the distribution of certain variates in the sample such as age, sex, social class, corresponds to the distribution of these same variates in the population. Neither of the empirical selection methods, however, is often used in biology.

Probability sampling is characterized thus: each unit of the population has a determinate probability of being selected, the sampling is performed on a chance basis compatible with this particular probability, and the formulae for evaluating the parameters of the population from the sample are related to this same particular probability.

3.2.2. Sampling Errors

Before considering the main types of probability sampling, and in order to grasp the respective merits of the various possibilities, it is important to define what is meant by the accuracy of a sample. The accuracy of an estimate $\hat{t}$ based on a sample of a parameter T of the population is measured as the difference between these two values and called the error or total error of the sample:

$$\text{Total error} = \hat{t} - T.$$

The total error is generally made up of (1) the random sampling error, (2) the bias, sometimes called the systematic error or distortion, and (3) the random error of measurement or of observation, called the measuring error.

The random sampling error (1) is the error produced by the replacement of the population by a sample.

An estimate $\hat{t}$ is said to be non-biased if the expected value or mathematical expectation $E(\hat{t})$ of all the estimates obtainable by this method of sampling is equal to the true value T of the parameter to be estimated. Otherwise the method of estimation is said to be biased. Here the bias or systematic error is given by the expression

$$E(\hat{t}) - T.$$

The precision of an estimate $\hat{t}$ is the difference between the estimate and its expected value $\hat{t} - E(\hat{t})$.

In probability sampling it is possible to determine mathematically the sample distribution of the estimate $\hat{t}$ and often of the systematic errors inherent in the sampling method.

The measuring error is the error committed in determining the value of the variate investigated in each unit of the population, for example, reading errors or imprecise determinations. An example illustrating the three components of the total error of a sample is a sample survey of the prevalence of pulmonary tuberculosis based on direct examination of sputum. Here a random sampling error is introduced because the population is replaced by a sample, a bias is induced by the limited sensitivity of direct bacteriological examination, and microscope technicians may well commit many reading errors.

In the following paragraphs only the random sampling error and the bias which are inherent in the sampling method employed will be considered; it is assumed that the variate can be correctly measured throughout the population.

3.2.3. Types of Probability Sampling

The simplest type is random sampling, called also unrestricted random sampling and this will be studied in greater detail in section 3.3. In this form of sampling the supposition is made that all the units of the population have previously been numbered; this is not always the case and can be a severe imposition. One then draws a certain number of numerals at random by the use of Tables of random numbers, for instance, or by throwing dice. The units corresponding to these numerals comprise the sample.

When this simple situation does not exist, or this method is felt to be inadequate, other, more complex sampling methods must be considered. It may be that the units of the population differ greatly in size, for instance, farmsteads of varying acreage. Instead of taking the sample strictly at random, one can pick the farmsteads from a map by pin-pointing them at random. One then speaks of sampling with probabilities proportional to the size of the unit. Size can be measured in different ways. The size of a farm may be indicated by the surface area of the cultivated land, or by the acreage sown, or by the head of cattle. In a more general way, a positive arbitrary probability can be assigned to each unit of the population, and one then speaks of sampling with arbitrary probabilities.

Whenever the variate investigated varies appreciably from one group of the population to another, for example from one region to another, a separate sample may be taken in each of these groups or strata. This approach is called stratified sampling.

When the population is scattered over a large territory, simple random sampling or sampling with arbitrary probabilities generally produce a geographically scattered sample. The cost of moving from one unit of the sample to another, either to identify the units selected or to record the value of the variate investigated, or even to supervise the work *in situ*, may be prohibitive.

One can try to limit the scatter of the sample by taking compact clusters of units (cluster sampling).

If these clusters contain a large number of units, or if the value of the variate is costly to record in relation to the delimitation of the clusters, or if the variate is largely constant throughout each cluster, there is often some advantage in taking only a fraction of the units contained within the clusters selected. This is called two-stage sampling or sub-sampling. This can be extended to three or more stages.

In systematic sampling units are selected at regular intervals, for example every kth unit. A number is selected at random between 1 and k, say i, and one then takes the units bearing the numerals i, $i+k$, $i+2k$, $i+3k$ and so on. This generally facilitates the choice of the sample. Moreover, systematic sampling is usually more precise than simple random sampling, given the same size of sample.

When the value of an auxiliary variate in close relation to the variate under investigation is known, a variety of sampling procedures and methods of estimation are available. The auxiliary variate may be, for instance, the value of the variate investigated at the time of a recent census. A ratio estimate method is applied when there are good reasons to believe that this relation is linear and passes through the origin. Otherwise a regression estimate is used. In estimates of this kind (ratio or regression) the mean value of the auxiliary variate in the population is assumed to be known, and the sample serves only to estimate the value of the ratio or coefficient of regression.

It may be that the mean value of the auxiliary variate in the population is not known. If it is relatively easy to record the auxiliary variate in relation to the one investigated, one can first take a sample that will produce an estimate of the mean of the auxiliary variate in the population, and then secondly another sample can be taken (generally a sub-sample of the first one) in which the relation between the auxiliary variate and the one under investigation is measured. This is called double sampling or two-phase sampling. The auxiliary variate may be, for instance, the weight of sugar beet, and the variate investigated may be the sugar content of the beet, in a survey to estimate the total sugar yield of the beet harvest.

These various methods are often combined. Thus one can have a stratified sample where in each stratum the units are selected by two- or three-stage sampling.

In a survey of the prevalence of pulmonary tuberculosis one may take a certain number of clusters of persons. For each person in each cluster three expectorations of sputum can be taken at daily intervals. The microscopist can then read off the number of bacillae in every tenth field in the microscope. There is here a complex superposition of different sample units.

Hence it will be seen that the sampling design chosen depends on the knowledge available as to the distribution of the variate investigated in the population. Section 3.4 outlines in detail how such knowledge is considered and used.

3.2.4. Situations in which Sampling is preferable to an Exhaustive Survey

Whenever sampling is preferred to an exhaustive survey, one or more of the following reasons is generally advanced:

(a) Sampling is the only possible method. This is the case where the variate cannot be studied without destroying the material.

(b) Sampling is less expensive than an exhaustive survey. This is generally so, although the actual savings from sampling may not be as large as supposed. Calculations of the cost of sampling should also include cost of determining the method of enquiry, of selecting the sample and of handling the results, which may necessitate lengthy calculations.

(c) Sampling produces more precise and more detailed results than an exhaustive survey. A sampling survey does not require a large staff, hence they may be selected, trained and supervised with greater care. One can also employ better qualified personnel and improved diagnostic and measuring procedures. For example in a population census only the blind can be counted, but in a sample survey opthalmologists can be asked to measure the degree of loss of vision and to determine the cause of blindness in each case.

(d) Sampling enables the results to be obtained more quickly. In censuses, for example, it often happens that in spite of the use of computers to process the results they are not forthcoming for several years. As a palliative one often does a quick tabulation of certain results from a sample of the data.

In view of the advantages, one may well wonder whether exhaustive surveys ought not to be systematically replaced by sample surveys. The answer is certainly in the negative. One of the aims of any census or exhaustive survey is to obtain results for every administrative (or other) unit of the population. A sample survey would be able to provide adequate information for the whole country and for the large administrative units, but not for each village or each urban or rural borough.

3.3. SIMPLE RANDOM SAMPLING

3.3.1. Definitions and Notation

Random sampling is probability sampling at its simplest level. To take a random sample of n different units from a population consisting of N units, is like drawing at random n tickets in a lottery of N tickets, or n balls from an urn containing N balls. This is sampling without replacement, for the drawn ball is not replaced in the urn before the next draw is performed.

We put y_i for the value of the variate (y) associated with the ith unit of the population being studied. The mean value $\bar{Y}$ of (y) in the population is written so:

$$\bar{Y} = \frac{y_1+y_2+ \dots +y_N}{N} = \frac{\sum_{i=1}^{N} y_i}{N}.$$

Assuming that the first unit drawn in the sample bears the numeral $a_1, \dots,$ and the ith the numeral a_i, etc., then (y) will assume in the sample the values $y_{a_1}, \dots, y_{a_i}, \dots, y_{a_n}$. But since this form of notation, although correct, is cumbersome, it is conventional to identify the units in the sample by the order in which they are drawn. Thus y_j is also the value of the variate in the jth draw and it can take all the possible values not already issued in the $(j-1)$ preceding draws. This definition may seem ambiguous, but it does not give rise to confusion if this convention is borne in mind.

The following notation is used:

	Population	Sample
Number of units	N	n
Mean value of the variate	$\bar{Y} = \frac{I}{N}\sum_{i=1}^{N} y_i$	$\bar{y} = \frac{I}{n}\sum_{i=1}^{n} y_i$
Total value of the variate	$Y = N\bar{Y}$	$y = n\bar{y}$

In the course of the exposition it will become clear that capital letters refer to the population, and small to the sample; if there is reason to suppose that the population derives from some hypothetically infinite universe, the parameters of this universe are indicated by Greek characters.

The mean value $\bar{y}$ of the sample is a non-biased estimate of the mean value $\bar{Y}$ of the population. In fact, the mathematical expectation E of the random variable $\bar{y}$ can be written so:†

$$E(\bar{y}) = E\left[\frac{1}{n}(y_1 + \dots + y_n)\right] = \frac{1}{n}[E(y_1) + \dots + E(y_n)].$$

By virtue of symmetry it is deduced that the probability of drawing a determinate unit in the jth draw when it is unknown which units have been drawn in the preceding tries is the same for every unit and equal to $1/N$.

Hence $E(y_j) = \frac{y_1}{N} + \frac{y_2}{N} + \dots + \frac{y_N}{N} = \bar{Y}$ for any $j \leqslant n$.

Therefore

$$E(\bar{y}) = \bar{Y}. \tag{3.1}$$

Most of the treatises on sampling written with a view to its application in descriptive surveys give the following expression for the variance of the values y_i about their mean $\bar{Y}$:

$$E[(y_i - \bar{Y})^2] = \frac{1}{N}\sum_{i=1}^{N}(y_i - \bar{Y})^2.$$

Nevertheless, in analytical surveys it is logical to regard the particular population being studied as a large sample of a hypothetically infinite universe. In such cases the mean $\bar{Y}$ is regarded as an estimate of the true mean μ of this universe. Likewise the variance σ^2 in this universe is estimated by

$$S^2 = \frac{1}{N-1}\sum_{i=1}^{N}(y_i - \bar{Y})^2 = \frac{N}{N-1}E[(y_i - \bar{Y})^2]. \tag{3.2}$$

This distinction is theoretically very important. Further most formulae take a simpler form when they are expressed in terms of S.

3.3.2. Properties of the Sample

One of the properties of the sample mean has already been given by means of formula (3.1). The precision of $\bar{y}$, as an estimate of $\bar{Y}$, is represented by its variance $V(\bar{y}) = E[(\bar{y} - \bar{Y})^2]$, which is not always known. It is therefore necessary to estimate $V(\bar{y})$. We shall consider the following theorem:

† See, for example, Deming (1950).

THEOREM 3.1. In unrestricted random sampling

(a) the mean $\bar{y}$ of the sample is a non-biased estimate of the mean $\bar{Y}$ of the population;

(b) the variance $V(\bar{y})$ of $\bar{y}$ is given by the formula

$$V(\bar{y}) = \frac{N-n}{Nn} S^2; \tag{3.3}$$

(c) the variance of the (y_i) in the sample

$$s^2 = \frac{1}{n-1} \sum_{i=1}^{n} (y_i - \bar{y})^2 = \frac{1}{n-1} \left(\sum_{i=1}^{n} y_i^2 - \frac{1}{n} \left(\sum_{i=1}^{n} y_i \right)^2 \right) \tag{3.4}$$

is a non-biased estimate of S^2, and therefore

(d)
$$v(\bar{y}) = \frac{N-n}{Nn} s^2 \tag{3.5}$$

is a non-biased estimate of $V(\bar{y})$.

Proof.

(a) Follows from the foregoing paragraph.

(b) It should be noted that $\bar{y} - \bar{Y} = \frac{1}{n} \sum_{i=1}^{n} (y_i - \bar{Y})$.

Therefore $V(\bar{y})$ may be written

$$V(\bar{y}) = E[(\bar{y} - \bar{Y})^2] = E\left[\frac{1}{n^2} \left(\sum_{i=1}^{n} (y_i - \bar{Y}) \right)^2 \right],$$

then, upon expanding the square and interchanging the operators E and Σ,

$$V(\bar{y}) = \frac{1}{n^2} \left[\sum_{i=1}^{n} E[(y_i - \bar{Y})^2] + \sum_{i=1}^{n} \sum_{j \neq i}^{n} E[(y_i - \bar{Y})(y_j - \bar{Y})] \right]. \tag{3.6}$$

It should be mentioned that $E[(y_i - \bar{Y})^2] = \frac{N-1}{N} S^2$ by virtue of (3.2).

Finally, for the products of the terms on the right-hand side of (3.6), we get

$$E[(y_i - \bar{Y})(y_j - \bar{Y})] = \frac{1}{N(N-1)} \sum_{i=1}^{N} \sum_{j \neq i}^{N} (y_i - \bar{Y})(y_j - \bar{Y})$$

$$= \frac{1}{N(N-1)} \left[\sum_{i=1}^{N} \sum_{j \neq i}^{N} (y_i - \bar{Y})(y_j - \bar{Y}) + \sum_{i=1}^{N} (y_i - \bar{Y})^2 - \sum_{i=1}^{N} (y_i - \bar{Y})^2 \right]$$

and then, on grouping the first two terms of the right-hand entry,

$$E[(y_i-\bar{Y})(y_j-\bar{Y})] = \frac{1}{N(N-1)}\left[\sum_{i=1}^{N}(y_i-\bar{Y})\right]^2 - \frac{1}{N(N-1)}\sum_{i=1}^{N}(y_i-\bar{Y})^2.$$

By definition the first term of the right-hand side is zero. By virtue of (3.2) the second is equal to S^2/N.

Upon substituting these expressions in (3.6), we obtain

$$V(\bar{y}) = \frac{S^2}{Nn}(N-1-n+1) = \frac{N-n}{Nn}S^2.$$

(c) s^2 can be written as

$$s^2 = \frac{1}{n-1}\sum_{i=1}^{n}[(y_i-\bar{Y})-(\bar{y}-\bar{Y})]^2,$$

and then upon expansion

$$s^2 = \frac{1}{n-1}\left[\sum_{i=1}^{n}(y_i-\bar{Y})^2-n(\bar{y}-\bar{Y})^2\right].$$

By taking the mathematical expectation of the foregoing expression, and by virtue of (3.2) and of (3.3), we get

$$E(s^2) = \frac{1}{n-1}\left(n\frac{N-1}{N}S^2-\frac{N-n}{N}S^2\right) = S^2.$$

(d) By virtue of (c),

$$E[v(\bar{y})] = \frac{N-n}{Nn}E(s^2) = \frac{N-n}{Nn}S^2 = V(\bar{y}).$$

The quotient n/N is called the sampling fraction f and the term $(N-n)/N = 1-f$ is called the finite population correction factor. Whenever n is small in relation to N, this correction factor can be replaced by 1 and then as a first approximation the formulae become

$$V(\bar{y}) = \frac{S^2}{n}, \quad v(\bar{y}) = \frac{s^2}{n}.$$

Thus, as a first approximation, the precision of a simple random sample depends only on the variance S^2 and on the size n of the sample, and not on the number of units N in the population.

For the estimate $\hat{Y}$ of the total value of the variate in the population Y, we have the following formulae:

$$\hat{Y} = N\bar{y} \tag{3.7}$$

$$V(\hat{Y}) = N^2 V(\bar{y}) \tag{3.8}$$

$$v(\hat{Y}) = N^2 v(\bar{y}), \tag{3.9}$$

where $V(\bar{y})$ and $v(\bar{y})$ are given by the formulae (3.3) and (3.5) respectively.

3.3.3. Calculating the Confidence Limits of the Mean

For the majority of distributions met with in biology, the distribution of the mean $\bar{y}$ of a random sample tends rapidly to a normal distribution of mean $\bar{Y}$ and of variance $V(\bar{y})$ as the size n of the sample increases.

The lower and upper limits of the confidence interval of the mean $\bar{Y}$ at the level of significance α are then estimated by

$$\left.\begin{aligned} \hat{\bar{Y}}_I &= \bar{y} - t_\alpha \sqrt{v(\bar{y})} \\ \hat{\bar{Y}}_S &= \bar{y} + t_\alpha \sqrt{v(\bar{y})} \end{aligned}\right\} \tag{3.10}$$

where $\hat{\bar{Y}}_I$ and $\hat{\bar{Y}}_S$ are the lower and upper limits of the confidence interval respectively, and t_α is the value of the Student's t distribution with $n-1$ degrees of freedom, corresponding to the level α of significance.

EXAMPLE 3.1. The fifth column of Table 3.1 shows the number (y) of discharges (and of deaths) of patients affected with arteriosclerotic heart disease in the $N = 26$ hospitals of Copenhagen and its suburbs during a twelve-month period beginning 1 April 1962. (The other columns of Table 3.1 are ignored in the present Example. They will serve to illustrate other more elaborate sampling techniques.)

Suppose that the number of discharges (y) of patients affected with arteriosclerotic heart disease is unknown. One can then take a sample of n hospitals at random and in each of them find the number of such discharges from the hospital records. Suppose that $n = 12$ is to be taken as the size of the sample.

A common method of selecting a sample at random is to use a table of random numbers. An extract from the tables of Fisher and Yates (1953) is given in the Appendix, with some examples of its use. Here the task is to

TABLE 3.1. NUMBERS OF BEDS, TOTAL NUMBER OF DISCHARGES, AND NUMBER OF DISCHARGES OF PATIENTS WITH ARTERIOSCLEROTIC HEART DISEASE IN THE HOSPITALS OF COPENHAGEN, DENMARK, AND ITS SUBURBS IN THE YEAR FROM 1 APRIL 1962 TO 31 MARCH 1963

Name of hospital	Number of beds M	$\sum_{i=1}^{h} M_i$	Discharges† total x	arteriosclerotic heart disease y	$\frac{100y}{M}$
1. De Gamles By	1460	1460	935	266	18.2
2. Rigs	1327	2787	22,878	266	20.0
3. Bispebjerg	1201	3988	25,253	691	57.5
4. Frederiksberg	1079	5067	17,144	580	53.8
5. Gentofte	1020	6087	25,078	770	75.5
6. Kommune	973	7060	19,046	840	86.3
7. Glostrup	780	7840	20,248	608	77.9
	7840		130,582	4021	
8. Blegdam	503	8384	6949	426	84.7
9. Sundby	402	8745	8144	235	58.5
10. Nørre	391	9136	475	30	7.7
11. Military	359	9495	5937	75	20.9
12. St. Joseph	330	9825	9534	239	72.4
13. Øresund	314	10,139	2514	64	20.4
14. Diakonisse	299	10,438	7247	190	63.5
15. St. Lukas	210	10,648	3857	78	37.1
16. Sønderbro	208	10,856	4074	150	72.1
17. Orthopaedic	200	11,056	3438	0	0.0
	3216		52,169	1487	
18. Finsen Institute	184	11,240	2213	14	7.6
19. Dronning Louises	167	11,407	1613	0	0.0
20. Rud. Berghs	160	11,567	1715	1	0.6
21. Fuglebakken	160	11,727	1317	0	0.0
22. St. Elisabeth	157	11,884	2698	75	47.8
23. Radium Institute	126	12,010	1516	0	0.0
24. Forchhammersv.	79	12,089	2286	0	0.0
25. Lyngby	75	12,164	187	1	1.3
26. Balders	68	12,232	130	17	25.0
	1176		13,675	108	
Total	12,232		196,426	5616	

† Including deaths.

Source: Medical Report II for the Fiscal Year 1962/1963. Published by the National Health Service of Denmark.

select, by means of the table, twelve different numbers between 1 and 26. This could be done by taking, in the table, successive two-digit numbers from a given reference point, retaining only those numbers up to and including 26, but for the sake of simplicity and in order to avoid wasting figures, it can be decided

(a) to consider all the numbers up to and including 30;
(b) to subtract 30 from all the numbers between 31 and 60;
(c) to subtract 60 from all the numbers between 61 and 90;
(d) to discard all numbers greater than 90 and every reduced number between 27 and 30.

Taking, for instance, line 6 of the table (p. 69) as the reference point and reading down, we obtain:

Number	Reduced number	Decision	y_i
03	3	Accepted	691
62	62 − 60 = 2	Accepted	266
08	8	Accepted	426
07	7	Accepted	608
01	1	Accepted	266
72	72 − 60 = 12	Accepted	239
88	88 − 60 = 28	Discarded, above 26	
45	45 − 30 = 15	Accepted	78
96		Discarded, above 90	
43	43 − 30 = 13	Accepted	64
50	50 − 30 = 20	Accepted	1
22	22	Accepted	75
96		Discarded, above 90	
31	31 − 30 = 1	Rejected, already accepted	
78	78 − 60 = 18	Accepted	14
84	84 − 60 = 24	Accepted	0
		Total $\sum_{i=1}^{12} y_i =$	2728

We then calculate:

(1) The mean $\bar{y}$

$$\bar{y} = \frac{2728}{12} = 227.3.$$

(2) The variance $v(\bar{y})$ (formulae 3.4 and 3.5)

$$s^2 = \frac{1}{11}\left(691^2+266^2+\ \ldots\ +0^2-\frac{2728^2}{12}\right) = 56{,}645$$

$$v(\bar{y}) = \frac{26-12}{12\times 26}56{,}645 = 2542$$

$$\sqrt{v(\bar{y})} = 50.4.$$

(3) $\hat{\bar{Y}}_S$ and $\hat{\bar{Y}}_I$ at the level of significance $\alpha = 0.05$ (formulae 3.10).

As $t_{0.05} = 2.201$,

$$\hat{\bar{Y}}_S = 227.3+2.201\times 50.4 = 338.2$$

$$\hat{\bar{Y}}_I = 227.3-2.201\times 50.4 = 116.4.$$

(4) It is also possible to estimate the total number of discharges of the arteriosclerotic patients Y, should this be of more interest or value than the mean number of discharges $\bar{y}$ by hospital:

$\hat{Y}$ as well as $\hat{Y}_S$ and $\hat{Y}_I$ (formula 3.7):

$$\hat{Y} = 26\bar{y} = \frac{26\times 2728}{12} = 5911$$

$$\hat{Y}_S = 26\hat{\bar{Y}}_S = 8793$$

$$\hat{Y}_I = 26\hat{\bar{Y}}_I = 3026.$$

The parameters of the population can be calculated from Table 3.1. Thus:

$$Y = 5616$$

$$\bar{Y} = 216$$

$$S^2 = 71{,}429.6$$

$$V(\bar{y}) = 3205.$$

The sampling error for the estimate of the mean $\bar{Y}$ is $\bar{y}-\bar{Y}$, i.e. $227-216 = 11$, which is a very low figure considering that the standard error of $\bar{y}$ is nearly five times greater.

As already mentioned, formulae (3.10) are correct as long as the y_i follow a normal distribution. They are useful as a first approximation even for small samples when the distribution of the y_i is reasonably close to normality. Otherwise, when the distribution is highly skew, and particularly when the sample is small, other formulae must be used or special tables consulted (Hald, 1951; Geigy, 1960).

3.3.4. Determining the Sample Size necessary to obtain the Desired Precision

The precision of an estimate is measured by the standard error of that estimate, hence the smaller the standard error the greater the precision. The precision is usually specified by the range $2i$ of the confidence interval of the mean of the sample at the level of significance α.

By virtue of (3.10),

$$2i = \hat{Y}_S - \hat{Y}_I = 2t_\alpha \sqrt{v(\bar{y})}.$$

Upon replacing $v(\bar{y})$ by its expression in (3.5), we have

$$i = t_\alpha \frac{s}{\sqrt{n}} \sqrt{\frac{N-n}{N}}. \tag{3.11}$$

If N is much greater than n, the expression $(N-n)/N$ on the right-hand side of this expression is very close to unity. Then, as a first approximation,

$$n' = \left(\frac{t_\alpha s}{i}\right)^2. \tag{3.12}$$

If n' is small enough in relation to N, for example of the order of no more than 10% of N, it is not necessary to proceed further. Otherwise it would seem indicated to take n directly from formula (3.11). In that case

$$n = \frac{1}{\left(\frac{i}{t_\alpha s}\right)^2 + \frac{1}{N}}. \tag{3.13}$$

Considering that on the right-hand side of (3.13) the term $(i/t_\alpha s)^2 = 1/n'$, we can write (3.13) so:

$$n = \frac{Nn'}{N+n'}, \tag{3.14}$$

where n' is given by (3.12).

The standard deviation S of the distribution of y may be known. If so, it is sufficient to replace s by S, and t_α by the standard normal deviate corresponding to the level α, in the foregoing formulae.

EXAMPLE 3.2. Determine the number of hospitals to be included in a sample so as to estimate the number Y of discharges of patients suffering from arteriosclerotic heart disease in the 26 hospitals of Copenhagen and its suburbs correct to within 2000 discharges and with a 5% level of significance.

A previous sample of 12 hospitals has already given an estimate $s^2 = 57{,}000$ (cf. Example 3.1).

$$26i = 2000 \qquad (i = 77).$$

Further, by virtue of (3.12),

$$n' = \frac{2.2^2 \times 57{,}000}{77^2} = 46.5 \cong 47.$$

This value is greater than $N = 26$; therefore the exact formula (3.14) should be used:

$$n = \frac{26 \times 47}{26+47} = \frac{1222}{73} = 16.7.$$

A sample of $n = 17$ hospitals should be taken.

In determining the size n of a sample thus it is presupposed that the degree of precision desired has already been decided. This is not always easy to do beforehand; neither is an estimate of S always available. In many cases, however, upper and lower limits can be assigned to S from consideration of the distribution under investigation or by analogy with other similar distributions. In general the choice of i is more difficult. If too large an interval is chosen the sample will not yield valid conclusions; if i is too small, the sample becomes too large and the cost prohibitive. As a general case, the choice of i is an application of the decision function theory. In practice, one often arrives at a satisfactory solution by visualizing different conceivable situations and considering the risks one is ready to accept of taking a wrong decision on the basis of the sample.

The foregoing calculations were performed with regard to the sampling error only, that is, on the implicit assumption that the accuracy of the results was not influenced by other sources of error such as errors of measurement, of interpretation, non-response, etc. We shall now consider a simple case where the data are subject to a constant bias due, for example, to defective calibration of the measuring device. All the data are affected by a systematic error b, and so the mean $\bar{y}$ is also affected. If μ is the true mean in the population, the total error of the estimate $\bar{y}$ of the mean μ can be written so:

$$\bar{y} - \mu = \bar{y} - E(\bar{y}) + E(\bar{y}) - \mu.$$

This expression contains, on the one hand, the sampling error $\bar{y} - E(\bar{y})$ and, on the other, the bias or systematic error $b = E(\bar{y}) - \mu$. By squaring both sides and taking the mathematical expectation, we get

$$E[(\bar{y} - \mu)^2] = E[(\bar{y} - E(\bar{y}))^2 + (E(\bar{y}) - \mu)^2]. \tag{3.15}$$

The mean square error of $\bar{y}$ is the sum of the random sampling variance and the square of the bias.

The following table shows the root mean square error of $\bar{y}$ as a ratio to the standard deviation S of the distribution for different sizes of samples, without bias ($b = 0$), and with bias of 10% and 30% of the standard deviation S. The calculations are carried out on the assumption that N is very large, so that the finite population correction factor is equal to unity.

By putting $b = kS$,

$$\sqrt{E[(\bar{y}-\mu)^2]} = S\sqrt{\frac{1}{n}+k^2}.$$

The table thus gives $\sqrt{\frac{1}{n}+k^2}$ for different values n and for $k = 0, 0.1$ and 0.3.

n	$b = 0$	$b = 0.1S$	$b = 0.3S$
10	0.32	0.33	0.44
50	0.14	0.17	0.33
100	0.10	0.14	0.32
200	0.071	0.12	0.31
500	0.045	0.11	0.30

Thus, with increasing size of sample, the total error is progressively controlled by the bias.

A further source of difficulty is generally encountered whenever several variates are investigated simultaneously. Each one will have its own distribution in the population, each with a different S^2. In addition, the required precision of the estimates of the mean may vary from one variate to another. There is a risk that some particular variate may require an unjustifiably large sample. In such instances one finds as many possible sizes n as there are variates and one must be selective and consider the relative importance of the various items.

3.3.5. Application to Qualitative Variates

The formulae to be discussed in this subsection are relevant in simple random sampling where for each unit of the population a particular attribute is either present or absent. For example, the attribute may be the response, positive or negative, to a serological test.

The N units of the population are distributed into two classes. In the first the attribute is present—say that A is the number of such units. The other class contains the units where the attribute is absent—here $N-A$ is their number. The notation is as follows:

	Population	Sample
Number of units	N	n
Number of units with the attribute	A	a
Proportion of units with the attribute	$P = A/N$	$P = a/n$

P or A must be estimated from the sample and the precision of this estimate calculated.

The derivation of formulae for proportions is simplified by the introduction of an auxiliary variate (y) which assumes the value 1 if the attribute is present in the unit, or 0 if it is absent. The variate y is a random variate to which the properties outlined in sub-section 3.3.2 apply. We then have:

$$A = \sum_{i=1}^{N} y_i \qquad a = \sum_{i=1}^{n} y_i$$

$$P = \overline{Y} = \frac{1}{N}\sum_{i=1}^{N} y_i \quad p = \bar{y} = \frac{1}{n}\sum_{i=1}^{n} y_i$$

Hence it is sufficient to apply the theorem of sub-section 3.3.2 to the variate (y) thus defined.

We then have the following corollary:

Corollary 3.1. A sample of n units is taken at random from a population of N units:

(a) The proportion p of sampled units possessing the attribute is a non-biased estimate of the mean value P of these units in the population.

(b) The variance $V(p)$ of p is given by the formula

$$V(p) = \frac{N-n}{N-1}\,\frac{P(1-P)}{n} \tag{3.16}$$

(c) The formula
$$v(p) = \frac{N-n}{N}\,\frac{p(1-p)}{n-1} \tag{3.17}$$

provides a non-biased estimate of $V(p)$.

This corollary is established without difficulty when one expresses S^2 and s^2 in the formulae (3.2) and (3.4) as a function of P and of p. The result is

$$S^2 = \frac{1}{N-1}\sum_{i=1}^{N}(y_i-\bar{Y})^2 = \frac{1}{N-1}\left[\sum_{i=1}^{N}y_i^2-N\bar{Y}^2\right]$$

$$= \frac{1}{N-1}(A-NP^2) = \frac{N}{N-1}(P-P^2),$$

whence, finally,

$$S^2 = \frac{N}{N-1}P(1-P).$$

Likewise

$$s^2 = \frac{n}{n-1}p(1-p).$$

For the confidence limits of P at the level of significance α,

$$\left.\begin{aligned}\hat{P}_S &= p+t_\alpha\sqrt{\frac{N-n}{N}}\cdot\sqrt{\frac{p(1-p)}{n-1}}\\ \hat{P}_I &= p-t_\alpha\sqrt{\frac{N-n}{N}}\cdot\sqrt{\frac{p(1-p)}{n-1}}\end{aligned}\right\} \quad (3.18)$$

When P is less than 0.20 or, conversely, exceeds 0.80, the distribution of p tends to become too skew, unless n is very large, and use is made of special tables, for example Geigy (1960).

EXAMPLE 3.3 (Hypothetical). In a population of 10,000 adult persons a serological test is made on a random sample of 100. Positive reactions are found in 15 tests. Find to within 5% the confidence limits of the proportion of subjects positive to the serological test in the population.

In this case $p = 15/100 = 0.15$,

$$v(p) = 0.99\,\frac{0.15\times 0.85}{99} = 0.001275$$

$$\sqrt{v(p)} = 0.0357$$

$$\hat{P}_S = 0.15+2\times 0.0357 = 0.22$$

$$\hat{P}_I = 0.15-2\times 0.0357 = 0.08.$$

The more precise formulae yield

$$\hat{P}_S = 0.24$$

$$\hat{P}_I = 0.09$$

It is seen that here the approximate formulae give acceptable approximations.

3.4. OTHER SAMPLING METHODS

3.4.1. General Considerations

As we have seen, simple random sampling is little used except for surveys of relatively homogeneous populations for which a detailed inventory or frame is at hand and where the units are not too scattered (if traveling is involved).

Choice of sample size has already been discussed for simple random sampling. The problem can be generalized. For each sampling method there is a corresponding degree of precision and a certain cost. It is therefore a question of using the available information to select the sampling method that will achieve the maximum precision for a given cost, or that will minimize the cost for a desired degree of precision. As we shall see, selection of the best method depends upon making judicious use of the information available.

In fact, practical considerations may impose restrictions on the selection of the sample and must be considered together with the optimum statistical requirements. For example, the financial budget of the organization carrying out the survey may not only set a maximum limit on the expenditure but also itemize expenses by sub-headings, for example traveling expenses, overtime, automatic data processing, etc. The survey may require costly field laboratory work, and if only one field laboratory can be purchased with the funds available, the sample may have to be concentrated. It might be necessary to restrict the survey in time, for example to a few days or weeks corresponding to a part of the reproduction cycle of some species of insect, in which case the budget would have to allow for several teams working simultaneously if the entire population is to be covered.

Various more elaborate techniques of sampling and estimation are presented in the sub-sections which follow.

3.4.2. Ratio Estimate Method

A ratio estimate is not in fact a new sampling method but rather a different technique of estimation that enables supplementary information to be taken into account.

Suppose that we are to find the number of diabetic cases treated for the first time at out-patient clinics during a period of one year. One may take a sample of n clinics out of the N in the country as a whole and then, for each of these clinics, find the total number y of diabetic patients appearing for treatment for the first time.

By virtue of the foregoing section, an estimate of the average number of first treatments per clinic is provided by the expression

$$\bar{y} = \sum_{i=1}^{n} y_i/n,$$

and the estimated total number of first treatments is

$$\hat{Y} = N\bar{y} = \frac{Ny}{n}.$$

In many countries out-patient clinics are required to report each year the total number of new out-patients attending for "all causes". From such reports the total number X and the average number $\bar{X}$ of new out-patients in all clinics are available. One may therefore inquire whether the ratio estimates

$$\bar{y}_q = \frac{\bar{y}}{\bar{x}}\bar{X} \quad \text{or} \quad \hat{Y}_q = \frac{\bar{y}}{\bar{x}}X$$

are not perhaps better estimates than $\bar{y}$ and $\hat{Y}$, where $\bar{x}$ is the sample mean of new out-patients "all causes".

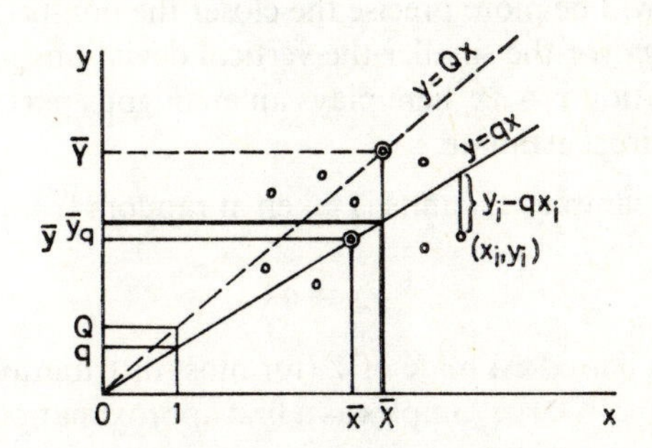

FIG. 3.1. Ratio estimate method.

One may surmise from Fig. 3.1 that if there is a close correlation between y and x and, as a first approximation, if y is proportional to x, then the ratio estimate $\bar{y}_q$ may be more precise than the direct estimate $\bar{y}$.

In a general way it can be assumed that (1) the two variates x and y are both measured in each sampled unit, and (2) we know the total value of all the x, X, or its average value $\bar{X}$ in the population. We thus have the following notation:

	Population	Sample
Number of units	N	n
Value of the variate investigated in unit i	y_i	y_i
Value of the auxiliary variate in unit i	x_i	x_i
Total value of the variate investigated	$Y = \sum_{i=1}^{N} y_i$	$y = \sum_{i=1}^{n} y_i$
Total value of the auxiliary variate	$X = \sum_{i=1}^{N} x_i$	$x = \sum_{i=1}^{n} x_i$
Mean value of the variate investigated	$\bar{Y} = \frac{Y}{N}$	$\bar{y} = \frac{y}{n}$
Mean value of the auxiliary variate	$\bar{X} = \frac{X}{N}$	$\bar{x} = \frac{x}{n}$
Ratio	$Q = \frac{Y}{X} = \frac{\bar{Y}}{\bar{X}}$	$q = \frac{y}{x} = \frac{\bar{y}}{\bar{x}}$

The results that follow are given here without proof. It is intuitively clear that the sampling will be more precise the closer the points (x_i, y_i) are to the straight line $y = qx$ (or the smaller the vertical deviations $y_i - qx_i$ are about this line). The deviation $y_i - qx_i$ here plays an analogous part to the deviation $y_i - \bar{y}$ used in the direct estimate.

THEOREM 4.1. A sample of n units is taken at random from a population of N units;

$$\bar{y}_q = q\bar{X} \tag{4.1}$$

where $q = \bar{y}/\bar{x}$ is a biased estimate of $\bar{Y}$ (for most distributions), but the bias is of the order of $1/n$. In large samples as a first approximation the variance of $\bar{y}_q$ is given by

$$V(\bar{y}_q) = \frac{N-n}{Nn} \sum_{i=1}^{N} \frac{(y_i - Qx_i)^2}{N-1} \tag{4.2}$$

and its estimate, from the sample, is given by

$$v(\bar{y}_q) = \frac{N-n}{Nn} \sum_{i=1}^{n} \frac{(y_i - qx_i)^2}{n-1}. \tag{4.3}$$

By expanding the squares in the term on the right-hand side of (4.3), it simplifies to

$$v(\bar{y}_q) = \frac{N-n}{Nn(n-1)}\left[\sum_{i=1}^{n} y_i^2 - 2q\sum_{i=1}^{n} x_i y_i + q^2 \sum_{i=1}^{n} x_i^2\right]. \tag{4.3'}$$

The formulae in $\hat{Y}_q = qX = N\bar{y}_q$ are obtained directly by analogy with the formulae (3.7), (3.8) and (3.9).

The term $y_i - Qx_i$, which appears on the right-hand side of (4.2), can also be written as $y_i - \bar{Y} - Q(x_i - \bar{X})$. Upon substituting this expression in (4.2), after a few simplifications we obtain

$$V(\bar{y}_q) = \frac{N-n}{Nn}\left[\frac{1}{N-1}\sum_{i=1}^{N}(y_i-\bar{Y})^2 - \frac{2Q}{N-1}\sum_{i=1}^{N}(y_i-\bar{Y})(x_i-\bar{X}) + \frac{Q^2}{N-1}\sum_{i=1}^{N}(x_i-\bar{X})^2\right].$$

The first term of the right-hand member is in fact $V(\bar{y})$, the variance of the arithmetic mean $\bar{y}$ of the sample.

One further puts

$$\frac{1}{N-1}\sum_{i=1}^{N}(y_i-\bar{Y})(x_i-\bar{X}) = \varrho S_x S_y,$$

where S_x and S_y are given by the expressions

$$S_x^2 = \frac{1}{N-1}\sum_{i=1}^{N}(x_i-\bar{X})^2, \quad S_y^2 = \frac{1}{N-1}\sum_{i=1}^{N}(y_i-\bar{Y})^2,$$

and ϱ is the coefficient of correlation between y and x. One obtains after a few transformations,

$$V(\bar{y}) - V(\bar{y}_q) = \frac{N-n}{Nn}(2Q\varrho S_x S_y - Q^2 S_x^2),$$

or, in other words, the ratio estimate is more precise than the direct estimate when

$$\varrho > \frac{1}{2}\cdot\frac{S_x/\bar{X}}{S_y/\bar{Y}} \tag{4.4}$$

When there are good reasons to doubt that the relation between X and Y passes through the origin, regression estimate can be tried:

$$\bar{y}_{\text{reg}} = \bar{y} + b(\bar{X} - \bar{x}). \tag{4.5}$$

One may surmise, as previously, that the deviations $y_i-[\bar{y}+b(x_i-\bar{x})]$ about the regression will influence the precision of the estimate and play a part similar to y_i-qx_i in formula (4.3).

EXAMPLE 4.1. In Example 3.1 of sub-section 3.3.3 a direct estimate $\bar{y}$ of the number of discharges of patients suffering from arteriosclerotic heart disease in one year was found. One can also calculate a ratio estimate with the total number (x) of discharges from "all causes" as the auxiliary variate because the total number of discharges is usually available for each hospital and because it can be assumed that there is a linear relation through the origin between (x) and (y).

Calculations give

$$\bar{X} = 7554.8 \qquad \bar{Y} = 216.0$$

$$S_x^2 = 69{,}748{,}975.4 \qquad S_y^2 = 71{,}429.6$$

$$\varrho = 0.882$$

$$\frac{1}{2}\,\frac{S_x/\bar{X}}{S_y/\bar{Y}} = 0.447.$$

Therefore, since this latter expression is less than ϱ, the ratio estimate will be more precise than the direct estimate by virtue of formula (4.4).

By way of illustration we shall calculate the ratio estimate of the number of discharged patients with arteriosclerotic heart disease in the twelve hospitals selected in Example 3.1.

The following twelve pairs are available from Table 3.1:

	x	y
	25,253	691
	22,878	266
	6949	426
	20,248	608
	935	266
	9534	239
	3857	78
	2514	64
	1715	1
	2698	75
	2213	14
	2286	0
Total	101,080	2728

Moreover,

$$\bar{x} = 8423.3 \qquad \bar{y} = 227.3 \qquad q = 0.026989$$

$$\sum_{i=1}^{12} x_i^2 = 1{,}752{,}698{,}618 \qquad \sum_{i=1}^{12} x_i y_i = 42{,}030{,}554 \qquad \sum_{i=1}^{12} y_i^2 = 1{,}243{,}256$$

$$\bar{y}_q = 203.9 \qquad \hat{Y}_q = 5301$$

$$v(\bar{y}_q) = 1024.7 \qquad v(\hat{Y}_q) = 692{,}709$$

The sampling error $\bar{y}_q - \bar{Y}$ becomes $203.9 - 216.0$, i.e. about -12, or in absolute magnitude as much as the random sampling error of Example 3.1.

It is more useful, however, to compare the standard errors or the variances of $\bar{y}$ and of $\bar{y}_q$. We have, respectively, $V(\bar{y}) = 3205$, $V(\bar{y}_q) = 712$. The ratio estimate thus has a variance more than four times as small as that of the direct estimate, and hence the confidence limits are about twice as small. The gain in accuracy is appreciable as could be expected.

3.4.3. Sampling with Probability Proportional to the Size of the Unit

In simple random sampling the standard deviation of the mean $\bar{y}$ becomes greater as the values y_i are more dispersed. This often happens when the sampling units are of different size and the value of the variate investigated is related to the size of the unit. Thus, for instance, the number of patients eospitalized in an establishment depends on the number of beds available, and the quantity of milk provided from a farm depends on the number of cows or of the grazing area.

Whenever the variate varies proportionally with the size of the unit, it seems logical (1) to reduce its variability by considering the quantities y_i/M_i, where M_i represents the size of the unit i, and (2) to arrange that units of greater size are correspondingly more likely to be included in the sample. It is then said that the units are sampled with probability proportional to the size of the unit. Unlike the sampling methods described in section 3.3 where the same unit could only appear once in the same sample, here any one unit can be included as many times as it is drawn. In this sense one speaks of sampling with replacement, analogous to drawing balls from an urn where each drawn ball is put back in the urn before the next draw.

The following notation applies:

	Population	Sample
Number of units	N	n
Value of the variate in unit i	y_i	y_i
Size of unit i	M	M
Sum of the sizes of the units in the population	$M = \sum_{i=1}^{N} M_i$	
Probability of drawing unit i		$\pi_i = \frac{M_i}{M}$

The following theorem can be formulated:

THEOREM 4.2. In a population of N units, n are drawn with replacement and with probability $\pi_i = M_i/M$ proportional to the size M_i of the unit. Here

$$\bar{y}_{pp} = \frac{M}{Nn} \sum_{i=1}^{n} \frac{y_i}{M_i} \tag{4.6}$$

is a non-biased estimate of $\bar{Y}$, and its variance is given by

$$V(\bar{y}_{pp}) = \frac{M}{nN^2} \sum_{i=1}^{N} M_i \left(\frac{y_i}{M_i} - \frac{Y}{M} \right)^2. \tag{4.7}$$

From the sample the estimate is

$$v(\bar{y}_{pp}) = \frac{M^2}{N^2 n(n-1)} \sum_{i=1}^{n} \left(\frac{y_i}{M_i} - \frac{N\bar{y}_{pp}}{M} \right)^2. \tag{4.8}$$

This formula can be presented in a form which is easier to compute:

$$v(\bar{y}_{pp}) = \frac{M^2}{N^2 n(n-1)} \left[\sum_{i=1}^{n} \left(\frac{y_i}{M_i} \right)^2 - \frac{1}{n} \left(\sum_{i=1}^{n} \frac{y_i}{M_i} \right)^2 \right]. \tag{4.9}$$

The concept of size is arbitrary in many cases. The size of a farm can be measured by the surface area of the fields, the acreage sown with grain, the number of cows, etc. The cultivated acreage is probably a good general measure of the size of a farm, but when milk production is being studied the number of cows or the head of cattle is preferable. In sample surveys several different variates are very often recorded in each unit and there is often no adequate single criterion of size. Furthermore, often only an estimate of the size is available. Hence, one can consider a sequence π_i of positive numbers, the sum of which over N units is equal to unity: $\sum_{i=1}^{N} \pi_i = 1$, π_i being analogous to the relative size of the unit M_i/M.

If the units are taken with probabilities π_i and with replacement, we have

$$\bar{y}_\pi = \frac{1}{nN} \sum_{i=1}^{n} \frac{y_i}{\pi_i}, \tag{4.10}$$

which is a non-biased estimate of $\bar{Y}$.

EXAMPLE 4.2. Estimate the total number of discharges of patients with arteriosclerotic heart disease during a period of 12 months using a sample of 12 hospitals selected with probabilities proportional to the number of beds. The sample may be selected by using cumulative sums ΣM_i. These sums are given in Table 3.1. Twelve numbers up to and including 12,232 are chosen from a table of random numbers. This is done in the present example by taking five-digit numbers and subtracting 20,000 from any number between 20,001 and 40,000; 40,000 from any number between 40,001 and 60,000; and so on. Those hospitals for which the sum ΣM_i is at least equal to or just greater than the random number selected are included in the sample. Starting immediately after the last random number drawn in Example 3.1 (line 26), and taking the first five digits, we obtain:

Number	Reduced number	Decision	100y/M
36,671	16,671	Rejected	
07,285	7285	Sample hospital No. 7	77.9
10,158	10,158	Sample hospital No. 14	63.5
55,196	15,196	Rejected	
53,812	13,812	Rejected	
51,863	11,863	Sample hospital No. 22	47.8
35,917	15,917	Rejected	
.	.	.	.
.	.	.	.
02,960	2960	Sample hospital No. 3	57.5
49,834	9834	Sample hospital No. 13	20.4
84,607	4607	Sample hospital No. 4	53.8
.	.	.	.
.	.	.	.
51,295	11,295	Sample hospital No. 19	0.0
21,313	1313	Sample hospital No. 1	18.2
29,012	9012	Sample hospital No. 10	7.7
.	.	.	.
90,846	10,846	Sample hospital No. 16	72.1
46,406	6406	Sample hospital No. 6	86.3
20,318	318	Sample hospital No. 1	18.2

We thus have

$$\sum_{i=1}^{n} \frac{y_i}{M_i} = 5.234, \qquad \sum_{i=1}^{n} \left(\frac{y_i}{M_i}\right)^2 = 3.237022;$$

hence

$$\bar{y}_{pp} = \frac{12{,}232}{26 \times 12} 5.234 = 205.2$$

$$v(\bar{y}_{pp}) = \frac{(12{,}232)^2}{26^2 \times 12 \times 11} (3.237022 - 2.282896) = 1599.9.$$

Calculation of $V(\bar{y}_{pp})$ gives a value of 1530, that is, a little less than half the variance of the direct estimate with random sampling. Although the gain in accuracy as compared to simple random sampling is appreciable, it is not as great as that obtained with the ratio estimate ($V(\bar{y}_q) = 712$).

3.4.4. Systematic Sampling

It can be a tiresome operation to draw a large number of units at random from a frame. A considerable amount of time is often saved by taking the sample systematically, that is, by drawing every kth name on a list. Equally one can foresee that a systematic draw that distributes the sample uniformly throughout the population will usually be more precise than random sampling. All the same one should check whether the variate in question is subject to periodic fluctuations, and whether the interval between two consecutively sampled units coincides with this period, or is a multiple thereof.

The mean value $\bar{y}_s$ of the sample is a non-biased estimate of the mean $\bar{Y}$ of the population. Unfortunately, there is no general formula to calculate an estimate $v(\bar{y}_s)$ of the variance $V(\bar{y}_s)$. In certain cases, when the sample is very large, it can be taken in the form of l independent, systematic sub-samples, the interval between two consecutive units obviously being l times greater in these sub-samples. The variance between the mean values $\bar{y}_{s_1}, \ldots, \bar{y}_{s_l}$ of the sub-samples can be taken to estimate the variance of the combined sample.

If k is the interval between two consecutive units, or, what is the same thing, the reciprocal of the sampling fraction, in order to draw the sample one chooses a number at random between 1 and k and thereupon takes each kth unit from this number onwards. For instance, if one wants a systematic sample of $\frac{1}{20}$th of the patients admitted to hospital in one year, choose any number between 1 and 20 (e.g. 14), and then take every twentieth line of the Admission Register of the hospital from the 14th line onwards, that is, the lines 14, $14+20 = 34$, $14+(2\times 20) = 54$, . . ., $14+(p\times 20)$ and so on. Alternatively one can take 5 samples of one-hundredth each by choosing any five numbers between 1 and 100 and then taking each 100th line from each of these five numbers onwards.

3.4.5. Stratified Random Sampling

It was seen in Section 3.3 that the variance of a sample is correspondingly greater the greater the dispersion of the values y_i about the mean. Clearly, therefore, if the total population can be broken down into homogeneous strata, the sampling will be more precise.

A stratification may be used on other grounds. It may be that separate estimates are wanted for different groups of the population, e.g. for each of the large administrative divisions of the country, or for the urban and rural populations separately, and that a given precision is imposed on each estimate. There is then an advantage in considering each of these groups as a stratum and in taking an independent and separate sample in each one. It may also be the case that no single sampling method, or no particular frame, is truly appropriate for the population as a whole, but that the population can be broken down into strata for which there are appropriate frames and sampling methods. For instance, the population of a country may be broken down into the settled urban population, the settled rural population and the nomad population. In the urban stratum the sampling unit may be a block of houses, in the rural stratum villages or parishes, and in the third stratum one could carry out area sampling where the sampling unit is some area delineated on a map or on aerial photographs.

The N units of the population are allocated to k strata of size $N_1, \ldots, N_k$. In each stratum one sample is taken; say n_h is the number of units sampled in stratum h. We then have the following notation:

	Population	Sample
Number of units	N	n
Number of strata	k	
Number of units in stratum h	N_h	n_h
Value of the variate in the *i*th unit of stratum h	y_{hi}	
Mean value of the variate in stratum h	$\bar{Y}_h = \frac{1}{N_h} \sum_{i=1}^{N_h} y_{hi}$	$\bar{y}_h = \frac{1}{n_h} \sum_{i=1}^{n_h} y_{hi}$

	Population	Sample
Total value of the variate in stratum *h*	$Y_h = \sum_{i=1}^{N_h} y_{hi}$	$y_h = \sum_{i=1}^{n_h} y_{hi}$
Total value of the variate in the population	$Y = \sum_{h=1}^{k}\sum_{i=1}^{N_h} y_{hi} = \sum_{h=1}^{k} Y_h$	$y = \sum_{h=1}^{k}\sum_{i=1}^{n_h} y_{hi} = \sum_{h=1}^{k} y_h$
Mean value of the variate	$\bar{Y} = \dfrac{Y}{N}$	$\bar{y} = \dfrac{y}{n}$
Variance of the variate in stratum *h*	$S_h^2 = \dfrac{\sum_{i=1}^{N_h}(y_{hi}-\bar{Y}_h)^2}{N_h-1}$	$s_h^2 = \dfrac{\sum_{i=1}^{n_h}(y_{hi}-\bar{y}_h)^2}{n_h-1}$

We then have the following theorem:

THEOREM 4.3. If, for any $h \leqslant k$, $\hat{\bar{Y}}_h$ is a non-biased estimate of the mean $\bar{Y}_h$ in stratum h, and if the k estimates are independent, then

$$\bar{y}_{st} = \frac{1}{N}\sum_{h=1}^{k} N_h \hat{\bar{Y}}_h \tag{4.11}$$

is a non-biased estimate of $\bar{Y}$. (The subscript "st" stands for stratified.)

Further, if $V(\hat{\bar{Y}}_h)$ is the variance of $\hat{\bar{Y}}_h$, estimated by $v(\hat{\bar{Y}}_h)$, then the variance of $\bar{y}_{st}$ is given by the expression

$$V(\bar{y}_{st}) = \frac{1}{N^2}\sum_{h=1}^{k} N_h^2 V(\hat{\bar{Y}}_h), \tag{4.12}$$

and its estimate is

$$V(\bar{y}_{st}) = \frac{1}{N^2}\sum_{h=1}^{k} N_h^2 v(\hat{\bar{Y}}_h). \tag{4.13}$$

It will be seen that the theorem is general. It is valid whatever sampling method is used to take the sample in each stratum, providing the estimates of the means in the k strata are independent and non-biased. For instance, in one stratum random sampling could be used, in another use could be made of sampling with probability proportional to the size of the unit, and so on.

Proof of Theorem 4.3 derives directly from the fact that the mathematical expectation is a linear operator and the k random variates $\hat{\bar{Y}}_1, \ldots, \hat{\bar{Y}}_k$ are independent; hence their covariances are zero.

If the sample is taken at random in each stratum, the following corollary is implied as a special case of Theorem 4.3:

Corollary 4.1. In stratified random sampling with n_h units taken in stratum h $(h = 1, \ldots, k)$

$$\bar{y}_{st} = \frac{1}{N} \sum_{h=1}^{k} N_h \bar{y}_h \tag{4.14}$$

is a non-biased estimate of $\bar{Y}$.

The variance is given by the expression

$$V(\bar{y}_{st}) = \frac{1}{N^2} \sum_{h=1}^{k} N_h(N_h - n_h) \frac{S_h^2}{n_h} \tag{4.15}$$

and it is estimated by

$$v(\bar{y}_{st}) = \frac{1}{N^2} \sum_{h=1}^{k} N_h(N_h - n_h) \frac{s_h^2}{n_h}. \tag{4.16}$$

In certain studies an estimate of the total Y may be more interesting than that of the mean $\bar{Y}$. The expression

$$\hat{Y}_{st} = N\bar{y}_{st} \tag{4.17}$$

is a non-biased estimate of Y, and the variances are given by the expressions

$$V(\hat{Y}_{st}) = N^2 V(\bar{y}_{st}) \tag{4.18}$$

$$v(\hat{Y}_{st}) = N^2 v(\bar{y}_{st}). \tag{4.19}$$

The confidence limits $\hat{Y}_S$ and $\hat{Y}_I$ at the level of significance α of the mean $\bar{y}$ are given by the expressions

$$\left.\begin{aligned} \hat{Y}_S &= \bar{y}_{st} + t_\alpha \sqrt{v(\bar{y}_{st})} \\ \hat{Y}_I &= \bar{y}_{st} - t_\alpha \sqrt{v(\bar{y}_{st})} \end{aligned}\right\}. \tag{4.20}$$

When all the n_h are large, say greater than 30, it is sufficient to read the t_α from a table of standard normal deviates. Otherwise, providing that the distribution of the y does not notably differ from a normal distribution in each of the strata, a table of Student's t-distribution can be used with a number of degrees of freedom somewhere between the smallest of the $n_i - 1$ and their sum (Satterthwaite, 1946).

In many surveys it is justifiable to assume that the variance between units is the same in all the strata, i.e. that only the central position of the distribution of the y varies from one stratum to another. The variances $S_1^2, S_2^2, \ldots, S_k^2$

are then regarded as k estimates of one and the same hypothetical variance σ^2_{comb} which one estimates in the sample by s^2_{comb} as follows:

$$s^2_{\text{comb}} = \frac{\sum_{i=1}^{k} \sum_{j=1}^{n_i} (y_{ij}-\bar{y}_i)^2}{\sum_{i=1}^{k} (n_i-1)}.$$

Therefore in stratified random sampling the estimate $v(\bar{y}_{\text{st}})$ of the variance of the mean of the y in the population is given by

$$v(\bar{y}_{\text{st}}) = \frac{s^2_{\text{comb}}}{N^2} \sum_{h=1}^{k} \frac{N_h(N_h-n_h)}{n_h} \tag{4.21}$$

whenever the variance of the y_{hi} is the same in each of the strata. This comes of replacing the s_h^2 by s^2_{comb} in formula (4.16). In this case the number of degrees of freedom of t_α in formulae (4.20) is of course $\sum_{h=1}^{k} (n_h-1) = n-k$.

Application to qualitative variates

On many occasions the investigator is interested in the proportion of units with some particular attribute. By setting $y_{hi} = 1$ or 0 according to whether the unit i of stratum h possesses, or does not possess, the attribute in question (cf. sub-section 3.3.5), we obtain the following notation which is supplementary to the one at the beginning of this sub-section:

	Population	Sample
Number of units with the attribute	A	a
Proportion of units with the attribute	$P = \frac{A}{N}$	$p = \frac{a}{n}$
Number of units with the attribute in stratum h	A_h	a_h
Proportion of units with the attribute in stratum h	$P_h = \frac{A_h}{N_h}$	$p_h = \frac{a_h}{n_h}$

Corollary 4.2 to Theorem 4.3. Given that the population is broken down into k strata of size $N_1, \ldots, N_k$, at random and without replacement we take n_1 units in the first stratum, ..., n_k units in the kth stratum. If $a_1, \ldots, a_k$ are

the number of sample units in the 1st, ..., kth stratum possessing the attribute, then

$$p_{st} = \frac{1}{N}\sum_{h=1}^{k}\frac{N_h}{n_h}a_h = \frac{1}{N}\sum_{h=1}^{k}N_h p_h \tag{4.22}$$

is a non-biased estimate of $P = A/N$.

The variance is given by

$$V(p_{st}) = \frac{1}{N^2}\sum_{h=1}^{k}\frac{N_h^2(N_h-n_h)}{N_h-1}\,\frac{P_h(1-P_h)}{n_h} \tag{4.23}$$

and estimated by

$$v(p_{st}) = \frac{1}{N^2}\sum_{h=1}^{k}N_h(N_h-n_h)\frac{p_h(1-p_h)}{n_h-1}. \tag{4.24}$$

These formulae derive directly from (3.16) and (3.17) and from (4.14), (4.15) and (4.16).

Allocation to strata

The formulae so far presented enable the mean of the variate in the population and the precision of the sampling to be estimated only after the sampling plan has been determined. The investigator is faced with the problem of (1) how to choose the strata, and (2) how many units to choose in each stratum.

Often the units fall naturally into various homogeneous groups and then these groups can be used to advantage as strata. In enumerative studies the advantage of stratification is correspondingly greater the smaller the variation of the variate within a stratum.

In sampling for analytical purposes, the strata will most often be distinct domains of study that one wants to compare. The epidemiologist may want to compare the prevalence of cancer of the stomach in rural, semi-urban and urban populations. The total number of cases of cancer, or their relative frequency, in the country as a whole is of less interest to him; he wants to discover whether this relative frequency is the same or significantly different from one domain of study or stratum to another. However, the administrator may be more interested in the total number of cancer cases in the whole country, for example to be able to plan adequate services.

In general, and if $\bar{y}_1, \bar{y}_2, \ldots, \bar{y}_k$ are the means of the samples of the k strata, the problem of the optimal allocation of the sample consists in finding the combination of sample sizes $n_1, n_2, \ldots, n_k$ for the k strata such that the variances of the differences $\bar{y}_i-\bar{y}_j$ are as small as possible in an analytical study, or such that the variance of the sum $N_1\bar{y}_1+N_2\bar{y}_2+\ldots+N_k\bar{y}_k$ is as small as possible in an enumerative survey for a given size $n = n_1+\ldots+n_k$, or for a given total cost, or for any other relevant relation between the n_i's. For any particular size n_0 it can be shown that, as a first approximation, the optimal allocation of the sample is proportional to the product S_hN_h in an enumerative survey

$$n_h = \frac{n_0 N_h S_h}{\sum_{h=1}^{k} N_h S_h},$$

and to S_h in an analytical survey designed to compare the differences of means between strata

$$n_h = \frac{n_0 S_h}{\sum_{h=1}^{k} S_h}.$$

EXAMPLE 4.3. Estimate the mean annual number of discharges of patients treated for arteriosclerotic heart disease using a stratified sample of $n = 12$ hospitals. These hospitals are assigned to three different strata according to the size of the hospital:

stratum 1—large hospitals (600 beds or more)
stratum 2—medium-sized hospitals (200–599 beds)
stratum 3—small hospitals (less than 200 beds)

The sampling is to be optimum, that is, the variance of the mean number of discharges for a given sample size n_0 is to be estimated as precisely as possible. The optimum allocation into three strata presupposes the knowledge of all the S_h^2, or at least an estimate of the S_h^2, by virtue of the formula given immediately above. These values can be calculated directly from Table 3.1. We have:

Stratum	N_h	S_h^2	S_h	N_hS_h	$\frac{12N_hS_h}{\Sigma N_hS_h}$	n_h
1	7	52,297	228.7	1600.8	6.2	6
2	10	16,466	128.2	1282.4	5.0	5
3	9	602	24.5	220.8	0.9	1
				3104.0		12

This allocation allots only one unit to stratum 3; the sampling error, therefore, cannot be calculated in this stratum. Stratum 1 contains seven hospitals from which six are to be selected for inclusion in the sample.

The following sample is obtained by random selection:

Stratum 1		Stratum 2		Stratum 3	
i	y_{1i}	i	y_{2i}	i	y_{3i}
1	266	10	30	26	17
2	266	12	239		
3	691	8	426		
4	580	9	235		
5	770	17	0		
7	608				
	3181		930		17

Calculation gives:

$$\hat{Y}_{st} = \tfrac{7}{6}\times 3181+\tfrac{10}{5}\times 930+9\times 17 = 5725$$

$$\bar{y}_{st} = \frac{5725}{26} = 220.2.$$

As pointed out previously, $v(\bar{y}_{st})$ cannot be calculated since the sample of the last stratum contains only one unit. In this example, the variance $V(\bar{y}_{st})$ may be calculated so as to compare it with the variance of a random sample. We get:

$$V(\bar{y}_{st}) = \frac{1}{26^2}\left(\frac{7}{6}S_1^2+\frac{50}{5}S_2^2+72S_3^2\right) = 397.65.$$

3.4.6. Two-stage Sampling (Sub-sampling)

As already mentioned, simple random sampling, and even stratified random sampling, raise certain difficulties whenever the sample is very dispersed geographically. First of all a detailed frame is needed. Then the dispersal itself complicates the organization and the verification of the work in the field, and travelling expenses, for example, are large. In order to simplify matters, the population is first broken into large units called primary units.

A definite number of primary units are then selected, from which a certain number of sampling units called secondary units are drawn. The primary units may be taken at random, or with probability proportional to their size. One may choose a direct method of estimation, or a regression estimate or a ratio estimate if these seem indicated. Supplementary stages can also be added. The technique is illustrated in Fig. 3.2.

The population is grouped in primary units, indicated by the letters A to E in the figure. The primary units A and E, encircled by a thicker line, were selected first, and then the secondary units represented by a dark circle were drawn from these two primary units. The primary unit E contains $N_E = 12$ units of which $n_E = 5$ have been drawn.

The following notation applies:

	Population	Sample
Number of primary units	N	n
Number of secondary units in the *i*th primary unit	M_i	m_i
Total number of secondary units	$M = \sum_{i=1}^{N} M_i$	$m = \sum_{i=1}^{n} m_i$
Average number of secondary units per primary unit	$\bar{M} = \frac{M}{N}$	$\bar{m} = \frac{m}{n}$
Value of the variate in the *j*th secondary unit of the *i*th primary unit	y_{ij}	
Total value of the variate in the secondary units of the *i*th primary unit	$Y_i = \sum_{j=1}^{M_i} y_{ij}$	$y_i = \sum_{j=1}^{m_i} y_{ij}$
Mean value of the variate in the *i*th primary unit	$\bar{Y}_i = \frac{Y_i}{M_i}$	$\bar{y}_i = \frac{y_i}{m_i}$
Total value of the variate	$Y = \sum_{i=1}^{N} Y_i$	$y = \sum_{i=1}^{n} y_i$
Mean value of the variate per primary unit	$\bar{Y}_i = \frac{Y}{N}$	$\bar{y} = \frac{y}{n}$
Mean value of the variate per secondary unit	$Y = \frac{Y}{M}$	$\bar{\bar{y}} = \frac{y}{m}$
Probability of drawing the *i*th primary unit		π_i

If there is an auxiliary variate X, the notations are analogous.

There are no special problems in proving the main formulae. However, the proof is long and must be repeated for each of the several different methods

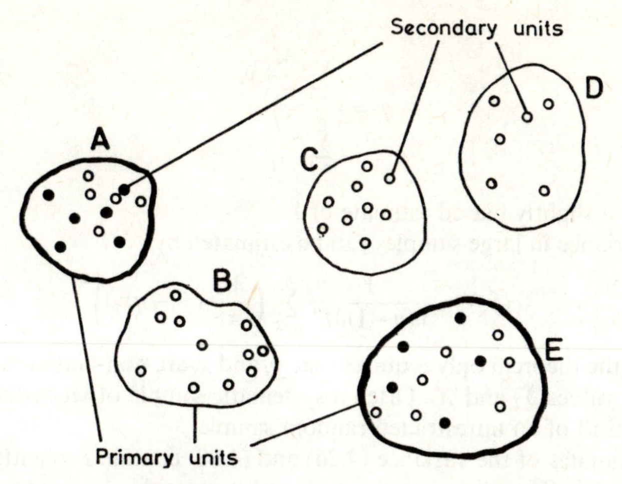

FIG. 3.2. Graphical representation of two-stage sampling.

of selecting the units at each stage. Instead, therefore, a fairly general theorem is given without proof and formulae for different selection schemes are derived as special cases.

THEOREM 4.4. Given a population consisting of N primary units, the ith primary unit comprising M_i secondary units.

A sample of n primary units is taken with replacement and with probabilities $\pi_1, \ldots, \pi_N$ of drawing the 1st, . . ., Nth primary units of the population respectively. From within each drawn primary unit a certain number of secondary units are selected at random and without replacement; m_i represents the number of secondary units selected in the ith primary unit. Further, (x) and (y) are two variates defined in each of the secondary units.

The direct estimate

$$\bar{\bar{y}}_\pi = \frac{1}{nM} \sum_{i=1}^{n} \frac{M_i \bar{y}_i}{\pi_i} \tag{4.25}$$

is a non-biased estimate of $\bar{\bar{Y}}$, and its variance estimated from the sample is given by

$$v(\bar{\bar{y}}_\pi) = \frac{1}{n(n-1)M^2} \sum_{i=1}^{n} \left(\frac{M_i \bar{y}_i}{\pi_i} - M \bar{\bar{y}}_\pi \right)^2 . \tag{4.26}$$

The ratio estimate

$$\bar{\bar{y}}_q = q\bar{\bar{X}},$$

where

$$q = \frac{\sum_{i=1}^{n} \frac{M_i}{\pi_i} \bar{y}_i}{\sum_{i=1}^{n} \frac{M_i}{\pi_i} \bar{x}_i}, \tag{4.27}$$

is usually a slightly biased estimate of $\bar{\bar{Y}}$.

The variance in large samples can be estimated by

$$v(\bar{\bar{y}}_q) = \frac{1}{n(n-1)M^2} \sum_{i=1}^{n} \left(\frac{M_i}{\pi_i} (\bar{y}_i - q\bar{x}_i) \right)^2. \tag{4.28}$$

In fact the theorem only requires that $\bar{y}_i$ and $\bar{x}_i$ are non-biased estimates of the mean values $\bar{Y}_i$ and $\bar{X}_i$. Often a systematic sample of secondary units is taken instead of an unrestricted random sample.

The estimates of the variance (4.26) and (4.28) contain explicitly only the differences between primary units. In fact the quantities $\bar{y}_i$ appearing in these formulae have the form

$$\bar{y}_i = (\bar{y}_i - \bar{Y}_i) + \bar{Y}_i .$$

The term $(\bar{y}_i - \bar{Y}_i)$ reflects the difference between secondary units within a primary unit; these differences are therefore taken into account in the formulae (4.26) and (4.28) even though they do not appear there explicitly.†

Finally, the foregoing formulae are directly applicable to proportions, conventionally assuming that y_{ij} becomes 1 or 0 according as the jth secondary unit of the ith primary unit possesses or does not possess the attribute being studied. In many applications one is interested not only in the proportion of units with a particular attribute, but also in the relation between the number of these units and the number with some other attribute. In a sample survey of trachoma, for instance, one may be interested in the number of persons with a trichiasis and in the relation of this number to the number of those with a cicatricial trachoma (Tr III and Tr IV of the Classification of McCallan). The ratio-estimate formulae are then applicable. In many cases, and more particularly in "artificial" populations such as the daily production of a laboratory, or a consignment of drugs, or "plots" in agricultural experi-

† In fact, whenever the primary units are equal in size and the variability between secondary units within the primary units is the same in every primary unit a fixed number $\bar{m}$ of secondary units is selected in each primary unit. One may put $y_{ij} = \bar{\bar{Y}} + a_i + e_{ij}$, and then, by analogy with analysis of variance, we have

$$V(\bar{\bar{y}}) = \frac{1}{n} \sigma_a^2 + \frac{1}{n\bar{m}} \sigma_e^2,$$

where σ_a^2 and σ_e^2 are the variances between and within the primary units respectively.

ments (Yates and Zacopanay, 1935), the population divides into definite numbers of primary units (e.g. bottles of drugs) each containing the same number of secondary units (e.g. tablets). *A priori* one may assume that the variability between secondary units will be the same within every primary unit and that therefore there is no reason to take a larger number of secondary units in one primary unit than in any other. With a consignment of drugs, one may be interested perhaps in the weight of the tablets, or in the number of defective tablets, or in the relation between the quantity of product a to that of product b in the content of each tablet. In this instance one puts in the formulae of Theorem 4.4:

$$\pi_i = \frac{1}{N},$$

$$M_i = \overline{M},$$

$$m_i = \overline{m}.$$

Scheme I. The population comprises N primary units, each with $\overline{M}$ secondary units.

Suppose that n primary units are drawn at random and with replacement and that in every primary unit $\overline{m}$ secondary units are drawn without replacement. We then have

$$\bar{\bar{y}}_{\mathrm{I}} = \frac{1}{n\overline{m}} \sum_{i=1}^{n} \sum_{j=1}^{\overline{m}} y_{ij} = \bar{\bar{y}}, \tag{4.29}$$

$$v(\bar{\bar{y}}_{\mathrm{I}}) = \frac{1}{\overline{m}^2 n(n-1)} \left[\sum_{i=1}^{n} y_i^2 - \frac{1}{n} \left(\sum_{i=1}^{n} y_i \right)^2 \right], \tag{4.30}$$

$$q_{\mathrm{I}} = \frac{\sum_{i=1}^{n} \sum_{j=1}^{\overline{m}} y_{ij}}{\sum_{i=1}^{n} \sum_{j=1}^{\overline{m}} x_{ij}} = \frac{y}{x}, \tag{4.31}$$

$$\bar{\bar{y}}_{q_{\mathrm{I}}} = q_{\mathrm{I}} \bar{\bar{X}},$$

$$v(\bar{\bar{y}}_{q_{\mathrm{I}}}) = \frac{1}{\overline{m}^2 n(n-1)} \left[\sum_{i=1}^{n} y_i^2 - 2q_{\mathrm{I}} \sum_{i=1}^{n} y_i x_i + q_{\mathrm{I}}^2 \sum_{i=1}^{n} x_i^2 \right]. \tag{4.32}$$

EXAMPLE 4.4 *(Hypothetical)*. A hospital has received 1000 bottles each containing 100 tablets. A sample of $n = 6$ bottles is taken at random and then from each bottle a sample of $\overline{m} = 20$ tablets is analysed for quality control. The following factors are to be controlled:

(1) the average weight $\bar{\bar{y}}$ of a tablet, (2) the proportion p of tablets with substandard consistency, and (3) the relation or quotient q between the content of two substances a and b in a tablet.

The following data are obtained for the six sampled bottles:

Bottle	Weight in g (sum of 20 tablets) y	Number of tablets with sub-standard consistency	Percentage content (sum of 20 tablets)	
			Substance a	Substance b
1	21.8	1	5.42	31.2
2	19.3	0	5.01	30.3
3	22.8	1	5.43	30.5
4	23.1	3	5.43	32.4
5	22.9	1	5.20	31.4
6	19.7	1	5.56	32.9
Total	129.6	7	32.05	188.7
Mean (per tablet)	1.08	0.058	0.267	1.572
Sum of the squares	2813.68	13	171.3999	5939.91
Sum of the products			1008.658	

Estimation of the variances

(1) weight $\bar{\bar{y}}_{\text{I}} = 1.08$ g per tablet

$$v(\bar{\bar{y}}_{\text{I}}) = \frac{1}{400 \times 30}(2813.68 - 2799.36) = 0.001193$$

$$\sqrt{v(\bar{\bar{y}}_{\text{I}})} = 0.035$$

(2) proportion $p_{\text{I}} = 0.058$ of tablets with sub-standard consistency

$$v(p_{\text{I}}) = \frac{1}{400 \times 30}(13 - 8.17) = 0.000403$$

$$\sqrt{v(p_{\text{I}})} = 0.020$$

(3) relation $q_{\text{I}} = 0.170$ of content of the substance a to the substance b as $q_{\text{I}} = \bar{\bar{y}}_{q_{\text{I}}}/\bar{X}$ one has

$$v(q_{\text{I}}) = \frac{1}{\bar{\bar{X}}^2} v(\bar{\bar{y}}_{q_{\text{I}}}) \cong \frac{1}{\bar{\bar{x}}^2} v(\bar{\bar{y}}_{q_{\text{I}}}) \quad \text{and then}$$

$$v(q_{\text{I}}) = \frac{1}{400 \times 30 \times 1.572^2}(171.3999 - 2q_{\text{I}} \times 1008.658 + q_{\text{I}}^2 \times 5939.91)$$

$$= 0.00000402$$

$$\sqrt{v(q_{\text{I}})} = 0.0020.$$

In most sample surveys where the units sampled are natural units such as towns, families, factories, etc. it is rare for them to be of identical size or even close enough in size for the differences to be ignored. Three possibilities present themselves:

(a) Stratify by size of primary units, for example, localities by numbers of inhabitants and take a random sample of primary units in each stratum.

(b) Regroup and dissect the units so as to obtain primary units of similar size, and select a random sample of such units for example, split the large villages and join hamlets to neighbouring villages so as to obtain blocks of about 500 inhabitants each.

(c) Take the units as they are and select the primary units at random, or better still with probability proportional to size.

By putting $\pi_i = 1/N$ the primary units are selected at random and we have *Scheme II*. The population consists of N primary units, the ith unit containing M_i secondary units. We draw n primary units at random and with replacement. In the ith drawn primary unit we take m_i secondary units at random and without replacement. Thus

$$\bar{\bar{y}}_{II} = \frac{N}{nM} \sum_{i=1}^{n} M_i \bar{y}_i \tag{4.33}$$

$$v(\bar{\bar{y}}_{II}) = \frac{N^2}{n(n-1)M^2} \sum_{i=1}^{n} \left(M_i \bar{y}_i - \frac{\sum_{i=1}^{n} M_i \bar{y}_i}{n} \right)^2 . \tag{4.34}$$

The variance of $\bar{\bar{y}}_{II}$ consists, except for a factor, in the sum of the squares of the deviations of the $M_i \bar{y}_i$ about their mean, hence usually the more disparate in size the primary units the larger the variance. One may then prefer a ratio estimate, usually biased, but more precise:

$$\bar{\bar{y}}_{q_{II}} = \frac{\sum_{i=1}^{n} M_i \bar{y}_i}{\sum_{i=1}^{n} M_i} \tag{4.35}$$

for which the variance is obtained by putting $\bar{x}_i = 1$ $(x_{ij} \equiv 1)$ in equation (4.28):

$$v(\bar{\bar{y}}_{q_{II}}) = \frac{N^2}{n(n-1)M^2} \sum_{i=1}^{n} [M_i(\bar{y}_i - \bar{\bar{y}}_{q_{II}})]^2 . \tag{4.36}$$

M_i appears in the formula for the variance only as a weighting factor of the deviations of the values $\bar{y}_i$ about their weighted mean, so therefore $v(\bar{\bar{y}}_{q_{II}})$ is not influenced directly by large differences in size.

If an auxiliary variate x is defined in every unit, a ratio estimate of the following type is obtained:

$$\bar{\bar{y}}_{q_x\text{II}} = \frac{\sum_{i=1}^{n} M_i \bar{y}_i}{\sum_{i=1}^{n} M_i \bar{x}_i}. \tag{4.37}$$

The variance of $\bar{\bar{y}}_{q_x\text{II}}$ can be derived without difficulty from equation (4.28).

If the primary units vary appreciably in size, a selection with probability proportional to size M_i is usually preferable to random sampling. In formulae (4.25)–(4.28) π_i then takes the value M_i/M. It is usual in these cases to take a constant number of secondary units from each primary unit selected; the sample then becomes self-weighted.

Scheme III. The population consists of N primary units, the ith unit containing M_i secondary units. We draw n primary units with probability proportional to size M_i of the unit. From each primary unit so drawn, a sample of $\bar{m}$ secondary units is taken at random and without replacement. With $x_{ij} \equiv 1$ and therefore $\bar{x}_i = 1$ the two estimates (4.25) and (4.27) are identical, and therefore

$$\bar{\bar{y}}_{\text{III}} = \frac{1}{n\bar{m}} \sum_{i=1}^{n} \sum_{j=1}^{\bar{m}} y_{ij} = \bar{\bar{y}}, \tag{4.38}$$

$$v(\bar{\bar{y}}_{\text{III}}) = \frac{1}{n(n-1)\bar{m}^2} \left[\sum_{i=1}^{n} y_i^2 - \frac{\left(\sum_{i=1}^{n} y_i \right)^2}{n} \right]. \tag{4.39}$$

For a ratio estimate of the type $y = qx$, we have

$$q_{\text{III}} = \frac{\sum_{i=1}^{n} \sum_{j=1}^{\bar{m}} y_{ij}}{\sum_{i=1}^{n} \sum_{j=1}^{\bar{m}} x_{ij}} = \frac{y}{x}, \tag{4.40}$$

$$\bar{\bar{y}}_{q\text{III}} = q_{\text{III}} \bar{\bar{X}},$$

$$v(\bar{\bar{y}}_{q\text{III}}) = \frac{1}{n(n-1)\bar{m}^2} \left[\sum_{i=1}^{n} y_i^2 - 2q_{\text{III}} \sum_{i=1}^{n} x_i y_i + q_{\text{III}}^2 \sum_{i=1}^{n} x_i^2 \right]. \tag{4.41}$$

EXAMPLE 4.5 *(Hypothetical)*. In a given district there are $N = 315$ schools attended by $M = 27{,}215$ pupils. A sample of 8 schools is taken with probability proportional to the number of pupils in attendance. In each of the 8 schools a sample of $\bar{m} = 50$ pupils is drawn. In each sample we determine the number of trachomatous pupils and the number of pupils with trachoma with abundant cicatrices. Estimate the proportion of trachomatous pupils in the school population and the proportion among them having trachoma with abundant cicatrices.

The following results are obtained:

School	Number of trachomatous pupils y_i	Number of trachomatous pupils with abundant cicatrices z_i
1	40	3
2	31	0
3	47	16
4	41	8
5	43	8
6	36	5
7	39	2
8	48	13
Total	325	55
Sum of the squares	13,421	591
Sum of the products	2426	

Calculation gives:

$$\bar{\bar{y}}_{\text{III}} = \frac{1}{8\times 50} 325 = 0.812$$

$$v(\bar{\bar{y}}_{\text{III}}) = \frac{1}{8\times 7\times 50^2}\left(40^2+\ldots+48^2-\frac{325^2}{8}\right) = 0.00156$$

$$\sqrt{v(\bar{\bar{y}}_{\text{III}})} = 0.039$$

$$q_{\text{III}} = \frac{z}{y} = \frac{55}{325} = 0.169$$

$$v(q_{\text{III}}) = v(\bar{\bar{z}}_{q_{\text{III}}})/\bar{Y}^2 \cong v(\bar{\bar{z}}_{q_{\text{III}}})/\bar{\bar{y}}^2$$

$$= \frac{1}{0.812^2\times 8\times 7\times 50^2}\left[\sum_{i=1}^{8} z_i^2 - 2\times\frac{55}{325}\sum_{i=1}^{8} z_i y_i + \left(\frac{55}{325}\right)^2\times\sum_{i=1}^{8} y_i^2\right] = 0.00167,$$

$$\sqrt{v(q_{\text{III}})} = 0.041.$$

Other schemes could be developed, for instance by taking π_i proportional to an estimate of the size of the unit.

In order to draw a two-stage sample, it is necessary:

(1) To define the primary units.

(2) To decide upon methods for drawing the units at each stage, particularly the primary stage.

(3) To apportion the sample: the decision must be made whether it is more important to draw the largest possible number of primary units, examining only one or two secondary units within each one, or, alternatively, to select only a few primary units which in turn are sampled intensively if not integrally.

(4) To use an auxiliary variate if necessary.

Determining the optimal distribution of the sample involves complicated calculations, and in practice one often proceeds by trial and error. Another stage may be added to the sampling, in which case a three-stage sampling plan is formed.

3.4.7. Two-phase Sampling

Another useful technique that is sometimes used and should be briefly mentioned is two-phase sampling, also called double sampling. In the first phase a sample is drawn on the basis of which certain data are collected. Then in a second phase another sample is taken, generally much smaller, which may even be a sub-sample of the first, and complementary data are collected. Several current applications will be considered.

In the stratified sampling described in sub-section 3.4.5 the sizes N_h of the different strata were assumed to be known. In fact the formulae contain N_h only indirectly in terms of the relative sizes $L_h = N_h/N$. A sample of size n' can be taken that enables the relative size $l_h = n'_h/n'$ to be estimated. In most cases this is relatively easy; for instance, the problem may be that of ascertaining the proportion of fields under a certain type of crop.

In the second phase the mean $\bar{y}_h$ of the variate investigated is determined in the different strata using a sample of n units.

In place of formula (4.11) of the stratified sample

$$\bar{y}_{st} = \frac{1}{N}\sum_{h=1}^{k} N_h \bar{y}_h = \sum_{h=1}^{k} L_h \bar{y}_h,$$

one substitutes

$$\bar{y}_d = \frac{1}{n'}\sum_{h=1}^{k} n'_h \bar{y}_h = \sum_{h=1}^{k} l_h \bar{y}_h. \qquad (4.42)$$

In the ratio and regression estimates considered previously, the population mean $\bar{X}$ or total X of the auxiliary variate were assumed to be known; this is far from always the case. Whenever it is relatively easy to determine the value of the auxiliary variate, a large sample of size n' is taken in order to estimate its mean value $\hat{\bar{X}}'$. A second sample is then taken to estimate the ratio or the coefficient of regression between the two variates. In place of the formula (4.1)

$$\bar{y}_q = q\bar{X},$$

one then has

$$\bar{y}_d = q\hat{\bar{X}}'. \tag{4.43}$$

For instance, in order to estimate the total sugar production from sugar beet, a large number of fields can be sampled from which the total beet yield $\hat{X}'$ is estimated by extraction and weighing. The sugar content of the beet is then determined on a smaller sample, namely the ratio q of the weight of sugar to the weight of the beet. The estimated total sugar yield is then given by the product $q\hat{X}'$.

When the number non-responses is high in a sample survey, the results are spoiled by imprecision. An ingenious way of overcoming this is to treat the sample as a stratified one consisting of two strata, namely, the stratum c of units for which a response was obtained and the stratum a of non-responses. One may then try at greater effort and cost to obtain a reply from a sub-sample of those who did not respond during the first interview or interviews. The first sample of size n consists partly of n_c units for which an estimate $\bar{y}_c$ of the variate has been obtained, and partly of n_a units for which no information is available. During the second phase a sub-sample of size n' of these n_a units is examined and the mean of this second sample is $\bar{y}'_a$.

If n_a is large, the estimate

$$\bar{y} = \frac{\bar{y}_c}{n_c}$$

is replaced with advantage by the double sampling estimate

$$\bar{y}_d = \frac{1}{n}(n_c\bar{y}_c + n_a\bar{y}'_a).$$

3.4.8. Conclusions

Various different sampling methods have been discussed in this section. In practice they are most often used in combination, e.g. a stratified two-stage sample with a ratio estimate. Choice of the best possible combination is generally a complicated matter and is based upon experience and common sense.

When the units are homogeneous and quite concentrated, random or systematic sampling is practicable.

When the units vary greatly in size, or when there are large differences in the variate investigated between different parts of the population, there is an advantage in stratification. If within each stratum there are still large differences in size, sampling with probability proportional to the size of the unit can be done (inasmuch as the value of the variate under investigation is associated with the size of the unit). When the population is geographically dispersed, there is an advantage in taking a two-stage sample, or a sample with several stages. When the variate is distributed rather uniformly in the population, or in each stratum, and when the cost of traveling from one primary unit to another is high, and when there is no difficulty and no major expense in determining the value of the variate once on the spot, there is an advantage in drawing only a relatively small number of primary units and studying them intensively. If some auxiliary variate is known from a previous study, and this variate is closely correlated with the variate under investigation, a ratio or regression estimate can be used.

3.5. PREPARATION AND ORGANIZATION OF SAMPLE SURVEYS

3.5.1. General Considerations

Anyone experienced in sample surveys will probably be surprised that so much space has been devoted to a description of sampling plans, whereas the important problems of selecting the information to be collected, collection procedures, populations concerned, etc., have not been treated. Those who are especially interested in socio-economic surveys will ask why the control of observation and measurement errors has not not yet been considered. In subsection 3.3.4 it was seen that a bias of the order of one-third of the standard deviation of the variate causes a very serious increase in the total error of the sample, and that in such cases there is no reason to select random samples of more than 50 units because the precision of the sample mean is controlled by the bias. In socio-economic surveys much larger systematic errors have been reported, for example of the order of the standard deviation. However, adequate control of these non-sampling errors usually requires experience and intimate knowledge of the subject matter, and techniques for such control cannot be summarized easily. It is well to remember that most of the non-sampling errors have their origin in faulty preparation of the survey. It is, therefore, proposed to review briefly the main problems encountered in preparing a survey.

3.5.2. Definition of the Aim of the Survey

As can be seen in the following examples, sampling procedure depends very much on the aim of the survey, which must therefore be precisely defined from the outset. In sub-section 3.4.5 it was pointed out that in stratified sampling the distribution of the sample was different for enumerative and analytical surveys. A survey may describe the situation at a given instant, or it may reveal the variations for example with time. A survey designed to estimate the prevalence of ischemic heart disease in a population necessarily involves an examination of a sample of the general population. On the other hand, a survey to establish trends in the prevalence of this disease could perhaps be conducted by examining hospital statistics or sampling medical practices.

3.5.3. Definition of the Population

The population to be covered by the survey should be so defined that any decision as to whether a unit belongs to the population or not is unambiguous. This is usually not difficult in population sample surveys, except perhaps in marginal cases such as hotel residents, military personnel, etc. Some surveys can pose population definition problems, however.

3.5.4. Frames

The compilation of a frame or directory, or even its up-dating, can be a very costly and lengthy operation. Sometimes an old frame has to be re-used and its deficiencies remedied by rectifying the sample after the draw is made.

Yates has listed the essential characteristics of a frame (1960). It must be accurate and not include units no longer belonging to the population, it should be complete and include all the units of the population, there should be no duplication, it should be adequate in referring exactly to the population under investigation, and it should be up-to-date.

For instance, a list of telephone subscribers is not an adequate inventory of the families of a district, as it only contains families enjoying this facility.

The directory may contain repetitions; one person may appear more than once, for example under both a private and a professional address. The directory may be incomplete, disregarding recent subscribers, or inaccurate in still including the names of persons who have passed away, or have left the district, or have canceled their subscription.

There are many different frames and, for a particular survey, it is a question of determining which is the most appropriate. A sample survey of the population of a country might be based upon a census, a population register, a cadastral map, a list of localities with information as to the number of their inhabitants, a geographical map, lists of gas or electricity subscribers, electoral registers, etc.

3.5.5. Selecting the Information to be collected

A list of the information to be collected in the sample survey is compiled. Usually this list will be too long. An overly-long questionnaire contains the following disadvantages:

(1) The attention of the interviewed wavers and the precision of the data suffers thereby.
(2) The collection of data takes a longer time and the survey thus becomes more expensive.
(3) Refusals and non-responses are usually more frequent.

A good technique to use when an overly-long questionnaire cannot be avoided is to select a certain number of sub-samples and to ask a different set (sub-list) of questions in each sub-sample.

3.5.6. Collection of Pertinent Supplementary Data

Collection of supplementary data is an important but usually neglected task. The literature should be searched for what is known about the variate under investigation and about other variates that may be in close relation to it. Sometimes variances are reported from previous studies with similar objectives; when this is the case the variance of the variate under investigation can be estimated for calculation of the sample size.

3.5.7. Collection Procedures

There are many media for the collection of the information desired, for example medical records, postal questionnaires, questionnaires delivered at the place of residence, interviews, an accounts ledger, a diary to be filled in by the head of the household, an examination or observation of the situation, direct measurements such as weighing, analysis of a blood specimen, etc. Wherever possible, an objective quantitative measure is preferable to a qualitative assessment. On the other hand, in some cases it is very difficult to describe some complex situation, for example the cleanliness of a house, by a simple quantitative measure.

In simple cases where the errors of observation are independent of the random errors, the measured value of the variate can be written in the ith unit:

$$y_i = \mu + a_i + e_i,$$

where a_i is the deviation between the true value of the characteristic in the unit i and in the population, and e_i is the measuring error. If σ_a^2 is the random sampling variance and σ_o^2 is the observation variance, the total error σ_t^2 per unit will be the sum of the variances

$$\sigma_t^2 = \sigma_a^2 + \sigma_o^2.$$

If the component due to the errors of observation is important in relation to σ_a^2 and if observation is relatively easy, there is an advantage is repeating the observations more than once, e.g. taking a blood specimen from a patient on three consecutive mornings, or making several independent readings from a radiography or from an electrocardiogram. If k independent determinations of the same measure are performed, the variance between the units then becomes

$$\sigma_t^2 = \sigma_a^2 + \frac{\sigma_o^2}{k}.$$

The elaboration of a questionnaire is an operation calling for considerable previous experience. Gallup (1947) suggested series of five questions for most public opinion polls, namely,

(1) To find out if the person is informed on the subject and has thought about it.

(2) The second question should elicit the person's general feelings on the subject.

(3) The third question should elicit more precise replies on particular points.
(4) The fourth reasons for the opinions expressed.
(5) The fifth the degree of conviction behind the views expressed.

Certain questions, even apparently the simplest, can easily be misunderstood. Desabie (1966) cites numerous examples. For example, in French the verb "to know" in the phrase "Do you know such and such an ointment or product?" is taken by educated persons to mean "Has someone told you about . . . ?", while others take it to mean "Have you used . . . ?".

3.5.8. Choice of the Type of Survey

The type of survey must be chosen after consideration of the variability of the variate under study. A morbidity survey limited to one month or one week of the year would, for example, probably give an adequate prevalence estimate of chronic diseases but an imprecise presentation of seasonal infections such as influenza and acute rhinopharyngitis.

3.5.9. Choice of the Survey Plan and Final Preparations

The survey plan should be decided upon including the choice of sampling units and the techniques of estimation then, having regard to the available budget and the desired accuracy, the number of units to be examined should be fixed.

One can then draw up the protocol of the survey and the instructions to the investigators and the inspectors, recruit the personnel, select the apparatus required, and work out the plans for handling and tabulating the data, etc. If necessary, a pilot survey can be carried out in order to check certain data on variability and cost, to test the field procedures and instructions and to complete the staff training.

APPENDIX. UTILIZATION OF TABLES OF RANDOM NUMBERS

In practice the units to be sampled in a population are selected by means of a table of random numbers. Numerous tables of random numbers are available. A list of the best known include:

The table of H. Burke Horton (1949) Interstate Commerce Commission, Washington.

The table of Fisher and Yates, contained in R. A. Fisher and F. Yates (1953), *Statistical Tables for Biological, Agricultural and Medical Research*, 4th edit., Oliver & Boyd, Edinburgh, one page of which is reproduced at the end of this Appendix.

The table of Kendall and Babington Smith, contained in M. G. Kendall and B. Babington Smith (1946) *Tracts for Computers* No. XXIV, 2nd edit., Cambridge University Press.

The table of Linder, contained in A. Linder (1953) *Planen und Auswerten von Versuchen*, Birkhäuser, Basle.

The table of the Rand Corporation.

The table of Tippett, contained in L. H. C. Tippett (1927) *Tables of Random Sampling Numbers, Tracts for Computers*, No. XV, Cambridge University Press.

A few examples will be given to illustrate the use of such tables.

EXAMPLE 1. Take at random four numbers ranging between 1 and 78. As these numbers have no more than two digits, two successive columns of the table are used. Reading from top to bottom and starting for instance from the left-hand corner (bottom of page 68) of the table reproduced in this Appendix we obtain the numbers 53–63–35–63. The number 63 appears twice. This can be perfectly admissible, but generally different numbers are preferred, in which case one more number is drawn from the table. The next is 98, but as this is greater than 78 it, too, is rejected. The fourth number is 02.

Depending upon whether the numbers are required to be different or not, we have

Different numbers	Numbers not necessarily all different
53	53
63	63
35	35
02	63

The table can be read in any other order, for example from left to right, or from bottom to top, etc.; but reading successive columns from top to bottom is the most logical and most direct.

It should be pointed out that it would be incorrect to start the sampling from the same reference point, say the top left-hand corner, every time. Starting points in the table can be assigned for each series of draws on a chance basis, or alternatively the last number used can be underlined and the next series of draws started at the next consecutive number.

The tables can of course be used to sample numbers of 1, 2, 3, 4 or more digits. It is sufficient merely to read a corresponding number of successive columns.

EXAMPLE 2. From a town of 18,500 inhabitants six inhabitants are to be selected at random from the Citizen Register. We assume that the Citizen Register numbers the inhabitants from 1 to 18,500, and we therefore select six five digit numbers from the table. Starting, for instance, at the seventh line, we draw the following numbers:

64 55 2
85 07 2
58 54 1
34 85 2
03 92 1
etc.

It will be found that many numbers must be drawn in order to obtain six that are no greater than 18,500. To avoid such "waste", it is usual to subtract a round number larger than the populations (in this instance, say, 20,000) as many times as possible from the number drawn and to use the remainder.

$$64{,}552-3\times 20{,}000 = 4552$$
$$85{,}072-4\times 20{,}000 = 5072$$
$$58{,}541-2\times 20{,}000 = 18{,}541 \quad \text{(rejected, too large)}$$
$$34{,}852-1\times 20{,}000 = 14{,}852$$
$$03{,}921 = 3921$$
$$62{,}953-3\times 20{,}000 = 2953$$
$$08{,}459 = 8459$$

If the Citizen Register does not in fact number each of the 18,500 inhabitants, this approach is cumbersome in that it implies that a numbering must be carried out. Other methods are however available. Suppose the inhabitants are listed on 370 pages of 50 names (lines) each. One could choose a page by taking a number at random between 1 and 370, and then select a person on the page by taking a number at random between 1 and 50. In this case five-digit numbers are selected: the first three digits to signify the choice of page and the following two the choice of line (name).

It is convenient to subtract 400 from any three-digit number between 401 and 800, and to subtract 50 from any two-digit number above 50.

Numbers above 800 are ignored, because if 800 were subtracted from such numbers to find a number below 370, one would in fact be giving three chances to pages numbered to 199 as compared to only two chances to the others, and an over-concentration of early pages in the sample would be obtained. Taking the first five digits downwards from line 7:

Page	Line
645 − 400 = 245	52 − 50 = 2
850	
585 − 400 = 185	41 = 41
348 = 348	52 − 50 = 2
039 = 39	21 = 21
629 − 400 = 229	53 − 50 = 3
084 = 84	59 − 50 = 9

We thus select the person named on the 2nd line of page 245, the person on line 41 of page 185, and so on.

This method is only accurate if the same number of persons is listed on each page. If page 185 should only contain 35 names, or if line 2 of page 245 gives the name of a deceased person, the number ought not to be replaced by another on the same page; a new "pair" must be sampled (page and line).

Perhaps the Citizen Register is a card-index comprising 37 drawers with 500 cards in each one. One could suppose all the cards stacked in one pile, a pile 5 metres high for instance, and then select cards at six different levels taken at random. Such a choice of cards is no longer rigorously random. Some of the cards will be more compressed than others, one could choose equally among several cards occurring at the selected level, a tattered card will be thicker than a new one, and so on. All the same, in practice the procedure is valid.

Finally, an example is given using a Register of a given number of pages, which do not necessarily contain the same number of names.

EXAMPLE 3. The 18,500 inhabitants of a town are recorded in a Register of 485 pages with at most 50 names per page. Six inhabitants are to be selected at random from the Register.

An example using cumulative sums has been considered previously (Example 4.2). In that example the number of beds in each hospital were pre-supposed added. A method has been proposed by Lahiri avoiding the calculation of cumulative sums. A number between 1 and 485 is taken at random; this number indicates the page of the Register. Another number is then taken at random between 1 and 50 namely between 1 and the maximum number of names per page. If the page indicated by the first number contains a number of names at least equal to this second number, then this page is retained and one selects the name on the line indicated by the second number. The following results are obtained by starting from the 14th line of the table and selecting numerical "pairs", that is, five-digit numbers consisting of one three-digit and one two-digit number. By convention 500 is subtracted from any number greater than 500 in the first three digits, and 50 from any

Pair	Page no.	Name no.	Number of names on page	Decision
070 85	70	35	31	Rejected
018 58	18	8	48	Accepted
728 47	228	47	44	Rejected
887 82	387	32	46	Accepted
451 77	451	27	46	Accepted
967 62	467	12	39	Accepted
433 16	433	16	41	Accepted
504 46	4	46	46	Accepted

number greater than 50 in the last two. Thus: The 8th name on page 18, the 32nd name on page 387, and so on, are selected.

The numbers taken at random may happen to be concentrated in one area, for example, two-thirds of the names may all be within the last 100 pages of the Register. Although the sample is random and taken quite correctly, the persons recorded on the last pages may for some reason or other have very different characteristics from those of the others in the Register, and there is a case for taking it again.

TABLE OF RANDOM NUMBERS†

(Extract)

53 74 23 99 67	61 32 28 69 84	94 62 67 86 24	98 33 41 19 95	47 53 53 38 09
63 38 06 86 54	99 00 65 26 94	02 82 90 23 07	79 62 67 80 60	75 91 12 81 19
35 30 58 21 46	06 72 17 10 94	25 21 31 75 96	49 28 24 00 49	55 65 79 78 07
63 43 36 82 69	65 51 18 37 88	61 38 44 12 45	32 92 85 88 65	54 34 81 85 35
98 25 37 55 26	01 91 82 81 46	74 71 12 94 97	24 02 71 37 07	03 92 18 66 75

† Fisher, R. A. and Yates, F. (1953), *Statistical Tables for Biological, Agricultural and Medical Research*, 4th edit., Oliver & Boyd, Edinburgh, Table XXXIII, Random Number II, p. 127.

02 63 21 17 69	71 50 80 89 56	38 15 70 11 48	43 40 45 86 98	00 83 26 91 03
64 55 22 21 82	48 22 28 06 00	61 54 13 43 91	82 78 12 23 29	06 66 24 12 27
85 07 26 13 89	01 10 07 82 04	59 63 69 36 03	69 11 15 83 80	13 29 54 19 28
58 54 16 24 15	51 54 44 82 00	62 61 65 04 69	38 18 65 18 97	85 72 13 49 21
34 85 27 84 87	61 48 64 56 26	90 18 48 13 26	37 70 15 42 57	65 65 80 39 07
03 92 18 27 46	57 99 16 96 56	30 33 72 85 22	84 64 38 56 98	99 01 30 98 64
62 95 30 27 59	37 75 41 66 48	86 97 80 61 45	23 53 04 01 63	45 76 08 64 27
08 45 93 15 22	60 21 75 46 91	98 77 27 85 42	28 88 61 08 84	69 62 03 42 73
07 08 55 18 40	45 44 75 13 90	24 94 96 61 02	57 55 66 83 15	73 42 37 11 61
01 85 89 95 66	51 10 19 34 88	15 84 97 19 75	12 76 39 43 78	64 63 91 08 25
72 84 71 14 35	19 11 58 49 26	50 11 17 17 76	86 31 57 20 18	95 60 78 46 75
88 78 28 16 84	13 52 53 94 53	75 45 69 30 96	73 89 65 70 31	99 17 43 48 76
45 17 75 65 57	28 40 19 72 12	25 12 74 75 67	60 40 60 81 19	24 62 01 61 16
96 76 28 12 54	22 01 11 94 25	71 96 16 16 88	68 64 36 74 45	19 59 50 88 92
43 31 67 72 30	24 02 94 08 63	38 32 36 66 02	69 36 38 25 39	48 03 45 15 22
50 44 66 44 21	66 06 58 05 62	68 15 54 35 02	42 35 48 96 32	44 52 41 52 48
22 66 22 15 86	26 63 75 41 99	58 42 36 72 24	58 37 52 18 51	03 37 18 39 11
96 24 40 14 51	23 22 30 88 57	95 67 47 29 83	94 69 40 06 07	18 16 36 78 86
31 73 91 61 19	60 20 72 93 48	98 57 07 23 69	65 65 39 69 58	56 80 30 19 44
78 60 73 99 84	43 89 94 36 45	56 69 47 07 41	90 22 91 07 12	78 35 34 08 72
84 37 90 61 56	70 10 23 98 05	85 11 34 76 60	76 48 45 34 60	01 64 18 39 96
36 67 10 08 23	98 93 35 08 86	99 29 76 29 81	33 34 91 58 93	63 14 52 37 52
07 28 59 07 48	89 64 58 89 75	83 85 62 27 89	30 14 78 56 27	86 63 59 80 02
10 15 83 87 60	79 24 31 66 56	21 48 24 06 93	91 98 94 05 49	01 47 59 38 00
55 19 68 97 65	03 73 52 16 56	00 53 55 90 27	33 42 29 38 87	22 13 88 83 34
53 81 29 13 39	35 01 26 71 34	62 33 74 82 14	53 73 19 09 03	56 54 29 56 93
51 86 32 68 92	33 98 74 66 99	40 14 71 94 58	45 94 19 38 81	14 44 99 81 07
35 91 70 29 13	80 03 54 07 27	96 94 78 32 66	50 95 52 74 33	13 80 55 62 54
37 71 67 95 13	20 02 44 95 94	64 85 04 05 72	02 32 90 76 14	53 89 74 60 41
93 66 13 83 27	92 79 64 64 72	28 54 96 53 84	48 14 52 98 94	56 07 93 89 30
02 96 08 45 65	13 05 00 41 84	93 07 54 72 59	21 45 57 09 77	19 48 56 27 44
49 83 43 48 35	82 88 33 69 96	72 36 04 19 76	47 45 15 18 60	82 11 08 95 97
84 60 71 62 46	40 80 81 30 37	34 39 23 05 38	25 15 35 71 30	88 12 57 21 77
18 17 30 88 71	44 91 14 88 47	89 23 30 63 15	56 34 20 47 89	99 82 93 24 98
79 69 10 61 78	71 32 76 95 62	87 00 22 58 40	92 54 01 75 25	43 11 71 99 31
75 93 36 57 83	56 20 14 82 11	74 21 97 90 65	96 42 68 63 86	74 54 13 26 94
38 30 92 29 03	06 28 81 39 38	62 25 06 84 63	61 29 08 93 67	04 32 92 08 09
51 29 50 10 34	31 57 75 95 80	51 97 02 74 77	76 15 48 49 44	18 55 63 77 09
21 31 38 86 24	37 79 81 53 74	73 24 16 10 33	52 83 90 94 76	70 47 14 54 36
29 01 23 87 88	58 02 39 37 67	42 10 14 20 92	16 55 23 42 45	54 96 09 11 06
95 33 95 22 00	18 74 72 00 18	38 79 58 69 32	81 76 80 26 92	82 80 84 25 39
90 84 60 79 80	24 36 59 87 38	82 07 53 89 35	96 35 23 79 18	05 98 90 07 35
46 40 62 98 82	54 97 20 56 95	15 74 80 08 32	16 46 70 50 80	67 72 16 42 79
20 31 89 03 43	38 46 82 68 72	32 14 82 99 70	80 60 47 18 97	63 49 30 21 30
71 59 73 05 50	08 22 23 71 77	91 01 93 20 49	82 96 59 26 94	66 39 67 98 60

REFERENCES

(a) Books, Reviews and Monographs

COCHRAN, W. G. (1963) *Sampling Techniques*, 2nd edit., John Wiley and Sons, New York.

DEMING, W. E. (1950) *Some Theory of Sampling*, 1st edit., John Wiley and Sons, New York.

DESABIE, J. (1966) *Théorie et Pratique des Sondages*, Dunod, Paris.

FISHER, R. A. and YATES, F. (1953) *Statistical Tables for Biological, Agricultural and Medical Research*, 4th edit., Oliver & Boyd, Edinburgh.

GEIGY (1960) *Wissenschaftliche Tabellen*, 6th edit., Geigy, Basle.

HALD, A. (1951) *Statistical Theory with Engineering Applications*, John Wiley and Sons, New York.

YATES, F. (1960) *Sampling Methods for Censuses and Surveys*, 3rd edit., Charles Griffin Co. Ltd., London.

(b) Original Papers

GALLUP, G. (1947) The quintadimensional plan of question design, *Public Opinion Quarterly*, **11**: 385–393.

SATTERTHWAITE, F. (1946) An approximate distribution of estimates of variance components, *Biometrics*, **2**: 110–114.

YATES, F. and ZACOPANAY, I. (1935) The estimation of the efficiency of sampling, with special reference to sampling for yield in cereal experiments, *J. Agric. Sci.*, **25**: 545–577.

CHAPTER 4

DATA

A. A. Weber

Lausanne, Switzerland

4.1. GENERAL CONSIDERATIONS

Until relatively recent times developments in medicine have, with rare exceptions, been the result of intelligent but quite intuitive and unplanned observation of life. Only at the end of the nineteenth century were the experimental and observational foundations of modern biology laid, consequent upon the works of Francis Galton and Karl Pearson, and only within the last few decades have these principles penetrated into medical research and experimentation. Today, however, developments in medicine are increasingly the result of analysis and interpretation of medical data collected according to strict criteria and scientific procedures.

Medical data cover a wide area of observation and measurement, including anatomical data, physiological measurements, determination of resistance or sensitivity to diseases or drugs, etc. These data are used as much by the medical practitioner for a clinical diagnosis (to determine the disorder or morbid condition affecting the health of his patient), as by the research worker to establish a group diagnosis. To make a group diagnosis the research worker describes as precisely and objectively as possible the state of health of a population group: the individual is of interest primarily insofar as he is representative of the population group to which he belongs. The research worker describes a population by eliciting the frequency distribution of a variate such as blood pressure, or the proportion of the population with a particular attribute, such as an electrocardiographic Q wave, or mycobacteria in sputum, etc. He constantly seeks results that are reproducible by other workers at other times.

The clinician sometimes has difficulty in enumerating the criteria on which his diagnosis is established. The research worker, for the requirements of his study, must always show that his medical cases clearly present one or a combination of several definite characteristics.

Data on mortality and morbidity are the basis of group diagnosis. For some diseases, such as rabies, tetanus, malignant tumors of the brain or stomach, the failure of treatment is often tantamount to the death of the patient. With other diseases, however, the mortality rate is neither a good index of the frequency of the disease, nor a good indication of the effectiveness of a specific course of treatment. This is so, for instance, in the case of pulmonary tuberculosis in Northern Europe.

Death is a determinate event and the time of death can be ascertained fairly accurately. Amongst young people death is almost always due to a single disease or a particular morbid process, even if several conditions are present at the time of death. But disease itself is far from having such precision. The distinction between sick and healthy persons is subjective and finely drawn. To some, sickness is a pathological process disabling or limiting appreciably the activity of the sufferer, to others it is as well any vague feeling of "not being well".

In defining many diseases this subjective element plays an important rôle —for example when headaches and diffuse muscular pains must be considered. It is not always easy to distinguish "Sick" or "Not sick", from "Hypochondriac" or "Stoic". The problem is complex, as in addition to the subjective element the very real phenomenon of the "iceberg", or sub-clinical forms of disease, is involved. These forms, usually unnoticed, added to the undiagnosed manifest forms of disease cause the classical determinations of disease frequencies in a population to be rather incomplete. Last (1963) has given an impressive account of this problem in the average medical practice in Great Britain.

Sickness is characterized no less by its severity than by the different clinical phases through which it may pass and by such components as complications and duration of illness. The onset may be insidious, and equally it can be very difficult to determine when the disease is cured and the person regarded as recovered.

More and more, in group diagnosis, data on mortality and morbidity are being supplemented by data of a quite different nature, such as information on the socio-economic conditions of the group, on environmental characteristics, on genetic constitution, or on the distribution of certain physiological variates in the group.

This chapter on data will, therefore, concern itself with a review and discussion of the types and use of medical data available in group diagnosis.

Primarily, in any investigation, the diagnostic criteria to be used must be decided upon and specified. The characteristics of good criteria are presented in section 4.2 together with a discussion of methods available to measure their adequacy.

Closely linked with the problem of criteria is that of classification of data. Classification should allow a meaningful description of the phenomena studied and an adequate presentation of the material collected. Commonly used classifications of medical data are presented in section 4.3.

When a topic is to be investigated, no source of information should be disdained *a priori:* Search of the literature and study of on-going projects will often sharpen the focus of the proposed investigation. A study of current morbidity and mortality statistics may suggest a relation between two variates. Such an indication, however, will only rarely be conclusive enough as to the nature of the relation. Costly clinical trials that may produce conclusive evidence, on the other hand, should only be undertaken when the hypotheses to be tested have been sufficiently documented from other sources. In practice, critical examination of current mortality and morbidity statistics often provides an indication of a relation that is then verified by a more or less elaborate *ad hoc* study.

Commonly available mortality and morbidity statistics are discussed in section 4.4. Current morbidity statistics are increasingly being supplemented by the results of general morbidity surveys carried out on a continuous basis or repeated after a certain number of years. These surveys have become in some countries an integral part of the system of routine health statistics. Such general morbidity surveys are therefore included in the discussions in section 4.4.

In working with mortality and morbidity data, the research worker is usually interpreting and analysing data already collected. When conducting an *ad hoc* survey or experiment, however, he will have to decide how the data are to be collected and recorded. This is the subject of section 4.5.

The *ad hoc* methods of investigation are then considered. Special surveys of a specific condition are treated in section 4.6. As even the interpretation of such surveys cannot provide conclusive evidence of causal relations there is a premium on conducting a trial as quickly as possible, that is, as soon as a trial is possible ethically and as long as it is justified logically and economically. Medical trials are considered in section 4.7.

4.2. DIAGNOSTIC CRITERIA

The successful use of medical data to determine the state of health of a population group depends on establishing adequate criteria for the diagnosis. Only when criteria are well defined does the study take form and only when they are well stated will the study results be meaningful to other workers.

This can perhaps be best illustrated by an example. The results of various surveys on the prevalence of ischemic heart disease among workers in Great Britain can be summarized as follows. The material is based upon 1848 workers aged 35 to 59 years (World Health Organization, 1963b).

Symptoms	Number	Prevalence per 100 subjects examined
A. Angina pectoris, or possible infarction, or intermittent claudication	155	8.4
B. Symptoms as A above and/or electrocardiographic Q/QS waves	192	10.4
C. Symptoms as B above and/or electrocardiographic ST depressions and T inversions	377	20.4

Depending upon the criteria adopted to define ischemic heart disease, prevalence rates vary between 8 and 20%. It must be borne in mind, further, that the range is probably even greater due to the subjectiveness of the definition of angina pectoris and intermittent claudication.

In group diagnosis four essential characteristics of a good diagnostic criterion can be defined. A criterion must be relevant to the aims of the study; it must be reproducible in the sense that as far as possible the results should not be conditional upon time, place or person; it must also be discriminating and adequately separate persons affected by the condition from those who are not; and finally any criterion should be easy and simple to apply.

Relevance and Simplicity of the Criterion

Some criteria may be appropriate for establishing an initial diagnosis, but not so at all for determining the continued existence of a disease after treatment. For example, in a survey of the prevalence of syphilis in a population group a positive qualitative serological test might be a relevant criterion to diagnose cases of syphilis (with the possible exception of early forms). It would not be a very useful test in a therapeutic trial as the antibody levels may remain high for a long period of time after cure. A quantitive test would be more relevant for such a study.

Further, the criterion however relevant must be defined in terms that can be handled. A prevalence survey based on field examinations should not use diagnostic criteria requiring bulky, elaborate equipment. Similarly, in a prevalence survey of ischemic heart disease in groups of apparently healthy adults, it would be impossible to base the diagnosis of arteriosclerosis on direct examination of the lumen of coronary arteries.

Reproducibility

It should be possible to reproduce the same or largely similar results whenever a specific measurement or observation is made on different persons at different times and with different equipment and observers. Reproducibility depends upon the variability of the observation, on the constancy of the variate under investigation and on the ability of the observer to record exactly what he has been instructed to measure. A subjective criterion is generally much less reproducible than an objective criterion. The precision of a test or of an observation connotes the reproducibility of very nearly identical results each time the test is repeated under "identical conditions". The precision of a test is therefore part of its reproducibility.

The *discriminatory power* of a test is measured in terms of (1) sensitivity to respond correctly in the presence of the disease or the phenomenon being studied, and (2) specificity not to respond when the disease (or phenomenon) is not present.

The discriminatory power of a test can be illustrated by a double-entry table (Table 4.1), filled in accordingly as the test produces a positive or negative result in the presence or absence of the disease or phenomenon.

The sensitivity of a test measures the capacity of the test to produce positive results in subjects affected by the disease under study. Generally it is given as a percentage (Thorner and Remein, 1961):

$$\text{Sensitivity} = \frac{a}{a+c} \times 100.$$

Table 4.1. Diagnostic Test Results in the Presence or Absence of the Disease

Disease / Test result	Present	Absent	Total
positive	a (true positive)	b (false positive)	$a+b$
negative	c (false negative)	d (true negative)	$c+d$
total	$a+c$	$b+d$	$N = a+b+c+d$

TABLE 4.2. NUMBER OF PERSONS EXAMINED BY SIZE OF REACTION TO AN INTRADERMAL 10 TU TUBERCULIN TEST (PPD RT XXII) AND PRESENCE OF PULMONARY CALCIFICATION, IN TWO AGE-GROUPS, DENMARK

Size of reaction (diameter of induration in mm)	Age—group									
	15–34 years					55 years or more				
	No. examined	Pulmonary calcification				No. examined	Pulmonary calcification			
		present		absent			present		absent	
		No.	%	No.	%		No.	%	No.	%
0–1	10,810	12	3.3	10,798	38.8	284	16	9.0	268	11.8
2–3	1831	1	0.3	1830	6.6	80	4	2.2	76	3.3
4–5	423	0	0.0	423	1.5	43	5	2.8	38	1.7
6–7	345	0	0.0	345	1.3	66	6	3.4	60	2.6
8–9	787	7	1.9	780	2.8	154	6	3.4	148	6.5
10–11	1405	21	5.7	1384	5.0	265	20	11.2	245	10.8
12–13	2065	24	6.5	2041	7.3	312	20	11.2	292	12.9
14–15	2514	61	16.5	2453	8.8	345	28	15.7	317	13.9
16–17	2186	62	16.8	2124	7.6	240	22	12.4	218	9.6
18–19	2265	69	18.7	2196	7.9	276	20	11.2	256	11.3
20–21	1343	47	12.7	1296	4.7	141	13	7.3	128	5.6
22–23	973	26	7.0	947	3.4	87	6	3.4	81	3.6
24–25	663	18	4.9	645	2.3	63	6	3.4	57	2.5
26&+	587	21	5.7	566	2.0	95	6	3.4	89	3.9
Total	28,197	369	100.0	27,828	100.0	2451	178	100.0	2273	100.0

The specificity of a test, on the other hand, measures its capacity to produce negative results in subjects not affected by the disease. It also is given as a percentage:

$$\text{Specificity} = \frac{d}{b+d} \times 100.$$

By analogy with the theory of tests of hypotheses, Collen *et al.* (1964) use the term "error of the first kind" in cases where the test fails to produce a positive result when the disease is present, and the term "error of the second kind" where the test fails to produce a negative result in the absence of disease. Thus,

$$\text{Error of the first kind, \%} \quad \frac{c}{a+c} \times 100.$$

$$\text{Error of the second kind, \%} \quad \frac{b}{b+d} \times 100.$$

In general, the sensitivity and specificity of a test will vary from one population to another with the relative frequency in the population of persons producing false positive and false negative reactions. An illustration of these concepts is provided by material from the Danish Tuberculosis Index (1955) reporting tuberculin test results in two age-groups of the Danish population.

Table 4.2 and Fig. 4.1 show the distributions by size of reaction to the tuberculin test among persons with and without pulmonary calcifications in two age-groups, 15–34 years and 55 years or more. Among young adults the persons with pulmonary calcifications include a larger proportion of those with large tuberculin reactions than the persons without pulmonary calcifications.

To measure the efficiency of the tuberculin test in selecting those persons in the population with pulmonary calcifications, one could, for example, put the criterion of a positive test at 8 mm of induration and draw up a double-entry table such as Table 4.3.

There is a strong correlation ($\chi^2 = 288.89$, 1 d.f.) between the presence of pulmonary calcifications and a reaction of 8 mm or more of induration in the ages 15–34 years. The specificity of the test in this age-group, however, is barely 50%: that is, there are as many false positives as true negatives with this criterion.

Among persons aged 55 years or more, the high frequency of small tuberculin reactions among persons with calcification signifies that, in this age-group, the tuberculin test does not produce a real division of those infected from those uninfected ($\chi^2 = 0.32$, 1 d.f.).

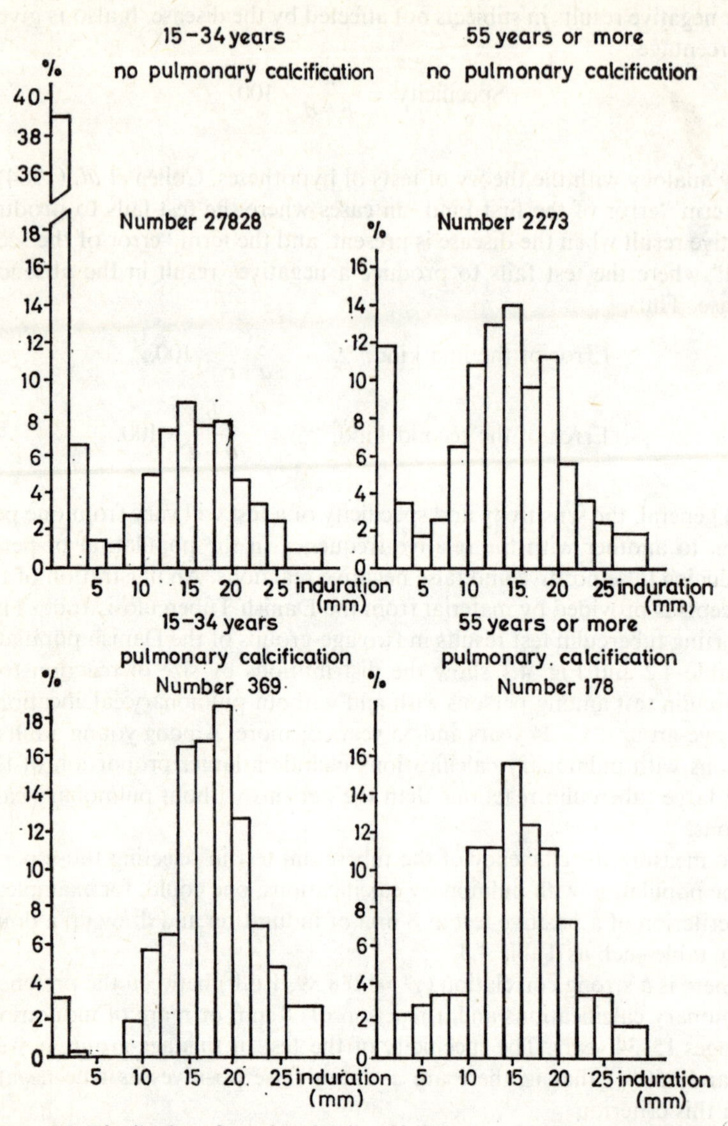

FIG. 4.1. Distribution of reactions by size of induration to an intradermal 10 TU tuberculin test among persons in two age-groups with and without evidence of pulmonary calcification.

TABLE 4.3. TUBERCULIN TEST RESULTS AMONG PERSONS IN TWO AGE-GROUPS WITH AND WITHOUT PULMONARY CALCIFICATION (DANISH TUBERCULOSIS INDEX, 1955)

Pulmonary calcification / Diameter of induration to intradermal 10 TU tuberculin test	Age-group					
	15–34 years			55 years or more		
	present	absent	total	present	absent	total
⩾8 mm	356	14,432	14,788	147	1831	1978
<8 mm	13	13,396	13,409	31	442	473
Total	369	27,828	28,197	178	2273	2451
Sensitivity	$\frac{356}{369} \times 100 = 96.5\%$			$\frac{147}{178} \times 100 = 82.6\%$		
Specificity	$\frac{13{,}396}{27{,}828} \times 100 = 48.1\%$			$\frac{442}{2273} \times 100 = 19.4\%$		
χ^2 1 d.f.	288.89			0.32		

If reactions of 12 mm or more of induration are regarded as positive, the sensitivity of the test decreases and the specificity increases accordingly:

	15–34 years	55 years or more
Sensitivity	88.9%	68.0%
Specificity	55.9%	34.0%

The above considerations can be applied to other tests or developed for combinations of tests. A diagnosis is very often established on the basis of a battery of tests. Collen *et al.* (1964) have indicated a method whereby possible combinations of test results are so grouped to provide the combination or combinations best suited to the discrimination required. This method, based on previous work by Neyman (1950), allows the selection of the combination of test results giving the highest specificity for a given sensitivity or

TABLE 4.4. RESPONSES OF DISEASED AND NON-DISEASED PERSONS TO FOUR EXAMINATION TESTS WITH CALCULATION OF ERRORS OF THE FIRST KIND (α) AND ERRORS OF THE SECOND KIND (β)

Test result				Number of diseased persons	Number of non-diseased persons	Q	Cumulative sum of the n_s from top	Cumulative sum of the n_{ns} from bottom	α	β
A	B	C	D	n_s	n_{ns}					
Neg	Neg	Neg	Neg	0	200	0.000	0	1000	0	1.000
Neg	Neg	Neg	Pos	1	300	0.003	1	800	0.01	0.800
Neg	Neg	Pos	Neg	1	200	0.005	2	500	0.02	0.500
Neg	Neg	Pos	Pos	1	100	0.01	3	300	0.03	0.300
Neg	Pos	Neg	Neg	1	50 }					
Neg	Pos	Neg	Pos	1	50 }	0.02	5	200	0.05	0.200
Neg	Pos	Pos	Neg	1	40	0.025	6	100	0.06	0.100
Pos	Neg	Neg	Neg	2	20	0.10	8	60	0.08	0.060
Neg	Pos	Pos	Pos	3	5	0.6	11	40	0.11	0.040
Pos	Neg	Neg	Pos	7	7	1.0	18	35	0.18	0.035
Pos	Neg	Pos	Pos	8	5	1.6	26	28	0.26	0.028
Pos	Neg	Pos	Neg	8	4	2.0	34	23	0.34	0.023
Pos	Pos	Neg	Pos	11	4	2.75	45	19	0.45	0.019
Pos	Pos	Neg	Neg	15	5	3.0	60	15	0.60	0.015
Pos	Pos	Pos	Neg	38	10	3.8	98	10	0.98	0.010
Pos	Pos	Pos	Pos	2	0	∞	100	0	1.00	0.000
Total				100	1000	—	—	—	—	—

vice versa. It can be briefly illustrated by a hypothetical example where the diagnosis of a disease is based on the results of four tests. For the sake of simplicity the response to each test will be assumed to be of the "yes" or "no" type.

Let us assume that in preparing a prevalence survey of ischemic heart disease results of a battery of four tests, designated by the letters A, B, C and D, have been obtained from a group of persons known to be suffering from ischemic heart disease and from a group of persons not suffering from the disease. A could be the answer to a questionnaire as to the presence of cardiac pain, B the result of an electrocardiogram (pathological or non-pathological), C an evaluation of blood pressure (indicated as high or normal) and D a determination of blood cholesterol similarly presented. The results of these tests among the diseased and non-diseased persons can be arranged in a table such as Table 4.4. There are $2^4 = 16$ possible combinations of test results, and for each combination there are a certain number of sick (n_s) and of well (n_{ns}) persons evidencing that pattern of results.

For each combination the ratio $Q = n_s/n_{ns}$ is found, and then the sixteen combinations are listed by increasing value of Q. The cumulative frequency of diseased persons for all the combinations corresponding to a Q of up to a given value is calculated, as well as the cumulative frequency of the non-diseased persons for all the combinations corresponding to a Q at least equal to this value. These sums in fact represent the errors of the first kind (α) and errors of the second kind (β): they appear in the last two columns of Table 4.4. With such a table at hand, a specificity $(1-\beta)$ or sensitivity $(1-\alpha)$ can be selected best suited to the requirements of the proposed study.

In the present example, if the combinations with a Q equal to, or greater than, 0.025 are considered as delimiting the positive region, i.e. the region where the test results are deemed to be positive, one commits an error of the first kind $\alpha = 0.05$ and an error of second kind $\beta = 0.100$. With this threshold of $Q \geqslant 0.025$, a model of Table 1 gives:

Test results $Q \geqslant 0.025$ \ Ischemic heart disease	Present	Absent	Total
Positive	95	100	195
Negative	5	900	905
Total	100	1000	1100

$$\text{Sensitivity } \frac{95}{100} \times 100 = 95\%.$$

$$\text{Specificity } \frac{900}{1000} \times 100 = 90\%.$$

Thus, if the diagnostic criterion of disease for the prevalence survey is taken to be any of the configurations under the horizontal dividing line, only 5% of diseased persons will be missed, but 10% non-diseased added.

4.3. SELECTION AND CLASSIFICATION OF MEDICAL DATA

In every study one of the first operations to be performed is to select the variates to be recorded and their classification. This selection and classification concerns not only the medical data, but also data characterizing and describing other aspects of the investigation such as the patients, the environment, the treatment or preventive measures applied, etc.

Disease can be classified in many ways, for instance by anatomical location, aetiology, and so on.

The classification in most common use is the International Classification of Diseases, the 8th Revision of which came into force in January 1968. It is revised every 10 years by an International Conference convened by the World Health Organization. The 8th Revision was published in two volumes (World Health Organization, 1967a). Volume 1 contains the so-called Classification proper, in the form of four- or three-digit lists, special lists for tabulation, rules for selecting the cause of death, international regulations, and so on. The other volume consists of an alphabetical Index.

The 8th Revision includes 670 three-digit categories of diseases and 182 three-digit categories of accidents, poisonings and violence by external cause, all grouped into 17 sections. They are sub-divided, for the most part, by means of an additional digit. For accidents, poisonings and violence there is also a classification by nature of injury. The Classification was conceived particularly for the compilation of mortality statistics, for which purpose it is widely used.

Clearly, such a classification is unable to meet directly the needs of every possible user. There are numerous adaptations for special purposes, for example those prepared by the U.S. Department of Health, Education and Welfare (1962), or by the Swedish Health Service (Kungl. Medicinalstyrelsen, 1964 and 1965), for classifying and indexing medical reports. The World Health Organization, as well as various medical societies, have also recommended a number of more detailed classifications which bear upon certain defined groups of diseases, and they have made recommendations on the criteria to be adopted for recording and classifying diagnoses both in field and in clinical studies. These recommendations are generally refinements of the International Classification in particular domains.

Each research worker should choose the classification that meets his diagnostic needs and criteria. Often he will find that he needs a more detailed classification than the International Classification of Diseases offers for the particular disease or diseases, or symptoms, that interest him. In any case, however, it will be to his advantage to select a classification compatible with the International Classification of Diseases, if only because it may be desirable to compare the statistics from his investigation with other statistics.

All too often insufficient attention is paid to the patients themselves, and to the population from which they are drawn, all effort being concentrated on describing the disease. Here too the inquirer will gain a certain advantage in using as far as possible generally accepted standards and criteria. For instance, age of subjects is often reported differently, and it is not rare to find statistics on the same persons presented with different age classifications.

It might be well to consider the main groupings and definitions in this respect. The age of a subject at the time of an event is defined as the interval of time from the moment of birth up to the time of the event being considered. Age is generally given in whole periods of time, such as number of completed days, weeks, months or years. If a person born on the 15th of February 1931 was the victim of an accident on the 28th of November 1966, the person is said to have met with the accident at the age of 35 years and nine months, or at the age last birthday of 35 years. A child having lived 36 hours is considered to be one day old.† The same principle is applied to express the duration of some phenomenon. A person dying 36 hours after an accident is said to have survived one day.

The following age classifications have been adopted by the World Health Assembly for the publication of mortality and morbidity statistics (World Health Organization, 1967a).

(a) for general purposes:

(i) under 1 year, single years to 4 years, 5-year groups from 5 to 84 years, 85 years and over;

(ii) under 1 year, 1–4 years, 5–14 years, 15–24 years, 25–34 years, 35–44 years, 45–54 years, 55–64 years, 65–74 years, 75 years and over;

† In some countries only the difference between the date of birth and the date of the event is considered, i.e. a child born on the 3rd of March and dying on the 4th of March is said to have lived for one day. Consistency, however, requires that only children who have lived for 24 hours but less than 48 hours should be regarded as having lived one day.

(iii) under 1 year, 1–14 years, 15–44 years, 45–64 years, 65 years and over.

(b) for special statistics of infant mortality:

(i) by single days for the first week of life (under 24 hours, 1, 2, 3, 4, 5, 6 days), 7–13 days, 14–20 days, 21–27 days, 28 days up to, but not including, 2 months, by single months of life from 2 months to 1 year (2, 3, 4, . . ., 11 months);

(ii) under 24 hours, 1–6 days, 7–27 days, 28 days up to, but not including, 3 months, 3–5 months, 6 months but under 1 year;

(iii) under 7 days, 7–27 days, 28 days but under 1 year.

If an age-distribution more detailed than one of the above groupings is used, the distribution should be such as to be reducible to one of the recommended groupings.

Various classifications and recommendations have also been established with regard to socio-economic criteria, such as place of residence, marital status, level of education and occupation. The list is to be found in the *Directory of International Standards for Statistics* published by the United Nations Statistical Office (1960). On occasion it is also useful to consult the classifications and definitions used in the latest population and housing censuses, as well as proposals for future censuses.

4.4. CURRENT MORTALITY AND MORBIDITY STATISTICS

Current mortality and morbidity statistics are sometimes disdained by research workers due to the incompleteness of these statistics or the inaccuracy or heterogeneity of the diagnoses concerned. However, as mentioned previously these statistics, in spite of their weaknesses, can be a useful guide to the investigator. Through the intelligent use of mortality data the association of several carcinogenic substances with elevated mortality from various neoplasms was suggested or corroborated. The drastic drop of mortality and morbidity from acute paralytic poliomyelitis that followed the mass vaccination programmes provided a necessary confirmation of the effectiveness of the vaccine used.

However, it is true that the interpretation of these statistics requires prudence and sound judgement. Artificial geographical or chronological variations can be produced by differences in diagnosis or in procedure or in the composition of the populations studied. It is necessary, therefore, to carefully assess the validity of morbidity and mortality statistics before using them.

Mortality statistics

In most of the so-called developed countries mortality statistics are now considered to be fairly complete, although the uniformity of diagnosis leaves much to be desired. It is probable, for instance, that in many countries the decreasing mortality from cancer of the stomach over the last 20 years may be partly attributed to a better diagnosis of the cancers of the other organs of the digestive apparatus, cancers previously often classified as cancers of the stomach. Possibly such artefacts may also affect comparisons between countries, and even between parts of the same country. But such diagnostic differences cannot fully explain the very considerable differences in reported mortality from cancer of the stomach found between various countries or between population groups in the same country (Backett, 1966). In short, mortality statistics should not be disregarded by the research worker. Their study is still a rich source of indices and ideas for a research programme.

Mortality statistics by cause are usually given as the number of persons dying from a specific cause in one year per 100,000 persons exposed to the risk of dying from that cause during that year. These rates are given by sex and age groups. The mortality from a particular cause in one population group is thus presented by up to forty specific rates by age and by sex. For comparisons of mortality from one country to another, or over different periods of time, to be unaffected by differences in the age and sex structure of the populations under consideration, use is made of various standardized rates, the definition and calculation of which is given in current works of reference (Swaroop, 1960; Benjamin, 1959).

The definitions and methods of calculation of several terms and rates used in mortality and morbidity statistics are given in the Appendix.

Morbidity statistics

Morbidity is more difficult to describe than mortality. Within a certain period of time a person may experience several episodes of sickness, whether induced by the same disease, e.g. repeated crises of asthma, or by different diseases. A person can as well suffer simultaneously from more than one disease. Morbidity statistics can record the total number of persons sick in a population group; or the total number of diseases present in the group, one patient possibly being counted several times; or the total number of episodes of sickness within a certain period of time. Statistics can be collected regarding persons contracting a disease or condition over a period of time (incidence rate) or regarding all persons ill at a given moment or within a limited period of time (prevalence rate). Exactly what the morbidity statistic represents must be clearly defined.

As with mortality statistics, specific morbidity rates are calculated by age and by sex whenever the cases considered can be referred to a definite population.

The main current sources of morbidity statistics are briefly reviewed below.

The statistics of cases of compulsorily notifiable disease are an important source of information for certain diseases. In the so-called developed countries cases of quarantinable diseases and of a few other conditions such as the paralytic forms of acute poliomyelitis, cerebrospinal fever, diphtheria and some other serious illnesses are notified to the authorities relatively well, perhaps even "over-notified" in the sense that an appreciable number of suspect cases are reported as well as confirmed ones. These statistics, however, are still being compiled for a certain number of diseases for which individual notification does not seem necessary any more, and so for these diseases far too few cases are notified to produce valid statistics. This applies, for instance, to cases of chicken pox and whooping cough which are still compulsorily notifiable in many countries. In a sample of households, the Canadian Sickness Survey, Lossing (1955) compared the recorded notifications with the cases found at the time of the morbidity survey, and estimated that only 14% of the cases of measles were notified, 11% of the whooping cough cases, 17% of the chicken pox cases and only 14% of the mumps.

Study of the internal consistency of statistics of compulsorily notifiable disease is generally more profitable than direct geographic comparisons, and is often the source, or point of departure, of interesting research. For example, a decrease in the number of cases of infectious hepatitis in Scandinavia was found to be accompanied by a greater average age of the notified cases, suggesting an actual decrease in the frequency of the infection. This situation is peculiar to Scandinavia and it is not found to occur in other European countries for which valid statistics are available on this disease (World Health Organization, 1963a).

Compulsory notification has lately been extended to certain diseases or chronic conditions such as tuberculosis and cancer where, generally, the information is recorded in a register kept more or less up to date. In some instances the register is coupled with a very thorough "follow-up" of the notified cases until recovery or death. In such instances it is possible to supplement the incidence rates furnished by the notification with prevalence, survival and recovery rates. Notification is currently being introduced for other conditions such as poisoning and congenital abnormalities.

Hospital morbidity statistics are available in many countries, some publishing the number of patients hospitalized on a given date, and others numbers of hospital admissions and discharges. Statistics based on discharges (including death) are potentially more accurate and complete. The hospital diagnostic facilities will have been employed to determine the sickness treated, and the discharged patient can be classified by type of treatment received, mode of discharge (home, transfer, etc.) and health condition at discharge (cured, stationary, etc.).

The population served by a hospital or a group of hospitals cannot always be ascertained with the necessary accuracy, and so it is not always possible to calculate population-based morbidity rates. It is now recognized that the relative frequency of hospitalizations depends to a large extent on socio-economic factors. For instance, the admission rates are almost always found to be higher amongst the population of large towns than in the semi-urban and rural areas of a country (Logan, 1964).

The value of hospital morbidity statistics depends a great deal on the quality of the medical recording and reporting in the hospital. Hospital medical records are frequently used for retrospective surveys and quick explorative studies. For these purposes, it is very convenient if a hospital index is maintained in which patients are classified by main diagnosis, one or two secondary diagnoses, surgery performed, and so on. It should always be borne in mind that the hospitalized population is a very selective population. For example, certain studies on hospitalized patients that suggested a statistical association between two different diseases have in fact only shown that physicians have more often sent to hospital complicated cases where the two diseases were present than simple cases when only one was found.

In recent years various morbidity statistics have been compiled from the files of general medical practitioners (Logan and Cushion, 1958) or from out-patient services (Dolejsi and Vacek, 1960). In countries with a highly developed health service, these statistics are very useful additions. The value of statistics obtained in this way is dependent upon the correctness of the diagnosis, which can however be evaluated, e.g. by a detailed examination of a sample of the files.

Other sources of morbidity statistics include social insurance statistics; these statistics are based on social insurance records which are primarily produced for purposes other than the collection of data on morbidity, but which remain a potentially important source; school hygiene statistics which are also a potentially useful source subject to the availability of a system such as health record cards enabling individual follow-up through the school years; and statistics of various other specialized services such as laboratory services, maternity and child welfare services, and so on.

The various health records on which these current statistics are based usually contain important additional information. If adequately stored they may permit additional analyses or investigations, some of which are described in the following subsections.

Various interesting experiments are at present being carried out to combine in a single file through record linkage the data relating to the health of a particular person obtained from different sources. Thus, for instance, in the Oxford Record Linkage Study all the data relating to birth, all hospitalization and death of each member of a population group of about 750,000 persons are combined on one card (Acheson, 1967, 1968). Reference may also be made to Newcombe *et al.* (1957) and the United Nations and World Hearth Organization (1962).

General morbidity surveys

Several countries have undertaken general morbidity surveys based on a sample of households (Linder, 1965). In these surveys, information has been obtained by questionnaire concerning the number of cases of sickness, the duration and degree of disability on the basis of the information provided by the household members. The value of this information was assessed on a sample basis by confrontation with the files of medical practitioners.

Clearly, such surveys are of interest for public health authorities. Nevertheless, more specific and limited surveys are to be preferred in the study of one disease or a definite group of diseases. Such a survey might be conducted in the general population or on restricted populations such as occupational groups. It would call upon a specialized staff and, whenever possible, employ objective criteria and measurements.

4.5. COLLECTION AND RECORDING OF DATA

When a research worker decides to collect *ad hoc* information for his investigation, he is faced with the double problem of how to collect the information and how and where to record it. The information may already be contained in various records such as medical histories of hospital patients, death certificates, autopsy protocols, etc., in which case collection is mainly the problem of retrieval. Otherwise he will have to plan the collection of his basic data.

Unless detailed instructions are laid down for the collection and recording of data there is a great risk that the study will be spoiled by inconsistencies and heterogeneity in the material collected. In large-scale studies, with several teams working independently, a detailed protocol for the collection and recording of information is a condition *sine qua non*, but even small studies conducted by a single worker should be carried out according to minutely detailed specifications as to how the data is to be collected and processed in order to avoid bias and inconsistency. A physician has every chance of changing, however slightly, his diagnostic criteria in the course of a study unless the criteria are clearly specified. Similarly, a medical secretary may alter the way or sequence in which data are recorded if there are no precise instructions, and an interviewer may be inclined to progressively modify the formulation of his questions, or interpretation of replies, unless he has a rigid scheme to adhere to. Such a protocol must include precise directions for checking procedures, coding, extraction and presentation. It is not rare to find that the handbook containing the protocol is larger than the final report of the study.

The scale of such a study can be very large, for instance the first American study of the relation between cigarette smoking and lung cancer was conducted on 200,000 persons (Hammond and Horn, 1954, 1958). In such cases a pilot study is almost invariably run to test the procedures of the study and especially the adequacy of the instructions for the collection and recording of data. The collection of data is specified by describing one by one the necessary handling operations. A data flow chart may be required indicating the different operations as they follow each other: recording, coding, sorting, collation of data sets, transcription, etc. Special care must be taken at every stage that bias is not introduced into the recordings.

If persons are to be examined more than once they must be identified precisely enough for relocation. If different cards are used for the same person there is an advantage in assigning to each individual an identification number to appear on each of his record cards. Wherever a national identity number is widely used and accepted it can be used with advantage. If specimens are collected for analysis in different laboratories, they should be accompanied by shift cards on which the test results can be recorded for later transfer to the basic card or document.

Care must be taken in the design of cards and other documents so that the recording of each subsequent piece of information is straightforward and not influenced by information already recorded such as the result of previous tests or type of treatment administered.

Similarly, instructions must be simple, expressed in a language understandable by lay members of the team as well as the professional staff. They must make clear the necessity of objective recording and include provision for editing and checking the data recorded.

If the information concerned is very simple and only a few subjects are involved, a notebook with one line or one page for each subject may be sufficient. If the data are more numerous, perhaps because the study refers to a large number of subjects, or because more data are recorded per subject, use is generally made of individual cards. The most convenient are cards of Bristol type, or sets of differently coloured cards, e.g. green for forwarding laboratory test results, light blue for radiological findings, etc. These cards can be processed directly by counting those displaying a particular characteristic. Much has been said of cards with marginal perforation, but the advantages in our opinion are fairly limited as regards extracting information.

If the data are numerous and if processing involves long and complicated operations, it is useful to provide for automatic data processing with conventional punch card equipment or with a computer.

If the information is relatively simple and can be recorded without difficulty in pre-coded form, there may be an advantage in using cards of the "mark-sensing" type which permit immediate machine-reading of the code inscribed on the card by a soft black lead pencil. "Mark-sensing" cards are particularly useful as shift cards carrying the results of an examination that are then automatically punched onto a master card or transferred onto magnetic tape. In general it is advisable not to use the same document for recording and automatic processing, but to transpose the data on the record cards onto punch cards and not try to make do with a single card. Whenever records are filled in in the field, record cards may become soiled and tattered, causing frequent trouble in their automatic processing.

On a large-scale survey, a computer is useful. In some instances the results of analysis can be transferred directly into the computer, e.g. the photoelectric readings of a blood analysis can be entered into the computer via an analog-to-digital convertor. Transformation of data can be performed automatically, for instance, data can be expressed in terms of standard units, or a rate of response can be integrated, before the computer automatically stores the data in a memory cell for future reference.

As regards utilization of the information, a computer can be used with advantage to screen the data for gross inconsistencies, to adjust models to the data and to reduce the data by means of statistical analysis. In "follow-up" investigations computers have been used to single out automatically the persons who should be contacted again, and the computer may even be instructed to issue the convening letters to those persons. See World Health Organization (1967b).

4.6. SPECIAL MORBIDITY SURVEYS

The purpose of special morbidity survey is to furnish accurate replies to a small number of specific questions on certain diseases or conditions. They are carried out on certain population groups using standardized techniques of diagnosis, such as clinical examinations according to a pre-arranged plan, observations, measurements and laboratory testing. Medical examination is not always necessary. Thus, urine specimens can be collected by laymen for glucose determinations in a survey of the distribution of the glucose levels, and tuberculin tests can be carried out by nurses in the schools to determine the distribution of tuberculin allergy in the school population.

The aim of a prevalence survey is to determine, in a well-defined population and at a given time, the proportion of individuals with a particular disease, symptoms or characteristics. For example, instances of suspect lung shadows on X-ray or blood pressure above a certain given level can be counted. Such surveys are generally concerned with a particular population group, such as schoolchildren, or persons belonging to a certain occupational group. The whole population group can be considered, or a sample of it. In any event, one records the state of this population at the time of the examination in relation to certain selected measures or parameters. For any one individual the examination may extend over several days; for instance, a person with a lung cavity detected by photo fluorography may have to undergo a tomography in order to locate the exact position of the cavity. But still, only the condition of the patient at the time of the examination is determined, hence the prevalence survey is ill-suited to a study of acute diseases.

A prevalence survey can be the starting point of a systematic follow-up —extending for a period of months or years—of subjects selected in the light of results of the initial survey, and may suggest possible risk factors or hypothesis as to the etiology of the disease.

The interpretation of the results of a prevalence survey strikes many reefs. The prevalence of a disease at a particular moment depends on its incidence in the past and on the duration of the disease up to the death or recovery of the patient. Thus, for instance, Nyboe (1957) has shown that the relation at a given moment between the prevalence of positive reactions to tuberculin and the age of the subjects tested may be the same in two populations with very different evolutions of the incidence rates of tuberculous infection by age during previous years.

The study of the variation with age of arterial pressure has been taken as a second example of these difficulties. Figure 4.2 shows the mean diastolic pressure in men as given by Master *et al.* (1952, 1958). It is seen from the

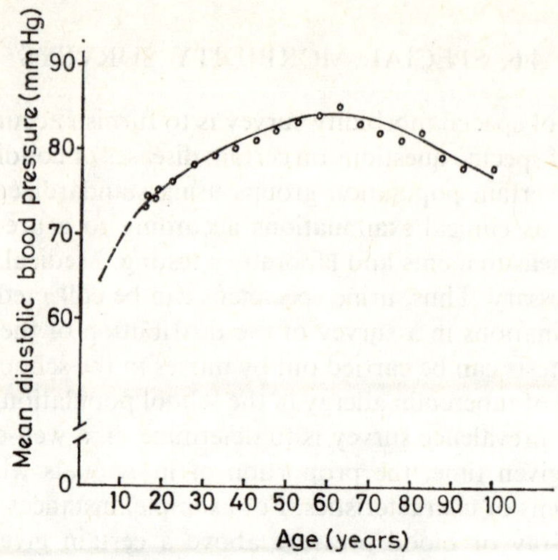

FIG. 4.2. Mean blood pressure among men, by age (Master *et al.*, 1952, 1958).

curve that the mean values drop from age 65 onwards. The following questions then present themselves: Is this drop genuine? If it is, what is the explanation of a drop in arterial blood pressure after age 65? Does this generation tend to suffer from higher blood pressure than older generations? Do persons with a higher arterial pressure die younger than those without? etc. The answers to these questions would usually require deeper studies than a prevalence survey provides.

Whenever the population under study is neither natural nor stable, but a special group, such as the population of a rapidly expanding town or the labour force of a factory, particular attention must be paid to possible effects on the actual prevalence of the disease of the "forces" that controlled the movements in and out of this population. Workers in some occupation groups may be found particularly free from disease not so much because this occupation is "healthy" but because workers in this occupation must be in very good health always and those unfit are progressively weened out.

With the exception of surveys that are the starting point for systematic surveillance of the persons examined, it might not at first seem necessary to collect extensive data on the identity of the subjects. Through a legitimate effort to limit the amount of data collected, one is inclined to retain only data enabling the disease and a few other relevant factors to be described. This introduces the risk of being unable to show possible distortions in the observed population or, conversely, being unable to show that the sample is in fact representative with respect to various characteristics. At the other extreme of course, the survey can be destroyed by an enormously large questionnaire. It is always necessary and comforting to check that the observed population conforms to the underlying population with respect to several criteria, viz. age, sex, socio-economic groups, and so on; or, if not, to describe the observed population according to these criteria. It is fundamental, for instance, to eliminate the influence of age when comparing the prevalence of ischemic heart disease between office workers and manual workers, if one wants to bring to light a possible association with physical activity. One needs to check that the groups are comparable as regards other factors which may influence the prevalence of the disorder, and if the groups are not comparable in these respects, both populations should be broken down into homogeneous sub-groups in relation to these factors before proceeding to the comparisons sub-group by sub-group.

As already mentioned, prevalence surveys often provide the basis for selection of special risk groups, or individuals, for a prospective study. Other sources of material for prospective or retrospective studies are registers of persons affected by a particular disease, files of hospital patients, birth certificates, etc. Medical certificates of the cause of death are a potentially important basis of retrospective studies. An excellent account of prospective and retrospective studies, by Dr R. Doll, is to be found in the work edited by Witts (1959) on medical surveys and clinical trials.

In a retrospective study the presence of a disease or a particular pathological condition is observed in certain persons. Then, working backwards, an attempt is made to determine whether or not some event or phenomenon suspected of influencing the disease occurred amongst these persons with the same frequency as in another group of persons free of the disease or condition and called controls. In a prospective study, one selects for study individuals displaying certain characteristics or exposed to a certain hazard believed to be a risk factor in contracting a disease or in modifying adversely its course. These individuals are then followed for a period of time to see whether or not the anticipated effect is produced.

The possible effect of radiations during pregnancy on the development of leukemia in the child has been the object of a retrospective study by Stewart *et al.* (1958) and Stewart (1961). For this purpose she commenced by studying medical certificates of the cause of death of infants that had attributed death to a leukemia, and then continued by questioning the parents as to any radiological treatments or investigations which the mother may have had during pregnancy. A similar set of questions was put to parents whose children had not had leukemia. The frequency of radiological treatments was then compared in the two groups.

The same effect was studied prospectively by MacMahon (1962). He followed 734,243 children born alive in 37 large maternity hospitals in the United States between 1947 and 1954. The exposure to X-rays during pregnancy was estimated by reviewing systematically 1% of the pregnancy and delivery records for these 734,243 births. Those few children who did die from cancer were traced by scrutinizing death certificates in the area concerned. The intra-uterine X-ray exposure of these children was drawn from the hospital records and compared with exposures found in the 1% sample.

In retrospective studies one proceeds from consideration of a group of subjects manifesting a particular morbid condition, to one or several that are similar in all respects except as regards the presence of the condition in question. One is sometimes content to compare the initial group of subjects with the total population directly. In each of the groups the frequency—or the intensity—of exposure to a preceding "risk" is determined, and then this frequency or intensity is compared in the different groups.

The choice of the case group, and of the control group, is of great importance to the success of the study. If the control groups differ greatly from the group being studied in several respects, then the chances are greater that any differences in morbidity found are explicable by differences in the composition of the groups, regardless of any possible effect of the risk factor being studied.

A method commonly used in forming control groups is the method of matched pairs. For each patient or person having the observed characteristic, a person is selected who does not have this characteristic, but who is similar in other basic factors. In fact, one can take as many "healthy" control subjects for a patient as are wanted to make up different control groups.

In a survey of the possible influence of removal of the tonsils on the risk of contracting an acute anterior poliomyelitis, for each case of poliomyelitis another pupil was taken of the same sex and age and studying in the same class (Medical Research Council, 1955). Stewart *et al.* (1958), in their study of leukemia, took for each deceased infant another whose birth had been recorded in the same Register of Births, of the same age within six months, of the same sex and living as close as possible to the place of residence of the deceased leukemic child. Schwartz and Denoix (1957), in a study of the correlation between cigarette smoking and lung cancer, selected patients hospitalized with lung cancer as their case group, and for each of these patients selected four hospitalized patients as controls. They were matched with the lung cancer patients as regards age, sex, hospital in which treatment had been received, and date of interview. The controls were selected as follows:

1st group—patients suffering from a cancer of another site of the respiratory tract, or from a cancer of the digestive system;
2nd group—patients suffering from a cancer of any other site;
3rd group—patients with a disease other than cancer;
4th group—accident cases.

The smoking habits of the patients were noted at an interview.

By forming these four groups the writers were able to show that the smoking–cancer relation was specific to lung cancer, and much less so to cancers of other sites of the respiratory tract and of the digestive system. Another precaution which is always useful, and even necessary, is to verify that the group are comparable as regards other factors, for instance: education, social background, alcohol consumption and so on; and, if they are not comparable, to determine whether the difference is great enough to explain the differences of morbidity or mortality between the groups. Even when the exposure to risk is different between the case and control groups and no differences in their composition can be found, a cause and effect relation can never be deduced with certainty between the factor investigated (tonsil operation, irradiation, smoking) and the disease (poliomyelitis, leukemia or lung cancer). A fortuitous correlation may be induced by the fact that the case group was not representative of all persons suffering from the disease, or because the controls were ill-chosen, or because whatever precautions were taken to see that the collection of data was similar in all the groups, a bias was introduced into the replies. Such a bias may well be due to the patients themselves. (For instance, a patient who knows that he has lung cancer may blame himself bitterly for the cigarettes he has smoked and exaggerate the number now that the smoking–cancer relation has been established with a quasi-certitude.)

The retrospective survey has the great disadvantage also that the magnitude of the risk run by exposure to the factor being studied cannot be measured.

In *prospective studies* persons exposed to a risk are followed up for periods of weeks, months or years in order to study an eventual development of the disease or pathological conditions supposedly inherent in the risk. A parallel study is carried out on a control group and then the frequency of the disease at the end of the period of observation is compared in the two groups. Under certain conditions, the control group may be the total population of a region or country. Often the persons exposed to the risk are distributed over a certain number of homogeneous sub-groups according to the degree or intensity of exposure to the risk. For example, in a prospective study of the relation between the amount of serum cholesterol in the blood and ischemic heart disease, subjects could initially be classed as:

controls	normal amounts of serum cholesterol,
sub-group 1	serum cholesterol level moderately elevated,
sub-group 2	high serum cholesterol level

It may also well be that persons exposed to a very great risk should be excluded from the study for reasons of medical ethics, e.g. cases where the amount of serum cholesterol is exceedingly high.

The risk of developing the disease even in the most exposed group may be very slight, not exceeding one case in a thousand per annum, or perhaps the differences in incidence between the different groups may be of this order. In such cases the study must then include considerable numbers of subjects, up to hundreds of thousands for each group, to provide meaningful results.

But as there must be sufficient information from the start regarding the identity and characteristics of each subject, and relatively accurate information regarding the exposure to the risk involved, and as this information should be recorded in a uniform manner in each group and the subjects followed for months or even years, it is readily apparent that such a study can be a very large and a very costly undertaking that should only be contemplated if there are firm working hypotheses to be verified.

As compared with retrospective studies, the prospective study has the following advantages: (1) the amount of exposure to "risk" run by the subjects in the different groups can be evaluated, whether this "risk" reflects a cause and effect relation or not, and (2) a bias of memory or reply in the different groups is probably much less serious than in retrospective studies.

On the other hand, a prospective study is generally more costly and takes more time than a retrospective study. Unfortunately, neither ever provides the certainty of a cause and effect relation between the risk under investigation and the disease or conditions that are found.

An objection often raised to prospective studies is the long waiting period before the results become available. In certain cases this delay can be shortened by taking a period prior to the beginning of the study as the initial period of observation. One such case is where the risk run by persons in a particular industry is being studied and where the Personnel Department can provide the names of persons who have been doing certain "dangerous" jobs during the last 10 or 20 years. The study can then be carried out on these persons and on suitable controls, e.g. employees in the same factories who have been doing other work over the same period of time. This, of course, is only possible if one can trace back all persons subjected to the risk factor under investigation. Often the increased risk of contracting a disease or condition is so great amongst those exposed that one can dispense with control subjects in the strict sense of the term. One ought then to have sound information as to the frequency of the disease in the population as a whole.

No matter how earnestly patients are followed up throughout the study and regardless of the efforts employed, there are always a certain number of drop-outs: some persons may lose interest, others just disappear without leaving word, others may emigrate, or die, or suddenly become subjected to a much greater risk than that which is being studied. So the period over which any one subject is followed will differ; and therefore, in general, it is necessary to express the results in terms of number of person-years of follow-up utilizing actuarial techniques, for instance, life-table techniques (Wiesler, 1962; Swaroop, 1960; Bahn and Bodian, 1964). In any event, however, it is always necessary to check that those who have disappeared or been lost from sight —and this applies to all studies with "follow-up"—have the same characteristics in the various groups and that their number is kept as small as possible.

4.7. THERAPEUTIC AND PROPHYLACTIC TRIALS

An altogether different type of study will now be considered, i.e. studies that satisfy the statistical principles of experiments as presented by Fisher (1951). Comparable groups or experimental units are drawn up and different treatments allocated at random to these units. With such a design it is possible to infer from differences observed between units after treatment a significant effect of the treatments. The trial must, further, satisfy certain minimum conditions to be compatible with the laws of medical ethics. These conditions are as follows (Schwartz *et al.*, 1960):

(1) Voluntary consent on the part of the subject and his own physician.
(2) Experiments should be beneficial to the community.
(3) It is impossible to foretell which of the treatments is the most advantageous.
(4) Control subjects should receive treatment that is recognized to be efficaceous, and if their condition worsens they should have the optimum treatment required. No person included in the trial should in any way suffer from it.

Some workers demand far more stringent conditions, some even go so far as to claim that the allocation of treatments at random, or the prescription of standard treatment, is contrary to ethics. However, a complete lack of experimentation is no less to be condemned. One can maintain that when the medical profession is faced with two treatments and no-one knows whether one is better then the other, the rules of medical ethics operate in favour of a therapeutic trial to compare the advantages of the two treatments rather than to remain in doubt and prescribe treatment which is perhaps not the best that the patient can receive, and which may even be of very limited value.

Therapeutic trials are comparative studies of the efficacy of treatments administered according to a strict experimental design on ambulatory or hospitalized patients. Only a passing reference is made to them here, as such trials are discussed at greater length in Chapter 15 of Volume 2. Readers will profit equally by consulting the works of Herdan (1955), Schwartz *et al.* (1960) and Witts (1959).

Prophylactic trials are concerned with comparing the efficacy of preventive treatments, or of prophylactic measures, administered according to a strict experimental design on persons who are apparently not sick. These trials answer the question whether persons who have received preventive treatment against a particular disease fall sick less often or less severely than those who have not received this treatment. The latter group is the control group.

There is a smaller premium in medical trials than in laboratory experiments in using complex experimental designs such as lattice, factorial design with confounding, etc. The greatest advantage of such complex design lies in the possibility of keeping a strict balance between experimental units throughout the trial. This is difficult if not almost impossible in medical trials where patients may refuse to cooperate at any point in the trial, may disappear, or may suddenly present a rather serious form of the disease that requires intensive treatment. An attempt to balance these losses would be quite difficult in such complex sophisticated trial designs.

Simple trial designs are usually selected depending on the type of disease under study. When vaccines are being tested, persons participating in the trial are generally distributed at random into as many groups as there are vaccines to be compared. Frequently one of the groups comprises persons receiving an inert substance or sometimes no treatment at all. There may moreover for one test treatment be several control groups such as one group receiving a standard treatment of known efficacy, one group a placebo and one group no treatment at all. Often the vaccines are allocated to persons on a "pseudo-random" basis, e.g. with reference to their date of birth (even number = vaccine, odd = no vaccine), or even allocated systematically (vaccine A, vaccine B, placebo, vaccine A, vaccine B, placebo, and so on).

In the trial of the Medical Research Council (1959) on B.C.G. vaccines, more than 50,000 children between 14 and $15\frac{1}{2}$ years of age were given a serially numbered identity card. Those who had a negative tuberculin reaction and whose number finished in certain digits, were vaccinated with the whole vaccine, others with correspondingly different digits were given a weakened vaccine, while other children were not vaccinated at all. Vaccination was performed between 1950 and 1952 and the test subjects were followed up for many years. The low rates of tuberculosis incidence in the age-group in question do in fact require a large initial population and a long period of follow-up.

In the U.S. test of the Salk vaccine and of a "placebo", more than 200,000 children were included in each group and were followed up for about eight months.

When the disease to be prevented is grave (smallpox, paralytic forms of acute anterior poliomyelitis, etc.), new cases are quickly recognized and recorded and so a less intensive follow-up is sufficient. But when less serious diseases are studied (whooping cough, common colds), the risks of omission are great, even when there is close supervision, and so very intensive surveillance of the persons included in the survey is necessary.

It is also necessary to be sure that the groups subjected to the various vaccines or preventive treatments, and the control groups as well, are similar, and this is done by comparing in the different groups the distributions of persons by age, sex and some other characteristics. Further, one must be certain that the intensity of follow-up is equivalent in the different groups, and that the subjects who "disappear", or refuse to cooperate, or who are eliminated for different reasons, are similar in the different groups. At any rate, their total should be quite small, otherwise great doubt is cast upon the value of the trial.

Since such trials are performed on volunteers, or at least on persons who have agreed to participate, a control group is indispensable, whether the controls receive a dummy treatment, placebo, or treatment of recognized efficacy. Bradford Hill (1955), in commenting upon the results of an excellent trial of Diehl *et al.* (1938), in connection with a vaccine against the common cold, fully illustrates this requirement. The subjects who volunteered for the vaccination had on average 1.6 colds each in the year following the vaccination, as compared with 5.9 during the previous year. This result could have seemed miraculous if the authors had not been prudent enough to administer a placebo to some of these subjects, for whom the average number of colds was 2.1 after the "treatment" compared with 5.6 previously.

The group may also be its own control. For instance, in the US fluoridation trial (Arnold *et al.*, 1956) two towns were chosen, partly because their water supply had only a very low fluor content. The dental condition of the child population of these two towns was noted and fluor was introduced into the water supply of one town, whereupon there was a large reduction in the number of caries, more particularly in regard to children born after fluoridation was started. Five years later fluor was also added to the water supply of the other town with similar results.

In the tests considered up to the present, the experimental unit has almost always been the individual person. Clinical trials have indeed been carried out on smaller units, e.g. comparisons of treatments administered to each of the two arms or the two eyes of the same patient (Linder, 1959), but larger units have also been utilized, such as the population served by a water adduction system.

Two extreme types of trials will be briefly considered at the end of this section: the field trial of mass treatments and laboratory experiments. In the first one the experimental unit is a population group forming an epidemiological entity for the disease considered; in the second one, animals or some biological preparation.

Field trials of mass treatments

Until now we have been concerned with what happens to someone when this person is subjected to a particular form of prophylactic or therapeutic treatment. One may equally enquire as to what may happen if a collective group is submitted to a particular treatment. One may, for instance ask what happens in a community when an attempt is made to break the chain of transmission in relation to a contagious disease. The experimental unit is no longer the person or individual alone, but the epidemiological entity. If the mode of infection is essentially familial, the family may become the experimental unit, and if the sources of infection go beyond the family circle, the unit becomes a block of houses, a residential estate, a village or a town as the case may be.

In a comparative study of different treatment of the seasonal epidemics of conjunctivities and trachoma in a rural population (Weber and Linder, 1960; Weber, 1960), one approach was to try to reduce the reservoir of pathogenic germ carriers at the start of the epidemic period to slow down the onset of the epidemic, and another was the massive reduction of the number of flies, presumed to be the important vector in the transmission of seasonal conjunctivitis. The village was therefore chosen as the experimental unit: it would have been difficult to produce a massive and lasting reduction of the fly population in only half of a village, and, besides, we were interested in eliminating the reservoir of germs as radically as possible in the village.

Laboratory experiments

These experiments, relating to animals or to various different biological preparations, culture media, etc., are generally undertaken at any stage of a research programme, in order to obtain very precise answers to specific questions. The questions usually raised, broadly speaking, are: Can a particular substance produce the desired reaction? What is the dose–effect relation? Can a particular substance produce acute or sub-acute intoxication?, etc.

Laboratory experiments at the introduction of a new drug or treatment may mean that fewer questions will need to be answered by therapeutic or preventive trials, and the questions themselves may be better specified.

Similarly, laboratory experiments may equally well be carried out at the end of an experimental programme in order to check certain specific hypotheses suggested by earlier trials, e.g. a possible side-effect not anticipated but suggested during clinical trials.

Two types of errors are common in this field. The first is to believe that it is sufficient to observe attentively, then conscientiously record what happened, and then analyse these findings by recognized methods. By proceeding in this way, interesting phenomena will almost always be overlooked by being immersed in the mass of observational data. The second error is to believe that the reactions of the human organism are identical to those observed in animals or on the experimental material.

It is essential to have good working hypotheses before undertaking such experiments and to determine exactly what type of data must be recorded. As far as the second error is concerned when dealing with animals, the type or types suitable should exhibit reactions which are comparable to those associated with the human organism as regards the action being studied. Moreover, animals are unable to complain, and so unless there is some anatomical lesion or apparent trouble, it is impossible to detect anything that may have been aggravated under a particular treatment.

These two types of trials—laboratory experiments and field trials of mass treatments—are not strictly speaking medical trials but have been included briefly in this chapter because the total picture of medical investigation is not complete without them.

APPENDIX. DEFINITION OF TERMS IN COMMON USAGE

The principal measures of ill health in a population are the incidence, prevalence and duration of the cases of disease. The incidence is a measure of the frequency of new cases over a given period. The prevalence is a measure of the frequency of the existing cases of disease, either at a given moment, which is called point prevalence, or at any time during a given period, which is called period prevalence. The duration of a disease, though not always easy to measure, is the interval of time separating its onset and its end (cure or death).

The terms are applicable to all events or processes having a definite duration, including any well-defined stage of a disease, such as the precicatricial phase of trachoma. For deaths, which theoretically have no duration, one can only measure the number occurring during a given period. Mortality rates are therefore rates of incidence of fatal cases of disease.

In order to compare the frequency of a disease in different population groups, rates are calculated by referring the number of cases at a given time or during a given period in each group to the population of the group.

For morbidity rates, the denominator is usually the average population exposed to the risk in question over the period of time (or the point in time) under consideration. For instance, the following hypothetical figures might be considered for an epidemic of the common cold in a hotel over a period of one week. It is assumed that no sick persons are admitted to the hotel and that the sick are transferred to a clinic within a day.

Day	Number of customers present	Customers affected (new cases)
Sunday	93	3
Monday	178	1
Tuesday	162	3
Wednesday	157	2
Thursday	151	0
Friday	85	1
Saturday	126	2
Total	952	12

There are thus 12 cases per 952 person-days, or per $\frac{952}{7} = 136$ person-weeks, i.e. a rate of incidence at the hotel equal to $\frac{12}{136} = 0.09$ cases per person remaining one week, or 9 cases per 100 persons per week.

Suppose three isolated cases of food poisoning occur during a summer season at a boarding house with the following numbers of residents:

Length of stay	Number of residents
1 week	12
2 weeks	45
3 weeks	18
4 weeks	5
Total	80

The number of person-weeks is:

$$(1\times 12)+(2\times 45)+(3\times 18)+(4\times 5) = 176.$$

The incidence of food poisoning is therefore $\frac{3}{176} = 0.02$ cases per person-week, or about 2 cases per 100 tourists spending one week at the boarding house.

A rate is general, or crude, if it is calculated in relation to the population as a whole, or it may be specific in reference to a particular group of the population or a particular disease. Thus, for instance, we have a specific mortality rate for subjects 40–44 years of age, and a specific prevalence rate for workers of a given age-group in a factory. A rate is generally given per 1000 persons (or per 10,000 or 100,000 persons) exposed to the risk under consideration during a period of time or at a point in time. Hence

$$\text{Crude rate} = \frac{\text{Number of events in the population } P \text{ during the period } T}{\text{Average number of persons at risk in the population } P \text{ during the period } T}\times 1000.$$

$$\text{Specific rate} = \frac{\text{Number of events in group } G \text{ of population } P \text{ during the period } T}{\text{Average number of persons at risk in group } G \text{ of population } P \text{ during the period } T}\times 1000.$$

For point prevalence rates, one considers events at a given moment M, rather than during the period T.

Mortality rates

The crude annual mortality rate is defined as

$$\begin{array}{c}\text{Crude annual}\\ \text{mortality rate}\end{array} = \frac{\text{Number of deaths in the population } P \text{ during one year}}{\text{Mean population } P \text{ during the year}}\times 1000.$$

In stable populations the mean population is taken to be the mid-year population.

The crude mortality rate permits only very general comparisons. In order to compare the mortality of different populations in relation to particular diseases, specific mortality rates by causes are generally calculated as per say 100,000 population as follows:

$$\text{Specific annual mortality rate due to cause of death } A = \frac{\text{Number of deaths attributed to cause } A \text{ in the population } P \text{ during one year}}{\text{Mean population } P \text{ during the year}} \times 100{,}000.$$

Certain rates have very little value for analytical studies, and they could be much improved by the use of more explicit denominators. For example, mortality from road traffic accidents would be more meaningful and much more useful in comparative studies if referred to, say, 1000 vehicles licensed per year or 100,000 passenger-miles.

Likewise, specific rates are defined for given groups, e.g. persons of the same sex and of the same age. In these rates the number of persons in the group exposed to the risk or an estimate of this number should be taken as the denominator; thus, in order to calculate the maternal mortality rate in the 25 to 29 year age-group, all pregnant women in this age-group should comprise the denominator. This figure is not generally known, however, and this rate is expressed instead as the annual number of maternal deaths at age 25 to 29 years in relation to the number of children born alive during the year of mothers aged from 25 to 29 years.

Applying the same principle, the number of persons suffering from a certain disease in the population can be regarded as the population exposed to the risk of dying from this disease. If the number of deaths is related to 100 or 1000 cases of disease, clinical mortality rates, now commonly called fatality rates, are obtained. They are generally calculated as follows:

$$\text{Fatality rate} = \frac{\text{Number of deaths attributed to disease } A \text{ in the population } P \text{ during one year}}{\text{Number of new cases of disease } A \text{ in the population } P \text{ during the year}} \times 1000.$$

In follow-up studies mortality is often expressed as the probability of dying within a given number of years after the beginning of follow-up per 100 or 1000 persons exposed to risk at the start of the period, and not in terms of the average number exposed to the risk during the period. The denominator of such a rate is then the number of person-years of exposure. These rates are analogous to the mortality probabilities of Life Tables (Wiesler, 1962).

Morbidity rates

Morbidity is a more complex factor than mortality. A distinction can be drawn between diseases existing at a given instant or at any time during a given period (prevalence) and diseases that start during a period (incidence). One should also state whether the unit counted is

(1) persons sick,
(2) cases of disease,
(3) episodes of disease (distinguishing relapses of one disease from new attacks after recovery from the preceding attack).

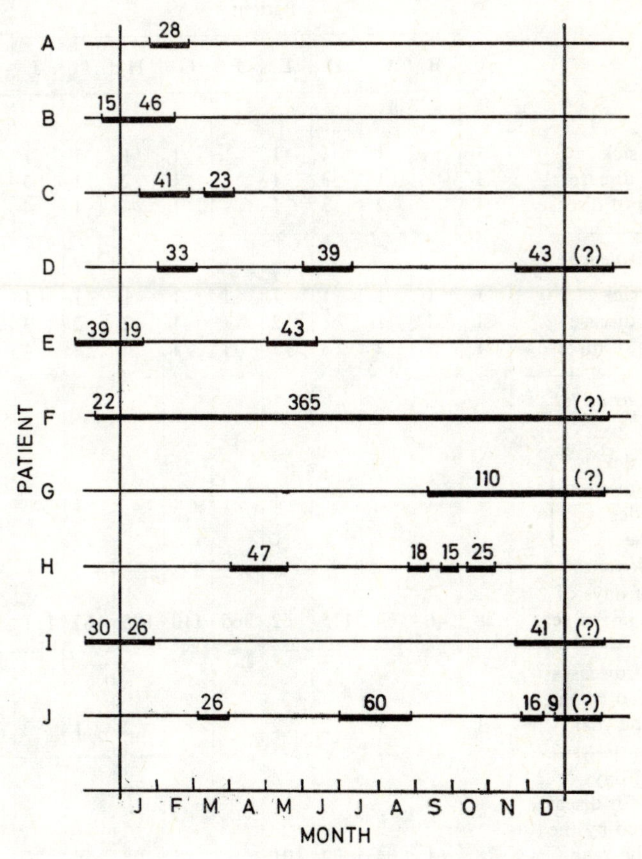

Fig. 4.3. Distribution of episodes of disease for ten persons during one year.

In order to grasp these concepts better, suppose that a population of 100 persons is followed for one year and that 10 have suffered from a certain disease during this year. Cases of other diseases have been disregarded for the sake of the illustration. The 10 persons are represented in Fig. 4.3, a separate horizontal line being assigned to each patient and a period of sickness being indicated by a thickened line. The figures are the number of sick days for each episode of disease. Any episode of sickness occurring less than one month after the end of the previous episode is assumed quite arbitrarily to be a relapse of the same disease, while episodes occurring one month or more later are assumed to be new cases.

The data can be summarized as follows:

	Patient										Total
	A	B	C	D	E	F	G	H	I	J	
Incidence:											
persons sick	1		1	1	1		1	1	1	1	8
cases of disease	1		1	3	1		1	2	1	3	13
episodes of disease	1		2	3	1		1	4	1	4	17
Annual prevalence:											
persons sick	1	1	1	1	1	1	1	1	1	1	10
cases of disease	1	1	1	3	2	1	1	2	2	3	17
episodes of disease	1	1	2	3	2	1	1	4	2	4	21
Prevalence as of 1st January:											
persons sick, cases of disease, or episodes of disease		1			1	1			1		4
Number of days of sickness in the year	28	46	64	115	62	365	110	105	67	111	1073
Number of diseases terminated by the end of the year	1	1	1	2	2			2	1	2	12
Number of days of sickness for diseases terminated by the end of the year	28	61	64	72	101			105	56	86	573

The annual prevalence rate of cases of disease is:

Annual prevalence rate (cases) $= \frac{17}{100} \times 1000 = 170$ per 1000.

The point prevalence rate at the beginning of the year per 1000 persons is:

$$\frac{4}{100} \times 1000 = 40 \text{ per } 1000.$$

Similar rates can be defined with reference to persons sick or episodes of disease.

Two further expressions should also be introduced:

The *mean duration of the cases of disease terminated by the end of the year*, defined as the ratio of the total sum of days of sickness of diseases terminated by the end of the year to the total number of such diseases, is

$$\frac{573}{12} = 47.8 \text{ days};$$

the *mean number of days of sickness during the year* is defined as the ratio of the total sum of days of sickness in the period to the population exposed to the risk:

$$\frac{1073}{100} = 10.7 \text{ days per person.}$$

For a more detailed account, readers should refer to the works of Benjamin (1959), Taylor and Knowelden (1958) and Swaroop (1960).

REFERENCES

(a) Books, Reviews, and Monographs

ACHESON, E. D. (1967) *Medical Record Linkage*, Oxford University Press, London.

ACHESON, E. D. (1968) *Record Linkage in Medicine*, E. & S. Livingstone Ltd., London and Edinburgh.

BENJAMIN, B. (1959) *Elements of Vital Statistics*, George Allen & Unwin Ltd., London.

FISHER, R. A. (1951) *The Design of Experiments*, Oliver & Boyd, Edinburgh, 6th ed.

HERDAN, G. (1955) *Statistics of Therapeutic Trials*, Elsevier, Amsterdam.

HILL, A. B. (1955) *Principles of Medical Statistics*, Lancet, London, 6th ed.

LINDER, A. (1959) *Planen und Auswerten von Versuchen*, Birkhäuser, Basle, 2nd ed.

LOGAN, W. P. D. and CUSHION, A. A. (1958) *Morbidity Statistics from General Practice*, vol. 1 (Studies on Medical and Population Subjects, No. 14), H.M. Stationery Office, London.

MASTER, A. M., GARFIELD, C. I. and WALTERS, M. B. (1952) *Normal Blood Pressure and Hypertension*, Lea & Febiger, Philadelphia.

NEWCOMBE, H. B., AXFORD, S. J. and JAMES, A. P. (1957) *Family Linkage of Vital and Health Records*, Report Atomic Energy of Canada Ltd. No. 470, Atomic Energy of Canada Ltd., Chalk River, Ontario.

NEYMAN, J. (1950) *First Course in Probability and Statistics*, Henry Holt, New York.

SCHWARTZ, D., FLAMANT, R., LELLOUCH, J. and ROUQUETTE, C. (1960) *Les Essais Thérapeutiques Cliniques*, Masson, Paris.

SWAROOP, S. (1960) *Introduction to Health Statistics*, E. & S. Livingstone Ltd., Edinburgh.

[SWEDEN] KUNGL. MEDICINALSTYRELSEN (1964) *Klassifikation av Sjukdomar*, del 2, Civiltryck, Stockholm.

[SWEDEN] KUNGL. MEDICINALSTYRELSEN (1965) *Klassifikation av Sjukdomar*, del 1, Civiltryck, Stockholm.

TAYLOR, I. and KNOWELDEN, J. (1958) *Principles of Epidemiology*, Churchill, London.

THORNER, R. M. and REMEIN, Q. R. (1961) *Principles and Procedures in the Evaluation of Screening for Disease* (Public Health Monograph No. 67), U.S. Department of Health, Education, and Welfare, Public Health Service, Washington.

UNITED NATIONS AND WORLD HEALTH ORGANIZATION (1962) *Use of Vital and Health Statistics for Genetic and Radiation Studies*, New York.

UNITED NATIONS STATISTICAL OFFICE (1960) *Directory of International Standards for Statistics*, ST/STAT/SER.M/22, REV. 1, New York.

U.S. DEPARTMENT OF HEALTH, EDUCATION, AND WELFARE, PUBLIC HEALTH SERVICE (1962) *International Classification of Diseases, Adapted for Indexing Hospital Records by Diseases and Operations*, vols. 1 and 2, Washington.

WEBER, A. A. (1960) *Problèmes de Statistique Mathématique Posés par les Programmes de Santé Publique*, Geneva. (Thesis, Faculty of Sciences of the University of Geneva.)

WITTS, L. J. (Editor) (1959) *Medical Surveys and Clinical Trials*, Oxford University Press, London.

WORLD HEALTH ORGANIZATION (1967a) *Manual of the International Statistical Classification of Diseases, Injuries and Causes of Death*, 1965 rev., vols. 1 and 2, Geneva.

(b) Original Papers

ARNOLD, F. A., DEAN, H. T., JAY, P. and KNUTSON, J. W. (1956) Effect of fluoridated public water supplies on dental caries prevalence. *Publ. Hlth Rep. (Wash.)*, **71:** 652–658.

BACKETT, E. M. (1966) The epidemiology of gastric cancer. (Unpublished WHO working document EURO-350 (1)/3.)

BAHN, A. K. and BODIAN, C. (1964) A life table method for studying recurrent episodes of illness or care. *J. chron. Dis.*, **17:** 1019–1031.

COLLEN, M. F., RUBIN, L., NEYMAN, J., DANTZIG, G. B., BAER, R. M. and SIEGELAUB, A. B. (1964) Automated multiphasic screening and diagnosis. *Amer. J. publ. Hlth*, **54:** 741–750.

DANISH TUBERCULOSIS INDEX (1955) The relation of tuberculin sensitivity to pulmonary calcifications as an index of tuberculosis infection. *Bull. Wld Hlth Org.*, **12:** 261–275.

DIEHL, H. S., BAKER, A. B. and COWAN, D. W. (1938) Cold vaccines. An evaluation based on a controlled study. *J. Amer. med. Ass.*, **111:** 1168–1173.

DOLEJSI, V. and VACEK, M. (1960) Méthodes pour l'étude de la morbidité en Tchécoslovaquie. *Santé publ. (Buc.)*, **3:** 219–233.

HAMMOND, E. C. and HORN, D. (1954) The relationship between human smoking habits and death rates. A follow-up study of 187, 766 men. *J. Amer. med. Ass.*, **155:** 1316–1328.

HAMMOND, E. C. and HORN, D. (1958) Smoking and death rates—Report on forty-four months of follow-up of 187,783 men. I. Total mortality. II. Death rates by cause. *J. Amer. med. Ass.*, **166:** 1159–1172, 1294–1308.

LAST, J. M. (1963) The iceberg. "Completing the clinical picture" in general practice. *Lancet*, **2:** 28–31.

LINDER, F. E. (1965) National health interview surveys, pp. 78–111. In: *Trends in the Study of Morbidity and Mortality* (WHO Public Health Papers No. 27), Geneva.

LOGAN, R. F. L. (1964) Studies in the spectrum of medical care, pp. 3–51. In: *Problems and Progress in Medical Care*, McLachlan, G. (Ed.), Oxford University Press, London.

LOSSING, E. H. (1955) Reporting of notifiable diseases. *Canad. J. publ. Hlth*, **46:** 444–448.

MACMAHON, B. (1962) Prenatal X-ray exposure and childhood cancer. *J. nat. Cancer Inst.*, **28:** 1173–1191.

MASTER, A. M., LASSER, R. P. and JAFFE, H. L. (1958) Blood pressure in white people over 65 years of age. *Ann. intern. Med.*, **48:** 284–299.

MEDICAL RESEARCH COUNCIL COMMITTEE ON INOCULATION PROCEDURES AND NEUROLOGICAL LESIONS (1955) Poliomyelitis and tonsillectomy. *Lancet*, **2:** 5–10.

MEDICAL RESEARCH COUNCIL TUBERCULOSIS VACCINES CLINICAL TRIALS COMMITTEE (1959) B. C. G. and vole bacillus vaccines in the prevention of tuberculosis in adolescents. *Brit. med. J.*, **2:** 379–396.

NYBOE, J. (1957) Interpretation of tuberculosis infection age curves. *Bull. Wld Hlth Org.*, **17:** 319–339.

SCHWARTZ, D. and DENOIX, P. F. (1957) L'enquête française sur l'étiologie du cancer broncho-pulmonaire. Rôle du tabac. *Sem. Hôp. Paris*, **33:** 3630–3643.

STEWART, A. (1961) Aetiology of childhood malignancies. Congenitally determined leukaemias. *Brit. med. J.*, **1:** 452–460.

STEWART, A., WEBB, J. and HEWITT, D. (1958) A survey of childhood malignancies. *Brit. med. J.*, **1:** 1495–1508.

WEBER, A. A. and LINDER, A. (1960) Auswertung der Ergebnisse eines Versuches zur Trachombekämpfung. *Biometrische Zeitschrift*, **2:** 217–229.

WIESLER, H. (1962) The investigation of mortality. *Annals of Life Insurance Medicine*, **1:** 1–89.

WORLD HEALTH ORGANIZATION (1963a) Quelques caractéristiques épidémiologiques de la Région européenne. (Unpublished WHO working document EUR/RC13/8.)

WORLD HEALTH ORGANIZATION (1963b) Survey of the prevalence of ischaemic heart diseases in certain European countries: Technical meeting. (Unpublished WHO document EURO-179.3 (Pr).)

WORLD HEALTH ORGANIZATION (1967b) Symposium on the use of electronic computers in health statistics and medical research. (Unpublished WHO document EURO-341.)

CHAPTER 5

PREPARATION AND REDUCTION OF DATA—GRAPHS

A. L. Delaunois

Gent, Belgium

5.1. PURPOSES OF STATISTICAL ANALYSIS

Accurate measurements of a quantity produce results which are scattered around a value called the "mean value". This scattering is caused by "errors" due to many different causes such as the experimenter, the instruments, the measurement method, the object, etc.

A biological object is particularly subject to disturbances, some of which may be traced to known or partially known causes but most of which are often unaccountable. These disturbances constitute sources of potential error hiding the answer to the questions in which the experimenter is interested. Because of the lack of information on the biological or medical object, it will generally be necessary to analyse the results obtained in the course of experiments and the best way of doing this is to make use of the mathematical method of "statistical analysis". Statistical analysis has two purposes:

(a) to reduce the often incomprehensible and complex results to a few easily understood quantities containing most if not all of the information relevant to the object under investigation;

(b) to reveal the meaning and the importance of these quantities while making due allowance for the errors caused by the disturbing influences.

This chapter deals with the first objective.

5.2. CLASSIFICATION OF DATA

Results obtained during experimental and clinical research are usually available in a chronological or alphabetical sequence. In that form they generally provide little or no clear information on the wanted variable. To acquire a first idea of the change of the variable under study, a "classification" should be made in which the data are arranged in columns and rows. Such a table will be clearer and easier to understand when properly designed and when it contains all the information on the meaning of the columns and the rows.

5.2.1. Small Number of Data

A simple and often occurring example of such a classification is the case where there is only one variable and the number of results is small. The results are classified according to their magnitude and tabulated in a first column (array). To the right of this one will be placed the corresponding calculated values as deviations and squared deviations derived from the numbers of the first column (see 5.5.1.1, Example 4).

5.2.2. Classification in the Form of a Frequency Distribution

Statistical data are also often classified in the form of a "frequency distribution". They may then be arranged according to:

(a) Their *qualitative properties* (for example, mortality from different causes, symptoms after administration of drugs, etc.). The properties are placed in a first column and next to this in a second column is indicated how often the property occurs.

EXAMPLE 1. Mortality from different causes in Sweden (1963)

Causes of death	Number of deaths
Tuberculosis (all forms)	443
Other infectious diseases	255
Cancer and other malignant tumors	15,014
Allergic and endocrinic diseases	1626
Neurological diseases	10,536
Diseases of the cardiocirculatory system	30,291
Diseases of the respiratory system	4832
Diseases of the digestive system	3177
Diseases of the uro-genital system	2042
Groups XI–XVI (WHO)	3054
Accidents	3330
Suicide	1406
Murder	63
Total	76,460

(b) The *value of the variable:*

(i) It may happen that among the values of the observations one figure occurs more than once. The different figures are arrayed in a first column, and in a second column the corresponding frequencies with which each value occurs are placed (see 5.5.1.1, Example 5).

(ii) In the case of many results it is a common rule to classify them in "groups" or "classes". One therefore divides the difference between the highest and the lowest found value (the "range") into a number of equal classes within well-defined upper and lower limits. These classes are arrayed in a first column. In a second column is marked the frequency of occurrence of a variable having a magnitude lying within the "class-interval". The "class-mid-value" or "class-mean", found by dividing the sum of the upper and lower limits of each class by 2, will be used together with the frequency of occurrence to calculate the mean value of the variable, the deviations, the squared deviations, etc. These calculated values are placed in contiguous columns (see 5.5.1.1, Example 5).

5.3. PARAMETERS

By classifying the results one way or another, we may see that many of the values found concentrate near the "mean" and that the number of values diminish as we depart from the mean. We say that the valuest show a "distribution" around the mean.

The measure for the central tendency of a random variable with a continuous distribution is defined, for a population, by the equation:

$$\mu = \frac{\int_{-\infty}^{+\infty} x\,\varphi(x)\,dx}{\int_{-\infty}^{+\infty} (x)\,dx} = \int_{-\infty}^{+\infty} x\,\varphi(x)\,dx$$

$\varphi(x)$ being the "probability density".

For a discrete distribution the mean value is given by:

$$\mu = \frac{\sum_{i=-\infty}^{\infty} x_i P(x_i)}{\sum_{i=-\infty}^{\infty} P(x_i)} = \sum_{i=-\infty}^{\infty} x_i P(x_i)$$

with $P(x_i)$ the probability of occurrence of discrete values of x.

The measure of dispersion, by analogy with that of central tendency, can be defined for a continuous distribution as:

$$\sigma^2 = \int_{-\infty}^{+\infty} (x-\mu)^2 \varphi(x)\, dx$$

and for a discrete distribution:

$$\sigma^2 = \sum_{i=-\infty}^{\infty} (x_i-\mu)^2 P(x_i).$$

The distribution exhibited by biological or clinical measurements is usually the "normal distribution". The curve of this distribution is bell-shaped and is well known as "Gauss curve" or "normal curve of error". If, however, the data have not a normal distribution it is possible, in many cases, to study them as such, or to transform or so adapt them that they can be considered as having a normal distribution.

A very important property of the Gauss curve is that a normally distributed population is completely known if two well defined values of the curve can be determined:

(a) The abscissa of the top of the curve. This is a measure of "central tendency", and is called the "mean" (μ).

(b) The distance from this mean value to the symmetrical inflexion points of the curve. This is a measure of the "dispersion" and is called the "standard deviation" (σ). These "parameters" are seldom fixed by hypothesis, but have to be estimated from the experimental data.

The theoretical mathematical forms of the distributions are of course reached only in the case of a theoretically infinite number of possible observations. Such an infinity group is called a "population".

For economic and practical reasons, biological and clinical experiments are necessarily limited in scope. In practice, one takes a relatively small number of observations out of the theoretical population and this "sample" will only roughly represent the actual distribution of the population. Therefore, a distinction has to be made between a parameter of the population and the estimate of the statistics, obtained from the experimental data.

The population, characterized by parameters, represents statistically the "hypothesis" (or its consequences); the sample, on the other hand, represents the "observations". By means of the hypothesis one tries to explain the resulting observations as well as the results of any other possible experiment of the kind under consideration.

It is seldom that a statistical hypothesis is so precisely formulated that it can be tested before the results supply, by inductive reasoning, some quantity necessary for their complete specification.

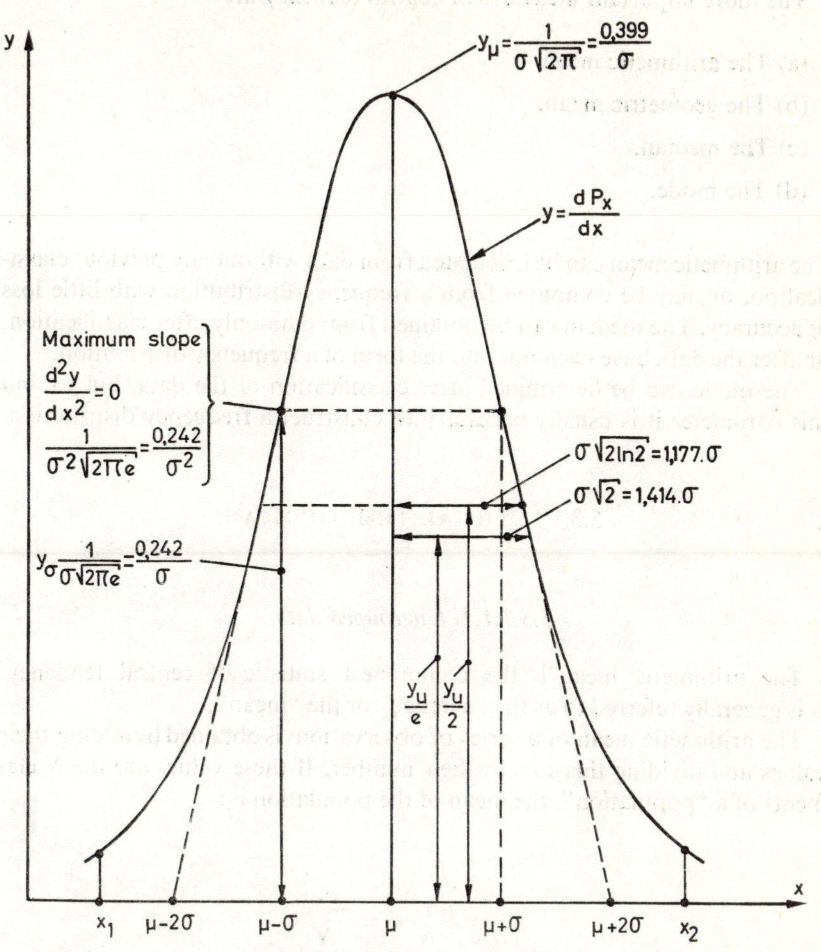

FIG. 5.1. Normal distribution for $\mu = 100$ and $\sigma = 10$.

From that point of view statistics, as mathematics of experiments, contrast with other branches of mathematical science concerned with deducing the logical consequences of a given set of postulates. Statistics are the mathematics of inductive reasoning, and they provide conclusions of general validity from results of restricted experiments.

5.3.1. Statistics of Central Tendency

The more important measures of central tendency are:

(a) The arithmetic mean.

(b) The geometric mean.

(c) The median.

(d) The mode.

The arithmetic mean can be calculated from data without any previous classification, or may be computed from a frequency distribution with little loss of accuracy. The median can be obtained from data only after classification, or after the data have been put into the form of a frequency distribution.

The mode can be determined after classification of the data, but to find this parameter it is usually necessary to construct a frequency distribution.

5.3.1.1. The Arithmetic Mean

5.3.1.1.1. Ungrouped data

The arithmetic mean is the commonest statistic of central tendency. It is generally referred to as the "average" or the "mean".

The arithmetic mean of a series of observations is obtained by adding their values and dividing this sum by their number. If these values are the N elements of a "population", the mean of the population is:

$$\mu_x = \frac{\sum_{i=1}^{N} x_i}{N} = \frac{\Sigma x_i}{N}.$$

In the case of a "sample" of n elements the mean of the sample, $\bar{x}$, is:

$$\bar{x} = \frac{\sum_{i=1}^{N} x_i}{n} = \frac{\Sigma x_i}{n}.$$

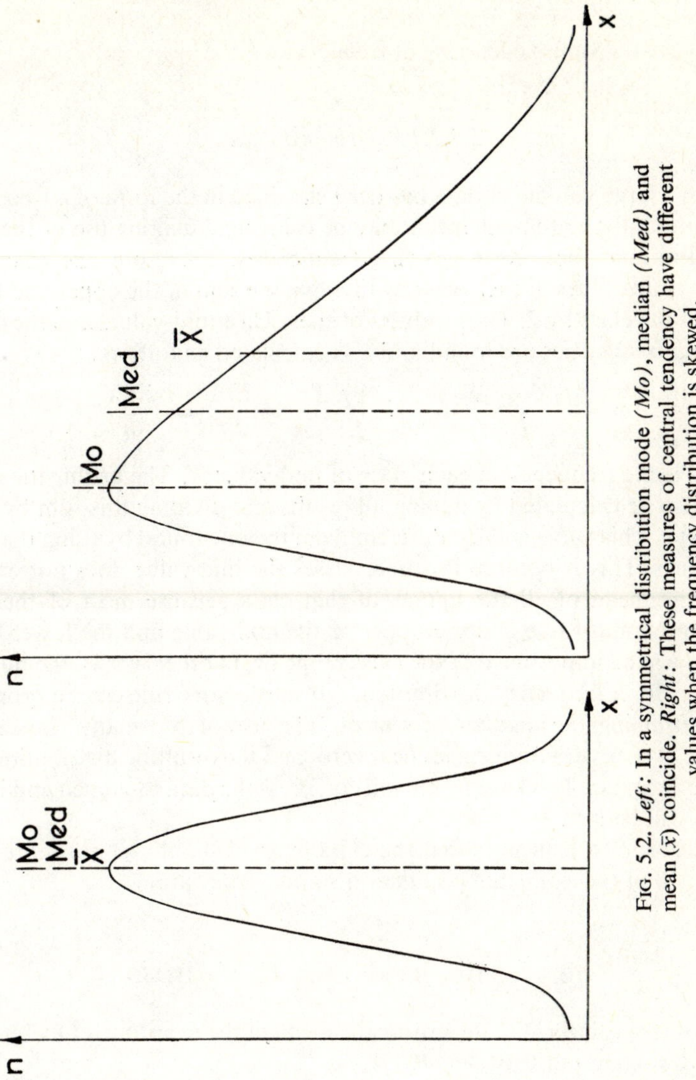

FIG. 5.2. *Left:* In a symmetrical distribution mode *(Mo)*, median *(Med)* and mean ($\bar{x}$) coincide. *Right:* These measures of central tendency have different values when the frequency distribution is skewed.

and if the data are available in a frequency distribution form, $\bar{x}$ becomes:

$$\bar{x} = \frac{f_1x_1+f_2x_2+ \ldots +f_nx_n}{f_1+f_2+ \ldots +f_n} = \frac{\Sigma fx}{\Sigma f} = \frac{\Sigma fx}{n} \quad \text{as} \quad \Sigma f = n$$

where x are the values occurring at frequencies f.

5.3.1.1.2. *Grouped data*

When a large volume of data has been classified in the form of a frequency distribution, the arithmetic mean may be calculated making use of the class mid-values and their corresponding frequencies. Therefore one calculates the class mid-values of each class by dividing the sum of the upper and lower limits of the class by 2. The products of these class mid-values and the corresponding frequencies are found and $\bar{x}$ is calculated as follows:

$$\bar{x} = \frac{f_1X_1+f_2X_2+ \ldots +f_nX_n}{f_1+f_2+ \ldots +f_n} = \frac{\Sigma fX}{\Sigma f} = \frac{\Sigma fX}{n},$$

where f is the frequency in each class of mid-value X. The arithmetic mean could also be calculated by adding all results and dividing this sum by their number, n. This value may be different from the one found by using the class mid-values. This is because in some classes the mid-value does not exactly reflect the mean of all the figures of that class because most of them lie between the mid-value and the upper or the mid-value and the lower limit. This shows the importance of the class range or, in other words, the number of classes for a frequency distribution. No satisfactory rule can be proposed for determining the number of classes. The use of too many classes will produce classes with frequencies near zero, and the resulting distribution will become irregular. Too few classes will compress the data too much and information will be lost.

Snedecor (1946) suggests that the class interval should be not more than one-fourth of the estimated population standard deviation.

5.3.1.2. The "harmonic mean" (H.M.)

This is the reciprocal of the arithmetic mean of the reciprocals of a series of values (Croxton and Cowden, 1939)

$$\text{H.M.} = \frac{1}{\frac{1}{n}\left(\frac{1}{x_1}+\frac{1}{x_2}+ \ldots +\frac{1}{x_n}\right)} = \frac{n}{\frac{1}{x_1}+\frac{1}{x_2}+ \ldots +\frac{1}{x_n}}$$

5.3.1.3. The geometric mean (G.M.)

The geometric mean is the Nth root of the product of the values

$$\text{G.M.} = \sqrt[N]{x_1 \cdot x_2 \cdot x_3 \ldots x_N}$$

or, using logarithms,

$$\log \text{G.M.} = \frac{\log x_1 + \log x_2 + \log x_3 + \ldots + \log x_N}{N},$$

and $$\text{G.M.} = \text{antilog}\, \frac{\log x_1 + \log x_2 + \log x_3 + \ldots + \log x_N}{N}.$$

The geometric mean is preferable to the arithmetic mean if the series of observations contain one or two abnormally large values. Especially where enormous variation in numerical values is common (as for example in parasitology and microbiology) the use of the geometric mean is recommended.

Example 2 (Misra and Ray, 1957). The geometric mean is calculated from the parasite counts per 100 fields of thick smear relating to 42 monkeys with plasmodic infection. The values found are:

7, 8, 3, 14, 2, 1, 440, 15, 52, 6, 2, 1, 20, 1, 1, 25, 12, 6, 9, 2, 1, 6, 7, 3, 4, 70, 20, 200, 2, 50, 21, 15, 10, 120, 8, 4, 70, 3, 1, 103, 20 and 90.

$$\log \text{G.M.} = \tfrac{1}{42}(\log 7 + \log 8 + \ldots + \log 90) = 0.974317$$

$$\text{G.M.} = \text{antilog}\, 0.974317 = 9.43.$$

The arithmetic mean found in this case is 34.6 and is a greatly inflated figure because of the presence of a few abnormally high values, e.g. 440, 200, 120 etc.

5.3.1.4. The median

The median is a value dividing a distribution so that an equal number of observations occurs on either side of it. With an odd number, $(2n+1)$, of values arrayed in order of magnitude the median will be found by taking the $(n+1)$th value, and with an even number of $2n$ values the median will be found between the nth and the $(n+1)$th value. By applying this rule in the case of grouped data only, the class in which the median will be found can be determined. To find the median from data in frequency-distribution form, one has to interpolate within the class in which the value of the median is expected to be found. Therefore, the sum of all frequencies is divided by 2:

$$\frac{\Sigma f}{2} = \frac{n}{2},$$

and starting from either end of the distribution one interpolates the class at $n/2$ to determine the value of this point.

(a) Proceeding upwards from the class at $n/2$ one finds:

$$\text{Median} = \text{lower limit of the class} + \frac{\frac{n}{2} - \Sigma f\text{-down}}{f_{\text{class}}} \cdot \text{class interval.}$$

Σf-down is the sum of the frequencies in the classes below the class in which the interpolation is done.

(b) Proceeding downwards from the class at $n/2$ the median becomes:

$$\text{Median} = \text{upper limit of the class} - \frac{\frac{n}{2} - \Sigma f\text{-up}}{f_{\text{class}}} \cdot \text{class interval.}$$

Σf-up is the sum of the frequencies in the class above that in which $n/2$ lies.

5.3.1.5. The Mode

The mode of a series is the value on which the data concentrate or, in other words, the value or values having the highest frequency. A distribution may have more than one mode (bi- or multimodality), but this happens infrequently. The curve of such a distribution shows two or more points of concentration, which are not necessarily of equal prominence. Bi- or multimodality is often due to the failure to separate different groups (males—females, children–adults, sick–healthy individuals), or may be present fortuitously. For a small number of observations when each value occurs only once no true mode exists.

Croxton and Cowden (1939) described a method of calculating the value of the mode M_0 to which the data concentrate within a class of a frequency distribution

$$\Delta f_1(M_0 - l_1) = \Delta f_2(l_2 - M_0)$$

$$\Delta f_1 M_0 + \Delta f_2 M_0 = \Delta f_2 l_1 + \Delta f_1 l_2;$$

but as

$$l_2 = l_1 + b$$

$$M_0(\Delta f_1 + \Delta f_2) = \Delta f_2 l_1 + \Delta f_1 l_1 + \Delta f_1 b$$

$$M_0 = \frac{l_1(\Delta f_1 + \Delta f_2)}{\Delta f_1 + \Delta f_2} + \frac{\Delta f_1 b}{\Delta f_1 + \Delta f_2}$$

$$M_0 = l_1 \frac{\Delta f_1}{\Delta f_1 + \Delta f_2} \times b$$

where Δf_1 = the frequency difference between classes I and II,
Δf_2 = the frequency difference between II and III,
b = the interval of class II (modal class).

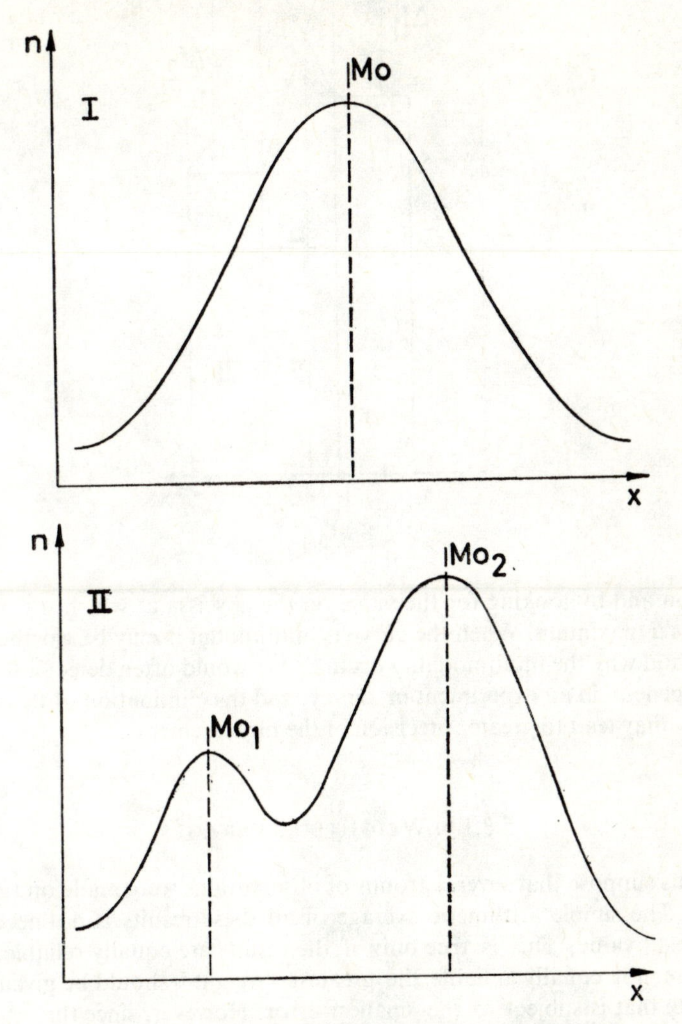

FIG. 5.3. I: Unimodal distribution. II: Bimodal distribution.

From this formula it is evident that the mode depends on the class-interval b. This interval is, as mentioned before, somewhat arbitrary. Using the graphical method, one may find the mode by making the histogram or frequency

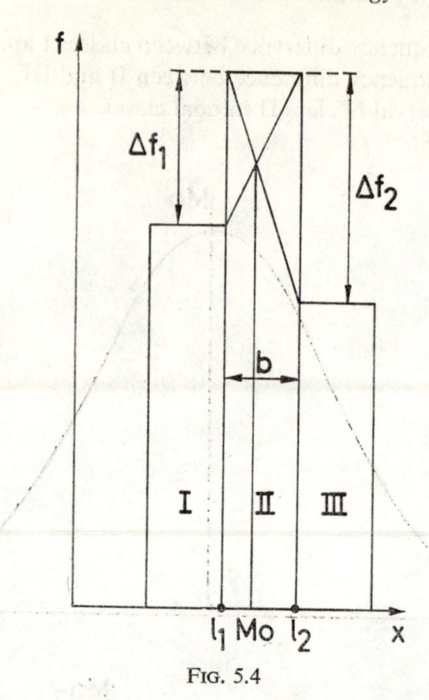

FIG. 5.4

polygon and by looking for the value on the abscissa at which the ordinate reaches a maximum. When the curve is multimodal it may be worthwhile to try to find why the multimodality occurs. One would often detect a source of heterogeneity in an experiment or survey, and the elimination of this heterogeneity may lead to greater precision of the measurements.

5.3.1.6. WEIGHTED AVERAGES

Let us suppose that several groups of observations were made on the same object. The simple arithmetic average of all these results is not necessarily the "best" value. This is true only if the results are equally reliable. But if they are not equally reliable the greatest "weight" should be given to the estimate that is subject to the smallest error. However, since the other estimates, though less reliable, may contribute additional information to the overall approximation of the value x, it would be wrong to neglect them. Considering all the estimates, one has to take into account their individual weight, which can be shown mathematically to be inversely proportional to the variance of the estimate. Therefore, to obtain the "weighted mean" one multiplies each individual estimate by a "weighting factor" equal to the reciprocal of its variance, these weighted values are then summed, and the sum is divided by the sum of the weighting factors, thus:

$$\bar{x}_w = \frac{\sum_{i=1}^{n} w_i x_i}{\sum_{i=1}^{n} w_i}.$$

EXAMPLE 3. Blood sugars are determined for three groups of men undergoing fast. The results for each group (mg/100 ml blood) are:

(1) 101.7 $s_1 = 21.0$ $s_1^2 = 441$ $1/s_1^2 = w_1 = 0.00225$

(2) 126.4 $s_2 = 19.7$ $s_2^2 = 338$ $1/s_2^2 = w_2 = 0.00262$

(3) 132.0 $s_3 = 23.5$ $s_3^2 = 552$ $1/s_3^2 = w_3 = 0.00181$

$$\text{Weighted mean} = \frac{0.00225 \times 101.7 + 0.00262 \times 126.4 + 0.00181 \times 132.0}{0.00225 + 0.00262 + 0.00181}$$

$$= 119.59.$$

5.3.2. Statistics of Dispersion

The second statistic which must be generally estimated from the experimental data is the "dispersion".

5.3.2.1. THE RANGE

The simplest measure of dispersion is the "range", that is the difference between the largest and the smallest values of the series of data. The range, being determined by only those two values, gives no information on the arrangement of the data *between* the two extremes, and is obviously influenced by the number of data of a series, as well as by extraordinarily large and small extreme values.

5.3.2.2. FRACTILES, QUARTILES, DECILES, AND PERCENTILES

A second way of measuring the dispersion is the use of "quartiles", "deciles", and "percentiles". The fractile (quartile, decile, or percentile), α, of an arrayed series of data is the value x_α below which a fraction, α, of the number of data is found. For example $\alpha = 0.1$ means that 10% of the data are below the value x. $x_{0.10}$ is the tenth "percentile". For $\alpha = 0.25$ the fractile is $x_{0.25}$, and it is called the "first quartile". The fractile $x_{0.75}$ is the "third quartile", and $x_{0.50}$ is the "second quartile" (or the median). A measure for dispersion employing the first and third quartiles is known as the "semi-interquartile range" and can be determined from $x_{0.75}$ and $x_{0.25}$, thus:

$$Q_{1.3} = \frac{x_{0.75} - x_{0.25}}{2}.$$

The value of x is not always directly available from a series of data. In that case one can easily find x_α by interpolation.

5.3.2.3. The average deviation (A.D.)

The "average deviation" or "mean deviation" is a measure of dispersion computed from the expression:

$$\text{A.D.} = \frac{\Sigma|x_i - \bar{x}|}{n}.$$

The average deviation takes into consideration the values of all the observations x_i of the distribution. It is the sum of the absolute values of the differences of each observation (value x_i) and the mean $\bar{x}$ divided by n, the number of observations. It has to be noted that the differences $(x_i - \bar{x})$ may have positive or negative values and that $\Sigma(x_i - \bar{x})$ is usually zero or very nearly zero. By using in the sum $\Sigma|x_i - \bar{x}|$ only the absolute values, all the deviations, without regard to sign, are added. Dividing them by n produces a mean or average. The average deviation is much less useful than the standard deviation.

5.3.2.4. The variance and the standard deviation

The "variance" of the distribution of a variable x, considering the whole population of N elements, is defined as

$$\sigma^2 = \frac{\sum_{i=1}^{N} (x_i - \mu)^2}{N} = \frac{\Sigma(x_i - \mu)^2}{N},$$

in which the squared deviations $(x_i - \mu)$ are summed and divided by N.

The squaring of $(x_i - \mu)$ is a more complex way of removing any difficulty introduced by the sign of the deviations. The variance is important for the further treatment of the results. The square root of σ^2 is called the "standard deviation":

$$\sigma = \sqrt{\sigma^2} = \sqrt{\frac{\Sigma(x_i - \mu)^2}{N}}.$$

σ^2 as well as σ measures the dispersion of the observations around the mean μ. σ is expressed in the same units as the data.

The variance of a *sample* of n observations of the variable x is given by

$$s^2 = \frac{\sum_{i=1}^{n} (x_i - \bar{x})^2}{n-1} = \frac{\Sigma(x_i - \bar{x})^2}{n-1}.$$

It may be noted that the divisor is not n but $n-1$.

5.3.2.5. DEGREES OF FREEDOM

The choice of $n-1$ is justified by the following considerations

(a) To evaluate the spread of a variable we need at least two values, and those two values will supply only one difference; with three values we obtain two independent differences; and, generalizing, n values give $n-1$ independent differences. If the mean is fixed by hypothesis an nth difference will be obtained by subtracting the observed mean from the theoretical mean. But if the theoretical mean is unknown the n observations contribute only $n-1$ independent differences or "degrees of freedom" to the estimate of the standard deviation (Mather, 1960).

(b) The use of $n-1$ as a divisor may also be justified by considering the case of only one observation. Knowing the mean, as fixed by hypothesis, this single observation will supply one difference for the estimation of s. If, however, the mean is unknown, the observation is the best estimate of the mean and the difference between the value of the mean and the value of the observation is inevitably zero, as they are the same. Using $n-1$ as a divisor the variance becomes $0/0$ which is indeterminate. But if one uses n as divisor the variance would be $0/1$, which is ridiculous (Mather, 1960).

(c) The dispersion of an observation is in fact due to the deviation of the value from the calculated mean, characterized by s, and the deviation of the calculated mean $\bar{x}$ from the theoretical value μ. This means that the variance

s^2 has to be replaced by

$$s^2 - \frac{s^2}{n},$$

$\frac{s^2}{n}$ being the variance of the mean (see 5.3.2.6).

Since
$$\frac{\Sigma(x_i - \bar{x})^2}{n} = s^2 - \frac{s^2}{n} = \left(\frac{n-1}{n}\right) \cdot s^2,$$

$$s^2 = \frac{\Sigma(x_i - \bar{x})^2}{n-1}$$

and
$$s = \sqrt{\frac{\Sigma(x_i - \bar{x})^2}{n-1}}.$$

When n is very great, $n-1 \approx n$, and it makes little or no difference whether one uses $n-1$ or n as the divisor. The equations of σ^2 or s^2 are often used in another form more convenient for computing purposes. Indeed, the sum of squares $\Sigma(x_i - \mu)^2$ can be transformed thus:

$$\Sigma(x_i - \mu)^2 = \Sigma([x_1 - \mu)^2 + (x_2 - \mu)^2 + \ldots + (x_N - \mu)^2]$$
$$= \Sigma[(x_1^2 - 2x_1\mu + \mu^2) + (x_2^2 - 2x_2\mu + \mu^2) + \ldots + (x_N^2 - 2x_N\mu + \mu^2)]$$

in which

$$x_1^2 + x_2^2 + \ldots + x_N^2 = \Sigma x_i^2$$

and
$$-2\mu(x_1 + x_2 + \ldots + x_N) = -2\mu\Sigma x_i.$$

Thus
$$\Sigma(x_i - \mu)^2 = \Sigma x_i^2 - 2\mu\Sigma x_i + N\mu^2.$$

Further, if
$$\mu = \frac{\Sigma x_i}{N},$$

$$\Sigma(x_i - \mu)^2 = \Sigma x_i^2 - 2\frac{\Sigma x_i}{N}\Sigma x_i + N\frac{\Sigma x_i}{N} \cdot \frac{\Sigma x_i}{N} = \Sigma x_i^2 - \frac{\Sigma x_i}{N} \cdot \Sigma x_i.$$

$$\sigma^2 = \frac{\Sigma x_i^2 - \frac{(\Sigma x_i)^2}{N}}{N} = \frac{\Sigma x_i^2}{N} - \left(\frac{\Sigma x_i}{N}\right)^2.$$

$\frac{(\Sigma x_i)^2}{N}$ is called the "correction term", and for s^2

$$s^2 = \frac{\Sigma x_i^2 - \bar{x}\Sigma x_i}{n-1}.$$

5.3.2.6. The variance of the mean

We have by definition $\sigma^2 = \dfrac{\Sigma(x_i-\mu)^2}{N}$

$$= \frac{\Sigma}{N}[(x_1-\mu)^2+(x_2-\mu)^2+\ \dots\ +(x_n-\mu)^2]$$

$$= \frac{\Sigma}{N}[(x_1^2-2x_1\mu+\mu^2)+(x_2^2-2x_2\mu+\mu^2)+\ \dots$$

$$+(x_N^2-2x_n\mu+\mu^2)]$$

and

$$\sigma^2 = \frac{\Sigma x^2}{N}-2\mu\frac{\Sigma x}{N}+\frac{N\mu^2}{N}$$

or:

$$\sigma^2 = \frac{\Sigma x^2}{N}-2\mu^2+\mu^2$$

$$= \frac{\Sigma x^2}{N}-\mu^2.$$

Extracting Σx^2 from this equation,

$$\Sigma x^2 = (\mu^2+\sigma^2)N.$$

On the other hand, one finds that

$$(\Sigma x)^2 = (x_1^2+x_2^2+\ \dots\ +x_N^2)+(2x_1x_2+2x_1x_3+\ \dots)$$

in which equation the first part is equal to $\Sigma x^2 = N(\mu^2+\sigma^2)$ and the second part contains $N(N-1)$ terms having an expected value of $\mu\cdot\mu = \mu^2$.

After transformation,

$$(\Sigma x)^2 = N\mu^2+N\sigma^2+N(N-1)\mu^2$$

$$= N\mu^2+N\sigma^2+N^2\mu^2-N\mu^2$$

$$= N\sigma^2+N^2\mu^2,$$

and after dividing $(\Sigma x)^2$ by N^2 we get

$$\frac{(\Sigma x)^2}{N^2} = \bar{x}^2 = \mu^2+\frac{\sigma^2}{N}.$$

The dispersion of $\bar{x}$ is given by:

$$\sigma_{\bar{x}}^2 = \frac{\Sigma}{N}(\bar{x}-\mu)^2 = \frac{\Sigma(\bar{x}^2)}{N}-\mu^2$$

$$= \mu^2+\frac{\sigma^2}{N}-\mu^2 = \frac{\sigma^2}{N},$$

and

$$\sigma_{\bar{x}} = \sqrt{\frac{\sigma^2}{N}} = \frac{\sigma}{\sqrt{N}}.$$

Although $\bar{x}$ varies from sample to sample, its variance is only $1/N$ of the original population variance. By taking larger samples the variance of $\bar{x}$ can be made small (Li, 1964).

To distinguish $\sigma_{\bar{x}}$ from the standard deviation of the population (σ) and the standard deviation of the sample (s), $\sigma_{\bar{x}}$ is called the "standard error of the mean".

5.4. THE OPERATOR *E*

A very useful and convenient procedure is to use, especially in theoretical work, the "operator *E*" (Li, 1964). The symbol *E* preceding a variable or a part of an equation simply means "the long-range average value of ..." or more formally "the mathematical expectation of ..." or "the expected value of ...". So, if one wants to define the population mean one may write $E\{x\} = \mu$, which is in fact a shorter way to express

$$\frac{\sum_{i=1}^{N} x_i}{N}.$$

Similarly, the variance of a population can be expressed as

$$E\{x-\mu\}^2 = \sigma^2,$$

which means also "the average value of all squared deviations in the population". Since an "averaging process" consists of adding all possible values, weighted by their probabilities or their relative frequencies of occurrence, the operator *E* can essentially be handled in the same manner as the sign Σ. Three elementary rules have to be kept in mind when using the operator *E*:

(A) $E\{c\} = c$.

The expected value of a constant c is c.

(B) $E\{cx\} = cE\{x\}$.

The expected value of the product of a constant c and a variable x is the value of the constant multiplied by the expected value of the variable.

(C) $E\{x_1+x_2+ \ldots +x_n\} = E\{x_1\}+E\{x_2\}+ \ldots +E\{x_n\}$.

The expected value of a sum is equal to the sum of the separate expected values.

From these simple rules one can see that

(a) $E\{a_1x_1+a_2x_2+ \ldots +c\} = a_1E\{x_1\}+a_2E\{x_2\}+ \ldots +c$.

(b) $E\{ax\}^2 = E\{a^2x^2\} = a^2E\{x^2\}$.

(c) $E\{x_1\} = E\{x_2\} = \ldots = E\{x_n\} = \mu$.

(d) $E\{x_1+x_2+ \ldots +x_n\} = n\mu$.

(e) $E\{x_1x_2\} = E\{x_1\}\cdot E\{x_2\} = \mu\cdot\mu = \mu^2$.

(f) From the definition of σ^2 one may find the expected value of x^2:

$$\begin{aligned}\sigma^2 = E\{x-\mu\}^2 &= E\{x^2-2\mu x+\mu^2\}\\ &= E\{x^2\}-2\mu E\{x\}+\mu^2\\ &= E\{x^2\}-2\mu\cdot\mu+\mu^2\\ &= E\{x^2\}-\mu^2\end{aligned}$$

and $$E\{x^2\} = \mu^2+\sigma^2,$$
which shows that the expected value of the square of a variable is always larger than that of the product of two statistically independent variables with the same mean value.

(g) From (f) one can see that

$$E\{x_1^2+x_2^2+ \ldots +x_n^2\} = E\{\Sigma x^2\} = n(\mu^2+\sigma^2).$$

(h) The expected value of $(\Sigma x)^2$ can also be found

$$E\{x_1+x_2+\ldots+x_n\}^2 =$$

$$E\{\underbrace{x_1^2+x_2^2+ \ldots +x_n^2}_{n \text{ square terms}}+\underbrace{2x_1x_2+2x_1x_3+ \ldots}_{\substack{n(n-1) \text{ product terms}\\ \text{of the type } x_1x_2.}}\}$$

The expected value of the square terms is $\mu^2+\sigma^2$, and that of the product terms is μ^2.

Hence
$$E\{(\Sigma x)^2\} = n(\mu^2+\sigma^2)+n(n-1)\mu^2$$
$$= n^2\mu^2+n\sigma^2 .$$
$$E\left\{\frac{(\Sigma x)^2}{n}\right\} = E\{n\bar{x}^2\} = n\mu^2+\sigma^2.$$
$$\frac{E\{(\Sigma x)^2\}}{n^2} = E\{\bar{x}^2\} = \mu^2+\frac{\sigma^2}{n}.$$

(i) In repeated sampling the value of the mean $\bar{x}$, which is considered as a constant when only one sample is available, varies from sample to sample. The long-range average value of $\bar{x}$ can be found thus
$$E\{\bar{x}\} = \frac{1}{n}E\{x_1+x_2+ \ \ldots \ +x_n\} = \frac{1}{n}(n\mu) = \mu.$$

$\bar{x}$ is an "unbiased estimate" of the population mean.

(j) As $\bar{x}$ varies from sample to sample, for each $\bar{x}$ there must be a corresponding deviation $(\bar{x}-\mu)$. The average value of the variance of $\bar{x}$ can be calculated:
$$\sigma_{\bar{x}}^2 = E\{\bar{x}-\mu\}^2$$
$$= E\{\bar{x}^2\}-\mu^2$$
$$= \mu^2+\frac{\sigma^2}{n}-\mu^2 = \frac{\sigma^2}{n}.$$

(k) The expected value of s^2 can be derived from the preceding equations
$$s^2 = \frac{\Sigma(x-\bar{x})^2}{n-1}.$$

The numerator of these equations can be split into two parts
$$\Sigma(x-\bar{x})^2 = \Sigma x^2-\frac{(\Sigma x)^2}{n}.$$

Thus:
$$E\{\Sigma(x-\bar{x})^2\} = (n\mu^2+n\sigma^2)-(n\mu^2+\sigma^2)$$
$$= (n-1)\sigma^2 ,$$

and
$$E\left\{\frac{\Sigma(x-\bar{x})^2}{n-1}\right\} = E\{s^2\} = \sigma^2 .$$

This means that, if a large number of random samples of n are taken, the long-range average value of s^2 is equal to the population variance σ^2. s^2 is called an "unbiased estimate" of σ^2.

5.5.1 The Practical Determination of s^2, s and $\sigma_{\bar{x}}$

5.5.1.1. Ungrouped data

The sum of all the values $x_1, x_2, x_3, \ldots, x_n$ is made and this sum is divided by n (the number of values) to obtain the mean $\bar{x}$:

$$\frac{\Sigma x_i}{n} = \bar{x}.$$

From each value is subtracted the mean $\bar{x}$. This gives the "deviations" $(x_i - \bar{x})$, and these deviations are squared, $(x_i - \bar{x})^2$. The values x_i, $(x_i - \bar{x})$ and $(x_i - \bar{x})^2$ are arrayed.

$n \begin{cases} x_1 \\ x_2 \\ \cdot \\ \cdot \\ \cdot \\ x_n \end{cases}$	$x_1 - \bar{x}$ $x_2 - \bar{x}$ $\cdot$ $\cdot$ $\cdot$ $x_n - \bar{x}$	$(x_1 - \bar{x})^2$ $(x_2 - \bar{x})^2$ $\cdot$ $\cdot$ $\cdot$ $(x_n - \bar{x})^2$	x_1^2 x_2^2 $\cdot$ $\cdot$ $\cdot$ x_n^2	$\bar{x} = \frac{\Sigma x_i}{n}$
Σx_i		$\Sigma(x_i - \bar{x})^2$	Σx_i^2	

The sum of the $(x_i - \bar{x})^2$ is $\Sigma(x_i - \bar{x})^2$, and this sum is divided by $n - 1$.

This gives
$$s^2 = \frac{\Sigma(x_i - \bar{x})^2}{n-1}.$$

The square root of s^2 gives:

$$s = \sqrt{\frac{\Sigma(x_i - \bar{x})^2}{n-1}}.$$

Dividing s by $\sqrt{n}$, one obtains $\sigma_{\bar{x}} = \dfrac{s}{\sqrt{n}}$.

Example 4 (from Emmens, 1948). The measurements of 20 uterine weights of members of a group of young female rats are arrayed.

	x (mg)	$x_i-\bar{x}$	$(x_i-\bar{x})^2$	x_i^2
	9	−12	144	81
	14	−7	49	196
	15	−6	36	225
	15	−6	36	225
	16	−5	25	256
	18	−3	9	324
	18	−3	9	324
	19	−2	4	361
	19	−2	4	361
	20	−1	1	400
$n = 20$	21	0	0	441
	22	1	1	484
	22	1	1	484
	24	3	9	576
	24	3	9	576
	26	5	25	676
	27	6	36	729
	29	8	64	841
	30	9	81	900
	32	11	121	1024
	$\Sigma x_i = 420$	$\Sigma(x_i-\bar{x}) = 0$	$\Sigma(x-\bar{x})^2 = 664$	$\Sigma x^2 = 9484$

From the values of x_i one determines the mean thus:

$$\bar{x} = \frac{\Sigma x_i}{n} = \frac{420}{20} = 21.$$

With this mean $(x_i-\bar{x})$ is calculated.

The sum of the deviations $\Sigma(x_i-\bar{x}) = 0$.

The deviations $(x_i-\bar{x})$ are squared: $(x_i-\bar{x})^2$ and the sum $\Sigma(x_i-\bar{x})^2$ is made.

Then:

$$s^2 = \frac{\Sigma(x_i-\bar{x})^2}{n-1} = \frac{664}{19} = 34.94,$$

$$s = \sqrt{s^2} = \sqrt{\frac{664}{19}} = 5.91,$$

and

$$\sigma_{\bar{x}} = \frac{s}{\sqrt{n}} = \frac{5.91}{\sqrt{20}} = 1.32.$$

Application of the equation

$$s = \sqrt{\frac{\Sigma x_i^2 - \bar{x}\Sigma x_i}{n-1}}$$

leads to less computing work.

The values of x_i are added and divided by n to obtain the arithmetic mean $\bar{x}$. The values of x_i are also squared and summed (Σx_i^2):

$$s^2 = \frac{\Sigma x_i^2 - \bar{x}\Sigma x_i}{n-1} = \frac{9484 - 21 \times 420}{19} = \frac{9484 - 8820}{19} = \frac{664}{19} = 34.94$$

$$s = \sqrt{34.94} = 5.91.$$

When a value occurs more than once one may calculate s^2, s and $\sigma_{\bar{x}}$ by using equations on page 134.

EXAMPLE 5. Cholesterol determinations carried out in the serum of normal children (C. Hooft, Department of Pediatrics, University of Gent).

x_i (mg%)	f	x^2	$f_i\, x_i$
155	1	24,025	24,025
160	2	25,600	51,200
165	1	27,225	27,225
170	1	28,900	28,900
175	2	30,625	61,250
180	2	32,400	64,800
190	2	36,100	72,200
195	2	38,025	76,050
200	3	40,000	120,000
204	1	41,616	41,616
205	1	42,025	42,025
210	2	44,100	88,200
215	2	46,225	92,450
220	2	48,400	96,800
230	2	52,900	105,800
235	2	55,225	110,450
240	2	57,600	115,200
245	1	60,025	60,025
250	2	62,500	125,000
255	1	65,025	65,025
260	2	67,600	135,200
270	3	72,900	218,700
285	1	81,225	81,225
315	2	99,225	198,450
325	1	105,625	105,625
330	1	108,900	108,900
375	1	140,625	140,625
455	1	207,025	207,025
$\Sigma x_i =$ 10,729	$\Sigma f_i = n = 46$		$\Sigma f_i x_i^2 =$ 2,663,991

One computes:

$$\Sigma x_i = 10{,}729,$$
$$\Sigma f_i = n = 46,$$
$$\Sigma f_i x_i^2 = 2{,}663{,}991.$$

The arithmetic mean is:

$$\bar{x} = \frac{\Sigma x_i}{n} = \frac{10{,}729}{46} = 233,$$

and $$s^2 = \frac{\Sigma f_i x_i^2 - \bar{x}\Sigma x_i}{n-1} = \frac{2{,}663{,}991 - (233 \times 10729)}{46-1} = 3647.42,$$

$$s = \sqrt{3647.42} = 60.4,$$

$$\sigma_{\bar{x}} = \frac{60.4}{\sqrt{46}} = 9.0.$$

5.5.1.2. Grouped data

When a large number of data have to be treated, one divides them into classes so as to reduce the computational work considerably. The range is divided into a number of equal classes of interval b. The frequency of the values found between the upper and lower limit of each class and the class mid-value or class-mean are determined. Those values are arrayed as shown in the following table and ΣX_i, Σf_i, $\Sigma X_i f_i$, $(\Sigma X_i f_i)^2$ and $\Sigma X_i^2 f_i$ calculated.

Class mid-values X_i	Frequency f_i	X_i^2	$X_i f_i$	$X_i^2 f_i$
X_1	f_1	X_1^2	$X_1 f_1$	$X_1^2 f_1$
X_2	f_2	X_2^2	$X_2 f_2$	$X_2^2 f_2$
X_3	f_3	X_3^2	$X_3 f_3$	$X_3^2 f_3$
.	.	.	.	.
.	.	.	.	.
.	.	.	.	.
X_n	f_n	X_n^2	$X_n f_n$	$X_n^2 f_n$
ΣX_i	$\Sigma f_i = n$		$\Sigma X_i f_i$	$\Sigma X_i^2 f_i$

s^2 is then calculated,

$$s^2 = \frac{\Sigma X_i^2 f_i - \dfrac{(\Sigma X_i f_i)^2}{n}}{n-1},$$

$$s = \sqrt{\frac{\Sigma X_i^2 f_i - \dfrac{(\Sigma X_i f_i)^2}{n}}{n-1}}.$$

$$\sigma_{\bar{x}} = \frac{s}{\sqrt{n}}.$$

EXAMPLE 6 (C. Hooft, Department of Pediatrics, University of Gent). Cholinesterol determination are made in the serum of children having lipoid nephrosis. The values are expressed in mg %, and are arrayed in classes having an interval of 100 mg.

(a) The values arrayed and divided into classes of $b = 100$ give the results shown in the table.

X	x_i	f_i	X	x_i	f_i	X	x_i	f_i
	335			425			500	
	360			425			505	
$X_1 = 350$	370	$f_1 = 5$		430			530	
	385		$X_2 = 450$	430	$f_2 = 8$	$X_3 = 550$	530	$f_3 = 9$
	395			460			550	
				470			568	
				480			570	
				495			575	
							590	
	610			700			800	
	615			708			800	
	620			710			810	
	625			715			810	
	640			720			820	
	640			720			820	
	650			740			820	
$X_4 = 650$	650	$f_4 = 15$	$X_5 = 750$	742	$f_5 = 17$	$X_6 = 850$	830	$f_6 = 16$
	660			745			850	
	665			760			850	
	667			770			860	
	675			775			860	
	680			780			875	
	685			780			880	
	690			780			880	
				780			895	
				798				

X	x_i	f_i	X	x_i	f_i	X	x_i	f_i
$X_7 = 950$	925	$f_7 = 4$	$x_8 = 1050$	1005	$f_8 = 7$	$X_9 = 1150$	1100	$f_9 = 3$
	940			1020			1125	
	940			1030			1190	
	960			1040		$X_{10} = 1250$	1200	$f_{10} = 2$
				1050			1220	
				1060				
				1080				

(b) After classification of the class mid-values, the frequencies, X_i^2, $X_i f_i$, and $X_i^2 f_i$, $\Sigma f_i = n$, $\Sigma X_i f_i$, and $\Sigma X_i^2 f_i$ are calculated.

Class mid-value X_i	Frequency f_i	X_i^2	$X_i f_i$	$X_i^2 f_i$
$X_1 = 350$	5	122,500	1750	612,500
$X_2 = 450$	8	202,500	3600	1,620,000
$X_3 = 550$	9	302,500	4950	2,722,500
$X_4 = 650$	15	422,500	9750	6,337,500
$X_5 = 750$	17	562,500	12,750	9,562,500
$X_6 = 850$	16	722,500	13,600	11,560,000
$X_7 = 950$	4	902,500	3800	3,610,000
$X_8 = 1050$	7	1,102,500	7350	7,717,500
$X_9 = 1150$	3	1,322,500	3450	3,967,500
$X_{10} = 1250$	2	1,562,500	2500	3,125,000
	$\Sigma f_i = 86 = n$		$\Sigma X_i f_i = 63,500$	$\Sigma X_i^2 f_i = 50,835,000$

These figures give the following results:

$$\bar{x} = \frac{\Sigma X_i f_i}{n} = \frac{63,500}{86} = 738.37$$

$$s^2 = \frac{\Sigma X_i^2 f - \dfrac{(\Sigma X_i f_i)^2}{n}}{n-1} = \frac{50,835,000 - \dfrac{4,032,250,000}{86}}{85}$$

$$= \frac{50,835,000 - 46,826,628}{85} = \frac{3,948,372}{85} = 46,451,$$

and $$s = \sqrt{46{,}451} = 215.5$$

$$\sigma_{\bar{x}} = \frac{s}{\sqrt{n}} = \frac{215.5}{\sqrt{86}} = \frac{215.5}{9.27} = 23.24.$$

5.5.2. Sheppard's Correction

If large numbers of data are treated in grouped form, it may happen that the standard deviation as computed from these data is different from the standard deviation that would be found by treating the data in ungrouped form. The reason is that the mid-values of any one class do not always agree with the arithmetic means of the values from that class.

For frequency distributions when we deal with continuous variables (or discrete variables, if there are no gaps in the data) the mid-values of classes for values greater than the mode tend to be too large (Croxton and Cowden, 1939). As the mid-values from the classes are used to compute the deviations from the arithmetic mean $\bar{x}$ of the data, the absolute values of these deviations are too great. On squaring the deviations, the errors do not offset each other, but are cumulative. The result is that the value of s for a frequency distribution is usually larger then the value of s for the same data in ungrouped form.

Sheppard (1898) proposed a correction which can be introduced in those cases where

(a) the distribution relates to a continuous variable,
(b) the number of observations decreases gradually at the two ends of the curve,
(c) the number of observations is sufficiently large.

This correction is about $\frac{1}{12}$ (0.0833) of the square of the class-interval, and is subtracted from s^2. If the correction is applied when it is not appropriate it may over-correct, producing an s-value which is too small. In general, Sheppard's correction should be used only when one is reasonably sure that it is applicable.

5.5.3. Computation by Means of a Working Scale

When the number of data is very large it is usually somewhat difficult to compute the class values. The reduction of these values to more easily handled digits greatly simplifies the computation; this can be done by using a "working scale" instead of the class mid-values.

As an origin for the working scale one chooses a "working mean", $\bar{x}_m$, which is generally the class mid-value of the class having the highest frequency. The deviation of the class mid-values of the classes from the arithmetic mean $\bar{x}$ can then be expressed as the difference of two quantities:

(a) a deviation of the class mid-value from the mean $(X-\bar{x}_m)$, and
(b) a deviation of the working mean from the arithmetic mean $(\bar{x}-\bar{x}_m)$.

Thus $(X-\bar{x}) = (X-\bar{x}_m)-(\bar{x}-\bar{x}_m)$

and $(X-\bar{x})^2 = (X-\bar{x}_m)^2-2(X-\bar{x}_m)(\bar{x}-\bar{x}_m)+(\bar{x}-\bar{x}_m)^2$.

The difference $(X-\bar{x}_m)$ is called the "working unit". For n equations one obtains

$$\begin{aligned}\Sigma(x-\bar{x})^2 &= \Sigma(x-\bar{x}_m)^2-2\Sigma[(x-\bar{x}_m)(\bar{x}-\bar{x}_m)]+\Sigma(\bar{x}-\bar{x}_m)^2\\ &= \Sigma(x-\bar{x}_m)^2-2(\bar{x}-\bar{x}_m)\Sigma(x-\bar{x}_m)+n(\bar{x}-\bar{x}_m)^2.\end{aligned}$$

$\bar{x}$ and $\bar{x}_m$ are constant, and since

$$\bar{x} = \frac{\Sigma x}{n}$$

we get

$$\begin{aligned}n(\bar{x}-\bar{x}_m) &= n\bar{x}-n\bar{x}_m\\ &= \Sigma(x)-n\bar{x}_m\\ &= \Sigma(x-\bar{x}_m).\end{aligned}$$

Replacing this value in the previous equation:

$$\begin{aligned}\Sigma(x-\bar{x})^2 &= \Sigma(x-\bar{x}_m)^2-\frac{2}{n}[\Sigma(x-\bar{x}_m)]^2+\frac{1}{n}[\Sigma(x-\bar{x}_m)]^2\\ &= \Sigma(x-\bar{x}_m)^2-\frac{[\Sigma(x-\bar{x}_m)]^2}{n}.\end{aligned}$$

Replacing x by the class mid-values X and introducing the class-frequencies, the variance is

$$s^2 = \frac{\Sigma f(X-\bar{x})^2}{n-1} = \frac{\Sigma f(X-\bar{x}_m)^2}{n-1}-\frac{[\Sigma f(X-\bar{x}_m)]^2}{n(n-1)},$$

and the standard deviation

$$s = \sqrt{\frac{\Sigma f(X-\bar{x})^2}{n-1}} = \sqrt{\frac{\Sigma f(X-\bar{x}_m)^2}{n-1} - \frac{[\Sigma f(X-\bar{x}_m)]^2}{n(n-1)}}.$$

The values are arrayed as shown in the table.

Class mid-value X	Frequency	Deviation $(X-\bar{x}_m)$ working unit	$f(X-\bar{x}_m)$	$f(X-\bar{x}_m)^2$
X_1	f_1	$X_1-\bar{x}_m$	$f_1(X_1-\bar{x}_m)$	$f_1(X_1-\bar{x}_m)^2$
X_2	f_2	$X_2-\bar{x}_m$	$f_2(X_2-\bar{x}_m)$	$f_2(X_2-\bar{x}_m)^2$
X_3	f_3	$X_3-\bar{x}_m$	$f_3(X_3-\bar{x}_m)$	$f_3(X_3-\bar{x}_m)^2$
.	.	.	.	.
.	.	.	.	.
.	.	.	.	.
X_n	f_n	$X_n-\bar{x}_m$	$f_n(X_n-\bar{x}_m)$	$f_n(X_n-\bar{x}_m)^2$
ΣX	$\Sigma f = n$		$\Sigma f(X-\bar{x}_m)$	$\Sigma f(X-\bar{x}_m)^2$

The correction term is calculated: $\frac{[\Sigma f(X-\bar{x}_m)]^2}{n}$, then the variance

$$s^2 = \frac{\Sigma f(X-\bar{x}_m)^2 - \frac{[\Sigma f(X-\bar{x}_m)]^2}{n}}{n-1},$$

and the standard deviation

$$s = \sqrt{s^2} = \sqrt{\frac{\Sigma f(X-\bar{x}_m)^2 - \frac{[\Sigma f(X-\bar{x}_m)]^2}{n}}{n-1}}.$$

EXAMPLE 7 (Romero Alvarez and Miranda Franco, 1959). The reliability of measuring water-dispersible DDT by means of a volumetric method was studied. The frequency distribution of 1022 weight measurements is arrayed in classes having a class interval of 10 g.

Class limits (g)	Class mid-values (g)	Working unit $\frac{X-\bar{x}_m}{b}$	Frequency f	$\frac{f(X-\bar{x}_m)}{b}$	$\left(\frac{X-\bar{x}_m}{b}\right)^2 f$
460–469	465	−11	5	− 55	605
470–479	475	−10	22	−220	2200
480–489	485	− 9	24	−216	1944
490–499	495	− 8	13	−104	832
500–509	505	− 7	15	−105	735
510–519	515	− 6	20	−120	720
520–529	525	− 5	46	−230	1150
530–539	535	− 4	44	−176	704
540–549	545	− 3	73	−219	657
550–559	555	− 2	47	− 94	188
560–569	565	− 1	49	− 49	49
570–579	575	0	91	0	0
580–589	585	1	104	104	104
590–599	595	2	88	176	352
600–609	605	3	77	231	693
610–619	615	4	57	228	912
620–629	625	5	73	365	1825
630–639	635	6	101	606	3636
640–649	645	7	44	308	2156
650–659	655	8	15	120	960
660–669	665	9	3	27	243
670–679	675	10	3	30	300
680–689	685	11	2	22	242
590–699	695	12	2	24	288
700–709	705	13	4	52	676
			$\Sigma f = n$ $= 1022$	$\Sigma\left[\frac{(X-\bar{x}_m)f}{b}\right]$ $= 705$	$\Sigma\left[\left(\frac{X-\bar{x}_m}{b}\right)^2 f\right]$ $= 22{,}171$

First of all, the working mean is chosen, for example $\bar{x}_m = 575$. Knowing the value of $\bar{x}_m$, the working units can be calculated:

$$\frac{X-\bar{x}_m}{b} = \frac{\text{class mid-value} - 575}{\text{class interval } (= 10)}.$$

They are multiplied by f, giving the value

$$\frac{f(X-\bar{x}_m)}{b}$$

and summed:

$$\Sigma \frac{f(X-\bar{x}_m)}{b}.$$

The arithmic mean of the working scale is

$$\frac{\Sigma\left[\frac{f(X-\bar{x}_m)}{b}\right]}{\Sigma f} = \frac{705}{1022} = 0.69.$$

In the working scale 575 g was taken as the zero point. The arithmetic mean lies 0.69 working units above 573. The working unit is 10 g, which means that the arithmetic mean $\bar{x}$ will be 6.9 g above the working mean: 575+ 6.9 = 581.9. One calculates

$$\frac{\Sigma\left[f\left(\frac{X-\bar{x}_m}{b}\right)\right]^2}{n} = \frac{705^2}{1022} = 486.3,$$

$$s'^2 = \frac{\Sigma_f\left(\frac{X-\bar{x}_m}{b}\right)^2 - \frac{\left(\Sigma f\frac{X-\bar{x}_m}{b}\right)^2}{n}}{n-1}$$

$$= \frac{22{,}171-486.3}{1021} = \frac{21{,}684.7}{1021} = 21.24,$$

and $$s' = \sqrt{21.24} = 4.61.$$

s' is the standard deviation on the working scale. To bring this value to the real scale 4.61 has to be multiplied by the working unit:

$$s = bs' = 10\times 4.61 = 46.1 \text{ g},$$

and

$$s^2 = b^2s'^2 = 100\times 21.24 = 2124.$$

As the working mean is arbitrary, one can choose $\bar{x}_m = 0$. The equations simplify and become:

$$\Sigma(x-\bar{x})^2 = x^2-\tfrac{1}{2}(\Sigma x)^2,$$

$$s^2 = \frac{\Sigma(x-\bar{x})^2}{n-1} = \frac{\Sigma X^2}{n-1} - \frac{(\Sigma X)^2}{n(n-1)},$$

$$s = \sqrt{\frac{\Sigma(X-\bar{x})^2}{n-1}} = \sqrt{\frac{\Sigma X^2}{n-1} - \frac{(\Sigma X)^2}{n(n-1)}}.$$

5.6. DIAGRAMS

Statistical data, usually used in arrayed form, can also be presented as diagrams, graphs or charts. Rather than aiming to give detailed information, they are designed to present a general picture.

Diagrams "prove" nothing, but bring outstanding features readily to the eye; they are therefore no substitute for such critical tests as may be applied to the data, but are valuable in suggesting such tests, and in explaining the conclusions founded upon them (Emmens, 1948). Statistical diagrams are of various kinds, the use of each type depending largely on the nature of the data and the purpose of the diagram.

The commonest types of diagrams are:

(1) Line or curve charts:
  (a) with arithmetic scale,
  (b) with semi-logarithmic or logarithmic scale,
  (c) with other special scales.
(2) Bar diagrams and histograms.
(3) Pie-charts.
(4) Statistical maps.
(5) Correlation diagrams.

There are also organization and procedure charts, but these have no statistical value and are not treated here (Brinton, 1939). As a general rule, a diagram should be made as simple and as clear as possible so that it can be directly understood without having to study it closely. Choosing the simplest form appropriate to the problem is of supreme importance. The various elements to be included in a diagram are:

(a) The heading or title.
(b) A prefatory note explaining the nature of the data in a clear and concise form.
(c) The coordinate lines; they must be kept to a minimum, but should clearly indicate the units used. They should not interfere with the curve, which must stand out sharply against the background.
(d) Source of data; in order to make the diagram self-contained the source of data should appear under the graph.

Diagrams should produce a truthful impression on the reader. They should also be attractive, and show good draughtsmanship. Their construction should be easily checked and should clearly show the feature that has been emphasized.

ordinates). Plotted points are connected by a solid line or by If several series have to be compared, they may be drawn on rt, using a separate symbol line for each series.

GARITHMIC, SEMI-LOGARITHMIC AND LOG $(x+1)$ SCALES

grams using a logarithmic vertical scale are mainly used for series ich show a geometric progression. The diagrams are called "semi-c" or "ratio" charts. They are important to show the existence of atio increases or decreases.

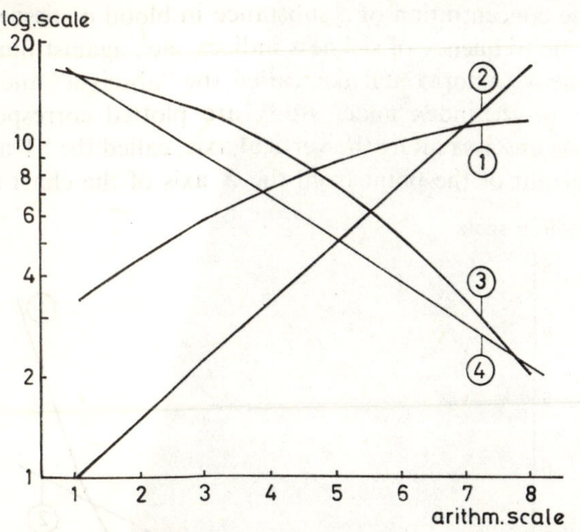

FIG. 5.6. Line diagrams having one axis arithmetic and the other logarithmic show straight lines for *constant* rate of increase or decrease (2, 4), but curved lines for *accelerated* or *diminishing* rate of increase or decrease (1, 3).

the ratio changes are constant, the line which represents the phenomenon ill be a straight line on the semi-logarithmic chart. If, however, the ratio hanges are not constant the line will remain curvilinear.

A semi-logarithmic chart shows:

(a) a constant ratio of change as a straight line,

(b) the magnitude of constant ratio by the slope of the line,

(c) a constant rate of increase by an ascending straight line and a constant rate of decrease by a descending line, and

(d) equal rates of change in corresponding series as parallel lines.

Besides the fact that logarithmic scales indicate whether the curve exhibits a constant ratio of increase or decrease, there is one more feature of consider-

5.6.

5.6.1.1. AR

Line diagrams are usually drawn
horizontal and vertical. Such diag
which are in arithmetic progression.
sists in plotting the values of a variab
or a man, the concentration of a subs
population, the frequency of sickness in
plotted on the X or horizontal axis call
the quantity or the index under study
position on the abscissa along the vertical
nate". The height of the point from the X

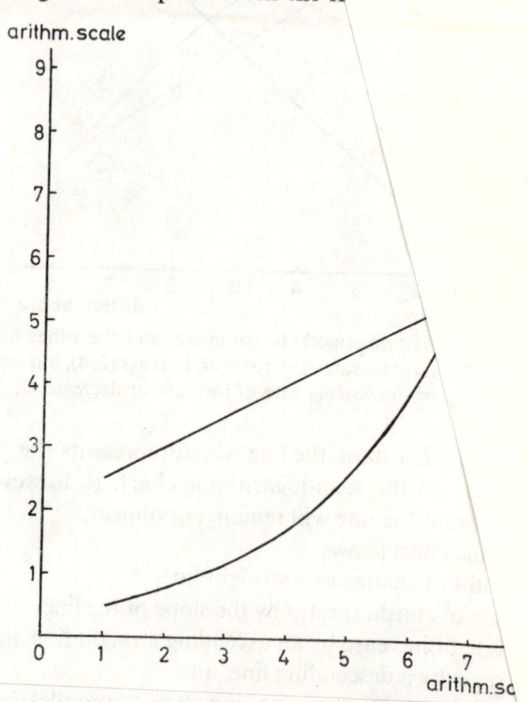

FIG. 5.5. Line diagrams with arithmetic scales for both ordinate a
show a straight line for constant *amount* of increase or decrease (2),
lines for constant *rate* of increase or decrease (1).

value (x, y c
symbol lines
the same ch

5.6.1.2. L

Line dia
of data wh
logarithm
constant r

If
w
c

able value. Heights of peaks are sensibly reduced while very low figures are somewhat raised above the abscissa. This feature makes it possible to study very low figures with greater ease and makes the logarithmic scale especially popular for charting figures of cases and deaths. Some non-symmetrical frequency curves can be made to appear symmetrical by changing the horizontal scale from an arithmetic to a logarithmic one. An asymmetric curve is called "log-normal" if it can be converted to a normal curve by this change. Log-normal frequency curves are characterized by a longer tail on the right-hand side of the peak (Bernstein and Weatherall, 1952). One difficulty arises as soon as figures approach zero level. Theoretically it is impossible to plot a zero figure on a logarithmic scale because the logarithmic value of zero corresponds to minus infinity, which cannot be shown on charts. In some fields of medical research, for example in eradication campaigns, the interest ultimately lies in ascertaining when the absolute zero level is attained and whether this level is subsequently maintained. The difficulty can be overcome if to each actual figure one unity is added, and thus instead of plotting $\log x$, $\log(x+1)$ is plotted. Since log 1 is zero, as soon as the zero level is attained for the values of x the plotted point will be situated on the abscissa. One precaution is necessary in interpreting the $\log(x+1)$ scale. The percentage decrease recorded between small values of x becomes reduced to some extent by the use of this method. However, the difference is insignificant when the values of x are large (Emmens, 1948).

Figures		Percentage decrease in successive values of	
x	$x+1$	x	$x+1$
10,000	10,001		
5000	5001	50	50
1000	1001	80	80
500	501	50	50
200	201	60	60
100	101	50	50
50	51	50	50
40	41	20	20
30	31	25	24
20	21	33	32
10	11	50	48
5	6	50	45
4	5	20	$16.7 \approx 17$
3	4	25	20
2	3	33	25
1	2	50	33
0	1	100	50

5.6.2. Bar Diagrams

In the construction of a bar diagram frequencies are represented by the lengths of bars. The bars may have an arbitrary width, though for the sake of clearness an appropriate width is necessary. The bars are placed on a common abscissa and there is a small space between two adjacent bars. For precise comparison the bars may be placed side by side, each bar having a length proportional to the absolute value of the component it represents. The number of bars which stick together should be limited to two or three, as more than that number tends to give a confusing picture. A marked contrast in shading the bars will improve the clearness of the picture.

5.6.3. Histograms

A special application of the bar-diagram is the histogram (Fig. 5.7). The histogram is the graphical representation of a frequency distribution. The x-values are placed along the abscissa and the frequency of each class is plotted in ordinates. The bars are in this case arranged side by side. Their width is determined by the upper and lower class limits, their height by the frequency of the values in the represented class. As the class interval is more or less determined (see 5.3.1.1.2), the units of the ordinate must be selected properly. A good compromise for this selection is a bar height for the class with the greatest number of values equal to two-thirds or three-quarters of the length of the total range of the observations represented on the abscissa. When the class intervals are unequal the heights of the bars have to be determined carefully. The frequencies of classes with unequal intervals cannot be compared simply by height. A method often used in these cases to represent the frequency is to use the area instead of the height.

5.6.4. Frequency Polygon

From a histogram of a frequency distribution with equal class intervals the frequency polygon can be derived. For this method the class mid-values are marked on the top side of the bars and these points are joined by straight lines. If the number of observations is large, and if the observations are classified in a larger number of narrower classes, the points of the frequency polygon will get nearer to each other. Theoretically one can imagine an infinitely large number of infinitely narrow classes. The points of the polygon

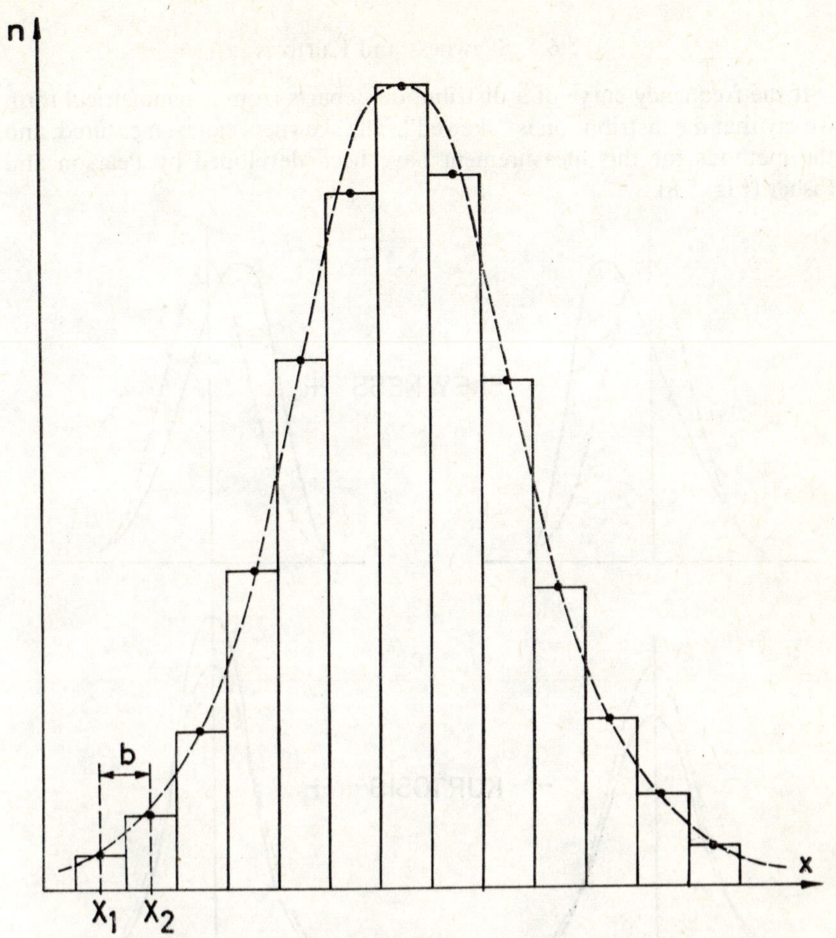

FIG. 5.7. Histogram of a frequency distribution (bars). Dotted line: the ideal normal distribution curve. $b = x_2 - x_1$: class interval. x_1 and x_2: class mid-values.

will then lie infinitely close and produce a smooth curve, called the "frequency curve" (Fig. 5.7). Although the data are classified in finite class intervals, it often occurs that the frequency polygon can indicate what the shape of the theoretical frequency curve can possibly be.

5.6.5. Skewness and Kurtosis

If the frequency curve of a distribution departs from a symmetrical form we say that the distribution is "skewed". The skewness can be measured, and the methods for this measurement have been developed by Pearson and Fisher (Fig. 5.8).

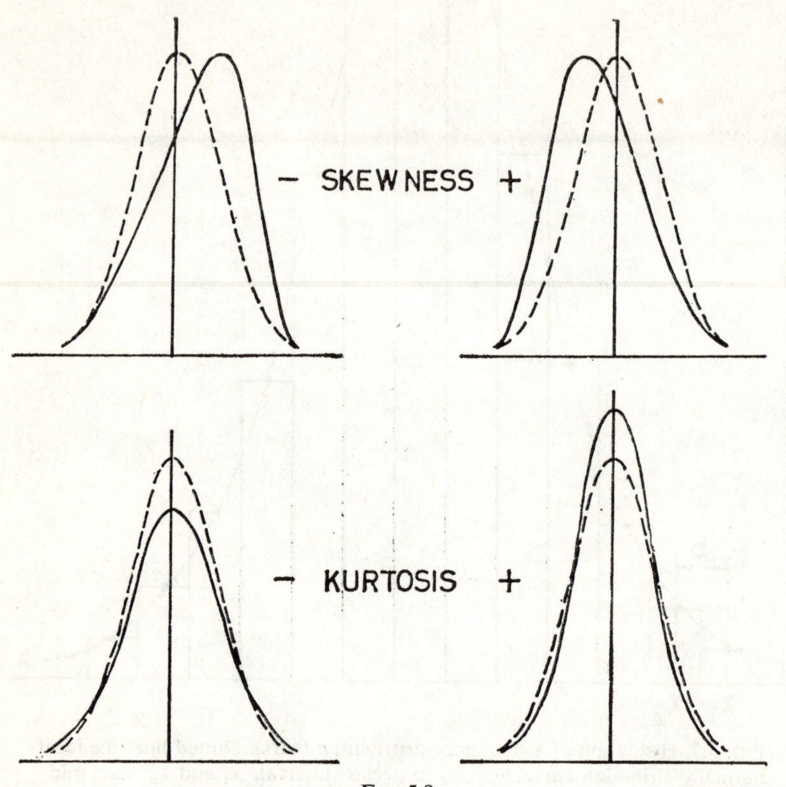

Fig. 5.8

The "Pearsonian measure of skewness" is given by the equation

$$Sk = \frac{\bar{x} - Mo}{s}.$$

But as the methods of locating the value of the mode are not entirely satisfactory, Pearson suggested an empirical formula for determining the mode. For moderately skewed distributions of continuous variables the mode

$Mo = \bar{x} - 3(\bar{x} - Med)$ [Med = median]

$$Sk = \frac{\bar{x} - Mo}{s} = \frac{\bar{x} - [\bar{x} - 3(\bar{x} - Med)]}{s} = \frac{3(\bar{x} - Med)}{s}.$$

For frequency distributions without skewness $Sk = 0$. Values of Sk_{max} approach ± 3 as a limit, but series observed data are not likely to show values exceeding ± 1. The method of R. A. Fisher uses "moments about the mean" and "cumulants" and is treated in Chapters 6 and 7. The direction of the skewness is obtained from the sign of the third moment about the arithmetic mean of a frequency distribution. A positive sign indicates skewness to the right, a negative sign skewness to the left. The degree of skewness can be calculated from the third and second moments about the mean.

The frequency curve may be perfectly symmetrical and still differ from a normal curve by having a narrower modal portion and higher tails, or a broader modal portion and lower tails. Such a curve has "kurtosis" and in the first case is called a "leptokurtic" curve, in the second case a "platykurtic" curve. Kurtosis is reflected in the fourth powers of the deviations of the values of a frequency distribution from their arithmetic mean (Fig. 5.8).

5.6.6. Cumulative Frequency Curve

The frequencies for each class of a frequency distribution can be cumulated by a process of successive summation. A cumulative frequency distribution is then established. If the cumulative totals are plotted in ordinate for the corresponding class mid-values in abscissa, and the obtained individual points joined by straight lines, an S-shaped curve will be obtained. With a normal distribution, and drawing a smooth line through the points, the resultant curve is a "sygmoid" (Galton) (Fig. 5.9). In this case the inflection point corresponds with the mid-point of the total frequency (median). Cumulative frequency curves are often used in biological research, especially for dose–response and LD_{50} determinations (probit analysis).

5.6.7. Correlation Diagram

Correlation theory makes it possible to calculate the degree to which two or more variables are associated. Sometimes the correlation is obvious, and this happens when there is a high degree of correlation. But it is often doubtful whether there is any correlation at all. In many cases the truth is between those two extremes. To make a correlation diagram one uses millimeter paper, and the value of one variable is plotted on the ordinate (y) and that

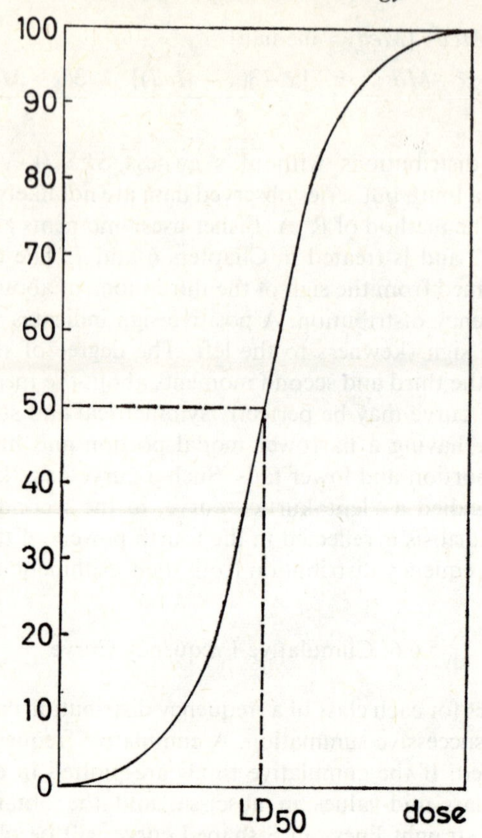

FIG. 5.9. Cumulative (sigmoid) frequency curve.

of the second variable, corresponding with the value *y*, is plotted on the abscissa (*x*). The point obtained (*x*, *y*) is called a "dot". The points found will occupy the area in a more or less "scattered" way. A correlation of $r = 1$ (maximum correlation) is found when all the points are on a straight line. No correlation ($r = 0$) is indicated by a great scattering of the points in the *x–y* area. Different degrees of correlation are indicated by the tendency to give to the dots a common direction in the *x–y* area (Fig. 5.10).

5.6.8. Pie-diagrams (Fig. 5.11)

A pie-diagram can be used to show the various component parts of a whole. It consists of a circle cut into slices which correspond in area with the quantities. This kind of diagram is seldom used in statistics.

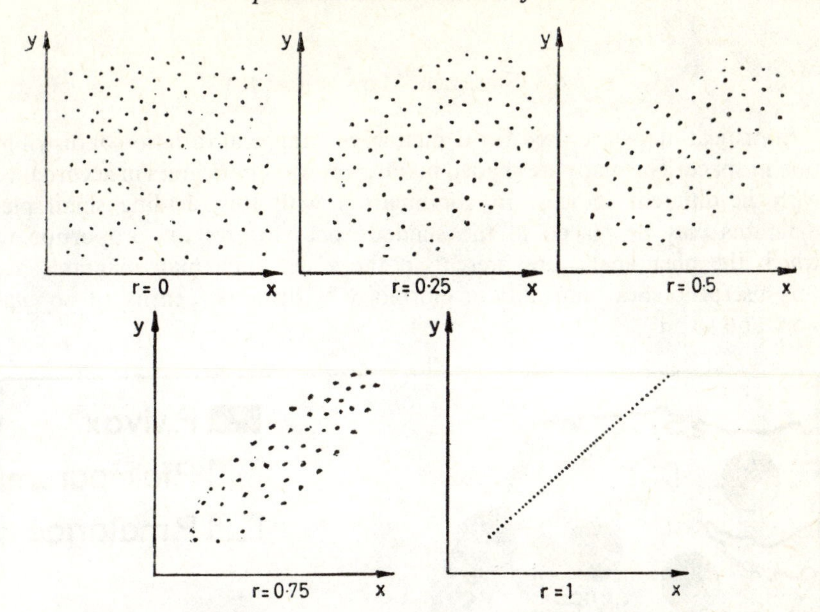

FIG. 5.10. Correlation diagrams of x, y values of two series of experiments and their corresponding correlation coefficients r.

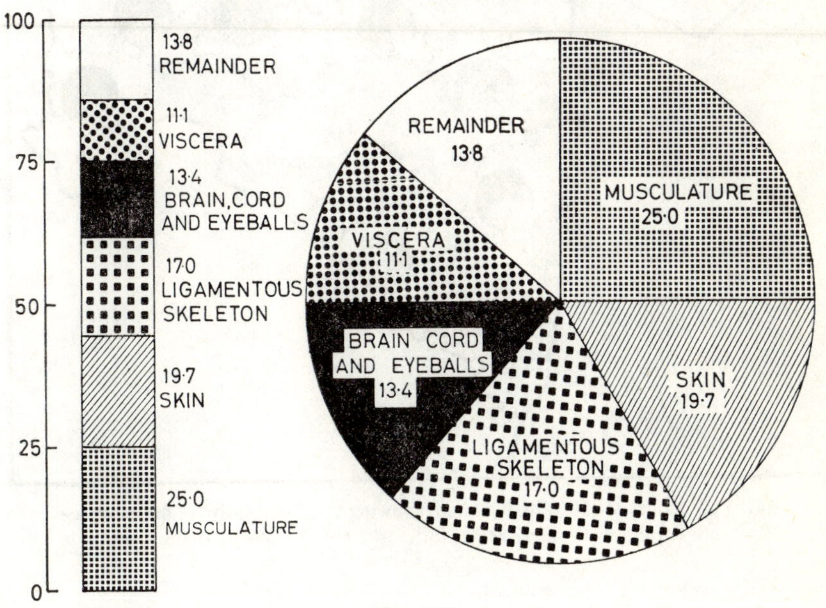

FIG. 5.11

5.6.9. Statistical Maps (Fig. 5.12)

Statistical maps are used for depicting geographical variation or distribution in space. The maps are shaded in different ways or degrees in accordance with the different indices. In combination with this shading small pie-diagrams may be placed in the shaded zones to give the proportion to which the phenomena are present in the zone. Such map diagrams are very useful to show mortality or morbidity by districts, density of population, and so on.

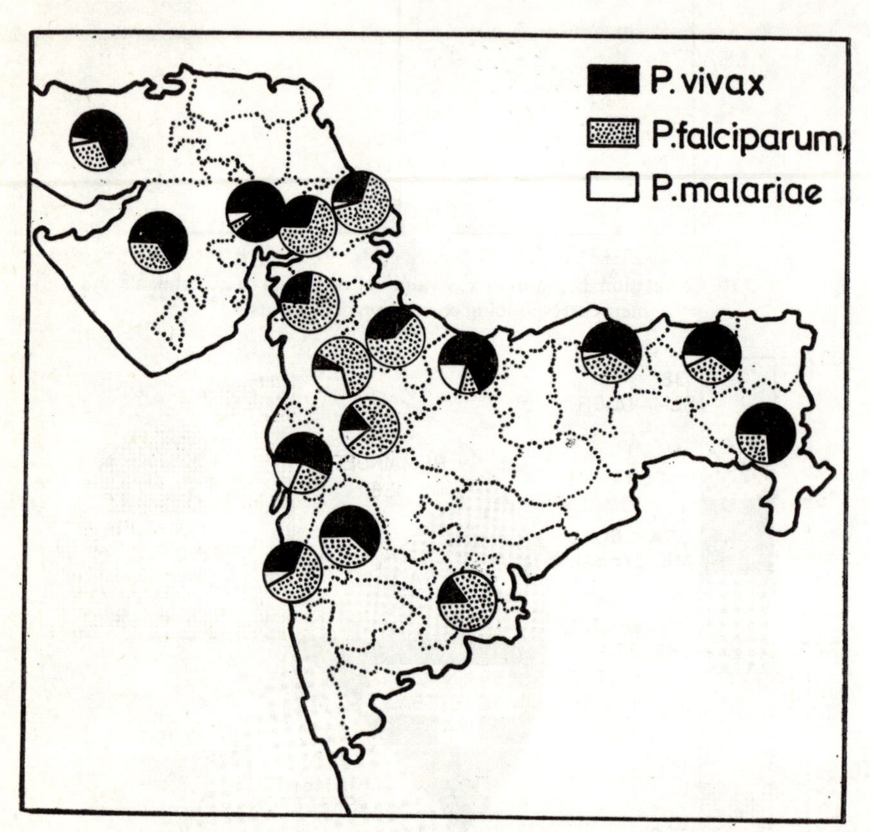

FIG. 5.12. Example of a statistical map showing the distribution of human plasmodia in Bombay State, India in 1950 (Swaroop, 1959).

5.6.10. Pictographs

Instead of using bars to show proportions or figures, it is sometimes more visually effective to use small pictures. These graphs are often used by insurance companies to show evolution of age, disease, etc. (Misra and Ray, 1957).

5.6.11. Nomograms (Fig. 5.13)

Among the aids to computation of fairly complicated mathematical formulae, graphical methods occupy a special place. These graphical charts, called "nomograms" or "nomographs", are devices for solving certain types

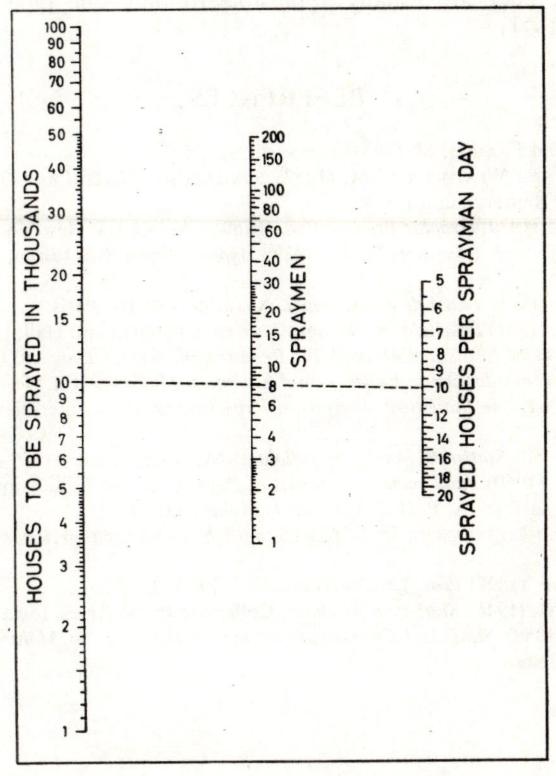

FIG. 5.13. Example of a nomogram for computing insecticide requirement on the basis of a sprayman's output (Swaroop, 1959).

of equations, primarily those containing three variables. If any two of these variables are known, the third can directly be determined from the conditioning equation between the three parameters. It can be shown mathematically that the conditioning equation defines three curves, whose points can be labeled in terms of three basic parameters (Menzel, 1960). Knowing the value on two of these curves, a straight line from the appropriate points will intersect the third curve at the value that satisfies the equation. For four variables the nomograms are identical in form with those of three, except that one or more of the scales are shifted, the necessary flexibility being obtained using a sort of slide rule index. Sometimes one of the four variables is treated as a constant. For each value of this constant a nomogram is calculated, and by interpolation between the various curves the appropriate point is selected. Using the nomograms one can immediately read fairly accurate values which would otherwise require complicated mathematical calculations. They are usually applicable to only one type of equation (D'Ocagne, 1921).

REFERENCES

ALVAREZ, R. and FRANCO, M. (1959) See Swaroop (1959).

BERNSTEIN, L. and WEATHERALL, M. (1952) *Statistics for Medical and Other Biological Students*, Edinburgh and London.

BRINTON, W. C. (1939) *Graphic Presentation*, Brinton Associates, New York.

CROXTON, F. E. and COWDEN, D. J. (1939) *Applied General Statistics*, Prentice-Hall, New York.

D'OCAGNE, M. (1921) *Traité de Nomographie*, Gauthier-Villars, Paris.

EMMENS, C. W. (1948) *Principles of Biological Assay*, Chapman and Hall, London.

FISHER, R. A. (1958) *Statistical Methods for Research Workers*, Oliver and Boyd, London.

LI, C. C. (1964) *Introduction to Experimental Statistics*, McGraw-Hill, New York.

MANDEL, J. (1964) *The Statistical Analysis of Experimental Data*, Interscience Publishers, New York.

MATHER, K. (1960) *Statistical Analysis in Biology*, Methuen, London.

MENZEL, D. H. (1960) *Fundamental Formulas of Physics*, Dover Publications, New York

MISRA, B. G. and RAY, A. P. (1957) *Indian J. Malar.*, **11:** 249.

MODLEY, R. and LOWENSTEIN, D. (1952) *Pictographs and Graphs*, Harper and Brothers, New York.

SHEPPARD, W. F. (1898) *Proc. London Math. Soc.*, **29:** 353.

SNEDECOR, G. W. (1946) *Statistical Methods*, Collegiate Press, Ames, Iowa.

SWAROOP, S. (1959) *Statistical Considerations and Methodology in Malaria Eradication*, W.H.O., Geneva.

CHAPTER 6

DISTRIBUTIONS AND TRANSFORMATIONS[†]

Leopold J. Martin
Brussels, Belgium

6.1. THE CONCEPT OF PROBABILITY

6.11. Probability in the Discrete Domain

Introduction. Objectively the probability of a discrete event can be regarded (1) from the experimental or frequential viewpoint as the limit of the relative frequencies of the event *observed* over a large number of trials in the *past*, or (2) from the point of view of combinatorial analysis of the realizations of the event in a fundamental probability set, thus permitting evaluation of the probability of a *future* event.

Both these viewpoints are valid for the conditional probability as well as for the unconditional, or absolute, probability. When only the event in itself is considered, i.e. disregarding the contingencies associated with the realization of the event, one evaluates unconditional (absolute) probabilities, usually referred to quite simply as the probabilities. The calculation is performed from consideration of some simple dichotomy, for instance, a favorable as against an unfavorable outcome, or survival versus death. But whenever the event is regarded from both viewpoints simultaneously, two dichotomies are involved, for instance, whether or not vaccinated and either surviving or deceased. In this case the probability of survival can be calculated both in the vaccinated group and in the non-vaccinated group. Here one determines the conditional probability—the probability of survival conditional upon having been vaccinated.

† *Translator's note:* Various queries have been discussed with Dr. Davies at the Department of Mathematical Statistics, Birmingham University, Birmingham 15, U.K., and he has kindly consented to assist me in investigating any "knotty points" if any more should remain.

We shall commence by considering the concept of absolute probability in section 6.111, and then pass on to its frequential, or *a posteriori*, aspect (6.111.1) and its classical, or *a priori*, aspect (6.111.2), called by some writers "Boolean Algebra" or "Algebra of Events". The inter-relation between these two concepts is treated in section 6.112, giving an example of independent events (6.112.1) and of non-independent events (6.112.2).

6.111. Absolute or Unconditional Probability

One is in the sphere of absolute or unconditional probability whenever the events being considered are dissociated from the contingencies of their realization, and the chance element is present.

We now consider the frequential aspect of past series of events which are thus of an experimental, or at least observational, nature.

6.111.1. Frequential aspect of statistical probability

The main characteristic of objective methods in general is their reproducibility. An experimental and easily reproducible event is the toss of a coin which can either fall "tails" (event E_1) or "heads" (event E_2). These two events exhaust all the possibilities, for it is very unlikely, and even impossible, for a circular coin to come to rest edgewise. If the coin is well balanced mechanically, statistically it is said to be *unbiased*, and the play is fair. In such cases the player subjectively expects to observe E_2 as often as E_1 since E_1 and E_2 are the only two possible realizations of the event E. In order to confirm this impression, the player repeats the toss n times and observes the event with an *absolute frequency* r (r is the initial letter of "realization of event" or "result favorable"). Table 6.111.1A shows the results obtained after three series of $n = 10$, 100 and 1000 throws.

The absolute frequencies of E_1 increase with n in chance fashion, but largely linearly. For a useful comparison of the results of tests made up of different numbers of trials, it is natural to reduce to unity the number r of successes observed, hence the relative frequency given in column 3 of Table 6.111.1A.

$$\mathcal{P} = \frac{r}{n}, \tag{a}$$

measured as a fraction of unity or as a percentage (column 4).

It is seen that the relative frequency of success in n trials is fairly constant in the series of trials from $n = 10$ to $n = 1000$. The relative frequency is the

TABLE 6.111.1A

Number of tosses n	Absolute frequency of success (favourable reactions) r	Relative frequency as a fraction of unity $\mathcal{P} = r/n$	As a percentage $100\tau\%$
10	4	0.400	40.0
100	55	0.550	55.0
1000	497	0.497	49.7

proportion of favorable realizations over n trials. The observation that the relative frequency is fairly constant implies some "permanency" in these physically independent trials.

Other types of permanency also exist; for instance, the relative proportion of boys (0.515) and girls (0.485), the proportion of men and women having a certain blood group in a well-defined human population.

6.111.2. *Probability of success in a future trial (The* a priori *probability defined in a fundamental probability set)*

6.111.21. Simple case. When the mechanism underlying the realization of the expected successful outcome is known, pure reason or logical analysis can lead to the prediction of a *theoretical* frequency, homologous to an observed permanency, of the event E_1 as the outcome of a game of heads or tails, and this quantity is called the ***probability of success in a trial*** in the game of toss. The probability p of tails in a trial is the ratio of the number (1) of favorable cases to the number (2) of possible cases, assuming that these latter are equally possible (using a coin without bias):

$$p = \tfrac{1}{2} = 0.5.$$

This is the simplest case of determination of the probability of an event by denumerating the number of equally possible cases on the one hand, and the number of favorable cases on the other.

In a pack of 32 playing cards there are 16 red cards. After having mixed the cards, the probability of selecting a red card is

$$P(\text{red}) = \tfrac{16}{32} = \tfrac{1}{2}.$$

Since the cards have been well shuffled before a card is drawn, it is presumed that the choice of a red card is as probable *a priori* as the choice of a black card. It is also known that there are four Kings, hence the probability of selecting a King after shuffling is

$$P\,(\text{King}) = \tfrac{4}{32} = \tfrac{1}{8}.$$

In throwing a perfectly cubical dice, the probability of scoring a 4 is 1/6. Here too the six eventualities are deemed to be equally possible by making the assumption that the construction of the dice is faultless.

These intuitive ideas will now be classified in terms of sets, sub-sets and "probabilistic urn".

6.111.22. Fundamental set, expected sub-sets of favorable events. This type of *a priori* probability implies the existence of a perfect (or supposedly perfect) mechanism which on the ideal plane generates equiprobable realizations of the random process. Such mechanisms are of two types, viz. (a) perfect objects like well-balanced coins, cubic or icosahedral† dice, playing cards, balls in an urn, and (b) sets of discrete elements which are idealizable by means of integers or letters of the alphabet, and can be seen to include the sub-set of favorable groupings. These discrete elements are sometimes compared to colored and/or numbered balls in an urn or box.

Whether perfect objects or ideal sets of discrete elements are concerned, one has to consider the *fundamental set of all the possible* events or occurrences or, better still, all the events which are equally possible by hypothesis or by construction. In this context one may employ the term "sample space" of von Mises (Feller, 1957, p. 14 and p. 91), or "fundamental probability set" (Neyman, 1952, p. 3), or "sample description space" (Parzen, 1960, p. 17).

Legal tender coin. Here the fundamental set is formed by two eventualities, namely, tails (designated 1), or heads (designated 0), only one of which is expected before tossing the coin. By construction the two events of the fundamental set have the same *a priori* weight, i.e. are equally possible *a priori*. The probability of tails is $\tfrac{1}{2}$.

Perfect cube. The fundamental set is formed from six events, being the scores 1, 2, 3, 4, 5 and 6 which by construction are equally possible *a priori*. A favorable event is, for instance, to throw the maximum score 6 with a probability 1/6, or an even score (2, 4 or 6) with probability $\tfrac{3}{6} = \tfrac{1}{2}$. These unbiased cubic dice with six square faces (not loaded or faked) are an idealization of the astragal used since ancient times, the origin and development of which has been recounted and illustrated by Florence David (1962).

† *Icosahedral*, having twenty faces.

Perfect icosahedron. One such dice, cut in rock crystal, is held in the Louvre museum in Paris, more particulars being given in the cited work of Florence David (p. 12). In Japan there are triplets of icosahedral dice in three colors and with their 20 triangular faces bearing double numbering from 0 to 9. These dice are cast in plastic material.

For an icosahedral dice the fundamental set formed by the scores of the 20 faces is the double series:

0 1 2 3 4 5 6 7 8 9

0 1 2 3 4 5 6 7 8 9

The favorable event, for instance, a throw of 5, can occur in two equally well-balanced ways owing to the assumed perfect construction of the dice. In the throw of the dice the event is a single number which, clearly, may come from the top or the bottom row. So the position is then as follows:

Event: read-out of a score:

Fundamental set		Power† =	20
Types of sub-sets	0,0	Power	2
	1,1	Power	2
	2,2	Power	2
	—		—
	—		—
	9,9	Power	2

In terms of sets the probability of the score of 5 becomes the probability of obtaining the sub-set (5,5) of power† 2 in the fundamental set of individual scores of power 20.

Further, the three icosahedral dice of the triplet are colored red, yellow and blue respectively. If numbers are read off in color sequence, the probability of a number such as 357 occurring is 1/1000 for there is a single favorable case in the 1000 equally possible cases of the fundamental set, i.e. first dice shows 3, second shows 5, third shows 7.

These dice are very useful for writing out quickly series of random numbers having probability 1/10 (1 dice), 1/100 (two dice) and 1/1000 (three dice).

Probabilistic urn. In the case of events where the outcome can be compared to drawing balls from an urn or box ("lucky dip") containing N balls, the fundamental set depends on the type of event being considered.

† The power of a finite set of elements is the (cardinal) number of elements constituting the set.

Suppose that the urn contains the total number of balls $N = B+R$, where B is the number of blue balls and R is number of red balls. If the event should result from drawing a single ball, the fundamental set is formed from the N balls comprising the two sub-sets of balls designated as B and R.

The problem now is to evaluate the probability of a blue ball being produced in a single draw. In other words, what is the probability of the ball belonging to the sub-set B?

If the event consists in the draw of two balls, the fundamental set consists of all the pairs of N balls that can be extracted. The power of this fundamental set is the number of possible combinations of 2 objects from a set of N objects, i.e.

$$^{N}C_2 = \frac{N(N-1)}{1.2} = \binom{N}{2}.$$

6.111.23. General definition of a priori *probability*. In short, providing the chance event results from the throw of a fixed and perfect object, or from the draw of balls from an urn, the question of probability of an event boils down to the probability of encountering an element of the favorable sub-set A in the fundamental set F of all the equally probable elements.

We set n_A for the power of the sub-set of realizations of the favorable event, and n_F for the power of the fundamental set F. By definition, the probability P of some event A is the quotient n_A/n_F of the number of favorable realizations over the number of equally probable realizations of this type of event, i.e.

$$P(A) = \frac{n_A}{n_F}. \qquad \text{(b)}$$

Mainly we use n_F and not n or N to indicate that the denominator is the power of the *fundamental* set F, and not the actual value n of a certain number of trials performed, or the actual value N of a finite real population. Only in the case where the urn contains N ideal balls which may be drawn with equal probability is n_F equal to N.

If the ideal object or system (coin, dice, urn) is an *anatomical entity*, the fundamental set for a defined event is a *functional entity*, the power of which depends on the type of event being considered.

The expected event A always defines a sub-set of the functional set F. Some examples are given in Table 6.111.23A.

The relationship linking a sub-set A to F is designated so:

$$A \subset F \qquad \text{(c)}$$

Probability is a concept having meaning only if it is defined for sub-sets belonging to the fundamental set.

TABLE 6.111.23A

Type of event	Fundamental probability set F	n_F = power of F	Event required A	n_A = power of A	Probability of A
Toss of coin	2 faces	2	heads	1	$\frac{1}{2}$
Throw of a cubic dice	6 faces	6	score 4	1	$\frac{1}{6}$
			even score	3	$\frac{3}{6} = \frac{1}{2}$
Throw of an icosahedral dice	20 faces	20	score 9	2	$\frac{2}{20} = \frac{1}{10}$
			even score	10	$\frac{10}{20} = \frac{1}{2}$
Throw of two cubic dice	36 configurations $36 = 6^2$	36	score 5 per the two-part terms: 41, 32, 14, 23	4	$\frac{4}{36} = \frac{1}{9}$
Throw of 3 cubic dice	216 configurations $216 = 6^3$	216			
Drawing from urn actual $N = 20$ balls of which 5 blue	Contents of urn	$n_F = N = 20$	draw of 1 blue ball	5	$\frac{5}{20} = \frac{1}{4}$

6.111.3. *Relation between the relative frequency r/n and probability p*

The connection between the *experimental plane* of series of identical trials in a homologable game of heads or tails and the *theoretical plane* where the probability of tails in *one* trial is defined can be seen from Fig. 6.111.3I.

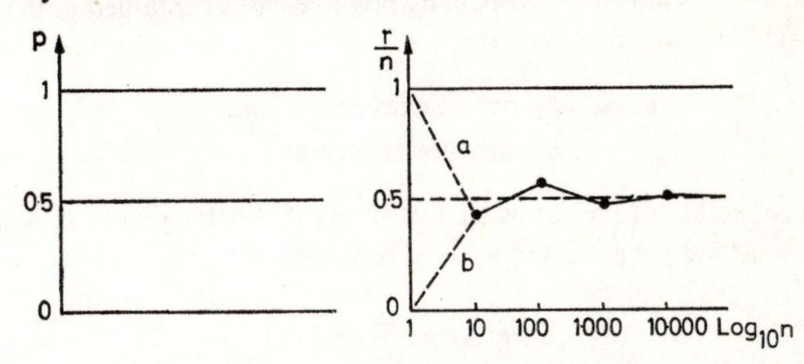

FIG. 6.111.3 I. Relation between a sequence of relative frequencies and probability. Part *a* starts from 1; part *b* from 0.

According to Fig. 6.111.3I, in an increasingly large series of trials the favorable event is realized with a relative frequency r/n which tends (in the sense of the theory of probability) towards the probability p of the event in one trial. Part (a) of the diagram schematically represents the course of the relative frequency when the first trial produces a success, whilst part (b) represents the course of r/n when the first trial results in a failure. This obvious difference, which is also maximal, decreases in successive series when the increase of the number of trials is plotted on a logarithmic scale.

The position on the right-hand side of the figure can be written so:

$$\frac{r}{n} \xrightarrow[\text{prob.}]{\text{in}} p = 0.5 \quad \text{as} \quad n \to \infty .$$

Thus, when the number of identical trials tends to infinity, the relative frequency of success in past trials tends probabilistically (according to Cantelli) towards the probability p of success in a future trial. This tendency in probability oscillates from one side to the other about the fixed limit p; furthermore, the oscillation amplitude tends to zero when the discrete operational variable n tends to infinity.

6.111.4. *Limits of the probability of an event*

The lower limit of probability of an event is $p = 0$, corresponding to impossible events in empty sub-sets:

$$\frac{0 \text{ favorable case}}{n_F \text{ equally possible cases}} = 0. \tag{a}$$

The upper probability limit of an event is the *certainty* of this event if the expected event is *any* of the n_F equally possible cases contained in the fundamental set F. In this case we have

$$\frac{n_F \text{ equally possible favorable cases}}{n_F \text{ equally possible cases}} = 1. \tag{b}$$

The probability of the realization of an event A belonging to F, written as $A \subset F$, if there are n_A favorable cases, is written so:

$$P(A) = \frac{n_A}{n_F} < 1, \tag{c}$$

$$A \subset F \tag{c'}$$

The probability of A thus satisfies the relations

$$0 \leqslant P(A) \leqslant 1. \tag{d}$$

The probability of a discrete event is a positive number less than or at most equal to unity and which is assigned to any sub-set of the fundamental probability set.

6.112. Conditional Probability

6.112.1. Conditional probability in the case of two independent events

We consider here some probabilities in two stages, namely, (1) extraction of a red card, i.e. (A) with a probability $P(\text{red}) = \frac{1}{2}$, and (2) extraction of a King knowing that the card is red, or even drawing a King on *condition* that the card is known to be red; or even drawing a King from a sub-*urn of red cards.*

The probability of such an event is a *conditional* probability which is written so:

$$P(\text{King} \mid \text{red}),$$

and reads: The probability of drawing a King on condition (or "given") that one has a red card.

For the calculation of the term $P(\text{King} \mid \text{red})$, the possible cases are 16 red cards and the favorable cases are 2 Kings. Hence $P(\text{King} \mid \text{red})$ is $\frac{2}{16} = \frac{1}{8}$.

In this case we have the following situation (Table 6.112.1A).

This latter probability (in to table) is confined to the red sub-set A, which influences the probability of a King.

An important relationship exists between the foregoing probabilities, and this relation can be stated explicitly by introducing the denominator $n_F = 32$ as used in the evaluation of absolute probabilities:

$$P(\text{King} \mid \text{red}) = \frac{2}{16} = \frac{\frac{2}{32}}{\frac{16}{32}} = \frac{P(\text{King and red})}{P(\text{red})}. \tag{a}$$

In this particular case the absolute probability of drawing a King is equal to the conditional probability of drawing a King, given that the card is red. As for the probability of drawing a King, nothing has been gained by fixing the prior condition that the card should be red. In this case the event "issue of a King" and the event "draw a red card" are *independent*. The

TABLE 6.112.1A

Set	Power
Fundamental set F Pack of 32 cards	$n_F = 32$
Red sub-set A	$n_A = 16$
Black sub-set B, $B = C\,A$	$n_B = 16$
Sub-set of Kings D	$n_D = 4$
Sub-set of red Kings R	$n_R = 2$
Absolute probability P(King) of a King being drawn =	$\frac{n_D}{n_F} = \frac{4}{32} = \frac{1}{8}$
Absolute probability P(King red) of a red King being drawn from the set of the pack =	$\frac{n_R}{n_F} = \frac{2}{32} = \frac{1}{16}$
Conditional probability P(King red) of a King being drawn from the red sub-set A =	$\frac{n_R}{n_A} = \frac{2}{16} = \frac{1}{8}$

assignment of the condition *at the same time* reduces by half the fundamental set, the power of which becomes 16 instead of 32, and the required sub-set with power of 2 instead of 4.

Double-entry table. Unlike the absolute probability which requires only a single-entry table, the conditional probability rests upon a double-entry or two-way table in A (issue of a King or not) and in B (red or black condition). In experiments and observations, use is made of the notion of factors, the factor A being the type of card (King or not a King), whilst the factor B is the color of the card (red or black).

TABLE 6.112.1B

		Color of the cards		Total
		B_1	B_2	
A	A_1 King	2	2	4
	Not-King A_2	14	14	28
		16	16	32

In Table 6.112.1B the four classes are re-grouped under the headings A and B (each in two states). In particular, the sub-set of red Kings is the intersection (or product) of the sub-set B_1 of red cards with the sub-set A_1 of Kings. For the intersection of A_1 and B_1, the notation is A_1 and B_1, or $A_1 \cap B_1$.

The mathematical definition of the probability of the event A_1 (draw a King) on condition that the card should be red (B_1), can now be written in terms of A and B:

$$P(A_1|B_1) = \frac{n_{(A_1 \cap B_1)}}{n_{B_1}}. \qquad \text{(b)}$$

Dividing top and bottom by n_F,

$$P(A_1|B_1) = \frac{P(A_1 \text{ and } B_1)}{P(B_1)}, \qquad \text{(b')}$$

providing that $P(B_1)$ is non-zero.

This definition is a corollary of the definition of probability in a fundamental domain restricted by the conditioning factor.

6.112.2. *Conditional probability in the general case of two non-independent events*

It is now convenient to leave the classical domain of gaming and consider some more specifically medical questions. Anticipating a later chapter (see χ^2-test), the first question is whether a strict diet may give a better survival rate after a first infarction of the myocardium.

Morrison (1960) has taken a group of 100 subjects having suffered a myocardial infarct. These are distributed at random into two groups of 50. One of these groups receives a strict diet B_1 and the other a free diet B_2 (diet factor). After 8 years' observation the survivals and deaths are counted in each group (factor of survival versus death).

The result of this prospective epidemiological enquiry is given in Table 6.112.2A where, as an idealization, the patients are comparable to balls of two colors, white for the survivors (A_1) and black for the deceased (A_2). In addition each ball is marked B_1 or B_2 according to the type of diet followed; this factor is supposed to act as the conditioning factor in the probability of survival.

The *fundamental set* F is of power $n_F = n_{..} = 100$ patients idealized. The *conditioning sub-set* B_1 is defined by the strict diet and is of power $n_{1.} = 50$.

The *sub-set being studied* is formed by the survivors considered from two points of view:

TABLE 6.112.2A. RELATION BETWEEN SURVIVAL AND THE TYPE OF DIETARY REGIME

		B Diet followed				
		$i = 1$ B_1 strict		$i = 2$ B_2 free		
A	$j = 1$ A_1 survive	n_{11}	28	n_{21}	12	$40 = n_{.1}$
	$j = 2$ A_2 death	n_{12}	22	n_{22}	38	$60 = n_{.2}$
		$n_{1.} = 50$		$n_{2.} = 50$		$100 = n_{..}$

Factors defining the table 2×2 of general term n_{ij}:

Factor A—type of reaction after 8 years; Factor B—type of diet.

Notation n_{ij}:

i – strict diet ($i = 1$), free diet ($i = 2$)
j – for survival ($j = 1$), death ($j = 2$)
$n_{11}+n_{12} = n_{1.}$ – marginal total of patients under B_1
$n_{21}+n_{22} = n_{2.}$ – marginal total of patients under B_2
$n_{11}+n_{21} = n_{.1}$ – marginal total of patients surviving (A_1)
$n_{12}+n_{22} = n_{.2}$ – marginal total of patients deceased (A_2)

N.B. (1) Summation following different subscripts 1 and 2 is designated by substituting a "dot" for these subscripts.

(2) If the summation is continued, one obtains the "mute" subscripts $n_{..}$ of the overall total.

(a) The overall sub-set of survivors of whatever diet, of power $n_{.1} = 40$ and leading to the absolute probability of survival

$$P(\text{survive}) = \frac{n_{.1}}{n_{..}} = \frac{40}{100} = 0.40. \qquad \text{(a)}$$

This overall probability of survival is also referred to as the *marginal*, or *unconditional*, probability of survival. In calculating it, one in effect neglects the type of diet which can condition or influence the probability of survival.

(b) The sub-set of survivors conditioned by the fact of having followed the strict diet (of power $n_{11} = 28$), and yielding the conditional probability of survival after 8 years' strict diet:

$$P(\text{survive} \mid \text{strict diet}) = \frac{n_{11}}{n_{.1}} = \frac{28}{50} = 0.56. \qquad \text{(b)}$$

The appreciable difference between the absolute probability of survival (0.40) and the conditional probability of survival after the strict diet (0.56) shows that the events, or the factors of survival and diet, are not independent. This question is only considered here from the standpoint of the definition of the conditional probability. The contingent association between these two factors can be verified by means of the χ^2-test on the basis of a random sample.

6.112.3. Use of the notion of conditional probability

Some authors consider the conditional probability to be of greater fundamental importance than the absolute probability. Without discussing this point further, we are able to list some of the objectives attainable using the conditional probability.

(1) The probability of the intersection or combined occurrence of two events can be written without previously knowing whether these events are independent or not.

(2) The conditional probability concerns relevant conditions in the occurrence of a phenomenon in the form

$$P(A|B).$$

(3) The conditional probability permits the probability of some event B occurring in the play, given any evidence A:

$$P(B|A).$$

(4) It prepares for the application of *statistical tests* of independence between characteristics such as B (diet followed) and A (survival or death).

(5) It clarifies the notion of fundamental "mortality rate" or "intensity of mortality" in demography, in disease statistics and in the follow-up.

6.113. Probability of the Result of Operations on One or Several Sets of Events

6.113.1. Probabilities of complementary events

Two events A and B are said to be complementary when the occurrence of one (success) necessarily precludes the occurrence of the other (failure). For instance, on tossing a coin, the outcome tails (A) excludes the heads (B). In this case the complement of event A is denoted as B = non-A. On the toss of a coin

$$P(A) = \tfrac{1}{2},$$

$$P(\text{non-}A) = P(B) = \frac{2-1}{2} = 1-\frac{1}{2} = \frac{1}{2}.$$

Consider now a more general example. In the throw of a dice, $P(A_6)$ is the probability of the realization of the event A_6, i.e. the maximum score; it is an eventuality occurring in one way only. The other event non-A_6, any non-maximal score (i.e. a score less than six), can be realized in $6-1 = 5$ ways. Hence

$$P(\text{non-}A_6) = \frac{5}{6} = \frac{6-1}{6} = 1-\frac{1}{6} = 1-P(A_6).$$

Thus the probability of the event non-A is the complement to unity of the probability of A.

The complementary set of A has been written by other writers as $\sim A$, $\tilde{A}$ (A tilde), A', $\bar{A}$ or A^c. In the present paper we shall use the notation non-A, or the notation $C\,A$ which signifies "complement of A".

One therefore has $B = CA$ with the necessary relation

$$A \text{ or } CA = F, \tag{a}$$

also written

$$A \cup C\,A = F$$

which states that the union of the complementary sets A and $C\,A$ will form the fundamental probability set F. Hence

$$n_A + n_{CA} = n_F, \tag{b}$$

$$n_{CA} = n_F - n_A. \tag{c}$$

We pass on to the probabilities. By definition,

$$P(A) = \frac{n_A}{n_F},$$

$$P(CA) = \frac{n_{CA}}{n_F} = \frac{n_F - n_A}{n_F} = 1-P(A),$$

therefore

$$P(CA) = 1-P(A). \tag{d}$$

The complement of the complement of A is A:

$$C(CA) = C^2A = A.$$

On dice, if A is the set of odd scores, then $B = CA$ is the set of even scores. The symbolic operator $C^2 = 1$ is a unitary operator which does not change the original set. Therefore $CB = C^2A$ is the set of *odd points*, say A.

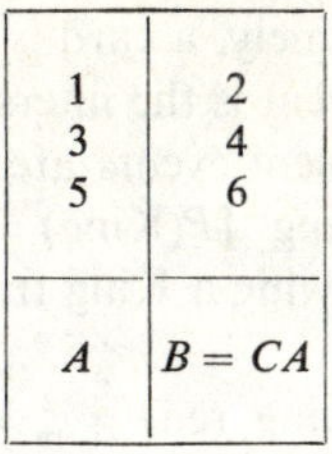

1 3 5	2 4 6
A	$B = CA$

F

In general, in the fundamental set F, two complementary events A and B enjoy the following properties:

(1) $B = CA$
(2) $CB = A$

(3) A or $CA = F$, also written $A \cup CA = F$; the union of the two complementary components is the fundamental probability set.
(4) A and CA, also written $A \cap CA = \phi$, where ϕ represents the empty set and is used in place of 0 to avoid any confusion.
The intersection of two complementary sets is empty. Also two complementary events are said to be *mutually exclusive*.

Use of complementary sets. It is sometimes difficult to write out the probability of an event A which consists of the occurrence of a determined occurrence on one or more occasions. By contrast there is often no difficulty in writing the probability $P(CA)$ of the eventuality in question not occurring, i.e.

$$P(A) = 1 - P(CA).$$

Biologists use complementary events in observations of the "All-or-nothing" type, e.g. death or survival, germination or non-germination, sterile or fertile tube, defective or non-defective vial. In studies of the propagation of excitations in random nets of neurons, Rapoport (1952) calculated the probability that a given neuron is missed by all the others; and hence the probability that the neuron would receive at least one input. These mutually exclusive events can be numbered 0 (absence of reaction) and 1 (reaction present), and by so doing one introduces the concept of the probability of random variables assuming the value 0 or 1—called the law of simple alternative.

6.113.2. Probability of the intersection of two events

6.113.21. Intersection of two independent events. The drawing of a red King realizes a composite event, namely, a card is drawn which is red as well as being a King. The required event is the intersection of the set of Kings with the set of reds. These constituent events are independent for the absolute probability of drawing a King [$P(\text{King}) = \frac{4}{32} = \frac{1}{8}$] is the same as the conditional probability of drawing a King from the conditioning set of red cards alone (6.112.1), i.e.

$$P(\text{King} \mid \text{red}) = \tfrac{2}{16} = \tfrac{1}{8}.$$

It is convenient to express the probability of the intersection *P* (King and red) from consideration of the probability of the constituent events: *P*(King) on the one hand, and *P*(red) on the other.

TABLE 6.113.21A

	Inter-section *A* and *B* red and King	*A* King	*B* red
Number of favorable cases	2	4	32
Number of possible cases	32	32	16
Probability	$\frac{2}{32} = \frac{1}{16}$	$\frac{1}{8}$	$\frac{1}{2}$
In general	$P(A \text{ and } B)$	$P(A)$	$P(B)$

From this table we have

$$\tfrac{1}{16} = \tfrac{1}{8} \times \tfrac{1}{2} \qquad \text{(a)}$$

or

$$P(\text{King} \cap \text{red}) = P(\text{King}) \times P(\text{red}) \qquad \text{(b)}$$

In section 6.112.1 it was seen that event *A* (drawing a King) was independent of the event *B* (drawing a red card), for the occurrence of *A* does not influence the outcome of *B*. We shall now show that the same result as in (a) or (b) is obtained from the standpoint of the conditional probability $P(A \mid B)$.

The extraction of a red King reduces to considering the intersection of the set A_1 of Kings with the set B_1 of red cards, say the set $A_1 \cap B_1$

$$P(\text{King} \cap \text{red}) = \tfrac{2}{32} = \tfrac{1}{16}. \tag{a}$$

We introduce the actual value 16 of the conditional sub-set, the red cards, multiplying by 16 top and bottom in the probability (King and red):

$$\tfrac{2}{32} \times \tfrac{16}{16} = \tfrac{16}{32} \times \tfrac{2}{16}$$

$$P(\text{King} \cap \text{red}) = P(\text{red}) \cdot P(\text{King} \mid \text{red}). \tag{b}$$

Events: $\quad A_1 \cap B_1 \quad B_1 \quad A_1 \mid B_1$

It is thus seen that the same result is obtained as in (a) by introducing the conditional probability $P(\text{King}|\text{red})$, which in fact is only $P(\text{King})$ by virtue of the fact that the event "King" and the event "red" are independent events.

By definition of independence, we thus have

$$\begin{aligned} P(A \mid B) &= P(A), \\ P(B \mid A) &= P(B). \end{aligned} \tag{c}$$

In this case of A being independent of B (and B of A), the probability of the intersection $A \cap B$ is simply the product of the probabilities of A and B respectively

$$P(A \cap B) = P(A) \cdot P(B). \tag{d}$$

6.113.22. Probability of intersection of two non-independent events. For two non-independent events, $P(A \mid B)$ is different to $P(A)$. As we shall show, one may in general write

$$P(A \mid B) \neq P(A)$$

$$P(A \text{ and } B) = P(B) \cdot P(A \mid B). \tag{a}$$

It is convenient to write $P(B)$ in first position, since the occurrence of B with probability $P(B)$ logically and chronologically precedes the extraction of the event "A given B", which depends on the occurrence of B.

6.113.221. Simple case, table 2×2. Before facing the general case, we will consider the table relating to two events (factors) A and B which are each present in two mutually exclusive modalities (Table 6.113.221A).

TABLE 6.113.221A

State A after 8 years	B = type of diet		
	B_1 = strict diet	B_2 = free diet	
A_1 survival	$n_{11} = 28$	$n_{21} = 12$	$n_{.1} = 40$
A_2 deceased	$n_{12} = 22$	$n_{22} = 38$	$n_{.2} = 60$
	$n_{1.} = 50$	$n_{2.} = 50$	$100 = n_{..}$

We consider the case which interests us most, i.e. the intersection A_1 and B_1, or the set of subjects who followed a strict diet and who lived 8 years after embarking on the test. For simplification, the actual relative frequencies are converted to probabilities as previously.

The probability of the intersection $A_1 \cap B_1$ is

$$P(A_1 \cap B_1) = \frac{n_{11}}{n_{..}} = \frac{28}{100} = 0.28. \quad \text{(a)}$$

The probability of belonging to group B_1 is $\frac{1}{2}$ by virtue of the chance allocation in the population:

$$P(A_1) = \frac{n_{1.}}{n_{..}} = \frac{50}{100} = 0.50. \quad \text{(b)}$$

The probability of survival on condition that the strict diet was obeyed is

$$P(A_1 | B_1) = \frac{n_{11}}{n_{1.}} = \frac{28}{50} = 0.56. \quad \text{(c)}$$

The general relation between the probabilities (a), (b) and (c) is obtained without difficulty by multiplying the factors of (a) top and bottom by the actual power $n_{1.}$ of the conditioning set in its strict modality:

$$P(A_1 \text{ and } B_1) = \frac{n_{11}}{n_{..}} = \frac{n_{1.}}{n_{..}} \times \frac{n_{11}}{n_{1.}}.$$

Using (b) and (c) the second term becomes

$$P(A_1 \cap B_1) = P(B_1) \cdot P(A_1 | B_1). \quad \text{(d)}$$

By this example we verify the general relation (d), yielding the probability of the intersection $(A_1 \cap B_1)$, say observing a subject who followed the strict diet and who survived:

$$P(A_1 \cap B_1) = P(B_1) \cdot P(A_1 | B_1) \quad \text{(e)}$$
$$0.28 = 0.50 \times 0.56.$$

It should now be asked whether the event A (survival or decease) is independent or not of event B (having followed a strict or a free diet). One therefore has to calculate the unconditional or marginal probability $P(A_1)$ of survival independently of the regime:

$$P(A_1) = \frac{40}{100} = 0.40. \quad \text{(f)}$$

It is thus seen that the conditional probability of survival in a strict diet $P(A_1 | B_1) = 0.56$ is distinctly superior to the unconditional probability of survival whatever the diet, i.e. $P(A_1) = 0.40$. The difference 0.56–0.40, or the quotient 0.56 : 0.40, provides an *index of non-independence* of the events B (diet followed) and A (surviving or deceased).

This epidemiological example is useful because it treats a chronological sequence, viz. B then afterwards A, i.e. initiation of the criterion B (diet), then afterwards observation of criterion A (death or survival) after 8 years. In this *prospective* case the probability $P(A | B)$ has a biological significance. One can also consider *retrospective* standpoints, i.e. the event A having been observed, what is the probability of intervention of B considered as cause?

In this latter case $P(B | A)$ is *relevant*:

$$P(B_1 | A_1) = \frac{P(B_1 \text{ and } A_1)}{P(A_1)} = \frac{0.28}{0.40} = 0.70. \quad \text{(g)}$$

Hence, whenever it is known that a patient subjected to this clinical test is alive, the conditional probability of the strict type of diet having been a benefic factor is 0.70. This example will be taken up further in section 6.115.1. The analysis of this situation is, however, a very delicate task, requiring the use of clinical documentation carefully compiled according to a rigorous plan.

6.113.222. General case, table $a \times b$. More generally we will now consider an event A which can occur in the mutually exclusive states $A_1, A_2, A_i, \ldots, A_a$, and an event B which can display mutually exclusive states $B_1, B_2, B_j, \ldots, B_b$. The frequency of the intersection (A_i and B_j) is written as n_{ij} ($i = 1, 2,$

$\ldots, a; j = 1, 2, \ldots, b$). By straight extension of formulae written in 6.113.222, we can write the following probabilities

$$P(A_i \cap B_j) = \frac{n_{ij}}{n_{..}} \qquad \text{(a)}$$

$$P(A_i) = \frac{n_{i.}}{n_{..}} \qquad \text{(b)}$$

$$P(A_i \mid B_j) = \frac{n_{ij}}{n_{.j}} \qquad \text{(c)}$$

6.113.23. Extension to three events. For three events, we have the formula

$$P(A \cap B \cap C) = P(A) \cdot P(B \mid A) \cdot P(C \mid A \cap B). \qquad \text{(a)}$$

6.113.3. Probability of the union of two events

The case of mutually exclusive events must be distinguished from the non-exclusive case.

6.113.31. Union of mutually exclusive events

6.113.311. Union of two mutually exclusive events (empty intersection). Consider the event consisting in dealing a King (A) or a Queen (B) from a pack of 32 cards. Clearly these two events are mutually exclusive for no card can be a King and a Queen at the same time!

The union of event A (deal a King) *with* event B (deal a Queen) consists in the arrival of a King *or* of a Queen. The expected event is nothing other than the arrival of one partner of a marriage. Clearly,

$$P(\text{King}) = P(A) = \tfrac{4}{32} = \tfrac{1}{8},$$
$$P(\text{Queen}) = P(B) = \tfrac{4}{32} = \tfrac{1}{8}.$$

The event A or B (King or Queen) also written $A \cup B$ is realized by $4+4 = 8$ cards, the arrival of each of which is equally probable *a priori*:

$$P(\text{King} \cup \text{Queen}) = P(A \cup B) = \tfrac{8}{32} = \tfrac{4}{32} + \tfrac{4}{32} = \tfrac{1}{4}. \qquad \text{(a)}$$

Relation (a) is the particular case of the general law in the case of two events which are mutually exclusive:

$$A \cap B = \phi$$
$$P(A \cup B) = P(A) + P(B) \qquad \text{(b)}$$

For two mutually exclusive events the probability of the union A or B is the sum of the probabilities of the constituent events, say $P(A)+P(B)$.

6.113.312. Union of several mutually exclusive events. Let there be an event E which can be realized by n mutually exclusive events $E_1 \ldots E_n$:

$$E = E_1 \cup E_2 \cup \ldots \cup E_n, \qquad \text{(a)}$$

where $\qquad E_i \cap E_j = \phi \; (i, j = 1, 2, \ldots, n; \; i \neq j).$

By extension of formula 6.113.311 (a), we get

$$P(E) = P(E_1) + P(E_2) + \ldots + P(E_n) \qquad \text{(b)}$$

6.113.313. Union of an event E with an exhaustive stratification of the fundamental set. An exhaustive stratification of the fundamental set F is defined as consisting in the decomposition of F by sub-sets F_i with the following properties:

(1) $F_i \subset F \; (i = 1, 2, \ldots, n)$
(2) Two sets F_i and F_j are always disjoint if

$$F_i \cap F_j = \phi \qquad (i \neq j) \;\; i, j = 1, 2, \ldots, n \qquad \text{(a)}$$

(3) The sub-sets F_i should exhaustively reconstitute the fundamental set F (for notation, see 6.113.322 k).

$$F = \bigcup_{i=1}^{n} F_i. \qquad \text{(b)}$$

From our definition of the product of two sets, the relation for any $E \subset F$ is

$$E = E \cap F. \qquad \text{(c)}$$

By the use of formulae (b) and (c) the probability of E can be found from consideration of components of the type $E \cap F_i$ which are more easily handled.

Consider now a simple example. On throwing two cubic dice, what is the probability of the same score with each one? We let F be the fundamental set of 36 realizations (i, j) $i, j = 1, 2, \ldots, 6$ of the scores ($n_F = 36$). The required event E is here constituted from six pairs (i, i) $i = 1, 2, \ldots, 6$ $(n_E = 6)$. Hence the probability is

$$P(E) = \frac{n_E}{n_F} = \frac{6}{36} = \frac{1}{6}.$$

The same problem will now be treated by introducing an exhaustive stratification of the fundamental set F, stratification defined by the total r of the scores. Accordingly,

$$r = i+j \ (i, j = 1, 2, \ldots, 6), \quad r = 2, \ldots, 12.$$

Figure 6.113.313 I shows that F_2 contains the score (1, 1), that F_3 contains the scores (2, 1) and (1, 2), and so on and so forth. In this way the set F is decomposed into 11 sub-sets F_r $(r = 2, \ldots, 12)$ satisfying the conditions (a). It should be noted that the subscript i of F_i that ranges from 1 to 11 is replaced by the subscript $r = i+1$ which ranges from 2 to 12.

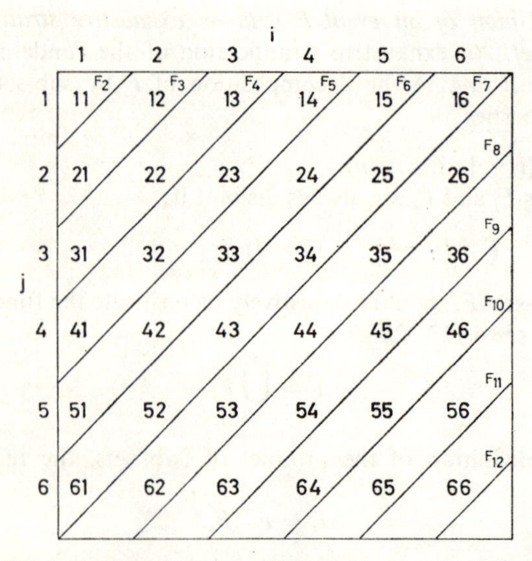

FIG. 6.113.313 I. Fundamental probability set for the score of two cubic dice.

We can therefore write

$$F = \bigcup_{r=2}^{12} F_r. \tag{d}$$

Substituting this result into (c),

$$E = E\cap\left(\bigcup_{r=2}^{12} F_r\right), \qquad \text{(e)}$$

i.e.
$$E = E\cap(F_2\cup F_3\cup \ldots \cup F_{12}),$$
or
$$E = (E\cap F_2)\cup(E\cap F_3)\cup \ldots \cup(E\cap F_{12}).$$

Since $F_2, F_3, \ldots, F_{12}$ are mutually exclusive, the intersections $E\cap F_2$, $E\cap F_3, \ldots, E\cap F_{12}$ are also mutually exclusiv.

Formula (b) will now be applied to several mutually exclusive events:

$$P(E) = P(E\cap F_2)+P(E\cap F_3)+ \ldots +P(E\cap F_{12}). \qquad \text{(f)}$$

From preliminary analysis of the problem, the power of the required set E, with its elements (1, 1)...(6, 6), is 6. As there are 11 sub-sets $E\cap F_r$ ($r = 2, 3, \ldots, 12$), we have to find 5 empty ones to discover the sub-sets $E\cap F_{r'}$ where r' is odd ($r' = 3, 5, 7, 9, 11$). In fact, by definition of F_r, each element of an $F_{r'}$ has an odd sum and so it cannot lead to two equal scores. Therefore

$$P(E\cap F_{r'}) = 0, \qquad r' = 3, 5, 7, 9, 11. \qquad \text{(g)}$$

Only the sub-sets $F_{r''}$ with even r'' lead to a non-empty intersection $E\cap F_{r''}$ of power 1. For $r'' = 6$ there is only the single element (3, 3) which, out of the elements of F_6, satisfies the property defining E ($i = 3; j = 3$).

Hence the probabilities of the intersections $E\cap F_{r''}$ are each $\frac{1}{36}$:

$$P(E\cap F_{r''}) = \tfrac{1}{36}, \qquad r'' = 2, 4, 6, 8, 10, 12. \qquad \text{(h)}$$

Since the union of the $F_{r'}$ and $F_{r''}$ reconstitutes F, we have

$$P(E) = \sum_{r'} P(E\cap F_{r'})+\sum_{r''} P(E\cap F_{r''}). \qquad \text{(i)}$$

By virtue of (h) and (g) we get

$$P(E) = 0+0+0+0+0+\tfrac{1}{36}+\tfrac{1}{36}+\tfrac{1}{36}+\tfrac{1}{36}+\tfrac{1}{36}+\tfrac{1}{36} = \tfrac{6}{36} = \tfrac{1}{6}, \qquad \text{(j)}$$

as obtained by direct analysis.

We again take formula (f) explicitly in reference to $P(E\cap F_r)$, on condition that F_r is realized previously:

$$P(E\cap F_r) = P(F_r)\cdot P(E|F_r). \qquad \text{(k)}$$

According to (h), only the probabilities in $F_{r''}$ are non-zero, r'' here being even. In Table 6.113.313A we calculate the factors of the second term of (k) with reference to the fundamental set F represented in Fig. 6.113.313 I above.

TABLE 6.113.313A

1	2	3	4
r''	$P(F_{r''})$	$P(E\mid F_{r''})$	$P(E\cap F_{r''})$
2	$\frac{1}{36}$	$\frac{1}{1}$	$\frac{1}{36}$
4	$\frac{3}{36}$	$\frac{1}{3}$	$\frac{1}{36}$
6	$\frac{5}{36}$	$\frac{1}{5}$	$\frac{1}{36}$
8	$\frac{5}{36}$	$\frac{1}{5}$	$\frac{1}{36}$
10	$\frac{3}{36}$	$\frac{1}{5}$	$\frac{1}{36}$
12	$\frac{1}{36}$	$\frac{1}{1}$	$\frac{1}{36}$
			$\frac{6}{36}$

The elements of column 4 lead to the same results as in (h).

Finally, the power of the set of cases studied is regarded as a sum of random variables. We have

$$n_E = \sum_r n_{E\cap F_r}.$$

Separating the contribution of the odd subscripts (r') from that of the even (r''), we get

$$n_E = \sum_{r'\text{ odd}} n_{E\cap F_{r'}} + \sum_{r'\text{ even}} n_{E\cap F_{r''}} = 0+0+0+0+0+1+1+1+1+1+1.$$

It is seen that $n_{E\cap F_r}$ plays the rôle of an indicative variable of the sub-set coming from the stratification of the fundamental set F (see 6.2.11.1).

6.113.32. Union of non-mutually-exclusive events

6.113.321. Union of two non-mutually-exclusive events. A simple example is taken from the game of dice. It can be used for randomly distributing patients into 2, 3 or 6 groups.

Notation: A represents the occurrence of an even score 2, 4 or 6, whilst B is a score which is a multiple of 3, i.e. 3 or 6. The probabilities are respectively

$$P(A) = \tfrac{3}{6} = \tfrac{1}{2}, \quad P(B) = \tfrac{2}{6} = \tfrac{1}{3}.$$

The event D which is of interest to us is the union of A with B, i.e. drawing of an even score, or of a score divisible by 3 ($D = A \cup B$). Elements of D are 2, 4, 6, 3 in the fundamental set F (see Fig. 6.113.321 I).

1	2	3	4	5	6	*F*
	2		4		6	*A*
		3			6	*B*
	2	3	4		6	*D*

FIG. 6.113.321 I. Union of two non-mutually exclusive events.

In the six equally possible cases (1, 2, 3, 4, 5, 6) contained in F, event D is realized in the four modalities 2, 3, 4 and 6. Therefore

$$P(D) = P(A \cup B) = \tfrac{4}{6}.$$

Figure 6.113.321I shows that 6 is an element common both to A and to B; the element 6 is no other than the intersection ($A \cap B$). Score 6 is counted twice, i.e. once in A and once in B. On the other hand the probability of the intersection A and B (score 6) is $P(A \cap B) = \tfrac{1}{6}$.

From consideration of the probabilities $P(A)$, $P(B)$, $P(A \cap B)$, it is possible to calculate the probability of the union ($A \cup B$) as follows:

$$\underset{\frac{4}{6}}{P(A \cup B)} = \underset{\frac{1}{2}}{P(A)} + \underset{\frac{1}{3}}{P(B)} - \underset{\frac{1}{6}}{P(A \cap B)} \qquad \text{(a)}$$

which, on reduction to a common denominator, reads $\tfrac{4}{6} = \tfrac{3}{6} + \tfrac{2}{6} - \tfrac{1}{6}$. It should perhaps be noted that here the union ($A \cup B$) means "either A or B", and not "A or else B"; since ($A \cap B$) is not empty.

One now subtracts $P(A \cap B)$ because the element 6 has been counted twice, once in A and once in B.

For mutually exclusive events A and B, the probability $P(A \cap B)$ is zero and so the formula applicable to events of this kind is verified.

Now if B is deemed to be a conditioning event, we have

$$P(A \cap B) = P(B) \times P(A|B). \quad \text{(b)}$$

Whence $$P(A \cup B) = P(A)+P(B)-P(B) \times P(A|B) \quad \text{(c)}$$
$$= P(A)+P(B)[1-P(A|B)].$$

6.113.322. Union of several non-mutually-exclusive events. Consider some event E consisting in the union of A, B and C which are non-mutually-exclusive sub-sets of F:

$$E = A \cup B \cup C. \quad \text{(a)}$$

In order to reduce this case to that of two sets, as treated in the foregoing section, we put

$$B \cup C = D. \quad \text{(b)}$$

We then get

$$E = A \cup D. \quad \text{(c)}$$

Therefore, by virtue of relation (b) above, we have

$$P(E) = P(A \cup D) = P(A)+P(D)-P(A \cap D). \quad \text{(d)}$$

In this relation, by applying (b) once more, the term $P(D)$ becomes

$$P(D) = P(B \cup C) = P(B)+P(C)-P(B \cap C). \quad \text{(e)}$$

It remains to evaluate $P(A \cap D)$. We have

$$P(A \cap D) = P[A \cap (B \cup C)]. \quad \text{(f)}$$

By virtue of the property of distributivity of the intersection over the union of sets,† we have

$$A \cap (B \cup C) = (A \cap B) \cup (A \cap C).$$

Whence

$$P[A \cap (B \cup C)] = P(A \cap B)+P(A \cap C)-P[(A \cap B) \cap (A \cap C)]. \quad \text{(g)}$$

† Analogous to the distributivity of multiplication over addition, i.e. $x.(y+z) = x.y+x.z$.

In order to evaluate $P[(A\cap B)\cap(A\cap C)]$, we refer to the diagram of Euler–Venn (Fig. 6.113.322 I).

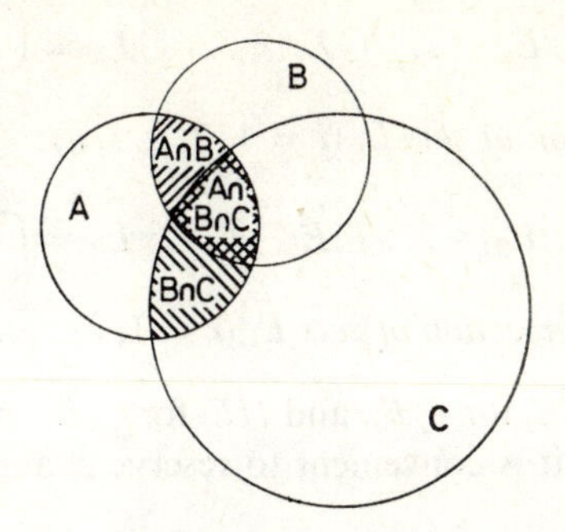

FIG. 6.113.322 I. Euler–Venn diagram.

Clearly,

$$(A\cap B)\cap(A\cap C) = A\cap B\cap C.$$

Therefore

$$P[(A\cap B)\cap(A\cap C)] = P(A\cap B\cap C). \tag{h}$$

Substituting the relations (e), (f) and (h) into (d),

$$\begin{aligned} P(E) &= P(A)+P(D)-P(A\cap D) \\ &= P(A)+P(B)+P(C)-P(B\cap C)-P(A\cap D), \end{aligned} \tag{i}$$

and so, finally, the classical formula for $E = A\cup B\cup C$ is

$$\begin{aligned} P(A\cup B\cup C) = P(A)+P(B)+P(C) \\ -P(B\cap C)-P(A\cap B)-P(A\cap C) \\ +P(A\cap B\cap C). \end{aligned} \tag{j}$$

It should be mentioned that $P(A\cap B\cap C)$ is given in the expression 6.113.23 (a).

This formula is due to Halley (Florence N. David and D. E. Barton, 1962, p. 25).

To write the general formula for the probability of union of k events $E_1, \ldots, E_k$, we introduce the two stenographic formulae:

$$E_1 \cup E_2 \cup \ldots \cup E_i \cup \ldots \cup E_k = \bigcup_{i=1}^{k} E_i$$

union of sets E_i $(i = 1, 2, \ldots, k)$;

$$E_1 \cap E_2 \cap \ldots \cap E_i \cap \ldots \cap E_k = \bigcap_{i=1}^{k} E_i \tag{k}$$

intersection of sets E_i $(i = 1, 2, \ldots, k)$.

Some writers put ΣE_i for $\cup E_i$, and ΠE_i for $\cap E_i$, but in order to avoid ambiguous notation, it is convenient to reserve Σ and Π for operations on numbers.

Thus, putting $\cup$ for the union or the sum of sets, and $\cap$ for the intersection or product of sets, for k sets one writes:

$$P\left(\bigcup_{i=1}^{k} E_i\right) = \sum_{i=1}^{k} P(E_i) - \sum_{i<j} P(E_i \cap E_j) + \sum_{i<j<l} P(E_i \cap E_j \cap E_l) - \ldots + (-1)^{k-1} P\left(\bigcap_{i=1}^{k} E_i\right). \tag{l}$$

After the writers mentioned, this latter formula is due to de Moivre. For $k = 3$ and the equalities $E_1 = A$, $E_2 = B$ and $E_3 = C$, the reader can readily verify that (l) leads to (j).

The probability of the intersection of k sets is found without difficulty by considering the complement of $\cup E_i$:

$$C\left(\bigcup_{i=1}^{k} E_i\right) = C(E_1 \cup E_2 \cup \ldots \cup E_k) = CE_1 \cap CE_2 \cap \ldots \cap CE_k = \bigcap_{i=1}^{k} (CE_i). \tag{m}$$

This relation declares that the complement of the union (sum) of k sets is the intersection (product) of the respective complements of these k sets (de Morgan).

The corresponding probabilities are written:

$$P\left[\bigcap_{i=1}^{k} (CE_i)\right] = P\left[C\left(\bigcup_{i=1}^{k} E_i\right)\right]. \tag{m'}$$

By definition of a complementary set and omitting the subscripts,

$$P[C(\cup E_i)] = 1-P(\cup E_i). \tag{n}$$

Substituting (m′) into (n), we obtain

$$P\left[\bigcap_{i=1}^{k}(CE_i)\right] = 1-\Sigma P(E_i)+\sum_{i<j} P(E_i\cap E_j)+ \ldots +(-1)^k P\left(\bigcap_{i=1}^{k} E_i\right). \tag{o}$$

Note. The operators C, $\cup$ and $\cap$ all apply to sets without parentheses except if C and $\cup$, or C and $\cap$, intervene simultaneously. The symbol $P(\ )$ is always accompanied by a parenthesis.

6.114. Application

Effects in all or nothing, Abbott's formula (1925)

Upon testing insecticide, a natural mortality with a probability $P(N)$ may be observed in the control preparation. In a well-planned and well-conducted experiment it can be supposed that the mortality of the control population applies equally to the preparations processed.

We set $P(I)$ for the probability of death by insecticide of a subject taken at random.

In fungostatic tests the non-germination effect of the control preparation (natural effect) is also supposed to happen in the preparations treated (toxic effect), and to the same extent as in the control preparation.

An analysis will now be made in terms of sets which leads to the classical formula of Abbott (1925). The observed probabilistic effect $P(R)$ results either from the natural effect of probability $P(N)$, or from the toxic effect $P(I)$, these two effects being regarded as independent by virtue of the correctly planned and performed testing:

$$P(N\cap I) = P(I|N)\cdot P(N) = P(N|I)\cdot P(I) = P(I)\cdot P(N). \tag{a}$$

However, these two effects are not necessarily mutually exclusive, for in general some of the subjects would have been killed by the toxicity if they had not died naturally, or vice versa.

Probability of observing a survival. The set of survivors S is composed of subjects which have survived both the natural effect (CN) and at the same time the toxic effect (CI), these two events being regarded as independent:

$$S = CN \cap CI. \tag{b}$$

Hence the probability of survival is

$$\begin{aligned} P(S) &= P(CN \cap CI) = P(CN)P(CI) \\ &= [1-P(N)][1-P(I)] \\ &= 1-P(N)-P(I)+P(N)\cdot P(I). \end{aligned} \tag{c}$$

The probability $P(R)$ of observing a death (from whatever cause) is the complement to 1 of the probability of survival. Using the notation of set theory, we have

$$R = CS, \tag{d}$$

hence the probability of observing a death is

$$P(R) = 1-P(S), \tag{e}$$

or, having regard to the last relation of equations (c),

$$P(R) = P(N)+P(I)-P(N)\cdot P(I). \tag{f}$$

Isolating $P(I)$ in the second term of (f),

$$P(R)-P(N) = P(I)[1-P(N)]. \tag{g}$$

Clearly, the excess of mortality over and above the natural mortality is equal to the product of the purely toxic mortality with the fraction of the population which is influenced by the toxic activity. So the probability of the purely toxic effect is

$$P(I) = \frac{P(R)-P(N)}{1-P(N)}. \tag{h}$$

This is *Abbott's formula* which shows that the effect properly due to the insecticide is measured by the mortality in excess of the natural mortality referred solely to the fraction of the subjects which have escaped natural mortality. Healy (1952) has produced tables for this correction.

Other derivation. One can also regard the set of observed deaths as the union of the sets N and I:

$$R = N \cup I. \tag{i}$$

Then, in the general case, the probability of at least one of the events (natural death or poisoning) is

$$P(R) = P(N)+P(I)-P(N\cap I). \tag{j}$$

In the *hypothesis of the independence of the natural and toxic effects*, we have

$$P(N\cap I) = P(N)\cdot P(I). \tag{k}$$

Hence (f) is obtained by substituting (k) into (j). It is emphasized that this hypothesis of independence is only tenable in the case of correctly planned and performed experiments.

Logical equivalence of the two approaches

We once again take formula (b), i.e.

$$S = CN\cap CI. \tag{l}$$

The probability of survival is

$$P(S) = P(CN\cap CI). \tag{l'}$$

The set of dead insects is the complement of S:

$$R = CS = C[CN\cap CI]. \tag{m}$$

By virtue of a theorem from set theory due to *de Morgan*, the complement of an intersection is equal to the union of the complements (see also 6.113.322 m). Therefore,

$$C[CN\cap CI] = C(CN)\cup C(CI). \tag{m'}$$

Since

$$CCN = N, \quad C\,CI = I, \tag{n}$$

one writes

$$R = N\cup I, \tag{o}$$

which again yields formula (i).

One may also proceed from this formula and apply the cited theorem, namely, the complement of a union of sets is equal to the intersection of the complements:

$$CR = C(N\cup I) = CN\cap CI; \tag{o'}$$

this immediately yields (b) as required since $S = CR$.

6.115. Probability of Causes

In section 6.113.313 a study was made of the *(a priori)* probability of some future event being associable with one or another cause from an exhaustive system. We shall now reverse the question and for some event which has actually occurred we require the *a posteriori* probability of a cause in some exhaustive system being set in play, given the *a priori* probability of the factor being involved.

We shall again begin with the simple case of two complementary causes (section 6.115.1) before proceeding to the general case.

6.115.1. Probability of one of two complementary causes entering into play

With reference to the example of survival after myocardial infarction, we consider Table 6.113.221A above and the associated formula (g):

$$P(B_1|A_1) = \frac{P(A_1 \cap B_1)}{P(A_1)}. \tag{a}$$

For the exhaustive system of types of diet (B_1 strict, $B_2 = CB_1$ free) we have

$$A_1 = (A_1 \cap B_1) \cup (A_1 \cap B_2).$$

Therefore

$$P(A_1) = P(A_1 \cap B_1) + P(A_1 \cap B_2) - P[A_1 \cap B_1) \cap (A_1 \cap B_2)], \tag{b}$$

or

$$P(A_1) = P(A_1 \cap B_1) + P(A_1 \cap B_2), \tag{c}$$

for B_1 and B_2 are complementary.

By substituting (c) into (a) one obtains the *a posteriori probability of the cause B_1 entering into play* for a survivor randomly selected in the fundamental set:

$$P(B_1|A_1) = \frac{P(A_1 \cap B_1)}{P(A_1 \cap B_1) + P(A_1 \cap B_2)}. \tag{d}$$

For this example the numerical values are: $P(A_1 \cap B_1) = 0.28$, $P(A_1 \cap B_2) = 0.12$, hence

$$P(B_1|A_1) = \frac{0.28}{0.28+0.12} = \frac{0.28}{0.40} = 0.70.$$

We now slightly change the notation by putting E for the event of survival in place of A_1 and retaining the exhaustive system of two (complementary) causes B_1 and B_2 with $B_2 = CB_1$. The relation (d) then becomes

$$P(B_1|E) = \frac{P(E \cap B_1)}{P(E \cap B_1)+P(E \cap B_2)} = \frac{P(E \cap B_1)}{P(E \cap B_1)+P(E \cap CB_1)}. \quad \text{(e)}$$

Such is the formula or criterion of the Reverend Th. Bayes in the particular case of an exhaustive system of two necessarily complementary causes (types of diet). Expanding the probability of the intersections $E \cap B_1$ and $E \cap CB_1$, we obtain

$$\begin{aligned} P(E \cap B_1) &= P(E|B_1) \cdot P(B_1), \\ P(E \cap CB_1) &= P(E|CB_1) \cdot P(CB_1), \end{aligned} \quad \text{(f)}$$

so that (e) becomes

$$P(B_1|E) = \frac{P(E|B_1) \cdot P(B_1)}{P(E|B_1) \cdot P(B_1)+P(E|CB_1) \cdot P(CB_1)}. \quad \text{(g)}$$

Thus the *a priori* probability $P(B_1)$ of B_1 becoming involved is utilized. This probability should be known in order to calculate the *a posteriori* probability of B_1 being a relevant factor, upon seeing the evidence E. In any actual case, through planning of the investigation and random allocation of the patients to B_1 or B_2, $P(B_1)$ is known, and therefore also $P(B_2)$,which equals $1-P(B_1)$.

The application of Bayes' formula involves no difficulty for one has:

$$P(B_1) = P(CB_1) = \tfrac{1}{2},$$
$$P(E|B_1) = 0.56, \quad P(E|CB_1) = 0.48.$$

Therefore

$$P(B_1|E) = \frac{0.56 \times \frac{1}{2}}{0.56 \times \frac{1}{2}+0.24 \times \frac{1}{2}} = \frac{0.28}{0.40} = 0.70.$$

But if the *a priori* probabilities are not known, one can proceed in two ways, i.e.

(a) The *a priori* entry of the causes into play can be declared equiprobable; and this is the case which has just been treated in regard to the value of $P(B_1) = P(B_2) = \frac{1}{2}$.

(b) One can subjectively balance or weigh the causes present, and therein lies all the difficulty.

6.115.2. Probability of the influence of one cause B_i belonging to an exhaustive system of n causes

Consider an event E which may occur for one or another cause of some exhaustive system. By hypothesis,

$$B_i \cap B_j = \phi, \quad \bigcup_{i=1}^{n} B_i = F. \tag{a}$$

The *a posteriori* probability of the cause being involved can be described by the extension of formula (g) in section 6.115.1, given the observation of E:

$$P(B_i|E) = \frac{P(E|B_i) \cdot P(B_i)}{\sum_{i=1}^{n} P(E|B_i) \cdot P(B_i)} \tag{b}$$

where $\sum_{i=1}^{n} P(B_i) = 1$.

So if $(n-1)$ of the probabilities $P(B_i), \ldots, P(B_{n-1})$ of the causes becoming involved are known, which is rare, it is possible to calculate (b), but otherwise the matter is controversial. Computer applications of the relation (b) in medical diagnosis are given *inter alia* by Ledley and Lusted (1959 and 1962), Overall and Williams (1965), Lincoln and Parker (1967). Wider applications of Bayes' theorem to clinical problems are given by Cornfield (1967).

6.12. Probability in the Continuous Domain

6.121. Continuous Probability in the Absolute or Unconditional Sense

6.121.1. Elementary continuous probability

Consider a game of skill consisting in throwing a constant mass along an OX axis, whilst remaining within the fixed limits (a, b)—see Fig. 6.121.1 I.

A priori, every point has an equal chance of being reached. It is considered to be a success if the mass is thrown into the interval (m, n) within (a, b). The *segmentary continuous probability* (see section 6.121.2) of seeing the thrown mass fall in (m, n) is the quotient of the measure of the set of these favorable cases over the measure of the fundamental set of the possible cases, i.e. (a, b).

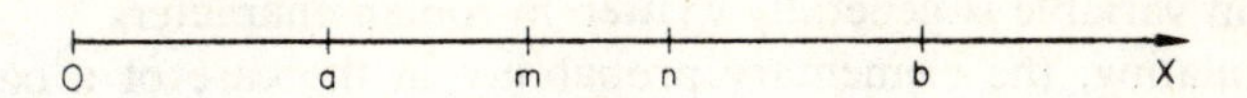

FIG. 6.121.1 I. Sub-set of the fundamental domain for a continuous variable.

For the continuous case it should be noted that the terminology changes. The concept of "measure of a continuous set" replaces the notion "power of a set of discrete elements". To define the measure of the fundamental set (a, b), with each point X of (a, b) we associate a function $f(X)$ called the probability density function and such that the measure of the fundamental set is unity.

Here the discrete probability p becomes the *local probability*:

$$dF(X) = f(X)\, dX, \tag{a}$$

or the elementary probability of finding the continuous random variable $\mathbf{X}$ in the infinitesimal interval centred on X and extending from $X - dX/2$ to $X + dX/2$. According to typographical convention we denote the random variable as $\mathbf{X}$.†

In fact the elementary probability is

$$\text{Prob}\left\{X - \frac{dX}{2} < \mathbf{X} < X + \frac{dX}{2}\right\} = \frac{\text{measure of the favorable segment}}{\text{measure of the fundamental probability domain}}. \tag{b}$$

The measure of the favorable segment on X is

$$dF(X) = f(X)\, dX, \tag{c}$$

† Dutch statisticians subline the random variable (van Dantzig, see de Jonge, 1963, I, p. 115), while Schlaifer (1959, p. 87) puts a tilde on the top.

on the *express condition* that the measure of the fundamental probability domain is equal to unity, i.e.

$$F(b)-F(a) = \int_b^a f(\mathrm{X})\, d\mathrm{X} = 1. \qquad \text{(d)}$$

The current variable is generally written in roman characters.

Recapitulating, the elementary probability in the case of a continuous random variable implies the coexistence of the three relations (a), (b) and (d) which *simultaneously* define both the elementary probability (pure number), and the probability density function $f(X)$ which is expressed in units inverse to those in which X is given. Thus, if X is given in cm, then

$$dF(X) = f(X)_{\mathrm{cm}^{-1}} \cdot dX_{\mathrm{cm}}. \qquad \text{(e)}$$

6.121.2. *Segmentary continuous probability*

Setting A for the favorable event consisting of the incidence of the thrown mass in the segment or sub-set (m, n), the set of favorable cases is measured by the integral of the probability density function $f(X)$ taken over the interval (m, n) of amplitude $i = (n-m)$, which is written

$$\int_m^n f(\mathrm{X})\, d\mathrm{X}.$$

The set of possible cases is measured by the integral of the probability density function $f(X)$ over the interval (a, b) of amplitude $R = b-a$, and this exhausts the set of possibilities.

Therefore the segmentary probability of seeing X fall in the segment or sub-set (m, n) within the fundamental domain (a, b) is written as

$$\text{Prob}\,(m < X < n) = \frac{\int_m^n f(\mathrm{X})\, d\mathrm{X}}{\int_a^b f(\mathrm{X})\, d\mathrm{X}}. \qquad \text{(a)}$$

EXAMPLE. We apply this general definition to throwing the mass m while calculating the probability that the mass m will fall within the segment (m, n)

of measure ($i = n-m$). Formula (a) can be used, providing the density function $f(X)$ is known. Intuitively, since every point has an equal chance of being reached, $f(X)$ is conceived to be constant, i.e.

$$f(X) = K. \tag{b}$$

Since the mass is certain not to fall outside the interval (a, b),

$$R = b-a, \tag{c}$$

we have the relation

$$\int_a^b f(X)\,dX = 1, \tag{d}$$

or

$$\int_a^b K\,dX = K(b-a) = KR = 1. \tag{e}$$

Therefore

$$K = \frac{1}{R}. \tag{f}$$

The set of favorable cases is measured by

$$\int_m^n K\,dX = K(n-m) = \frac{n-m}{R} = \frac{i}{R}. \tag{g}$$

Hence the segmentary probability of the point of impact X occurring in the interval (m, n) of measure i is

$$\text{Prob}\,[m < X < n] = \frac{n-m}{b-a} = \frac{i}{R}. \tag{h}$$

The continuous probability of observing X in the interval i within R whenever the density distribution function $f(X)$ is constant is represented by the fraction i/R contained between 0 and 1.

Geometric interpretation. Extending the notion introduced in section 6.111.23 for discrete random variables, Fig. 6.121.2 I shows that the probability of X occurring in i is measured by the area with base i and height $1/R$.

The probability of X being found at some point between a and b is equal to unity because

$$R\cdot\frac{1}{R} = 1.$$

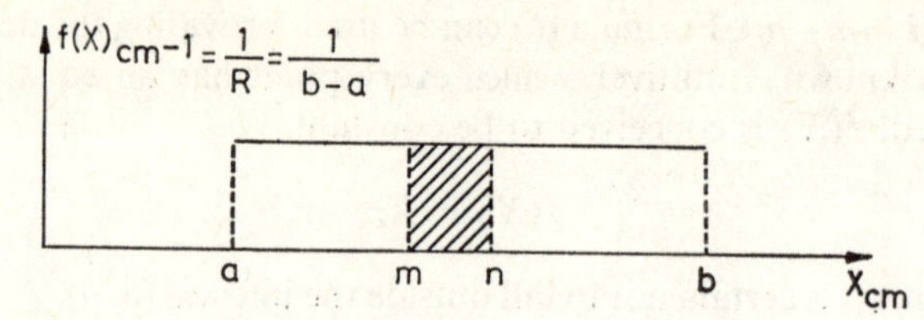

FIG. 6.121.2 I. Probability of a sub-set (m, n) in the fundamental probability domain (a, b).

It is seen that the ordinate of the rectangle which measures the segmentary probability of X in i is no other than the density $f(X)$, here being constant and equal to $1/R$ (uniform density). Since the segment OX is in this case a physical length, it is measured in cm. Therefore

$$f(X) = \frac{1}{R\,\text{cm}} = \frac{1}{R}\,\text{cm}^{-1} \tag{i}$$

is measured in cm^{-1}, this being a particular case of the general rule. If a continuous random variable X has a physical dimension D, the associated probability density function $f(X)$ is given in inverse units D^{-1} of D.

6.121.3. General concept of density and distribution for a continuous random variable

If instead of seeking the probability of X being incident in some interval $i = (m, n)$, it is required that X should fall at a point X exactly, the probability is clearly zero for i is reduced to 0 since m and n are merged at X. So rather than require from a theoretical point of view that X should be incident exactly at X, it is desired that X should in practice fall in a small interval dX centered on X, or having X as origin, while bearing in mind that this small interval should be contained within (a, b).

The *elementary* or *local probability* of X occurring in the interval dX is given in the particular case $f(X) = 1/R$ by applying the formula (6.121.2h) in the differential form

$$\text{Prob}\left\{X - \frac{dX}{2} < X < X + \frac{dX}{2}\right\} = dF(X) = \frac{dX}{R} \tag{a}$$

Figure 6.121.3 I shows that the probability $dF(X)$, thus defined, is measured by the area of an elementary rectangle of base dX centered on the point X and of height $f(X) = 1/R$. We thus obtain the simplest example of distri-

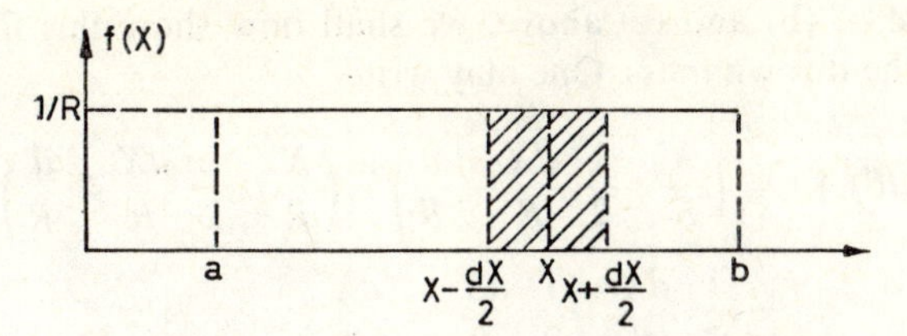

FIG. 6.121.3 I. Simple case of probability density.

bution of the continuous random variable taken at point X. The similar notation $dF(X)$ implies an infinitesimal increase of the *distribution* function $F(X)$.

In the example given, it is calculated without difficulty at any fixed point X. For the measure of the shaded rectangle, we have

$$\text{Base area}\left(a, X+\frac{1}{2}\,dX\right)-\text{Base area}\left(a, X-\frac{1}{2}\,dX\right)$$
$$=\frac{1}{R}\left(X+\frac{1}{2}\,dX-a\right)-\frac{1}{R}\left(X-\frac{1}{2}\,dX-a\right). \tag{b}$$

In this form the local probability, an infinitesimal quantity, becomes the difference of two finite quantities, i.e.

$$dF(X) = F(X+\tfrac{1}{2}\,dX)-F(X-\tfrac{1}{2}\,dX). \tag{c}$$

In this difference the first term

$$F\left(X+\frac{1}{2}\,dX\right) = \frac{1}{R}\left(X+\frac{1}{2}\,dX-a\right) \tag{d}$$

represents the total probability of finding X in the finite interval $(a, X+\frac{1}{2}\,dX)$ or Prob $\{X \leqslant X+\frac{1}{2}\,dX\}$.

In turn, the second term

$$F\left(X-\frac{1}{2}\,dX\right) = \frac{1}{R}\left(X-\frac{1}{2}\,dX-a\right) \tag{e}$$

represents the total probability of finding X in the interval $(a, X-\frac{1}{2}\,dX)$, or Prob $\{X \leqslant X-\frac{1}{2}\,dX\}$.

The difference between these two total probabilities is no other than the elementary probability $f(X)\,dX$ of finding X in the segment dX centered on X.

By the use of (b) and (c) above, we shall now show this firstly for the example of the thrown mass. One may write

$$dF(X) = \left(\frac{X}{R}+\frac{1}{2}\frac{dX}{R}-\frac{a}{R}\right)-\left(\frac{X}{R}-\frac{1}{2}\frac{dX}{R}-\frac{a}{R}\right)$$

$$= \frac{1}{2}\frac{dX}{R}+\frac{1}{2}\frac{dX}{R} = \frac{dX}{R}. \qquad \text{(f)}$$

In the general case one proceeds from

$$dF(X) = F(X+\tfrac{1}{2}\,dX)-F(X-\tfrac{1}{2}\,dX) \qquad \text{(g)}$$

and expands each term of the second item as a Taylor series limited to the first two terms, assuming that $F'(X)$ is derivable:

$$F(X+\tfrac{1}{2}\,dX) = F(X)+\tfrac{1}{2}\,dXF'(X)+\ldots \qquad \text{(h)}$$

$$F(X-\tfrac{1}{2}\,dX) = F(X)-\tfrac{1}{2}\,dXF'(X)+\ldots \qquad \text{(i)}$$

Whence $F(X)$ is eliminated like the term X/R in (f) and we are then left with

$$dF(X) = \tfrac{2}{2}F'(X)\,dX = f(X)\,dX \qquad \text{(j)}$$

on defining the probability density function as the limit of the quotient of the *local* probability over ΔX *centred at* X by the measure of the segment ΔX centred on X:

$$f(X) = \lim_{\Delta X \to 0} \frac{\text{Prob}\,\{X-\tfrac{1}{2}\,\Delta X < X < X+\tfrac{1}{2}\,\Delta X\}}{\Delta X}. \qquad \text{(k)}$$

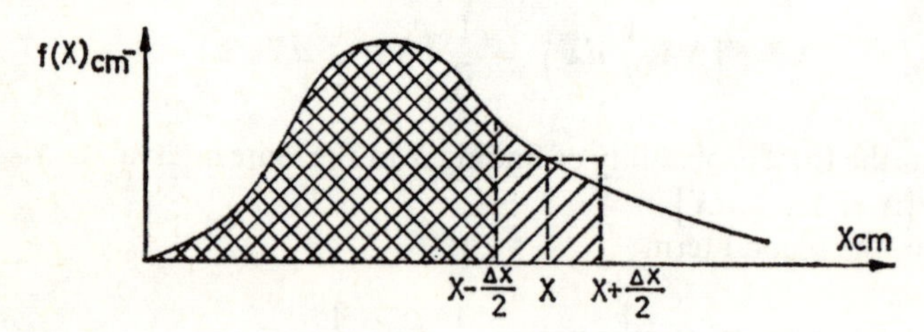

FIG. 6.121.3 II. Probability density $f(X)$ and distribution function $F(X)$ of a continuous variable X.

Figure 6.121.3 II shows the representation of $\Delta F(X)$ as a difference between shaded areas in the form of a trapezium of base ΔX and with curved upper limit. Apart from infinitesimals of order superior to arbitrary ΔX, one can approximate the trapezium by the hatched rectilinear rectangle of measure $f(X)\,\Delta X$. This approximation will be *a fortiori* better for smaller ΔX. If now one regards ΔX as corresponding to the interval I centered on X_i in a histogram (X_i, f_i'), where f_i' is the relative frequency, then the condition

$$\Sigma f(X)\,\Delta X \quad \text{centered at} \quad X_i \approx 1$$

will correspond approximately to the necessary relation

$$\Sigma f_i' = 1.$$

At the limit, after having covered the domain D or range of X by a large number of contiguous, but non-encroaching, intervals, one rigorously will have

$$\int f(\mathrm{X})\,d\mathrm{X} = 1,$$

which brings us back to the same definition of the probability density.

We now explicitly consider the notion of distribution. The function $F(X)$ corresponds to the relative frequencies *as accumulated at the end of the intervals* of the histogram, and of which it is an idealization. $F(X)$ is called the cumulative distribution of X. If X represents the stature, then $F(X)$ gives the probability of a stature equal to, or less than, X:

$$\text{Prob}\,(\boldsymbol{X} \leqslant X) = F(X). \tag{l}$$

Or, alternatively, the cumulative distribution function $F(X)$ gives the *(total) probability* of meeting with a stature at most equal to X.

Complementary distributions. Consider the complementary set of statures greater than X: the probability

$$\text{Prob}\,\{\boldsymbol{X} > X\} = 1 - F(X) \tag{m}$$

gives the *(total) probability* of observing a stature greater than X. It thus corresponds to the proportion of Saxon prisoners decapitated because they were taller in height than the sword of Dagobert, King of the Franks, was long in length. Another example: $1-F(X)$ corresponds to the proportion of male Tartar prisoners exterminated by Genghis Khan because their height exceeded the axle of his chariot ($\pm$1200).

In close analogy with (l), a random variable $\boldsymbol{X}$ is defined by Fisher (1959) if, for $0 \leqslant p \leqslant 1$, one has

$$\text{Prob}\,\{\boldsymbol{X} \leqslant X_p\} = p. \tag{l'}$$

So the quantile X_p is put into operation from the very beginning.

6.122. Conditional Probability in the Continuous Domain

Consider a continuous random variable $\boldsymbol{X}$ defined, for instance, in the theoretical domain $(0, \infty)$, and representing the duration X of survival of reticulocytes as from the instant $X = 0$. Here $f(X)$ is the probability density of X. The unconditional probability of survival time $\boldsymbol{X}$ lying in the interval $X, X+dX$ is

$$dF(X) = f(X)\, d(X) \tag{a}$$

This probability does not tell us anything about the reticulocytes *still* surviving at the instant X.

By definition, $F(X)$ is the probability of meeting with a reticulocyte, the survival of which is at the very most equal to X. There exists, therefore, as from the instant X, a proportion $1-F(X)$ of cells having a duration of survival greater than X:

$$\text{Prob}\,\{\boldsymbol{X} > X\} = 1-F(X). \tag{b}$$

For a reticulocyte still alive at the instant X, the conditional probability of its disappearance in the interval $X, X+dX$ is based no longer on the fundamental domain, as restricted by the condition which requires survival at the instant X, and extending onto $(X, +\infty)$. One therefore writes

$$\text{Prob}\,(X < \boldsymbol{X} < X+dX \,|\, \text{survival at } X) = \frac{f(X)\, dX}{1-F(X)}\,. \tag{c}$$

Such is the *imminence* $dJ(X)$ *of disappearance between* X *and* dX of a reticulocyte which is known to be still alive at the instant X; or, alternatively, the conditional probability of a reticulocyte, still alive at X, disappearing between X and dX (Martin 1960, 1962). In this context the *elementary probability* is said to be *instantaneous*, for X is a survival time even though the term *local probability* would be employed if X was a measure of space.

Just as a probability density function $f(X)$ at point X corresponds to $dF(X)$ (unconditional probability of proximity to X), so we have an "imminence rate $I(X)$ corresponding to $dJ(X)$ (conditional probability of proximity to X). So one has the series of relations:

$$\begin{aligned} dF(X) &= f(X)\, dX && 0 < \boldsymbol{X} < \infty \\ dJ(X) &= \frac{f(X)\, dX}{1-F(X)} && X < \boldsymbol{X} < \infty \\ &= I(X)\, dX. \end{aligned} \tag{d}$$

The imminence rate, measured here in hours^{-1}, therefore becomes

$$I(X)_{\text{hr}^{-1}} = \frac{f(X)_{\text{hr}^{-1}}}{1-F(X)}, \tag{e}$$

and so we have

$$I(X) > f(X). \tag{e'}$$

The imminence rate is also called "intensity" or hasard function. From (c) and (d), we see that the imminence rate $I(X)$ is related to the single distribution function $F(X)$ as follows

$$I(X) = -d \ln [1-F(X)], \tag{e''}$$

whence

$$F(X) = 1-\exp\left\{-\int_0^X I(X)\, dX\right\} = 1-\exp\{-J(X)\} \tag{e'''}$$

with $I(0) = 0$ since $F(0) = 0$, and $I(\infty) = \infty$ since $F(\infty) = 1$.

Having now expounded the concept of imminence, we will show the homology which exists with the conditional probability $P(A \mid B)$ seen previously in the discrete domain.

Noting A the chance observation of the death of a reticulocyte in the interval $(0, X+dX)$, i.e. seeing a cell with a life at the most equal to $X+dX$, we have

$$P(A) = \text{Prob}\,\{\boldsymbol{X} \leqslant X+dX\}. \tag{f}$$

We put B for the observation of a reticulocyte still alive at time X:

$$\begin{aligned} P(B) &= \text{Prob}\,(\boldsymbol{X} > X) \\ &= 1-F(X). \end{aligned} \tag{g}$$

Finally, for the event consisting in the observation of the death of a reticulocyte between X and $X+dX$ we put $E = A \cap B$, then by our definition of probability density we have:

$$\begin{aligned} P(E) = P(A \cap B) &= \text{Prob}\,(X < \boldsymbol{X} \leqslant X+dX) \\ &= f(X)\, dX. \end{aligned} \tag{h}$$

The conditional probability of observing between X and $X+dX$ the death of a reticulocyte still alive at X, and of which only a proportion $1-F(X)$ persists, is written so:

$$P(A \mid B) = \frac{P(A \cap B)}{P(B)} = \frac{f(X)\, dX}{1-F(X)}. \tag{i}$$

6.123. Probability of Events from Operations on Continuous Sets

6.123.1. *Complementary events in a continuous domain*

Consider once more the example of disappearing reticulocytes where there are two mutually exclusive states, viz. state 1 (survival) and state 0 (death). We put A for the set of vanished reticulocytes (state 0) at the instant X. From the point of view of waiting time, the probability of seeing some reticulocyte (taken at random) dead at the instant X (state 0), is also the probability that its lifetime $\mathbf{X}$ is less than X. The probability of A is therefore

$$P(A) = \text{Prob}\,(\mathbf{X} \leqslant X) = F(X). \tag{a}$$

The set of those which survive (state 1) after the instant X is the complement of A, and in this respect the probability is

$$P(C\,A) = \text{Prob}\,(\mathbf{X} > X) = 1 - P(A). \tag{b}$$

The evolution of the division of reticulocytes into vanished and survivors in the ratio $F(X)/1-F(X)$ is described in time X. One is thus concerned with a continuous series of instantaneous urns $U(X)$ where the proportion $p(X)$ of black balls (dead reticulocytes) is

$$p(X) = F(X). \tag{c}$$

In the same urn the proportion of white balls (reticulocytes still living) is

$$q(X) = 1 - F(X) \tag{d}$$

with at each instant

$$p(X) + q(X) = 1. \tag{e}$$

At each instant X the events A and B are contradictory or mutually exclusive, but it is no longer the same when the events A and B are defined on segments which overlap.

6.123.2. *Probability of the intersection of two continuous sets*

Consider now three events A, B and E in a domain such as $(0, \infty)$. The set E of points X which are common to A and B, is called the *intersection of A and B*:

$$E = A \cap B.$$

So as not to proliferate examples, we take the survival of reticulocytes again. Here A is the set of reticulocytes disappearing in the time period up to $X+\Delta X$, i.e. the lifetime $\mathbf{X}$ of which is less than $X+\Delta X$:

$$P(A) = \text{Prob}\,(\mathbf{X} \leqslant X+\Delta X); \tag{a}$$

B is the set of survivals at time X:

$$P(B) = \text{Prob}\,(\mathbf{X} > X) = 1-F(X). \tag{b}$$

The event E is defined as the disappearance of a survivor during the time interval ΔX common to the intervals defining A, on the one hand, and B on the other.

It is required to calculate the probability of the intersection $E = A \cap B$:

$$P(E) = P(A \cap B) = \text{Prob}\,(X < \mathbf{X} \leqslant X+\Delta X). \tag{c}$$

From the definition of the density function $f(X)$,

$$\text{Prob}\,(X < \mathbf{X} \leqslant X+\Delta X) = f(X)\,\Delta X. \tag{d}$$

From consideration of 6.113.22(a),

$$P(A \cap B) = P(B) \cdot P(A \mid B). \tag{e}$$

Since, by definition,

$$P(B) = 1-F(X),$$

the value of the conditional probability is again found from 6.122(i).

6.123.3. Probability of union of two continuous sets

It is convenient to distinguish the case of mutually exclusive (or non-overlapping) sets from the case of sets with non-empty intersection.

6.123.31. Probability of union of two mutually exclusive sets

This case is met with in bilateral tests of statistical hypotheses. It is given that $\mathbf{X}$ is a random variable of density $f(X)$ in the domain $(0, \infty)$. We use A

for the set of values of X less than a, and correspondingly B for the set of values of $X > b$, with $b > a$.

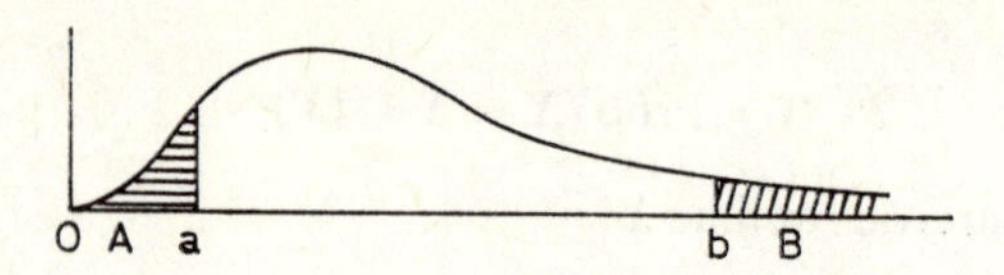

FIG. 6.123.31 I. Union of two mutually exclusive events for the continuous variable X.

The two sets A and B are mutually exclusive and so

$$P(A \cap B) = 0. \qquad \text{(a)}$$

Setting $\cup$ for the union of A and B, clearly $P(A \cup B)$ is the probability of finding X either in A or in B.

By virtue of the property of additivity of the definite integral, regarded as a function of sets, we have

$$P(A \cup B) = \text{Prob}\,[(0 < X < a) \cup (b < X < \infty)], \qquad \text{(a')}$$

$$\int_{(A \cup B)} f(X)\,dX = \int_0^a f(X)\,dX + \int_b^\infty f(X)\,dX = P(A) + P(B). \qquad \text{(b)}$$

Hence, if

$$P(A \cap B) = 0,$$

then

$$P(A \cup B) = P(A) + P(B). \qquad \text{(c)}$$

Special case: If $f(X)$ is symmetric about the origin, for instance, and if $a = -b$, then

$$P(A \cup B) = P(A) + P(B) = 2P(A). \qquad \text{(d)}$$

6.123.32. Probability of union, non-empty intersection

We take the above example limiting B to the upper bound Y (Fig. 6.123.32 I). The events A, B have an intersection $A \cap B$ of measure $f(X)\,\Delta X$. The union $A \cup B$ is the set of cases where either one observes A (a reticulocyte which disappears between 0 and $X + \Delta X$), or one observes B (a reticulocyte with lifetime somewhere between X and Y). The probability of $A \cup B$ is written so:

$$P(A \cup B) = \text{Prob}\,(0 < X < Y)$$

$$P(A \cup B) = \int_0^Y f(\mathrm{X})\, d\mathrm{X}. \qquad \text{(a)}$$

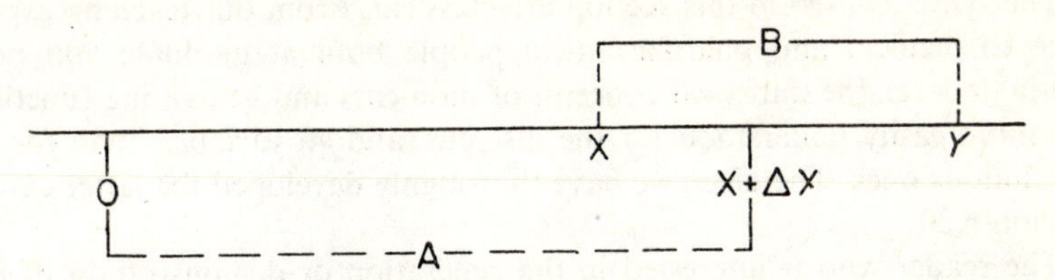

FIG. 6.123.32 I (see text).

Taking into consideration the point $X + \Delta X$, we have

$$\int_0^Y f(\mathrm{X})\, d\mathrm{X} = \int_0^{X+\Delta X} f(\mathrm{X})\, d\mathrm{X} + \int_{X+\Delta X}^{Y} f(\mathrm{X})\, d\mathrm{X}. \qquad \text{(b)}$$

The first term of the second entity is no other than $P(A)$. The second term breaks down as follows:

$$\int_{X+\Delta X}^{Y} f(\mathrm{X})\, d\mathrm{X} = \int_X^Y f(\mathrm{X})\, d\mathrm{X} - \int_X^{X+\Delta X} f(\mathrm{X})\, d\mathrm{X}$$

$$= P(B) - P(A \cap B). \qquad \text{(c)}$$

Hence,

$$P(A \cup B) = P(A) + P(B) - P(A \cap B), \qquad \text{(d)}$$

if $P(A \cap B) \neq 0$.

If, however, the intersection is empty, one again obtains the formula 6.123.31 (c). Finally, the formulae hold good for continuous events as well as for sets of discrete events. The notions of addition and subtraction remain the same in reference to *powers* (integers) of sets of events in the discrete domain, and also to *measures* (rational, irrational or transcendental numbers) in the case of events in the continuous domain.

6.2. PROBABILITY LAWS GOVERNING DISCRETE RANDOM VARIABLES

6.20. Notations

The symbols used in this section are classical. From our teaching experience to medical and pharmaceutical people both at graduate and post-graduate level, the statistical concepts of moments and generating functions are more easily understood for the discrete random variables than for the continuous ones. Therefore we have thoroughly developed the latter case in section 6.30.

The reader who is interested in the generation or demonstration of formulae in this section, which is mainly devoted to the *specification* of discrete probability laws, should refer to small but excellent books of, for example (in alphabetical order), Aitchison (1968), Aitken (1957) Bulmer (1965) and Weatherburn (1949), the very welcome and modern *Summa Statistica* produced by M. G. Kendall and Stuart (specially I, 1963) and finally the recent dictionary and bibliography of discrete distributions by Patil and Joshi (1968). In French, contributions are presented by Martin (1966, 1969). We now give the minimum set of essential notations in the discrete case.

6.201. Discontinuous or Discrete Random Variable, $\boldsymbol{r}$

$\boldsymbol{r}$ is used as an initial of reaction (r reactors on n exposed). As in section 6.211.1, the random variable $\boldsymbol{r}$ is indicated in bold italic type.

6.202. Discrete Probability Law or Discrete Probability Function, $P_{(\boldsymbol{r}=r)} = P_r$

The probability that the discrete random variable $\boldsymbol{r}$ takes the particular value r is contracted into P_r. On the domain D or range of r, we evidently have

$$\sum_D P_r = 1.$$

Distribution of the discrete variable, $\boldsymbol{r}$

$$\text{Prob}\,(\boldsymbol{r} \leqslant c) = P_0+P_1+\ldots+P_c = \sum_{r=0}^{r=c} P_r.$$

$$\varkappa'_{(1)} = \omega'(0) = \varkappa'_1 = \mu'_1$$
$$\varkappa_{(2)} = \omega''(0) = \mu'_{(2)} - \mu'^2_{(1)}$$
$$= \mu'_2 - \mu'^2_1 - \mu'_1 = \varkappa_2 - \varkappa'_1.$$

6.206. Probability generating functions for some discrete probability laws

In section 6.205 it is shown that all the generating functions may be derived from the probability generating function $G(t)$. Therefore we write it explicitly for the first seven cases and make reference to an urn model for the last three. The relevant section is also indicated.

	$G(t)$	Section
Simple alternative (Bernoulli)	$q+pt$	6.21
Positive binomial, $q+p = 1$	$(q+pt)^n$	6.22
Poisson	$e^{\mu(t-1)}$	6.23
Geometric	$p(1-qt)^{-1}$	6.24
Pascal, $q+p = 1$ (k integer and > 0)	$p^k(1-qt)^{-k}$	6.25
Negative binomial ($k > 0$), $q'-p' = 1$	$(q'-p't)^{-k}$	6.26
Logarithmic	$-\alpha \ln (1-\theta t)$	6.27
Eggenberger–Polya	Urn model: *a* white, *b* black initially; *c* balls added of the same color as that extracted	6.28, and Moran (1968), Johnson–Kotz (1969)
Hypergeometric, $r = 0$ included	Hypergeometric function. Urn model: $c = -1$	6.29, and Kendall–Stuart (1963, I), Moran (1968), Johnson–Kotz (1969, p. 143); other meaning p. 145, where $r = 0$ is excluded
Negative hypergeometric, $r = 0$ included	Urn model : $c = +1$	6.2.10, and Skellam (1948) (beta-binomial), Moran (1968, p. 77) (compound binomial)

6.21. Law of Simple Alternative

The realization of a success in a Bernoulli test.

6.211. Point-density Law of Simple Alternative

6.211.1. Definition

Consider *one* test where the probability of the realization of the event is p, whilst the probability of its non-realization is q. Therefore,

$$p+q = 1. \quad \text{(a)}$$

The law of simple alternative is defined by associating the values 1 and 0 of a random variable $\boldsymbol{r}$ with the respective probabilities p and q: this random variable assumes the value 0 (zero) in cases of failure or non-occurrence of the event (probability q), and the value 1 in cases of success or realization of this event (probability p). For any discrete *variable* according to convention (see p. 202), $\boldsymbol{r}$ is the random variable whose particular value is r. Throughout this section we abbreviate the notation of the probability that $\boldsymbol{r} = r$ to P_r,

$$\text{Prob}\,(\boldsymbol{r} = r) = P_r. \quad \text{(b)}$$

In set theory a function of this all or none type was introduced by de la Vallée Poussin (I, 1903, p. 221). The random variable $\boldsymbol{r}$ associated with (a) is called a *characteristic variable*, a *counting variable*, or an *indicator of an event* (Laboureur *et al.*, 1963, p. 318). The association between the values 0 and 1 assumable by the random variable, and the probability law in unique alternatives P_r, can be presented schematically as in Table 6.211.1A.

TABLE 6.211.1A

1 Event	2 r	3 P_r	4 rP_r	5 r^2P_r
Failure: non-realized	0	q	$0\times q$	$0\times q$
Success: realized	1	p	$1\times p$	$1\times p$
		$\sum P_r = 1$	$\sum rP_r = p$	$\sum r^2P_r = p$

The test may be the toss of a debased or non-debased coin; in the case of a non-debased coin $p = q = 0.5$. If the test is throwing a six-sided dice and the favorable event is the score "2", we then have $p = \frac{1}{6}$ and $q = \frac{5}{6}$. In the statistical control of industrial production, Japanese statisticians use dice with

20 sides (regular icosahedrons). The sides are numbered in two series of 0 to 9. Thus the probability of throwing "7" is $p = 2:20 = 0.1$, so therefore $q=0.9$. By means of these dice, colored red, yellow and blue, no difficulty arises in generating sequences of equiprobable random numbers ranging from 0 to 9. The simultaneous throw of two icosahedron dice of different color provides equiprobable sequences arranged between 00 and 99. An unequivocal reading is obtained if the red score is conventionally taken before the score of the blue dice. In this case $p = 0.01$ and $q = 0.99$. Finally, the complete throw of three dice of this kind, red, blue and yellow, covers the equiprobable range from 000 to 999 with $p = 0.001$ and $q = 0.999$ (see p. 159).

6.211.2. Parameters of reduction

In this most simple case the application of general formulae (No. 6.203) enables the parameters of reduction to be written without difficulty from consideration of the moments relative to the origin $\mu_1', \mu_2', \ldots$:

Mean: $E[\mathbf{r}] = \mu = \mu_1'$ ($E[\mathbf{r}]$ reads "expected value of" r),

$$E[\mathbf{r}] = \Sigma r P_r = 0 \times q + 1 \times p = p. \tag{a}$$

Important relation:

$$\mu_1 = E[(\mathbf{r} - E[\mathbf{r}])] \equiv 0. \tag{b}$$

The first central moment is identically zero.

Dispersion parameters: $V(\mathbf{r}) = \mu_2 = \sigma_r^2$.

Variance:

$$V(\mathbf{r}) = E[(r-\mu)^2] = \Sigma(r-\mu)^2 P_r. \tag{c}$$

After reduction we have

$$V(\mathbf{r}) = \mu_2 = \mu_2' - \mu_1'^2. \tag{c'}$$

The calculation yields

$$V(\mathbf{r}) = pq. \tag{c''}$$

By virtue of (a) and (c″) the variance is found to be less than the mean in the law of simple alternative:

$$V(\mathbf{r}) = pq = p - p^2 = E[\mathbf{r}] - (E[\mathbf{r}])^2 \tag{d}$$

Comparison of the mean with the variance is classical for probability laws governing discrete variables, for the mean and variance are then pure numbers with no dimension. The quotient

$$\frac{V(\mathbf{r})}{E[\mathbf{r}]} = \frac{\sigma_r^2}{\mu} \tag{e}$$

is called the *relative variance* by Clapham (1936), or the *variance to mean ratio* by Blackman (1942). In an earlier study of the type of distribution to match discrete data, systematic use was made of the *relative excess of variance* (Martin, 1961), i.e.

$$\delta_r = \frac{V(r)-E[r]}{E[r]} = \frac{\sigma_r^2-\mu}{\mu} = \frac{\sigma_r^2}{\mu}-1, \tag{f}$$

and this is no other than the variance/mean quotient decreased by 1. The parameter δ_r appears in one or another of the many forms in which the negative binomial law is written. That is why it is used consistently for all the discrete laws.

For the law of simple alternatives we have

$$\delta_r = \frac{V(r)-E[r]}{E[r]} = \frac{pq-p}{p} = q-1 = -p. \tag{g}$$

The relative excess of variance is negative for this particular case of the positive binomial law.

Parameters of shape:

The skewness parameter is (Fisher):

$$\gamma_1 = \frac{\mu_3}{\sigma_r^3} = \frac{pq(q-p)}{pq\sqrt{pq}} = \frac{q-p}{\sqrt{pq}}. \tag{h}$$

Pearson's β_1 has the value of γ_1^2.

Fisher's parameter of kurtosis

$$\gamma_2 = \frac{\mu_4}{\sigma^4}-3 \tag{i}$$

becomes

$$\gamma_2 = \frac{pq(1-3pq)}{pq\times pq}-3 = \frac{1}{pq}-6.$$

Pearson's β_2 has the value $\frac{\mu_4}{\sigma^4} = \gamma_2+3$.

6.212. Simple Alternative Distribution

Since the random variable can only assume the values $r = 0, 1$ with probabilities q and p respectively, it immediately follows that

$$\begin{aligned} &\text{Prob}\,(r = 0) = q, \\ &\text{Prob}\,(r \leqslant 1) = q+p = 1. \end{aligned} \tag{a}$$

6.22. Positive Binomial Law

6.221. Positive Point-density Binomial Law (Point-binomial)

The binomial law applies to independent phenomena which occur on the "all-or-nothing" principle according to two alternatives exclusively, e.g. success or failure, death or survival with a constant probability of success. Such trials are called *Bernoulli trials.*†

6.221.1. *Definition of positive point binomial density*

In the general case, consider n successive independent tosses of a non-debased coin, or a single toss of n numbered and independent coins, non-debased and thrown simultaneously. The probability of success E for one coin is constant and equal to p, whilst the constant probability of failure F has the value $q = 1-p$. After tossing, the *state of the system* is given by the number r of successes. This random variable may assume the $(n+1)$ values $r = 0, 1, 2, \ldots, n$. The domain extending from 0 to n is therefore finite.

A *combination* is of the type

$$E \quad E \quad F \quad F \quad E \quad F \quad E$$

n coins; r successes E out of n coins in one combination.

By virtue of the theorem of composite probabilities, the probability of a combination with r successes out of n coins is $p^r q^{n-r}$, whatever the sequencing of E, F.

The number of combinations with r successes out of n trials is written as:

$$\binom{n}{r} = {}^nC_r = \frac{n}{1} \cdot \frac{n-1}{2} \cdot \frac{n-2}{3} \cdot \ldots \cdot \frac{n-r+1}{r}. \tag{a}$$

The state r can be realized arbitrarily by any one of the *combinations with r successes*. Hence by virtue of the theorem of total probabilities, the probability of state r is

$$P_r = p^r q^{n-r} + p^r q^{n-r} + \ldots + p^r q^{n-r}.$$

† Bernoulli or Bernouilli *(Encycl. Britannica,* 1965, Vol. III, p. 528).

There are ${}^nC_r$ terms, hence P_r becomes

$$P_r = {}^nC_r \cdot p^r q^{n-r}, \quad \text{or} \quad P_r = \binom{n}{r} p^r q^{n-r} \qquad (r = 0, 1, 2, \ldots, n) \tag{b}$$

with the convention ${}^nC_r = \binom{n}{r}$.

The sequence of probabilities of state $P_0, P_1, \ldots, P_r, \ldots, P_n$ corresponds to the successive values $r = 0, 1, 2, \ldots, r, \ldots, n$, which the random variable representing the state of the system may exhibit. The sequence P_r is *generated* by the expansion of *Newton*'s binomial $(q+p)^n$, where the exponent n is a positive integer:

$$\begin{aligned} P_0 &= q^n \\ &\vdots \\ P_r &= {}^nC_r p^r q^{n-r} = \frac{n(n-1)\ldots(n-r+1)}{1\cdot 2\ldots r} p^r q^{n-r} \\ &\vdots \\ P_n &= p^n. \end{aligned} \tag{c}$$

Hence the name positive *binomial law* which associates with the discrete random variable $r = 0, 1, 2, \ldots, n$, the probabilities P_r, these being successive terms of the expansion of the binomial $(q+p)^n$ with $n > 0$.

Calculation of the successive terms. A recurrence law enables us to pass from P_r to P_{r+1}, i.e.

$$P_{r+1} = \frac{n-r}{r+1} \cdot \frac{p}{q} P_r. \tag{d}$$

The general nature of binomial points has been discussed by Hald (1952, pp. 668–9).

This binomial law applies only to *integer enumerations* of a random variable contained within the range from 0 to the *finite* positive upper bound n. The interval of variation R extends from 0 to n. It represents the range of possible values of r. On this domain the sum of the discrete probabilities is equal to unity:

$$\sum_{r=0}^{n} P_r = 1. \tag{e}$$

In fact, by reason of the mode of generation indicated above, we have

$$\sum_{r=0}^{n} P_r = (q+p)^n = 1^n = 1.$$

6.221.2. *Parameters of reduction*

Mean. The *mean* or expected value of r (or "expectation" of r) is given by the relation

$$E[r] = \mu = \sum_{r=0}^{r} rP_r. \tag{a}$$

The application of the $E[r]$ operation (take the expected value of r) is detailed in Table 6.221.2A.

TABLE 6.221.2A

1 Expected event	2 r	3 P_r	4 Particular value of r expected, rP_r
Non-realized:			
n failures out of n tests	0	${}^nC_0\, p^0q^n$	$0\times q^n$
Partially realized:	1	${}^nC_1\, p^1q^{n-1}$	$1\times npq^{n-1}$
r successes out of	2	${}^nC_2\, p^2q^{n-2}$	$2\times\dfrac{n(n-1)}{1\cdot 2}p^2q^{n-2}$
n tests, $r = 1, 2, \ldots, n-1$	r ⋮ $n-1$	${}^nC_r\, p^rq^{n-r}$	$r\times\dfrac{n(n-1)\ldots(n-r+1)}{1\cdot 2\ldots r}p^rq^{n-r}$
Realized *in every test:*			
n successes in n tests	n	${}^nC_n\, p^nq^0$	$n\times p^n$
		$\sum P_r = 1$	$\sum rP_r = np\dfrac{[q^{n-1}+(n-1)q^{n-2}p+\ldots+p^{n-1}]}{(q+p)^{n-1}}$

For the balance of column 4, we have

$$E[r] = \sum_{r=0}^{n} rP_r = np(q+p)^{n-1} = np, \tag{a'}$$

since $(q+p)^{n-1} = 1$. In this form the mean of a positive binomial in n Bernoulli tests (independent and of the same probability p) is the sum of n probabilities p that the indicator variable ("success") of simple alternative occurs in each test:

$$\mu_{\text{pos. bin.}} = n\mu_{\text{simple alt.}}. \tag{a''}$$

Parameters of variability. Absolute variability. As the measure of the *interval of variation* of amplitude $(0-n)$, we have

$$R = n. \tag{b}$$

The *variance* is $$V(r) = \sigma^2 = \Sigma(r-\mu)^2 P_r. \tag{c}$$
The calculation yields

$$V(r) = \sigma^2 = npq. \tag{c'}$$

It is seen here that for the positive binomial in n Bernoulli tests (independent and of constant probability p) the variance is the sum of n variances pq of the indicator variable of simple alternative in each test:

$$\sigma^2_{\text{pos. bin.}} = n\sigma^2_{\text{simple alt.}}. \tag{c''}$$

The *standard deviation* is

$$\sigma = \sqrt{npq}. \tag{d}$$

Relative variability. The *relative range,*

$$\frac{n}{\mu} = \frac{n}{np} = \frac{1}{p} \tag{e}$$

is equal to Gumbel's return period (1960, p. 23).

The *coefficient of variation* is

$$\gamma = \frac{\sigma}{\mu} = \frac{\sqrt{npq}}{np} = \frac{1}{\sqrt{n}}\sqrt{\frac{q}{p}}. \tag{f}$$

The *relative excess of variance*

$$\delta_r = \frac{V(r)-E[r]}{E[r]} = \frac{npq-np}{np} = q-1 = -p, \tag{f'}$$

is negative for the positive binomial as well as for the law of simple alternative; for either law the relative excess of variance is the same:

$$\delta_{r\ \text{pos. bin.}} = \delta_{r\ \text{simple alt.}}. \tag{f''}$$

Parameters of shape. Analysis yields the following results which are related to those obtained for the law of simple alternatives (6.211.112). *Skewness,* based on $\mu_3 = (q-p)npq;$ (g)

$$\gamma_1 = \frac{\mu_3}{\sigma^3} = \frac{q-p}{\sqrt{npq}} = \frac{q-p}{\sigma} \quad \text{(R. A. Fisher)}$$ (g′)

$$\beta_1 = \gamma_1^2. \quad \text{(K. Pearson)}$$

The positive binomial is symmetric for $p = q = \frac{1}{2}$, and non-symmetric for $p \neq q$.

Kurtosis, based on $\mu_4 = 3n^2p^2q^2 + npq(1-6pq);$ (h)

$$\beta_2 = \frac{\mu_4}{\sigma^4}. \quad \text{(K. Pearson)}$$ (h′)

From analysis,

$$\beta_2 = 3 + \frac{1-6pq}{npq}.$$

Using Fisher's notation,

$$\gamma_2 = \beta_2 - 3 = \frac{1-6pq}{npq} = \frac{1}{\sigma^2} - \frac{6}{n}.$$ (h″)

6.221.3. Relation between mean and variance for positive binomial distributions

Since $\mu = np$ and $\sigma^2 = npq$, therefore

$$\sigma^2 = (np)\cdot q = \mu q.$$ (a)

Since $q < 1$, therefore

$$\sigma^2 < \mu.$$ (b)

The variance of a positive binomial variable is less than its mean. Moreover, since $p = \mu/n$, therefore

$$\sigma^2 = \mu(1-p) = \mu - \mu\cdot\frac{\mu}{n} = \mu - \frac{\mu^2}{n}.$$

For a positive binomial variable, the variance and the mean are connected by the formula

$$\sigma^2_{(\text{bin.}>0)} = \mu - \frac{\mu^2}{n}.$$ (c)

This type of functional relation between parameters is very important as regards probability distributions, for it orients the inquirer in the search for some formula of adjustment, and also it may suggest an adequate representation or transformation of the data.

6.221.4. Example

Suppose that six homogeneous coins are tossed at the same time. The probabilities of obtaining 0, 1, 2, 3, 4, 5 or 6 Tails are given by the terms of the expansion

$$(q+p)^6 = 1\cdot q^6 + 6q^5p + 15q^4p^2 + 20q^3p^3 + 15q^2p^4 + 6q^1p^5 + 1\cdot p^6 .$$

The recurrence formula enables us to pass from one term to the next ($n = 6$):

$$P_{r+1} = \frac{6-r}{r+1}\cdot\frac{p}{q}\cdot P_r .$$

We proceed from $P_0 = q^6$,

$$P_1 = \frac{6-0}{0+1}\cdot\frac{p}{q}\cdot q^6 = \frac{6}{1}q^5p,$$

$$P_2 = \frac{6-1}{1+1}\cdot\frac{p}{q}\cdot\frac{6}{1}q^5p = \frac{6\cdot 5}{1\cdot 2}\cdot q^4p^2,$$

etc.

One recognizes in the coefficient of P_r the number of simple combinations of six elements taken two by two, say,

$$\binom{6}{2} = \frac{6\cdot 5}{1\cdot 2}.$$

Since $p = q = \frac{1}{2}$, therefore

$$q^6p^0 = q^5p^1 = q^4p^2 = q^3p^3 = q^2p^4 = q^1p^5 = p^0q^6 = \frac{1}{2^6} = \frac{1}{64}.$$

The exponent of p is the realization of the random variable r which gives the number of successes (Tails) in six tests, and the exponent of q shows the number of failures. No difficulty arises in verifying that the sum of the probabilities taken over the domain of variation of the random variable has the value of unity:

$$\sum_{r=0}^{6} P_r = \frac{1+6+15+20+15+6+1}{64} = \frac{64}{64} = 1.$$

For $r = 3$ the probability P_3 is maximum and unique. The value of the random variable such that the corresponding probability is maximum, is called the *mode of the binomial distribution*. If n is odd, one has two successive modal values. For instance, for $n = 3$ and $p = q = \frac{1}{2}$, one has the following situation ($p = \frac{1}{2}$):

r	0	1	2	3
P_r	$\frac{1}{8}$	$\frac{3}{8}$	$\frac{3}{8}$	$\frac{1}{8}$

The mean $\mu = np = 1.5$ is between the two modal values 1 and 2.

As a rule, there are two successive modal values if $np+p$ is an integer.

6.221.5. *Applications of the positive binomial to observations of presence or absence*

Events which are the subject of the simple alternative law ($q+p = 1$), are of the *all or nothing* type. The non-realized eventuality may be, for instance, survival in an insecticide test, whilst the realized event corresponds to the death of the insect during testing. By contrast, in the positive binomial law in n simultaneous or consecutive tests, providing they are representatively independent, the *degree of realization of the favorable event* has more graduation (column 1 of Table 6.221.2A). In fact one passes from total failure ($r = 0$) to partial failure (r between 1 and $n-1$), and then on to zero failure or integral success ($r = n$). The random variable r which indicates the realization of the event, now also represents the state E of the system.

In the analysis of physical or biological phenomena, however, it is common to find a clear-cut distinction between the outcome of events, i.e. failure or not a failure. From this ultra-simplified point of view which lumps together in a single non-fail event all the successes of degree other than 1 success out of n, or $(n-1)$ out of n, we are able to solve problems in scientific fields as far apart as genetics, cosmic radiation, neural, renal and sight physiology, epidemiology and numeration methods in bacteriology and hematology.

6.221.6. *Binomial models in biology*

In the theory of games of Heads or Tails the counting random variable r represents the number of possible successes in n independent tests where p is the constant probability of success. If the player receives a shillings for each success, the new random variable ra represents a random gain expressed in shillings. A more illuminating example, although being an extreme schematic representation, will now be taken from the field of human biometry.

Consider $n = 4$ environmental factors which may play a part in the post-natal growth of children, namely, temperature, degree of isolation, nourishment and air change (ventilation). Suppose that each of these factors acts independently of the others on an (0, 1) basis with elementary probability $p = \frac{1}{2}$. A factor may show itself to be unfavorable (0), in which case the effect on growth is zero, or else it can be favorable (1) and then its effect on growth is $c = 1$ cm.

If r factors act in the favorable sense (modality 1), the gain in height will be $r \times 1$ cm. Since r is contained between 0 and 4, the possible gain in height will vary from 0 to 4 cm according to the subjects. The interval 0–4 cm is the range of the discrete random variable, but dimensioned so that the gain in height is rc cm.

According to the chance manifestations of type 0 or 1, the environmental factors may realize different *complexions* (combinations). The *observable state* of the system is designated by r, which indicates the number of factors acting favorably in each complexion effecting this state, as expressed in terms of the observable gain in height:

$$X_r = rc \text{ cm.} \qquad \text{(a)}$$

The $2^4 = 16$ different and individually equiprobable complexions can thus be regrouped in five increasingly favorable states E_r as scaled by $r = 0, 1, 2, 3, 4$ for the random variable (see Table 6.221.6A).

TABLE 6.221.6A. THE SIXTEEN EQUIPROBABLE COMPLEXIONS REALIZED BY FOUR FACTORS ACTING ON A (0, 1) BASIS WITH $p = \frac{1}{2}$
(Complexions are read vertically)

I.	0	1000	1 1 1	0 0 0	1110	1
	0	0100	1 0 0	0 1 1	1101	1
	0	0010	0 1 0	1 0 1	1011	1
	0	0001	0 0 1	1 1 0	0111	1
II. r	0	1	2		3	4
III. State E_r	E_0	E_1	E_2		E_3	E_4
IV. $n_r = \binom{n}{r}$	1	4	6		4	1
V. P_r	$\frac{1}{16}$	$\frac{4}{16}$	$\frac{6}{16}$		$\frac{4}{16}$	$\frac{1}{16}$
VI. X_r, gain in cm	0	1	2		3	4

Notation:

I. Detail of the sixteen equiprobable complexions realized by four factors acting independently on the all-or-nothing principle (0, 1) with equal probability of action in the favorable sense ($p = q$). These $2^4 = 16$ complexions realize the fundamental probability set of power $n_F = 16$.

II. r represents the number of factors acting favorably.
III. State E_r.
IV. n_r: Power of the set of combinations realizing state E_r.
V. $P_r = n_r/n_p$ is the probability of E_r; $P_r = 1/2^{4n}C_r$ (see footnote, p. 247).
VI. X_r is the gain in cm of the children in state E_r.

Having introduced the quantum of increase $c = 1$ cm for all factors acting according to modality 1, we have to pass on from the random number r to the random increase of height:

$$X_r \text{ cm} = rc \text{ cm} = r \times 1 \text{ cm}. \tag{a}$$

We now introduce the *additive effect* of the factors on growth. The mean of this increase, i.e.

$$E[X] = \frac{\Sigma n_r X_r}{\Sigma n_r}, \tag{b}$$

presents itself in the form of a *weighted average of the* X_r, the weighting coefficients being the n_r.

The calculation yields

$$E[X] = \frac{1\times 0 \text{ cm} + 4\times 1 \text{ cm} + 6\times 2 \text{ cm} + 4\times 3 \text{ cm} + 1\times 4 \text{ cm}}{1+4+6+4+1} = \frac{32}{16} = 2 \text{ cm}.$$

The same result is obtained with the formula for the expected value:

$$E[X] = \Sigma P_r X_r$$

$$E[X] = \tfrac{1}{16}\times 0 \text{ cm} + \tfrac{4}{16}\times 1 \text{ cm} + \tfrac{6}{16}\times 2 \text{ cm} + \tfrac{4}{16}\times 3 \text{ cm} + \tfrac{1}{16}\times 4 \text{ cm} \tag{c}$$

$$= \tfrac{32}{16} \text{ cm} = 2 \text{ cm}.$$

This simple model thus leads to the prediction that the height of a group of children at birth (50 cm) will fan out between 50 and 54 cm. The sizes $(50+X_r)$ cm are given by the relative frequencies P_r from the binomial distribution.

The model is based on the concept of state of the child's environmental system in growth, and it can be extended to a larger number of factors present in the complexions. But if the number of factors n increases, the effect c cm must decrease in magnitude in such a way that the product nc cm, observed in the most favorable state, remains locally contained within biological limits. If the observed maximum increase *per annum* is about 5 cm, assuming that 50 factors are acting additively and independently, the general average effect produced factor by factor is of the order of 0.1 cm, or 1 mm, in magnitude.

Finally, biometry in genetics regards the phenotypic realization of height as the result of very many genes acting additively and independently according to equiprobable states (0, 1), the effect produced in c cm here varying from gene to gene. In this case the n genes realize combinations of the type

$$\overbrace{(0\ 1\ 1\ 0\ 0\ 1\ 1\ \ldots\ldots\ldots\ldots\ldots\ 1)}^{n}$$
$$\underbrace{(\downarrow\ \downarrow\ \ \ \ \ \downarrow\ \downarrow\ \downarrow)}_{r}$$

The combinations corresponding to the same r are regrouped in the state E_r, the probability of which (P_r) is given by the binomial distribution generated by

$$(q+p)^n \quad \text{with} \quad p = q = \tfrac{1}{2}.$$

The state E_r of the gene system (at microscopical level) is "projected" by averaging the quantity c onto the observable phenotypic level of the height.

X_r = original height$+rc$ cm (macroscopic level). Here too the maximum product rc must remain within limits compatible with biological obervation.

This conception was considered by Quetelet (1869 II, p. 38) for the height ($p = q = \frac{1}{2}$) and also for the weight ($p \neq q$), then became standard in genetic biometry (Fisher, 1918; Mather, 1949; and others).

6.221.7. *Binomial distributions and biological tests in all or nothing*

The outcome of an all-or-nothing test, or the *quantal response* (reaction), is of the type death or survival, germination or non-germination. As an example suppose that a group of $n = 50$ spores is exposed to a fungicide in a Petri dish and one observes a random response r of non-germination which varies with the dose of fungicide. It may be said that r is a random function of the dose D. The relation between these two magnitudes cannot be established in functional manner, but via the probability of r non-germinated spores being observed out of n (fixed) spores exposed to the dose D, i.e. the probability for which we set $P_r(D)$. By virtue of the independence of reaction of the spores, this probability is of binomial type:

$$[q(D)+p(D)]^n$$

with the following complementary relation for each dose:

$$p(D)+q(D) = 1.$$

The mean or expected value of the number of reactors to the dose D to be described out of the (fixed) n exposed, is

$$E[\mathbf{r}(D)] = np(D).$$

We shall now obtain a functional relation between $E[\mathbf{r}(D)]$ and D in terms of $p(D)$, which is the probability (across the domain D) of the non-germination of a spore taken at random in a group subjected to the dose D. No spores with "tolerance" equal to, or less than, D, will germinate. In general, the functional relation between $p(D)$ and D is of the asymmetrical sigmoid type, the two principal types of which are the log-Gaussian type (6.335) and the log-logistic type (6.336).

In short, the concept of biological quantal-response testing involves a discrete random variable $\mathbf{r}$ and its binomial probability distribution, a distribution of the continuous random variable $p(D)$, and finally two transformations (log-dose; probit or logit of the response) for rectifying the experimental sigmoid. This is to be seen from the internal unity of this section.

Finally, the parameters of the straight line of probit-log dose response are estimated by the method of maximum likelihood (C. I. Bliss, 1935; Sir R. A. Fisher, 1935), although the line of logit-log dose response is analysed by the method of minimum χ^2 as well (J. Berkson, 1949).

6.221.8. Representation of binomial points in a polar diagram, binomial probability paper

The simple alternative density where p is the probability of the favorable event A, has been represented elsewhere on a linear axis for which $r = 0.1$ (6.211). According to Sir R. A. Fisher (1947, p. 34), it is also possible to represent this density in a polar diagram (see Fig. 6.221.8I).

Out of n observed subjects, suppose that r individuals (reactors) exhibit a given characteristic A, and that the other $n-r$ subjects do not. In the above

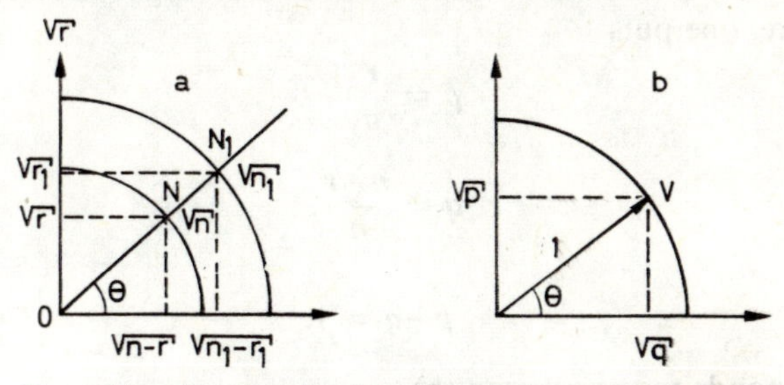

FIG. 6.221.8 I. Principle of the binomial probability paper.

diagram $\sqrt{r}$ is plotted along the ordinate axis, and $\sqrt{n-r}$ along the abscissas. Such are the *coordinates of the binomial point* $N(\sqrt{r}, \sqrt{n-r})$ (Fig. 6.221.8Ia).

By virtue of the theorem of Pythagoras,

$$(\sqrt{r})^2+(\sqrt{n-r})^2 = (\sqrt{n})^2, \tag{a}$$

or the obvious identity:

$$r+n-r = n. \tag{b}$$

This relation corresponds to the *observed* division of the subjects into classes A and non-A of effective values r and $n-r$ respectively. This division is represented by the point N of coordinates $(\sqrt{n-r}, \sqrt{r})$ and it results in the definition of an angle θ such that

$$\tan\theta = \sqrt{\frac{r}{n-r}}. \tag{c}$$

The point N_1 defines a division of n_1 subjects into r_1 and (n_1-r_1) individuals respectively, and in this connection the angle θ is correspondingly the same as for the n subjects divided into r and $(n-r)$. This division corresponds to the "split" of Mosteller and Tukey (1949). If, therefore, in several series of independent observations or experiments one finds n subjects divided between r and $(n-r)$ (or in the proportions p and q), and this remains invariant from one series to another, the binomial points $(\sqrt{n}, \sqrt{n-r})$ are arranged along a vector line at the same angle θ to the horizontal. Fisher's representation thus permits a convenient graphical test of the invariant nature of the division by series. This problem frequently arises in genetics, in psychology, in animal and human biometry, as well as in clinical and biological tests.

A quarter of a circle of radius n corresponds to each number n. It is possible to concentrate all these circle arcs in a single one by introducing the proportions p and q which play the part of reduced coordinates and lead to a unique law (distribution) in these reduced variables (Fig. 6.221.8Ib).

If, therefore, one puts

$$p = \frac{r}{n}$$
$$q = \frac{n-r}{n}, \tag{d}$$

then, clearly,

$$p+q = 1, \tag{e}$$

which corresponds to the relation (b).

With the point $V(\sqrt{p}, \sqrt{q})$ one associates an angle θ such that

$$\tan\theta = \sqrt{\frac{p}{q}} \quad \text{or} \quad \frac{\sin\theta}{\cos\theta} = \frac{\sqrt{p}}{\sqrt{q}}, \tag{f}$$

whence

$$\sqrt{p} = \sin\theta \quad \text{and} \quad \sqrt{q} = \cos\theta.$$

Upon squaring, we get

$$p = \sin^2\theta \quad \text{and} \quad q = \cos^2\theta. \tag{g) and (g'}$$

The binomial point V of reduced coordinates $(\sqrt{p}, \sqrt{q})$ thus possesses the *polar coordinates* $\varrho = 1$ and

$$\theta = \text{arc}\sin\sqrt{p}. \tag{h}$$

The relation (g) is represented in Fig. 6.221.8Ib. The angle θ varies between 0 and 90° as p runs the interval (0, 1). We shall again find the relation $p = \sin^2\theta$ under the name of *angular distribution for* θ varying between 0 and $\pi/2$ (6.314.11).

Masuyama (1951) proposed a more elaborate version of a stochastic binomial paper with very useful tables.

6.222. Positive Binomial Distribution

The positive binomial distribution is the probability up to an upper limit c, i.e. the probability of coming across at most c successes out of n independent experiments of the same probability p:

$$\text{Prob}\,(r \leqslant c) = P_0 + P_1 + \ldots + P_{c-1} + P_c, \tag{a}$$

or

$$F(c; n, p) = \sum_{r=0}^{r=c} \binom{n}{r} p^r (1-p)^{n-r}. \tag{b}$$

The probability of meeting with less than c successes is written so:

$$F(c-1; n, p).$$

For the probability, complementary to $F(c-1; n, p)$, of meeting with c or more successes, or even with at least c successes, we have

$$\text{Prob}\,(r \geqslant c) = \sum_{r=c}^{n} \binom{n}{r} p^r q^{n-r}, \tag{c}$$

or, according to Simons and Grubbs (1952),

$$P(c; n, p) = \sum_{r=c}^{n} \binom{n}{r} p^r q^{n-r}. \tag{c'}$$

The probability of having exactly c successes

$$P_c = \binom{n}{c} p^c q^{n-c} \tag{d}$$

is associated with $P(c; n, p)$ and $P(c+1; n, p)$ by the obvious relation

$$P_c = P(c; n, p) - P(c+1; n, p). \tag{d'}$$

Tables. Simons and Grubbs (1952) have tabulated the values $P(c; n, p)$ for $p = 0.1\ (0.01)\ 0.5$, and $n = 1\ (1)\ 150$ and $c = 1, 2, \ldots, n$. In order to spare the entries for $p > 0.5$, they considered (p. IV) the table of successive terms as noted by the number of successes r on the one hand, and the number of failures $(n-r)$ on the other hand.

Eisenhart (1950) has edited the binomial probability *Tables of the National Bureau of Standards* (1950). In addition to other methods, Borges (1970) gives a new simple approximation of the binomial by the Gaussian law to the order $1/n$ approximation instead of the usual order $n^{-1/2}$.

6.23 Poisson's Law

The Poisson law, its modifications and generalizations have been treated extensively in a monograph by Haight (1967).

6.231. Poisson's Point-density of Probability

6.231.1. Transition from the positive binomial to Poisson's law

In the study of a static Poissonian phenomenon, e.g. upon observing the number of red corpuscles in the squares of a hemocytometer, one may expect the following eventualities described by the random variable $\boldsymbol{r}$:

a vacant or non-occupied square observed: $r = 0$
a square with one corpuscle observed: $r = 1$
a square with two corpuscles observed: $r = 2$
and so on.

There is, of course, a physical limit to the accumulation of cells in a square, but in studying this law theoreticians ignore that this difficulty was recognized by Lancaster (1950) and Turner and Eadie (1957).

The foregoing events are designated by a discrete counting random variable r which can take the values 0, 1, ... But since the corpuscles are numerous (1600 to give an example) and are randomly distributed between numerous squares (400 small squares), the probability of being incident in one square is low ($p = \frac{1}{400}$). We are therefore concerned with a particular binomial variable where n is large, p is small, whilst the product $\mu = np = 1600 \times \frac{1}{400}$ remains finite and equal to 4. This is the average number of corpuscles per square of observation.

In general Poisson's law is the limit of the positive binomial law $(q+p)^n$ whenever the probability p of success tends to zero and while at the same time n increases sufficiently to make the product np finite and equal to the finite number μ, itself small. Thus in this process the steps of the range of the positive binomial conserve their equidistant spacing of 1 while the upper limit n increases indefinitely. More precisely Hoel (1948, p. 50) states that the binomial distribution may be replaced by the Poisson law when $\mu = np < 5$ even though the standard Tables of Molina (1942) cover the interval $0.001 \leqslant \mu \leqslant 100$.

For μ greater than 10, one can be content with the *Gaussian approximation* for a variable which remains Poissonian, i.e. a variable where variance equals the mean (6.231.2).

We now consider the steps involved in passing to the limit proceeding from the positive binomial in r and ending with the Poisson law for the same random variable r:

$$\lim_{\substack{n \to \infty \\ p \to 0 \\ np \to \mu}} \frac{n!}{r!\,(n-r)!} p^r q^{n-r} = e^{-\mu} \frac{\mu^r}{r!}. \tag{a}$$

In the binomial $(p+q)^n$ the general term is

$$P_r = \frac{n!}{r!\,(n-r)!} p^r q^{n-r}. \tag{b}$$

By virtue of the conditions imposed in (a) we substitute $p = \mu/n$ and $q = 1-\mu/n$ into the relation (b). Thus

$$P_r = \frac{n!}{r!\,(n-r)!} \left(\frac{\mu}{n}\right)^r \left(1-\frac{\mu}{n}\right)^{n-r}, \tag{c}$$

or

$$P_r = \frac{\mu^r}{r!}\left(1-\frac{\mu}{n}\right)^n \cdot \frac{n!}{(n-r)!\,n^r\left(1-\frac{\mu}{n}\right)^r}. \tag{d}$$

As $n \to \infty$, we have the following situations:

(1) $\mu^r/r!$ is constant, i.e. not dependent or n,

(2) $\lim\limits_{n\to\infty}\left(1-\frac{\mu}{n}\right)^n = e^{-\mu}$,

(3) $\lim\limits_{n\to\infty}\left(1-\frac{\mu}{n}\right)^r = 1$ for r remains finite and $\frac{\mu}{n} \to 0$.

On the other hand,

$$\frac{n!}{(n-r)!\,n^r} = \frac{1, 2, \ldots (n-r)(n-r+1) \ldots n}{(n-r)!\,n^r}. \tag{e}$$

Dividing top and bottom by $(n-r)! = 1, 2 \ldots (n-r)$, it remains to calculate

$$\lim_{n\to\infty} \frac{n-r+1}{n} \cdot \frac{n-r+2}{n} \cdots \frac{n}{n}.$$

This limit is the product of r factors tending towards 1 for $n \to \infty$. One writes thus the product

$$\left(1-\frac{r-1}{n}\right)\cdot\left(1-\frac{r-2}{n}\right) \cdots \left(1-\frac{1}{n}\right)\cdot 1,$$

which tends to unity as $n \to \infty$, since r is fixed in this process. Hence the limit of P_r for $n \to \infty$, or general term of the *Poisson point-density* at point r, is

$$P_r = \frac{\mu^r}{r!} e^{-\mu} \tag{f}$$

for $r = 0, 1, 2, \ldots, \infty$.

In the Poisson law r varies from 0 to ∞ since n (upper limit of r) tends towards ∞, whilst r varies from 0 to n finite in the case of the positive binomial law.

Interpretation of P_r. In the positive binomial distribution produced by $(q+p)^n$, P_r is the probability of observing r times a success of elementary probability p in a *finite* series of n independent, successive or simultaneous observations.

In the Poisson distribution, P_r is the probability of observing, r times per cell, an event of *low elementary probability p* in a *necessarily large* series of independent, successive or simultaneous observations. It is seen that the upper limit of the range, n, is no longer present explicitly in the Poisson point-density (f), although it is still present implicitly in the combination $\mu = np$. The Poisson probability point-density depends only on the single parameter μ.

Just the same as the binomial random variable, the Poisson random variable is essentially positive, but its range extends from 0 to ∞.

It is verified without difficulty that the sum of the $P_r = 1$:

$$\sum_{r=0}^{\infty} P_r = 1. \qquad \text{(g)}$$

In fact:

$$P_r = e^{-\mu}\frac{\mu^r}{r!},$$

$$\sum_{r=0}^{\infty} P_r = \Sigma e^{-\mu}\frac{\mu^r}{r!} = e^{-\mu}\Sigma\frac{\mu^r}{r!}$$

$$= e^{-\mu}\left(1+\frac{\mu}{1!}+\frac{\mu^2}{2!}+\frac{\mu^3}{3!}\ \dots\right).$$

Now inside the parenthesis is the series expansion of e^{μ}. Hence

$$\sum_{r=0}^{\infty} P_r = e^{-\mu}\cdot e^{\mu} = e^0 = 1. \qquad \text{(h)}$$

Graphical representation. Figure 6.231.1I shows the series of Poisson distributions for which the mean μ is equal to 0.5, 1 and 5 respectively. In principle the Poisson distribution is defined only for integer values of the random variable $r = 0, 1, 2, \dots$. The lines joining the points (r, P_r) are used to mark the general trend.

The maxima of the Poisson distribution of mean μ occur at $r = \mu-1$ and $r = \mu$ (see Fig. 6.231.1I) when μ is an integer.

$$P_{\mu-1} = P_{\mu} \qquad (\mu \text{ integer}). \qquad \text{(i)}$$

In fact, by making r equal to $\mu-1$ in (b), then multiplying top and bottom by μ:

$$e^{-\mu}\frac{\mu^{\mu-1}}{(\mu-1)!} = \frac{e^{-\mu}\mu^{\mu}}{(\mu-1)!}\times\frac{1}{\mu} = \frac{e^{-\mu}\mu^{\mu}}{\mu!}.$$

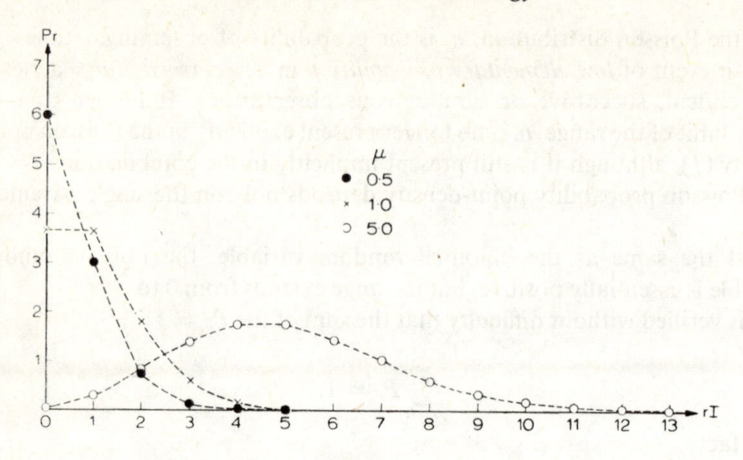

FIG. 6.231.1 I. Poisson point-density for $\mu = 0.5$, 1.0 and 5.0.

So therefore

$$P_{\mu-1} = P_{\mu},$$

and hence the horizontal plateau between $r = \mu - 1$ and $r = \mu$ for the Poisson density of mean μ. The same property is possessed by the positive binomial with integer $np+p$.

6.231.2. Descriptive parameters

Parameters of localization: the mean μ. It will now be seen that all the parameters of the Poisson distribution are given as a function of the single mean value μ (which is defined by the passage to the limit from the positive binomial).

One verifies that $E[\mathbf{r}]$ (expected value) is equal to μ:

$$E[\mathbf{r}] = \sum_{r=0}^{\infty} rP_r = \sum_{r=0}^{\infty} re^{-\mu}\frac{\mu^r}{r!}$$

Introducing $\mu e^{-\mu}$,

$$E[\mathbf{r}] = e^{-\mu}\mu \sum_{r=1}^{\infty} \frac{\mu^{r-1}}{(r-1)!}.$$

$$= e^{-\mu}\cdot\mu\cdot e^{+\mu} = \mu.$$

So as required

$$E[r] = \mu. \tag{a}$$

Parameter of variability: variance $\sigma^2 = \mu$. In the Poisson distribution the variance is equal to the mean, i.e.

$$V(r) = \sigma^2 = \mu. \tag{b}$$

This can be verified by evaluating the central second moment $\mu_2 = \sigma^2 = E[(r-\mu)^2]$ from the second moment from the origin $\mu_2' = E[r^2]$:

$$\mu_2' = E[r^2] = \Sigma r^2 P_r = \Sigma r^2 e^{-\mu} \frac{\mu^r}{r!}.$$

We now proceed as in the foregoing section by introducing $e^{-\mu}\cdot\mu$; but in addition we perform the operation $r/r! = 1/(r-1)!$. There thus remains

$$\mu_2' = e^{-\mu}\mu \sum_{r=0}^{\infty} r \frac{\mu^{r-1}}{(r-1)!}.$$

Breaking down the sum factor of the second entry in two terms:

$$\Sigma r \frac{\mu^{r-1}}{(r-1)!} = \Sigma(r-1)\frac{\mu^{r-1}}{(r-1)!} + \Sigma 1 \frac{\mu^{r-1}}{(r-1)!} = \mu e^{\mu} + e^{\mu}.$$

Hence

$$\mu_2' = \mu e^{-\mu}(\mu e^{\mu} + e^{\mu})$$

$$= \mu(\mu+1) = \mu^2 + \mu.$$

Here the second central moment becomes

$$\mu_2 = \sigma^2 = \mu_2' - \mu_1'^2$$

$$= \mu^2 + \mu - \mu^2 = \mu.$$

Thus the variance σ^2 is equal to the mean μ in the Poisson distribution.

Proceeding from the binomial distribution, it can also be shown that

$$\lim_{\substack{n \to \infty \\ p \to 0 \\ np \to \mu}} \sigma^2_{\text{bin.}} = \sigma^2_{\text{Poiss.}} \tag{c}$$

In the Poissonian case $\lim\limits_{\substack{n\to\infty \\ p\to 0}} np = \mu_{\text{finite}}$.

Since $p = \dfrac{\mu_{\text{finite}}}{n}$

$$\lim_{n\to\infty} q = \lim(1-p) = 1-\lim\frac{\mu}{n} = 1.$$

So the limit of the binomial variance is

$$\lim_{n\to\infty} np \cdot \lim_{n\to\infty} q = \mu\times 1 = \sigma^2_{\text{Poiss.}}.$$

Thus the same result is found as by the less direct method from the operator $E[(r-\mu)^2]$.

Since the variance of the Poisson random variable is equal to the mean μ as compared with the binomial, a measure of departure from an equality such as σ^2/μ is used, e.g. the relative variance (Clapham, 1936), the coefficient of dispersion (Blackman, 1942), the relative excess of variance (Martin, 1961).

$$\delta = \frac{\sigma^2-\mu}{\mu}, \tag{d}$$

which is zero in the Poisson case, negative in the positive binomial case ($\sigma^2 < \mu$), and positive for the negative binomial ($\sigma^2 > \mu$) (6.25 and 6.26).

The *standard deviation* is equal to the square root of the mean

$$\sigma = \sqrt{\mu}. \tag{e}$$

The coefficient of variation is written so:

$$\text{C.V.}\% = 100\frac{\sigma}{\mu}\% = 100\frac{\sqrt{\mu}}{\mu}\% = \frac{100}{\sqrt{\mu}}\%. \tag{f}$$

It decreases as the reciprocal of the square root of the mean.

6.231.3. Parameters of shape

Here we expound only the process of passage to the limit from the positive binomial distribution.

Skewness. This term is the limit of the parameter β_1 relative to the positive binomial law $(q+p)^n$ for $\substack{n\to\infty \\ p\to 0 \\ np\to\mu}$: $\beta_1 = \dfrac{(p-q)^2}{npq}$.

We write β_1 as a function of μ and of n before passing to the limit: since $q = 1-p$ and $p = \dfrac{\mu}{n}$, therefore

$$\beta_1 = \frac{(1-2p)^2}{n(1-p)p} = \frac{\left(1-2\dfrac{\mu}{n}\right)^2}{n\left(1-\dfrac{\mu}{n}\right)\dfrac{\mu}{n}}.$$

But since μ remains finite as $n \to \infty$,

$$\lim \left(1-\frac{2\mu}{n}\right)^2 = 1 \quad \text{and} \quad \lim \left(1-\frac{\mu}{n}\right) = 1;$$

the remaining relation is

$$\beta_1 = \frac{1}{\mu}. \tag{a}$$

Kurtosis. For Poisson's law the coefficient β_2 is the limit of the homologous term

$$\beta_2 = 3+\frac{1-6pq}{npq} \tag{b}$$

of the binomial law. In fact it is known that

$$\lim_{\text{bin.}} (npq) = \mu_{\text{Poiss.}},$$

whence

$$\lim_{n\to\infty} pq = \frac{\mu}{n}.$$

Setting these values in β_2, one obtains, from consideration of (b),

$$\beta_2 = 3+\frac{1-6\dfrac{\mu}{n}}{\mu}.$$

Since μ remains finite as $n \to \infty$, $\dfrac{6\mu}{n} \to 0$,

$$\lim_{n\to\infty} \beta_2 = 3+\frac{1}{\mu}. \tag{b'}$$

It is thus seen that in the Poisson distribution the variance and the shape parameters β_1 and β_2 all depend only on the mean μ. Thus *Poisson's law is completely determined by the value of* μ.

6.231.4. *Evolution of Poisson's law when the mean increases*

When the mean increases:

(1) The variance $\sigma^2 = \mu$ increases.

(2) The coefficient of variation $= \dfrac{100}{\sqrt{\mu}}$ decreases as $\dfrac{1}{\sqrt{\mu}}$.

(3) Skewness $\beta_1 = \dfrac{1}{\mu}$ decreases as $1/\mu$, $\gamma_1 = \dfrac{1}{\sqrt{\mu}}$ decreases as $\dfrac{1}{\sqrt{\mu}}$.

(4) *Kurtosis* $\beta_2 = 3+\dfrac{1}{\mu}$ decreases equally as $1/\mu$, whilst $\gamma_2 = \dfrac{1}{\mu}$ decreases equally as $1/\mu$. So when μ increases, and starting from $\mu = 10$, it may be said that the *Poisson law tends to a Gaussian law* for which $\beta_1 = 0$ and $\beta_2 = 3$, or $\gamma_1 = 0$ and $\gamma_2 = 0$ (6.322.213).

6.231.5. *Evaluation of Poisson distributions*

The principle is as follows: (1) *Calculate the observed mean*

$$\bar{r} = \frac{\Sigma r f_r}{\Sigma f_r} \tag{a}$$

as indicated in Table 6.231.5A. (2) *The successive terms of a Poisson law of mean* μ are:

$$P_0 = e^{-\mu}, \quad P_1 = \frac{\mu}{1} e^{-\mu}, \quad P_2 = \frac{\mu^2}{2!} e^{-\mu}, \ldots,$$
$$P_r = \frac{\mu^r}{r!} e^{-\mu}, \quad P_{r+1} = \frac{\mu^{r+1}}{(r+1)!} e^{-\mu} \ldots \tag{b}$$

(where $\bar{r}$ is an estimator of μ which may be designated $\bar{r} = \hat{\mu}$). This series of P_r is generated from P_0 by the simple recurrence formulae

$$P_{r+1} = \frac{\mu}{r+1} P_r \qquad r = 0, 1, 2, \ldots. \tag{c}$$

TABLE 6.231.5A. HEMOCYTOMETER (STUDENT, 1907), FREQUENCY f OF SQUARES CONTAINING r YEAST CELLS, ADJUSTMENT OF A POISSON DISTRIBUTION

1	2	3	4
r	f_r	rf_r	T_r
0	—	0	3.71
1	20	20	17.37
2	43	86	40.65
3	53	159	63.41
4	86	344	74.19
5	70	350	69.45
6	54	324	54.16
7	37	259	36.21
8	18	144	21.19
9	10	90	11.02
10	5	50	5.16
11	2	22	2.19
12	2	24	0.85
13	—	—	0.31
14	—	—	0.10
15	—	—	0.03
	400	1872	400.01

$$\bar{r} = \frac{1872}{400} = 4.68.$$

$$T_r = 400e^{-4.68}\frac{4.68^r}{r!} \quad \text{theoretical frequencies } T_r.$$

$$T_0 = 4\times 0.9279 = 4\times 0.928 = 3.712.$$

The application of relations (c) yields the theoretical frequencies $T_1, T_2, \ldots$ of column 4.

Calculation of P_0 and $T_0 = 400P_0$

(1) If one knows an estimator $\bar{r}$ of μ, the first term $e^{-\bar{r}}$ is found in a table of exponentials. For example, for Fig. 6.231.5I consult a table of exponentials e^{-x} where the independent variable is noted to the second decimal. If x is at the end of the table, the reciprocal of e^x should be taken because $e^{-x} = 1 : e^x$.

Now enter with $\bar{r} = 4.68$. Read $P_0 = e^{-\bar{r}} = e^{-4.68} = 0.009279$. If P_0 is to be expressed as a percentage, multiply by 100 and then $100P_0 = 0.9279\%$.

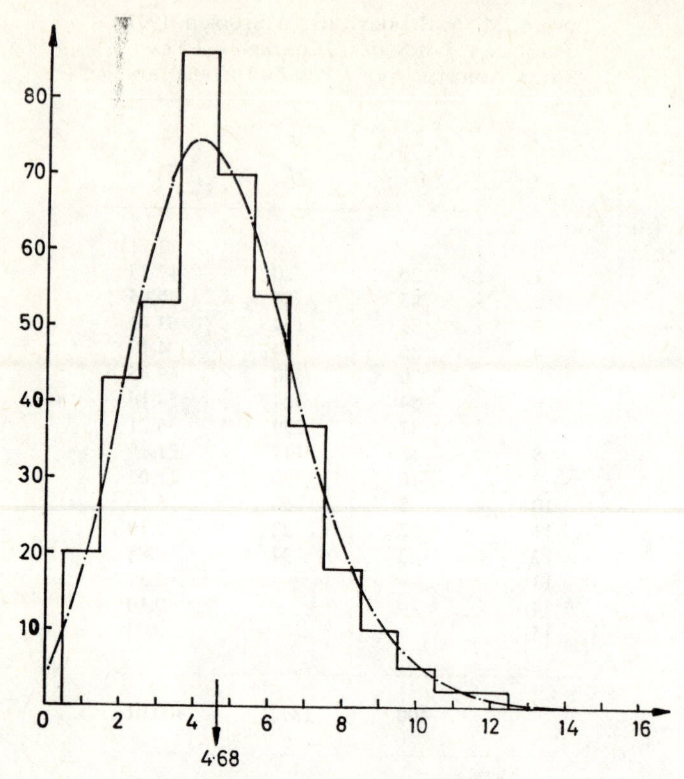

FIG. 6.231.5 I. Fitting a Poisson probability density.

Finally, the theoretical frequency of the vacant squares $T_0 = nP_0$ is

$$T_0 = 400 \times 0.009279 = 3.71.$$

Clearly, in a Poisson distribution of mean value $\mu = 4.68$ where the content of 400 individual squares is observed, the expected frequency of the vacant squares is about 3.71, i.e. somewhat less than 1%. This theoretical relative frequency is low and this in fact explains why one observes no vacant cell since $f_0 = 0$.

(2) If no table of exponentials is available, then $e^{-\bar{r}}$ is calculated by means of logarithms. In natural logs.,

$$\ln P_0 = \ln e^{-\bar{r}} = -\bar{r};$$

$\ln P_0$ is negative because P_0 is less than 1.

(3) But if no table of natural logarithms is to hand, use may be made of a

standard table of common logs. The change from $\log_{10}$ to ln is made by multiplying by the modulus ln 10 = 2.30103. It is convenient to use five decimal places:

$$\ln P_0 = 2.30103 \log_{10} P_0 = -\bar{r}.$$

Having calculated P_0 by one of these three methods, the recurrence formulae (c) are used to pass on successively to $P_1, P_2, \ldots$ and $T_1, T_2, \ldots$. This gives column 4 of the above table, and it remains to check that the sum again gives the total of 400 squares observed. The χ^2 test allows the comparison between the observed frequencies (f_r) on the one hand and the expected frequencies (T_r) on the other.

6.232. Poisson Distribution

6.232.1. Complementary distribution of Poisson

The complementary distribution of Poisson shows the probability of observing at least c elements per cell when the average is μ:

$$\text{Prob}\,(r \geqslant c \mid \mu) = \sum_{r=c}^{\infty} e^{-\mu} \frac{\mu^r}{r!}. \tag{a}$$

In order to arrive at a continuous distribution in μ, we derive the two terms of (a):

$$\frac{d\,\text{Prob}}{d\mu} = \sum_{r=c}^{\infty} \left[e^{-\mu} \cdot r \frac{\mu^{r-1}}{r!} - e^{-\mu} \frac{\mu^r}{r!} \right]$$

$$= \sum_{r=c}^{\infty} e^{-\mu} \frac{\mu^{r-1}}{(r-1)!} - \sum_{r=c}^{\infty} e^{-\mu} \frac{\mu^r}{r!}$$

$$\frac{d\,\text{Prob}\,(r \geqslant c \mid \mu)}{d\mu} = e^{-\mu} \frac{\mu^{c-1}}{(c-1)!}. \tag{b}$$

We integrate term by term between 0 and μ in the current variable x, thus obtaining the desired form of (a):

$$\text{Prob}\,(r \geqslant c \mid \mu) = \frac{1}{(c-1)!} \int_0^{\mu} e^{-x} x^{c-1}\, dx$$

$$= \frac{\Gamma_{\mu}(c)}{\Gamma(c)} \tag{c}$$

$$= I(\mu, c), \quad \text{or} \quad I_{\mu}(c)$$

using partial gamma function notation and in reference to section 6.334.12. We notice that x is a reduced gamma-c variate.

For the Poisson complementary distribution, Molina (1942, Table II) puts $P(c, \mu)$. (It should be noted that the symbol a of Molina is equivalent to our μ.) In this form

$$P(c, \mu) = \sum_{r=c}^{\infty} e^{-\mu} \frac{\mu^r}{r!},$$

and this is tabulated in six decimal places for $\mu = 0.001$ (0.001) 0.01 (0.01) 15 (1) 100 (Molina, 1942, Table II).

6.232.2. *Poisson distribution*

The Poisson distribution is written as a sum total of terms from zero to $(c-1)$, i.e.

$$\text{Prob}\,(r < c) = P_0 + P_1 + \ldots + P_{c-1} \qquad \text{(a)}$$

$$\text{Prob}\,(r < c) = 1 - \text{Prob}\,(r \geqslant c) \qquad \text{(b)}$$

$$= 1 - \frac{\int_0^{\mu} e^{-x} x^{c-1}\, dx}{(c-1)!} = \frac{(c-1)! - \int_0^{\mu} e^{-x} x^{c-1}\, dx}{(c-1)!}$$

$$= \frac{\int_0^{\infty} e^{-x} x^{c-1}\, dx - \int_0^{\mu} e^{-x} x^{c-1}\, dx}{(c-1)!}. \qquad \text{(c)}$$

The definition of $\Gamma(c) = (c-1)!$ is thus applied for c integer in this case.

Having regard to the additive property of the definite integral taken over two mutually exclusive ranges, the numerator becomes

$$\int_0^{\infty} e^{-x} x^{c-1}\, dx - \int_0^{\mu} e^{-x} x^{c-1}\, dx = \int_{\mu}^{\infty} e^{-x} x^{c-1}\, dx. \qquad \text{(d)}$$

Hence (b) is written as

$$\text{Prob}\,(r < c) = \frac{1}{(c-1)!} \int_{\mu}^{\infty} e^{-x} x^{c-1}\, dx. \qquad \text{(e)}$$

This formula is given in Feller (1952, I, p. 163).

Consider now the case $c = 2$. We therefore calculate P_0+P_1, say the sum total of the first two terms of a Poisson distribution of mean value μ:

$$\text{Prob}\,(r < 2) = \frac{1}{1!}\int_{\mu}^{\infty} e^{-x}x\,dx. \tag{f}$$

Integrating by parts, we get

$$\text{Prob}\,(r < 2) = -\left[xe^{-x}\right]_{\mu}^{\infty} + \int_{\mu}^{\infty} e^{-x}\,dx = -\{(0+0)-[\mu e^{-\mu}+e^{-\mu}]\}$$

$$= e^{-\mu}+\mu e^{-\mu} = P_0+P_1,$$

and, as expected we find the left member of (f)

$$\text{Prob}\,(r < 2) = P_0+P_1. \tag{g}$$

6.232.3. Poisson-sum probability paper

Graphical test of conformity to a Poisson law. Campbell (1923) and Miss Thorndike (1926) have drawn attention to a probability paper having the following characteristics:

(1) Abscissae: $\log \mu$,
(2) Ordinates: normal standard deviates u_p (6.322.222) or in probit scale $Y_p = 5+u_p$ (6.322.225.1).
(3) Network of equi-c curves showing the probability of observing *at least* c elements in a Poisson distribution of mean μ.

For the probability of observing at least c elements, we have

$$P_c+P_{c+1}+\ldots = 1-(P_0+P_1+\ldots+P_{c-1}). \tag{a}$$

Concisely (6.232.1)

$$\text{Prob}\,(r \geqslant c) = e^{-\mu}\sum_{r=c}^{\infty}\frac{\mu^r}{r!}, \tag{b}$$

which is related to the incomplete gamma function. The mean μ appears implicitly in the P_r as

$$P_r(\mu) = P_r = e^{-\mu}\frac{\mu^r}{r!}.$$

A modification of Miss Frances Thorndike's chart has been introduced by Dodge and Romig (1967, p. 35) by using abscissae $\log \mu = \log np$. This chart gives the probability of occurrence of c or less defectives in a sample of n pieces randomly selected from an infinite universe in which the fraction of defectives is p.

Another Poisson-sum probability paper was designed by Grimm (1962):

(1) Abscissae: $\sqrt{\mu}$ (note that $\sqrt{\mu}$ is the natural unit σ for Poisson's law).
(2) Ordinates: probit scale.
(3) Network of equi-c curves corresponding to the probability of finding *at most c* elements in a Poisson law of mean value μ. The equation of the curves is obtained from the approximation relation of Tseng Tung Cheng (1949):

$$\sum_{r=0}^{c} e^{-\mu}\frac{\mu^r}{r!} = \frac{1}{\sqrt{2\pi}}\int_{-\infty}^{u} e^{-u^2/2}\,du + \frac{1}{6\sqrt{2\pi\mu}}(1-u^2)e^{-u^2/2}+\delta.$$

The reduced deviation u is calculated from c in units $\sigma = \sqrt{\mu}$ and by applying the correction for continuity:

$$u = \frac{c+0.5-\mu}{\sqrt{\mu}}.$$

The residue δ is such that

$$|\delta| < 0.076\mu^{-1}+0.043\mu^{-3/2}+0.13\mu^{-2}.$$

Standard numeration of yeast cells on Poisson-sum probability paper. As an example, consider statistics of yeast cells in dilute suspension and set in the counting chamber of the Toma hemocytometer (Student, 1907) whose geometry is:

Base 1 mm² divided into 400 small squares; height 0.01 mm. Recording on Campbell–Thorndike probability paper in accord with the indications of Table 6.232.3A.

TABLE 6.232.3A. EXAMPLE OF STUDENT TREATED IN THE POISSON SUMS PROBABILITY SHEET OF THORNDIKE

1 r or c	2 f_r	3 Cumulative abs. frequency up to $c-1$, $f_0+f_1+\ldots+f_{c-1}$	4 Cumulative abs. frequency from the back $400-(f_0+\ldots+f_{c-1})$ $= f_c+f_{c+1}+\ldots$	5 Cumulative relative frequency from the back %
0	0	0	400	100
1	20	0	400	100
2	43	20	380	95.0
3	53	63	337	84.2
4	86	116	284	71.0
5	70	202	198	49.5
6	54	272	128	32.0
7	37	326	74	18.5
8	18	363	37	9.25
9	10	381	19	4.75
10	5	391	9	2.25
11	2	396	4	1.0
12	2 →	398	2	0.5
		400		

Example for $c = 2$, cumulative up to $2-1 = 1$. In column 3 one builds up to $c-1$. In column 4 the build-up is from the base. Verification: col. 3+col. 4 is equal to 400.

On the Poisson probability paper we mark off the point defined by (1) $c = 1, 2, \ldots$ (read off the equi-c curves), and (2) the relative cumulative frequency from the back read off on the ordinate axis (column 5). A series of points is thus obtained, each one of which is situated on an equi-c curve. If these points are situated on a vertical line, the data (r, f_r) follow a Poisson law with mean value μ graphically estimated by the fixed vertical marked off along the axis of the abscissae in log μ.

In fact the vertical is incident at the point 4.7, whereas the calculated mean value is $\bar{r} = 4.68$. Points which are oriented upwards to the left, correspond to a positive binomial and to other types derived therefrom. In the case where points are oriented upwards to the right, it is a question of a negative binomial (626), Polya–Eggenberger (628), or of distributions of the "contagious" type (e.g. Neyman, 1939; Neyman and Elisabeth Scott, 1957).

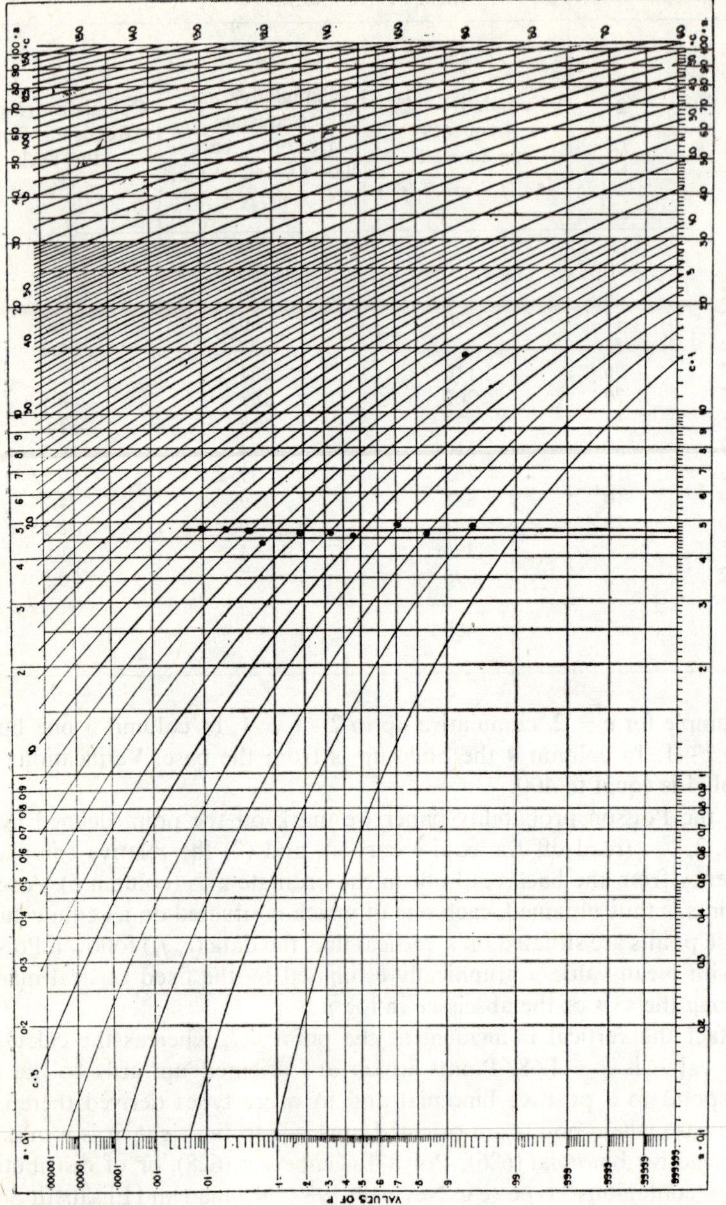

FIG. 6.232.3 I. Poisson-sum probability paper.

6.24. Geometric Law

6.241. Point Geometric Law

6.241.1. Definition

Consider a sequence of Bernoulli trials where we observe r failures of probability q prior to the first success of probability p which occurs in the trial of rank $(r+1)$. Setting 0 for a failure and 1 for a success, the following scheme is obtained:

Series	0	0 ...	0	1
	q	q	q	p
Rank	1	2	r	$r+1$

(a)

The probability of expecting r failures before the first success is

$$P_r = q^r \cdot p. \tag{b}$$

In particular, the probability of expecting 0 failures, and so of observing the first success at the outset, is

$$P_0 = q^0 \cdot p = p,$$

as expected.

The range of definition is $r = 0, 1, 2, \ldots$.

The sequence of probabilities

$$P_0 = p, \quad P_1 = qp, \quad P_2 = q^2p, \ldots \tag{c}$$

constitutes a geometric progression, the first term of which is p and the common ratio is q; hence the name *geometrical law.*

Over the range of definition $(0, \infty)$ the necessary relation is

$$\sum_{r=0}^{\infty} P_r = 1. \tag{d}$$

In fact,

$$\sum P_r = p \sum_{r=0}^{\infty} q^r. \tag{e}$$

The sum of the second term in this expression is the sum of the terms of a geometric progression of common ratio $q < 1$ and whose first term is 1:

$$\sum_{r=0}^{\infty} q^r = \frac{1}{1-q} = \frac{1}{p}. \tag{f}$$

Placing (f) in (e), it is seen that (d) is satisfied. The relation (f) is important for it enables the mean and the variance of r to be calculated in elegant fashion.

A recurrence relation is conveniently written so:

$$\begin{aligned} P_0 &= p \\ P_r &= qP_{r-1} \end{aligned} \quad (r = 1, 2, \ldots). \tag{g}$$

6.241.2. *Reduction parameters*

Mean. The expected value of r is written thus:

$$E[r] = \sum rP_r = p \sum_{r=0} rq^r = pq \sum rq^{r-1}. \tag{a}$$

The sum of the second term is the derivative of $\sum q^r$ with respect to q:

$$\sum rq^{r-1} = \frac{d}{dq} \sum q^r. \tag{b}$$

By virtue of formula 6.241.1 (f), we have

$$\frac{d}{dq} \sum q^r = \frac{d}{dq}\left(\frac{1}{p}\right) = \frac{d}{dq} \frac{1}{1-q} = -\frac{1}{(1-q)^2}(-1) = +\frac{1}{p^2}. \tag{c}$$

Placing (c) in (b) then (a), we have

$$E[r] = \frac{pq}{p^2} = \frac{q}{p}. \tag{d}$$

If $p = q = \frac{1}{2}$, *on the average* one expects a failure before a success.

Variance:

$$V(r) = \sum_{r=0} (r - E[r])^2 P_r = \sum \left(r - \frac{q}{p}\right)^2 P_r = \frac{1}{p^2} \sum (rp-q)^2 P_r. \tag{e}$$

After expansion of the sum of the second term and having regard to the second derivative of formula 6.241.1 (f), it can be shown that this sum is equal to q. Hence

$$V(r) = \frac{q}{p^2}. \tag{f}$$

The *standard deviation* of the geometric variable is

$$\sigma_r = \frac{\sqrt{q}}{p}. \tag{g}$$

The *coefficient of variation* is written so

$$\gamma_r = \frac{\sigma_r}{E[r]} = \frac{\frac{\sqrt{q}}{p}}{\frac{q}{p}} = \frac{1}{\sqrt{q}}. \tag{h}$$

The *relative excess of variance* is positive and equal to the mean:

$$\delta_r = \frac{\sigma_r^2 - \mu}{\mu} = \frac{\frac{q}{p^2} - \frac{q}{p}}{\frac{q}{p}} = \frac{\frac{q}{p}\left(\frac{1}{p}-1\right)}{\frac{q}{p}} = \frac{1}{p} - 1 = \frac{q}{p}.$$

Hence, according to (d),

$$\delta_r = E[r]. \tag{i}$$

The *parameters of shape* γ_1 and γ_2 are calculated by the central moments μ_3 and μ_4, as obtained from the moment-generating function which is written (Parzen, 1960, p. 219):

$$M(t) = \frac{p}{1-qe^t}. \tag{j}$$

Here we have

$$\mu_3 = \frac{q}{p^2}\left(1+2\frac{q}{p}\right),$$
$$\mu_4 = \frac{q}{p^2}\left(1+9\frac{q}{p^2}\right). \tag{k}$$

6.241.3. Relation between the variance and the mean

Comparing the relations (d) and (f), it is seen that

$$V(r) = \frac{q}{p} \cdot \frac{1}{p} = E[r] \cdot \frac{q+p}{p}. \tag{a}$$

Hence

$$V(r) = E[r](E[r]+1) = E[r]+(E[r])^2. \tag{b}$$

The variance of the geometric law is thus equal to the mean plus the mean squared. This relation is homologous to that which exists between the mean and the variance in the simple alternative law (6.211.1):

$$V(r) = pq = p(1-p) = p-p^2 = E[r]-(E[r])^2.$$

One thus senses the interest in studying the relation between the variance and the mean.

In Fig. 6.241.3I, it is apparent that the first quadrant of the (mean, variance) plane is divided by the bisector $\sigma^2 = \mu$ (Poisson) into two

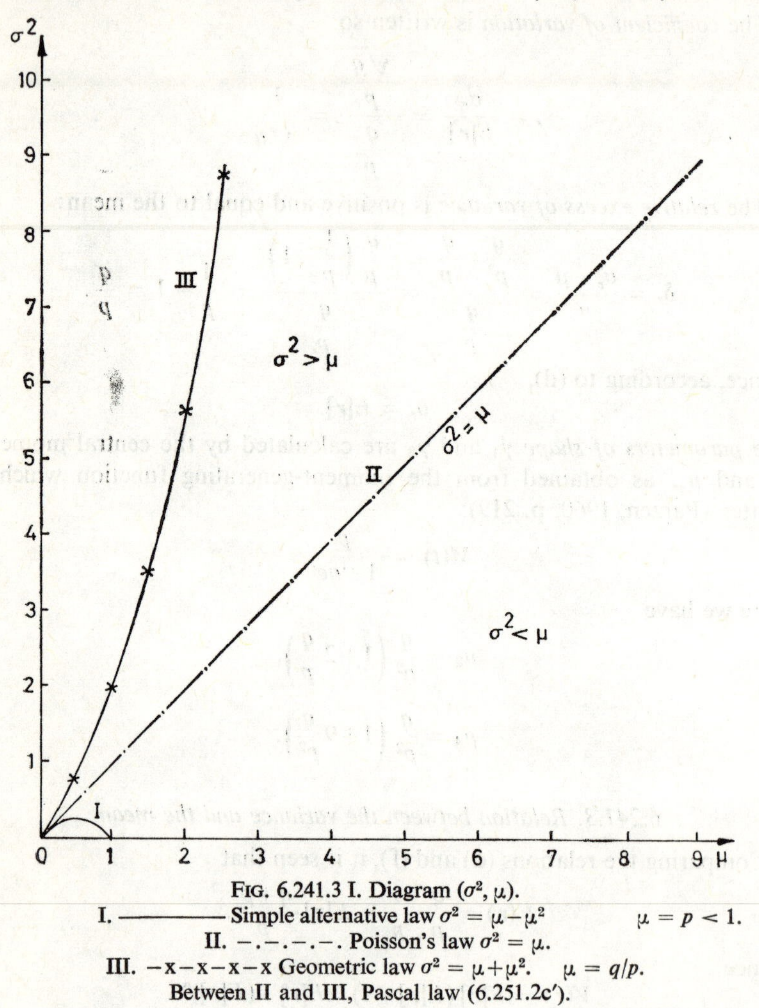

FIG. 6.241.3 I. Diagram (σ^2, μ).
I. ———— Simple alternative law $\sigma^2 = \mu-\mu^2$ $\mu = p < 1$.
II. –.–.–.–. Poisson's law $\sigma^2 = \mu$.
III. –x–x–x–x Geometric law $\sigma^2 = \mu+\mu^2$. $\mu = q/p$.
Between II and III, Pascal law (6.251.2c′).

regions, i.e. $\sigma^2 < \mu$ (law of simple alternative, positive binomial) and $\sigma^2 > \mu$ (geometrical, Pascal and negative binomial laws).

The points representative of other laws may equally be plotted in this diagram.

6.241.4. *Geometrical representation*

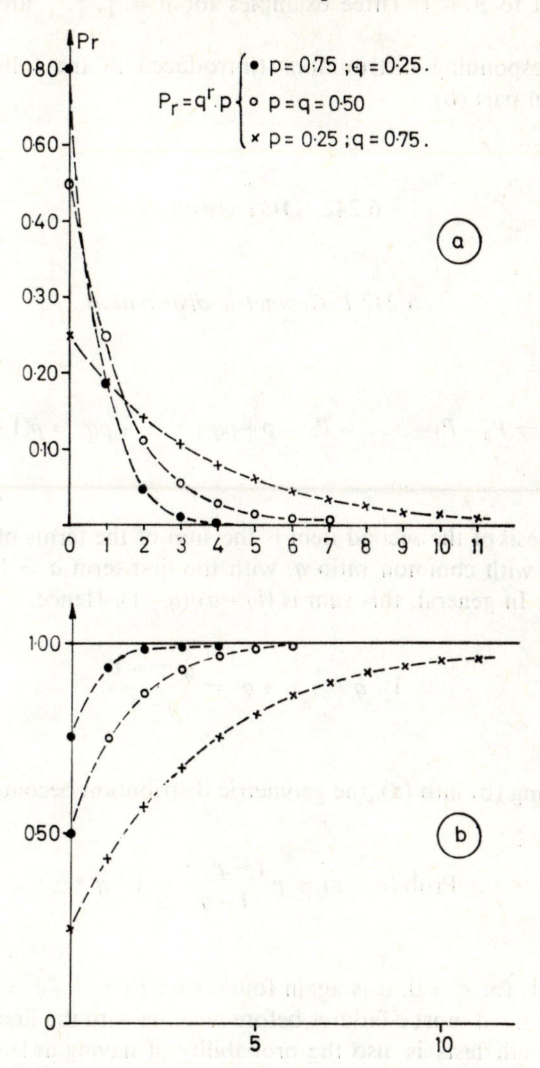

FIG. 6.241.4 I. Geometric point—density and distribution.

The point geometric law

$$P_r = q^r p \qquad (r = 0, 1, 2, \ldots)$$

is represented by a monotonically decreasing curve with ordinate at the origin equal to $p \leq 1$. Three examples for $p = \frac{3}{4}, \frac{1}{2}, \frac{1}{4}$ are given in Fig. 6.241.4I (a).

The corresponding distributions (introduced in the following section) are shown in part (b).

6.242. Distribution

6.242.1. Geometric distribution

Here

$$\text{Prob}\,(r \leq c) = P_0 + P_1 + \ldots + P_c = p + pq + \ldots + pq^c = p(1 + q + \ldots + q^c). \tag{a}$$

The parenthesis of the second item is the sum of the terms of a geometrical progression with common ratio q, with the first term $a = 1$ and with last term $l = q^c$. In general, this sum is $(lq-a)/(q-1)$. Hence,

$$1 + q + \ldots + q^c = \frac{q^{c+1}-1}{q-1}. \tag{b}$$

Substituting (b) into (a), the geometric distribution becomes

$$\text{Prob}\,(r \leq c) = p\,\frac{1-q^{c+1}}{1-q} = 1-q^{c+1}. \tag{c}$$

In particular, for $c = 0$, it is again found that $P_0 = 1-q = p$. This probability of having at most c failures before meeting with the first success out of $(c+1)$ Bernoulli tests is also the probability of having at least one success. The geometrical representation is given in Fig. 6.241.4I (b).

6.243. Another Form. Geometric Law in $n = 1+r$

If $n = r+1$ is the total number of tests to be performed before finally meeting with the first success, the following correspondence holds good:

Geometric law in $r = 0, 1, 2, \ldots$	Geometric law in $n = 1, 2, \ldots$	
$P_0 = p$	$P_1 = p$	
$P_1 = qp$	$P_2 = qp$	
$\cdot$	$\cdot$	
$\cdot$	$\cdot$	
$\cdot$	$\cdot$	
$E[r] = \dfrac{q}{p}$	$E[n] = 1+E[r]$	
	$= 1+\dfrac{q}{p} = \dfrac{1}{p}$	(a)
	Gumbel's return period (1960, p. 23)	
$V[r] = \dfrac{q}{p^2}$	$V[n] = \dfrac{q}{p^2}$.	(b)

6.244. Some Uses of the Geometric Law

1. In a sequence of Bernoulli trials, the geometric law indicates the probability of the "waiting time" (measured by failures) before the first success.
2. The geometric variable is the building block of the sequential test of a simple hypothesis and leads to the evaluation of the average sample number (A.S.N.) of items to get a decision (see, for example, Wald, 1959).

6.25. Pascal's Law

6.251. Pascal's Point-density Law

6.251.1. Definition

Pascal's law governs the probability of observing a random number $\boldsymbol{r}$ of failures before meeting with the kth success in a test of rank $(k+\boldsymbol{r})$. Since this eventuality of the kth success is fixed in rank with a probability p, only the preceding $(k+\boldsymbol{r}-1)$ tests are capable of providing $(k-1)$ successes. In this section the bold notation for the random variables $\boldsymbol{r}$ and $\boldsymbol{n}$ is particularly convenient, although it may have some drawbacks. The bold $\boldsymbol{r}$ and $\boldsymbol{n}$ are contrasted with the constant k which is not bold (see also section 6.254).

Consider one of the possible sequences:

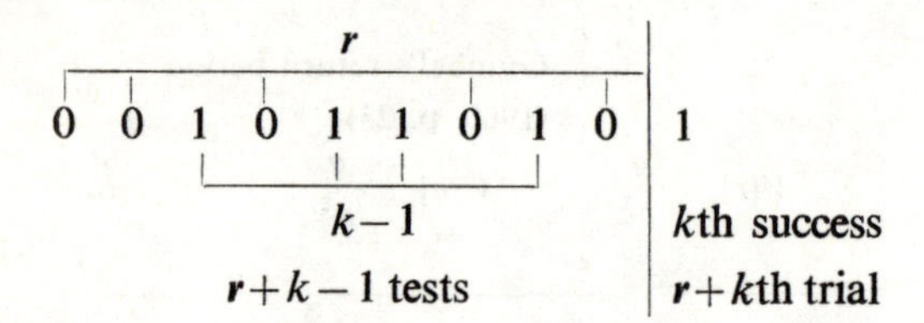

The probability of such a sequence is $q^{\boldsymbol{r}} \cdot p^{k-1} \cdot p$. But since it is a question of Bernoulli trials one may introduce the (random) total number of tests $\boldsymbol{n} = \boldsymbol{r}+k-1$ prior to the kth success, and then, *at this stage*, consider the value of $\boldsymbol{n} = \boldsymbol{r}+k-1$ then *fixed*. In the corresponding positive binomial the exponent of q is $\boldsymbol{r}$ and that of p is $(k-1) = \boldsymbol{n}-\boldsymbol{r}-1$, say $(\boldsymbol{r}+k-1-\boldsymbol{r})$. The probability of meeting with $\boldsymbol{r}$ failures *before* the kth success in $\boldsymbol{n}$ tests is thus

$$^{r+k-1}C_r q^r p^{k-1},$$

or, say,

$$^{r+k-1}C_r q^r p^{r+k-1-r}$$

upon introducing the exponent of p such that the sum of the exponents of both p and q is obviously $r+k-1$, the total number of tests performed in meeting with $(k-1)$ successes.

Finally, upon multiplying this probability by p (the probability of the kth success closing the series in question), the *Pascal probability of r failures before the kth success* is

$$P_r = {}^{r+k-1}C_r q^r p^{(r+k-1)-r} \cdot p, \tag{a}$$

or, more concisely,†

$$P_r = p^k \, {}^{r+k-1}C_r q^r .$$

The value

$$P_0 = p^k \tag{b}$$

is the probability of meeting with $r = 0$ failures before observing the kth success, e.g. an uninterrupted series of k successes in k Bernoulli tests:

$$\begin{aligned}
P_1 &= p^k \, {}^{1+k-1}C_r q = kp^k q, \\
P_2 &= p^k \, {}^{2+k-1}C_r q^2 = \frac{k(k+1)}{1\cdot 2} p^k q^2, \\
P_3 &= p^k \, {}^{3+k-1}C_r q^3 = \frac{k(k+1)(k+2)}{1\cdot 2\cdot 3} p^k q^3, \\
&\vdots \\
P_r &= p^k \, {}^{r+k-1}C_r q^r = \frac{k(k+1)(k+2)\ldots(k+r-1)}{1\cdot 2\cdot 3\cdot \ldots \cdot r} p^k q^r . \\
&\vdots
\end{aligned} \tag{b'}$$

Hence the recurrence law

$$P_r = \frac{k+(r-1)}{r} q P_{r-1} . \tag{c}$$

For $k = 1$, one again finds the rule for the geometric law (6.241.1 g).

We outline the combinations in the case of $k = 3$, where r is the number failed out of $(r+3)$ tests (see Table 6.251.1A).

† The reader will mentally dissociate the exponent k of p and the upper index of C.

TABLE 6.251.1A

	Before the 3rd success	3rd success	
$r=0$	1 1	1	$P_0 = p^3$
$r=1$	0 1 1	1	$P_1 = 3p^3q$
	1 0 1	1	
	1 1 0	1	
$r=2$	0 0 1 1	1	$P_2 = \frac{3\cdot 4}{1.2}p^3q^2$
	1 0 1 0	1	$= 6p^3q^2$
	1 0 0 1	1	
	1 1 0 0	1	
	0 1 0 1	1	
	etc.	etc.	

The successive terms of the Pascal law in (b′), written here more concisely as

$$p^k\left\{1, kq, \frac{k(k+1)}{1\cdot 2}q^2, \ldots\right\}, \tag{d}$$

where k is an integer, are generated by the expansion of the binomial with negative exponent, i.e.

$$p^k(1-q)^{-k}, \tag{e}$$

using the notation

$$\binom{-k}{r} = (-1)^r \frac{k(k+1)\ldots(k+r-1)}{1\cdot 2\cdot \ldots \cdot r} = {}^{-k}C_r. \tag{f}$$

In this sense Pascal's law is a special case of the negative binomial in which k has a positive value (see 6.261 e′):

$$\begin{aligned} P_r &= \binom{-k}{r} p^k(-q)^r \\ &= (-1)^r \binom{-k}{r} p^k q^r. \end{aligned} \tag{g}$$

6.251.2. Reduction parameters

From the cumulant-generating function, the first four cumulants are (Kendall and Stuart, 1963, I, p. 131):

Mean or expected value

$$E[r] = k\frac{q}{p} = \varkappa_1'. \tag{a}$$

Variance

$$V(r) = k\frac{q}{p^2} = \varkappa_2$$
$$= k\frac{q}{p}\cdot\frac{1}{p} = E[r]\left(\frac{1}{p}-1+1\right). \tag{b}$$

Now, since $\frac{1}{p}-1 = \frac{q}{p} = \frac{E[r]}{k}$, therefore

$$V(r) = E[r]\left(1+\frac{E[r]}{k}\right), \tag{c}$$

or, using the notation (μ, σ^2),

$$\sigma^2_{\text{Pascal}} = \mu+\frac{\mu^2}{k} \quad (k \text{ integer}). \tag{c'}$$

The relative excess of variance is therefore

$$\delta_r = \frac{V(r)-E[r]}{E[r]} = \frac{E[r]}{k} = \frac{q}{p}. \tag{d}$$

For Pascal's law δ_r is positive. Moreover, it is the same as for the geometric law:

$$\delta_{r\,\text{Pascal}} = \delta_{r\,\text{geom.}} \tag{d'}$$

This interesting property of independence of δ_r *vis-à-vis* k in Pascal's law bears a resemblance to the independence of δ_r *vis-à-vis* n in the positive binomial (6.221.2 f″).

From the expression

$$\varkappa_3 = \frac{kq(1+q)}{p^3}, \tag{e}$$

the *skewness* is calculated as

$$\gamma_1 = \frac{\varkappa_3}{\sigma^3} = \frac{kq(1+q)}{p^3} \cdot \frac{p^3}{kq\sqrt{kq}} = \frac{1}{\sqrt{k}} \cdot \frac{1+q}{\sqrt{q}}. \qquad \text{(f)}$$

The coefficient of skewness varies inversely as $\sqrt{k}$ and tends to zero as $\frac{1}{\sqrt{k}}$ if $k \to \infty$.

The *kurtosis* can be found from

$$\varkappa_4 = \frac{kq(1+6q^2)}{p^4}. \qquad \text{(g)}$$

6.251.3. *The plane* (μ, σ^2)

The representative point of a Pascal law (μ, σ^2) is situated in the region $\sigma^2 > \mu$ on the parabola in k and of equation

$$\sigma^2 = \mu + \frac{\mu^2}{k} \qquad (k > 1),$$

i.e. under the limit parabola $\sigma^2 = \mu + \mu^2$ corresponding to the geometric law (see Fig. 6.241.3I).

In the plane (μ, σ^2) the representative point of a Pascal law may be designated by the couple (μ, σ^2); but the couples (μ, k) or (σ^2, k) may also be convenient.

6.251.4. *Pascal variable regarded as the sum of k geometric variables*

We compare the respective formulae of the geometric and Pascal laws in Table 6.251.4A.

TABLE 6.251.4A

	Geometric law	Pascal's law
P_r	$P_r = q^r p$	$P_r = {}^{r+k-1}C_r q^r p^k$
$E[r]$	$\frac{q}{p}$	$k\frac{q}{p}$
$V(r)$	$\frac{q}{p^2}$	$k\frac{q}{p^2}$
$M(t)$	$\frac{p}{1-qe^t}$	$\left(\frac{p}{1-qe^t}\right)^k = (q'-p'e^t)^{-k}$

It may be seen that in a series of Bernoulli tests the Pascal variable which refers to the number of failures preceding the kth success, is the sum of k geometric variables referring to the number of failures preceding the first success. The same conclusion is obtained upon comparing the generating function $M(t)$ for the two laws; in this regard see the fourth line of the table (Parzen, 1960, pp. 218–19). In fact we have

$$M(t)_{\text{Pascal}} = [M(t)_{\text{geom.}}]^k. \tag{a}$$

The form of the Pascal law used in the table has been called the "binomial waiting time distribution" (Patil and Joshi, 1965, p. 217).

6.252. Pascal Distribution and Tables

In the tables for the negative binomial distribution in r, Grimm (1962, p. 241, formula 9) presents a formula, holding only for integer k, which is a special case of some $k > 0$ of the negative binomial, i.e. the Pascalian case. In Grimm's tables the entry is made using μ_1' and q' (which Grimm denotes as q).

In order to detect Pascalian cases, one has to look for integer k entries. Since $\mu_1' = kp' = k(q'-1)$, therefore the Pascalian entries correspond to the condition $k = \dfrac{\mu_1'}{q'-1}$ integer $\geqslant 1$.

It will also be seen that the Pascal distributions can be calculated from the positive binomial distributions (6.254).

6.253. Another Form of Pascal's Law

Pascal's law in $\boldsymbol{n} = k+\boldsymbol{r}$. In section 6.251.1 it was seen that

$$P_r = {}^{r+k-1}C_r q^r p^k \tag{a}$$

is the probability of having $\boldsymbol{r}$ failures before meeting with the kth success. We introduce the random variable

$$\boldsymbol{n} = k+\boldsymbol{r} \tag{b}$$

which realizes a *translation* of k units applied to r and which represents the *total number of tests carried out to meet with the kth success*. Expression (a) is then written in the new random variable $\boldsymbol{n}$:

$$P_n = {}^{n-1}C_r q^{n-k} p^k. \tag{c}$$

To eliminate r from within the sign C, we apply a result obtained from combinatorial analysis, i.e.

$$^{n-1}C_r = {}^{r+k-1}C_r = {}^{r+k-1}C_{r+k-1-r} = {}^{n-1}C_{k-1}. \qquad \text{(d)}$$

Hence the form sometimes called after Polya (Patil and Joshi, 1965, p. 297):

$$P_n = {}^{n-1}C_{k-1}\, q^{n-k} p^k, \qquad \text{(e)}$$

where $n = k, k+1, \ldots$.

The reduction parameters of the form in r (section 6.251.2) enable us to put

$$E[\boldsymbol{n}] = k+E[\boldsymbol{r}] = k+k\frac{q}{p} = k\left(1+\frac{q}{p}\right) = \frac{k}{p}, \qquad \text{(f)}$$

$$V(\boldsymbol{n}) = V(\boldsymbol{r}) = k\frac{q}{p^2}. \qquad \text{(g)}$$

For $k = 1$, formulae (f) and (g) become those of section 6.243 (right side column).

Relation (f) shows that the mean value of the total number of Bernoulli trials to be performed in order to meet with exactly the kth success is equal to k multiplied by the mean return period $1/p$ (Gumbel, 1960, p. 23). In particular, for $p = \frac{1}{2}$, we have

$$E[\boldsymbol{n}] = 2k. \qquad \text{(h)}$$

In a series of Bernoulli tests of probability $p = \frac{1}{2}$, the mean number of tests to be performed to meet finally with the kth success is $2k$. This result is expected intuitively.

6.254. Pascal Distribution in $\boldsymbol{n} = k+r$

The Pascal distribution in $\boldsymbol{n} = k+\boldsymbol{r}$ yields the probability of observing a number of tests $\boldsymbol{n} \leqslant n$ in order to obtain k successes, e.g. a sum of $n+1$ terms:

$$\text{Prob}\,(\boldsymbol{n} \leqslant n|k) = P_0+P_1+\ldots+\ldots+P_n \quad (n = k, k+1, \ldots), \qquad \text{(a)}$$

where the vertical line followed by k signifies "on condition k fixed", or "given k".

Such a cumulative Pascal probability can be obtained from *positive binomial distribution tables* (Schlaifer, 1959, p. 188; Patil, 1960) by reasoning thus:

(a) If one meets with *at least k successes* in the first n Bernoulli trials (binomial point of view), then one should perform *at most n trials* before meeting the first k successes (Pascalian point of view):

$$\text{Prob}_{\text{bin.}>0}\,(\boldsymbol{k} \geqslant k|n) = \text{Prob}_{\text{Pasc.}}\,(\boldsymbol{n} \leqslant n|k). \tag{b}$$

(b) If one meets with less than k successes in the first n trials (binomial viewpoint), more than n tests have to be performed to obtain the desired k successes:

$$\text{Prob}_{\text{bin.}>0}\,(\boldsymbol{k} < k|n) = \text{Prob}_{\text{Pasc.}}\,(\boldsymbol{n} > n|k). \tag{c}$$

A simple example will now be given. Consider the positive binomial $(\frac{1}{2}+\frac{1}{2})^5$ for which the successive terms are

$$\boldsymbol{r} = 0 \quad 1 \quad 2 \quad 3 \quad \textcircled{4} \quad 5$$

$$P_r = \frac{1}{32} \quad \frac{5}{32} \quad \frac{10}{32} \quad \frac{10}{32} \quad \frac{5}{32} \quad \frac{1}{32}.$$

We consider $k = 4$

$$\text{Prob}_{\text{bin.}>0}\,(\boldsymbol{k} \geqslant 4|5) = +\frac{5}{32}+\frac{1}{32} = \frac{6}{32}. \tag{d}$$

We calculate the successive terms of the negative binomial in $n = 4+\boldsymbol{r}$ for $\boldsymbol{n} \leqslant 5$, i.e. P'_4 and P'_5:

$$n \quad P'_n$$

$$4 \quad P'_4 = p^4 = \frac{1}{2^4} = \frac{1}{16} = \frac{2}{32}$$

$$5 \quad P'_5 = 4p^4q = \frac{4}{2^5} = \frac{4}{32}.$$

Therefore

$$\text{Prob}_{\text{Pasc.}}\,(\boldsymbol{n} \leqslant n|4) = P'_4+P'_5 = \frac{2}{32}+\frac{4}{32} = \frac{6}{32}, \tag{e}$$

and one again finds the value (d).

6.255. Application of Pascal's Law

Pascal's law is applicable to inverse sampling which treats Bernoulli counting events in the case where it is desired to count effectively a predetermined number k of *independent* elements observable with a *constant probability p*. These conditions are essential for the applicability of the law.

In as much as the technique of smear preparations is correct, and random observations are carried out, the counting of eosinophiles or of reticulocytes is one application of Pascal's law.

6.255.1. *Inverse sampling* (Tweedie, 1945; Haldane, 1945)

In a random number of tests it is required to find a pre-specified number of elements k defined according to the presence or absence of a particular characteristic. The number r of failures prior to finding the k desired elements is governed by a negative binomial law and, more especially, by Pascal's law (integer k).

Inverse sampling with ball return (or non-removal) is closely associated with the negative binomial law in cases of microscopical observations, etc. More generally, a factorial classification of the methods of sampling can be conceived (2×2), the first factor being the replacement of the unit drawn (Yes–No) (non-replacement is bound up with laws of the hypergeometric type 6.291), and the other factor being the direct or inverse characteristic.

Here from another quarter we again find the principle of formal classification of discrete distributions distinguishing, on the one hand, (a) *occurrence*, where one identifies and counts r elements out of a pre-specified total number n (in this case the random variable r is governed by the positive binomial law which applies thus to direct sampling) and, on the other hand, (b) waiting attitude where, noting the failures encountered, random sampling is continued while replacing the element drawn until the pre-specified number k of desired elements should be reached.

The random variable r follows the negative binomial in the Pascal form, which thus applies to inverse sampling.

When these two criteria are considered simultaneously, the following Table is obtained (Sibuya, Yoshimura and Shimizu, 1964, p. 421).

The technique and the statistical aspects of sampling are described in other sections. The only purpose of this table is to provide some factorial

TABLE 6.255.1A. PROBABILITY LAW ACCORDING TO TYPE OF SAMPLING

		Direct sampling	Inverse sampling
Replacement of the unit drawn	Yes	positive binomial	negative binomial
	No	positive hypergeometric or classical	negative hypergeometric

framework relating four important laws of discrete probability to four well-defined types of sampling. Quenouille (1958, pp. 18–19) has treated the two cases with replacement.

Kendall and Stuart (1963, I, p. 225) raise some objections to the term "inverse sampling" and use "sequential sampling (for attributes)".

6.255.2. *Counting eosinophiles on smears*

On a smear on a microscope slide, in order to enumerate the eosinophiles among the leucocytes, one performs after Tweedie (1945) *inverse binomial sampling*.

Fixing on a number of eosinophiles to be effectively observed, for instance, $k = 100$, one continues to count the leucocytes until the kth eosinophile should be observed. One thus proceeds in the expectation of the kth or 100th eosinophile, meanwhile the examination of the leucocytes continues and will only be stopped when the kth eosinophile has been observed (Haldane's labor-saving method of sampling, 1945). The number $\boldsymbol{r}$ of leucocytes examined before finding this kth eosinophile is a random variable which obeys the *negative binomial law* which applies in inverse sampling. The total number of examined leucocytes is $\boldsymbol{n} = k+\boldsymbol{r}$. Whilst the fraction $k/\boldsymbol{n}$ provides a biased estimate of the true proportion of eosinophiles, it can be shown that

$$\hat{p} = \frac{k-1}{\boldsymbol{n}-1} \tag{a}$$

gives a non-biased estimate. Here the numerator $k-1$ is fixed, while the denominator $\boldsymbol{n}-1$ is random.

Haldane (1945) gives an approximate formula for the standard error $s_{\hat{p}}$ of $\hat{p}$:

$$s_{\hat{p}} \approx \frac{1}{\boldsymbol{n}}\sqrt{\frac{k(\boldsymbol{n}-k)}{\boldsymbol{n}-1}} \tag{b}$$

$$\approx \hat{p}\sqrt{\frac{1-\hat{p}}{k-2}}. \tag{c}$$

When $\hat{p}$ is small, $s_{\hat{p}}$ is practically proportional to $\hat{p}$, and therefore the coefficient of variation of $\hat{p}$ in repeated samples of the same preparation is practically constant.

The exact formula is given by Finney (1949):

$$s^2_{\hat{p}} = \frac{\hat{p}(1-\hat{p})}{\boldsymbol{n}-2}, \tag{d}$$

where $\boldsymbol{n}$ is random.

We transform (d) by introducing *a priori* fixed k by the intermediary of $\hat{p}$:

$$\hat{p} = \frac{k-1}{n-1}.$$

Therefore $(n-1) = (k-1)/\hat{p}$, and

$$n-2 = \frac{k-1}{\hat{p}} - 1 = \frac{k-1-\hat{p}}{\hat{p}}.$$

We pass to the reciprocal:

$$\frac{1}{n-2} = \frac{\hat{p}}{k-1-\hat{p}}.$$

Therefore

$$s_{\hat{p}}^2 = \frac{1}{n-2} \cdot \hat{p}(1-\hat{p}) = \frac{\hat{p}}{k-1-\hat{p}} \cdot \hat{p}(1-\hat{p}),$$

or

$$s_{\hat{p}}^2 = \frac{\hat{p}^2(1-\hat{p})}{k-1-\hat{p}}. \tag{e}$$

Whence

$$s_{\hat{p}} = \hat{p}\sqrt{\frac{1-\hat{p}}{k-1-\hat{p}}}. \tag{f}$$

Upon comparing (f) with (c), it is seen that the denominator of Haldane's formula is consistently smaller than that of the exact formula and so it overestimates the variance of $\hat{p}$. These formulae also hold for the enumeration of reticulocytes (Dacie, 1956, p. 24).

6.26. Negative Binomial Law

6.261. Negative Binomial in r, Extension of Pascal's Law

This law has been the subject of an excellent monograph by Bartko (1961). The negative binomial in r with parameters k and p is an extension of Pascal's law for non-integer $k > 0$.

Again taking Pascal's law of 6.251.1 (a) written as

$$P_r = {}^{n+k-1}C_r p^k q^r = \binom{r+k-1}{r} p^k q^r = \frac{(r+k-1)!}{r!\,(k-1)!} p^k q^r, \tag{a}$$

if k is no longer an integer, then the factorials such as $(r+k-1)!$ and $(k-1)!$ are replaced by $\Gamma(r+k)$ and $\Gamma(k)$:

$$\binom{r+k-1}{r} = \binom{r+k-1}{k-1} = \frac{\Gamma(r+k)}{r!\,\Gamma(k)}.$$

The $r!$ is here since $r = 0, 1, 2, \ldots$ is always a positive integer. Hence the negative binomial form of Patil and Joshi (1965, p. 294) in a recent and important contribution by them:

$$P_r = \frac{\Gamma(r+k)}{r!\,\Gamma(k)}\, p^k q^r. \tag{b}$$

The terms of the positive binomial $P_{r|n} = \binom{n}{r} p^r q^{n-r}$ are generated by the expansion of the binomial sum $(q+p)$ raised to the positive exponent n, say $(q+p)^n$ where $q+p = 1$; where q and p are actual elementary constant probabilities of failure and success.

It will now be shown that the terms of the negative binomial

$$P_{r|k} = \binom{r+k-1}{r} p^k q^r \tag{b'}$$

are generated by the expansion of the binomial difference $(q'-p')$ raised to a negative power $-k$, i.e. $(q'-p')^{-k}$ with $q'-p' = 1$. The terms q' and p' are therefore not probabilities, but indeed simple functions of the probabilities q and p, shown to be equal to:

$$q' = \frac{1}{p}, \quad p' = \frac{q}{p}. \tag{c}$$

TABLE 6.261A

$(q+p)^n$ $q+p = 1$ $r = 0, 1, 2, \ldots$ n fixed number of possible successes has a *fixed upper bound* n. (p^n corresponds to $r = n$)	$(q'-p')^{-k}$ $q'-p' = 1$ $r = 0, 1, 2, \ldots$ k fixed $n = k+r = k, k+1, k+2, \ldots$ the total number n of tests for k successes has a *fixed lower bound* k (p^k corresponds to $r = 0$)

We place in parallel the formulae for the mean and variance in respect of the two types of binomial:

$$E[r] = np \qquad E[r] = k\frac{q}{p} = (-k)(-p') = kp'$$

$$= (-k)\left(-\frac{q}{p}\right) \tag{d}$$

$$V(r) = npq \qquad V(r) = k\frac{q}{p}\frac{1}{p} = (-k)(-p')q'$$

$$= kp'q'.$$

In these four lines one notes the following homologs, some of which have been indicated by Greenwood and Yule (1920, p. 279, note):†

q	homologous to	q'	
p	homologous to	$-p'$	
n	homologous to	$-k$	(d′)
np	homologous to	$(-k)(-p')$	
npq	homologous to	$(-k)(-p')q'$	

These homologs continue partly in the expansion of the binomials, for the positive binomial has $(n+1)$ terms, whilst the negative binomial has an unlimited number of terms:

$(q+p)^n$	$(q'-p')^{-k}$
$= q^n + \binom{n}{1}q^{n-1}p + \dots$	$= q'^{-k} + \binom{-k}{1}q'^{-k-1}(-p') + \dots$
$\binom{n}{r}q^{n-r}p^r + \dots$	$\binom{-k}{r}q'^{-k-r}(-p')^r + \dots$
p^n	$\dots.$

We show that the successive terms of Pascal's law are again found (see 6.251.1 b and b′). We have successively

$$q'^{-k} = \left(\frac{1}{p}\right)^{-k} = p^k$$

$$\binom{-k}{1}q'^{-k-1}(-p') = \frac{-k}{1}\left(\frac{1}{p}\right)^{-k-1} \cdot \left(\frac{-q}{p}\right)$$

$$= kp^k \cdot \frac{p}{p} \cdot q = kp^k q \tag{e}$$

† Homolog is not equal (see formula (c), p. 257).

$$\binom{-k}{r}q'^{-k-r}(-p')^r = \binom{-k}{r}\left(\frac{1}{p}\right)^{-k-r}\left(\frac{-q}{p}\right)^r$$

$$= \binom{-k}{r}\left(\frac{1}{p}\right)^{-k}\left(\frac{1}{p}\right)^{-r}(-q)^r\left(\frac{1}{p}\right)^r$$

$$= (-1)^r\binom{-k}{r}p^kq^r. \qquad \text{(e')}$$

Now since, for integer k,

$$(-1)^r\binom{-k}{r} = (-1)^{2r}\frac{k(k+1)\dots(k+r-1)}{1\cdot 2\dots(r-1)} \qquad \text{(f)}$$

and

$$(-1)^{2r} = 1,$$

we again find P_r as given in (b′).

The negative binomial has a proteiform aspect. Thus, in order to estimate parameters by the method of maximum likelihood, Katti and Shastri (1965, p. 36) have written it so:

$$P_r = \frac{k(k+1)\dots(k+r-1)}{r!}p'^r(1+p')^{-k-r}, \qquad \text{(g)}$$

where, clearly, $q' = 1+p'$ (using k' for the k used above).

From (e) the form (g) is manipulated as follows, since $q' = 1+p'$,

$$p'^rq'^{-k-r} = \left(\frac{q}{p}\right)^r\left(\frac{1}{p}\right)^{-k-r} = q^rp^k,$$

which again yields the form known and given in (b′) of section 6.251.1.

6.262. Different Types of the Negative Binomial Law

The multiple aspects of the negative binomial enable the following main types to be distinguished (Martin, 1966). The relevant parameters are shown in "boxes" of the left, whilst P_0, P_r, the reduction parameters and, if useful, the recurrence relation, are listed on the right-hand side.

q, p, k

$$P_0 = p^k$$

$q+p = 1$

$$P_r = \binom{k+r-1}{r} p^k q^r \qquad \text{(a)}$$

$$E[r] = k\frac{q}{p}, \quad V(r) = k\cdot\frac{q}{p}\cdot\frac{1}{p} = k\cdot\frac{q}{p^2}$$

$$P_r = \frac{k+(r-1)}{r} qP_{r-1}.$$

$\boxed{q', p', k}$

$$P_0 = \frac{1}{(1+p')^k}$$

$$P_r = \binom{k+r-1}{r}\frac{p'^r}{(1+p')^{r+k}} \qquad \text{(b)}$$

$q'-p' = 1$

$(q'-p')^{-k}$

Fisher (1953, p. 197), Martin and Shastri (1965, p. 36) (k' is denoted k)

$$\begin{cases} p' = \dfrac{q}{p} \\ q' = 1+\dfrac{q}{p} = \dfrac{1}{p} \end{cases}$$

$$E[r] = (-k)(-p') = kp'$$

$$V(r) = (-k)(-p')q' = kp'q' = kp'(1+p')$$

$$\begin{cases} p = \dfrac{1}{q'} \\ q = pp' = \dfrac{p'}{q'} \end{cases}$$

$$\delta_r = \frac{V(r)-E[r]}{E[r]} = \frac{kp'+kp'^2-kp'}{kp'} = p'$$

(δ_r is denoted δ).

$\boxed{\mu, k}$

From (a)

$$P_0 = \left(\frac{k}{\mu+k}\right)^k = \left(\frac{\mu+k}{k}\right)^{-k} = \left(1+\frac{\mu}{k}\right)^{-k}$$

$$P_r = \binom{k+r-1}{r}\left(1+\frac{\mu}{k}\right)^{-k}\left(\frac{\mu}{\mu+k}\right)^r$$

Anscombe (1950, p. 358) (c)

$$\frac{\mu}{k} = \frac{q}{p}$$

$$\mu p = k(1-p) = k-kp$$

$$\mu p + kp = k$$

$$p = \frac{k}{\mu+k}$$

$$q = \frac{\mu}{\mu+k}$$

In general

$$\frac{(k+r-1)!}{r!\,(k-1)!} = \binom{k+r-1}{r} = \frac{\Gamma(k+r)}{r!\,\Gamma(k)}$$

integer $k > 0$ $\qquad\qquad$ $k > 0.$

δ, k

$\delta = p' \quad 1+\delta = q'$ $\qquad$ $P_0 = \dfrac{1}{(1+\delta)^k} = (1+\delta)^{-k}$

$[(1+\delta)-\delta]^{-k}$ $\qquad$ $P_r = \dbinom{k+r-1}{r} \dfrac{1}{(1+\delta)^k} \left(\dfrac{\delta}{1+\delta}\right)^r$ $\qquad$ (d)

$1+\delta = \dfrac{1}{p} \quad p = \dfrac{1}{1+\delta}$

$q = \dfrac{\delta}{1+\delta}$

μ, δ

$k = \dfrac{\mu}{p'} = \dfrac{\mu}{\delta}$ $\qquad$ $P_0 = \left(\dfrac{1}{1+\delta}\right)^{\mu/\delta}$

$q = \dfrac{\delta}{1+\delta}$ $\qquad$ limit form of Eggenberger–Polya (1923) Bertrand and Th. Gayet-Haillon (1952), Martin (1961b).

$$P_r = \binom{k+r-1}{r} \left(\frac{1}{1+\delta}\right)^{\mu} \left(\frac{\delta}{1+\delta}\right)^r$$

$$P_r = \frac{\mu+\delta(r-1)}{r} \cdot \frac{1}{1+\delta} P_{r-1} \qquad \text{(e)}$$

$$\lim_{\delta \to 0} P_0(\mu, \delta) = e^{-\mu}.$$

R or X, k $\qquad$ $P_0 = (1-X)^k$

$X = \dfrac{p'}{q'} = \dfrac{p'}{1+p'}\ ; \quad \dfrac{\mu}{\mu+k} = q$ $\qquad$ $P_r = \dbinom{k+r-1}{r} (1-X)^k X^r$

$p = 1-X$ Anscombe (1950)

$$R = \frac{p'}{q'} = \frac{\mu}{\mu+k} \qquad P_0 = \frac{1}{q'^k} \qquad \text{(f)}$$

$$p = \frac{1}{q'}$$

The p and q of Anscombe and Bliss are our p' and q'

$$P_r = \binom{k+r-1}{r}\frac{R^r}{q'^k}$$

Bliss (1953, p. 177)

μ, μ_2

McKendrick (1926, p. 100) and revived by J. O. Irwin (1960).

$$\left[\frac{\mu_2}{\mu}-\left(\frac{\mu_2}{\mu}-1\right)\right]^{-b/c} \qquad \text{(g)}$$

McKendrick introduced a function of progression from the state r to the state $r+1$ on averaging constants b and c which are unrelated to those of section 6.281. This function of progression is:

$$f(r) = b+cr \qquad \text{(h)}$$

$$P_r = \frac{b}{c}\left(\frac{b}{c}+1\right)\ldots\left(\frac{b}{c}+r-1\right)\frac{\left(1-\frac{\mu}{\mu_2}\right)^r}{r!}\left(\frac{\mu}{\mu_2}\right)^{b/c}$$

$$= \binom{\frac{b}{c}+r-1}{r}\left(1-\frac{\mu}{\mu_2}\right)^r\left(\frac{\mu}{\mu_2}\right)^{b/c}$$

$$\frac{b}{c} = \frac{\mu^2}{\mu_2-\mu} = k$$

$$q' = \frac{\mu_2}{\mu} = 1+\delta$$

$$p' = \frac{\mu_2}{\mu}-1 = \delta.$$

6.263. Probability-generating Function (p.g.f.) of the Negative Binomial Law (Bartlett, 1960)

The generating function of the terms P_r of the binomial law is written as follows:

$$G(z) = (q'-p'z)^{-k} \tag{a}$$
$$= (1+p'-p'z)^{-k} \tag{a'}$$
$$= (1+p')^{-k}\left(1-\frac{p'}{1+p'}z\right)^{-k}. \tag{a''}$$

But since

$$\left(1-\frac{p'}{1+p'}z\right)^{-k} = \sum_{n=0}^{\infty}\frac{\Gamma(n+k)}{n!\,\Gamma(k)}\,\frac{p'^n}{(1+p')^n}z^n, \tag{b}$$

upon substituting (b) into (a''), we get

$$G(z) = \sum_{n=0}^{\infty}\frac{\Gamma(n+k)}{n!\,\Gamma(k)}\,\frac{p'^n}{(1+p')^{n+k}}z^n.$$

6.264. Estimation of the Parameters of the Negative Binomial

Adjustment. Developing an idea of Sir R. A. Fisher (1941), Anscombe (1949, 1950) studied the estimation of the parameters μ and k of the negative binomial. This method has been further developed by Bliss (1953) and Fisher (1953).

In short the following operations are carried out:

(α) An efficient estimate of μ is given by $\bar{r}$.

In order to estimate k, three procedures have been proposed.

(β_1) From $s_r^2 = (\sum(r-\bar{r})^2 f_r)/(\sum f_r - 1)$, the estimate of k is given by

$$\hat{k} = \frac{\bar{r}^2}{s_r^2-\bar{r}}\,. \tag{a}$$

We know in effect that

$$p' = \frac{\bar{r}}{k} = d,$$

where $d = (s^2-\bar{r})/\bar{r}$ which is the estimate of δ (relative excess of variance).

Therefore

$$\hat{k} = \frac{\bar{r}}{d} = \frac{\bar{r}^2}{s_r^2 - \bar{r}}.$$

In general, this method is to be used if $k > 1$.

(β_2) The relative frequency f_0/N of the observed zeros is homologable to the probability

$$P_0 = p^k.$$

Now

$$p = \frac{1}{q'} = \frac{1}{1+p'} = \frac{1}{1+\bar{r}/k}.$$

If, therefore, one writes

$$\frac{f_0}{N} = \left(\frac{1}{1+\bar{r}/k}\right)^k, \tag{b}$$

a transcendental relation in k is obtained. We take the log of the two numbers, changing the sign,

$$\log \frac{N}{f_0} = k \log \left(1+\frac{\bar{r}}{k}\right), \tag{b'}$$

obtaining an equation in k which can be solved by successive approximations (Bliss, 1953, p. 179). According to Anscombe (1949, p. 167), this technique is recommended for $k < 1$.

(γ) *Anscombe* also commends a transformation of the numerations r into a new variable y which depends on k, but not on $\bar{r}$. It can be considered whenever n is small, say

$$y = \log_{10} (r+\tfrac{1}{2}k), \tag{c}$$

or

$$y = \sinh^{-1} \sqrt{\frac{r+c}{k-2c}} \qquad (c = \text{const.}) \tag{d}$$

This is more complicated than the first two methods. Reference should be made to Anscombe (1949 and 1950) for the conditions of application.

(δ) In order to obtain a fully efficient adjustment from the values of first approximation given by the relations (a) or (b), Sir R. A. Fisher (1953) developed a method of giving for k a *solution of maximum likelihood*, discussed in detail and applied by Bliss (1953, p. 180). The formulae for the variance of $\bar{r}$ and of $\hat{k}$ in the method used are given by the cited writers.

(ε) In view of the heavy volume of calculations involved in these calculations, Martin and Katti (1965) have devised programs for computer approximation of the Poisson law, the negative binomial, as well as other distributions of "contagious" type. These programs have been applied to 35 examples presented in the published literature (Beall, 1940, and Bliss, 1953). Erna Weber (1961) has produced an excellent treatment of examples concerning the approximation of the negative binomial by the procedures α, β_1, β_2 and δ.

6.265. Negative Binomial Law Considered as a Compound Poisson Distribution

In a study of recurrent accidents and multiple episodes of diseases, Greenwood and Yule (1920) considered phenomena in two stages:

(1) A Poisson distribution of mean μ of the number r of accidents happening to a person:

$$p(r \mid \mu) = \frac{e^{-\mu}\mu^r}{r!} \qquad (r = 0, 1, 2, \ldots). \tag{a}$$

Here μ is a clinical, individual characteristic which is fixed on for an individual person by setting the mean annual level of accidents of this person. That is the reason why we use the notation $(r \mid \mu)$ for this probability of r accidents on the condition μ.

(2) This mean level μ is not constant in the population, but is assumed to vary from person to person according to a continuous probability density function $f(\mu)$.

After trying a Gaussian law, Greenwood and Yule (p. 274) retained a gamma law for his parameter r and constant c. But since we use these notations in a different sense, we transform r to k and c to γ, where k is the classical exponent of the negative binomial and $\gamma = 1/\delta$ will be shown to be the reciprocal of the relative excess of variance also called the index of contagion.

The gamma law which governs the variability of μ is written so (6.334.2)

$$f(\mu; k, \gamma) = \frac{\gamma^k}{(k-1)!} e^{-\gamma\mu}\mu^{k-1} \qquad (\mu \geqslant 0). \tag{b}$$

The denominator is written as $(k-1)!$ or $\Gamma(k)$ according to whether k is integer or not. For this gamma-k law (see 6.334.21) we have

$$E[\mu] = k/\gamma \qquad V(\mu) = k/\gamma^2. \tag{c}$$

Hence the *probability P_r of observing r accidents per annum in the population* is written by forming an explicit weighted mean for all the values of μ in the range $(0, \infty)$:

$$P\left(r \,\middle|\, \begin{matrix}\text{all } \mu \\ \text{in } 0, \infty\end{matrix}\right) = \int_0^\infty p(r, \mu) \cdot f(\mu; k, \gamma)\, d\mu. \tag{d}$$

It is seen that each Poisson probability is weighted by the infinitesimal weight $f\,d\mu$. After integration, μ disappears and then P_r depends only on the parameters k and γ and on the random variable $\boldsymbol{r}$.

Finally, the probability resulting from the composition of the Poisson variable $\boldsymbol{r}$ with parameter μ, with the gamma law for μ and having parameters k and γ, is

$$P\left(r \,\middle|\, \begin{matrix}\text{all } \mu \\ \text{in } 0, \infty\end{matrix}\right) = \binom{k+r-1}{r}\left(\frac{\gamma}{1+\gamma}\right)^k\left(\frac{1}{1+\gamma}\right)^r. \tag{e}$$

If in this form it is proposed to put

$$p = \frac{\gamma}{1+\gamma} \quad \text{and} \quad q = \frac{1}{1+\gamma}, \tag{f}$$

where $p+q = 1$, we get the formula (a) of section 6.262.

The reduction parameters are as follows:

$$E[r] = \frac{k}{\gamma}$$

$$V(r) = \frac{k}{\gamma} \cdot \left(1+\frac{1}{\gamma}\right) \tag{g}$$

$$\delta_r = \frac{1}{\gamma}.$$

6.266. Some Applications of the Negative Binomial Law

The field of application of the negative binomial, or of its equivalent forms, justifies the lengthy discussion which has been devoted to it. The negative binomial has gained for itself a veritable "freedom of the city" in applied statistics, in biology, medicine and epidemiology. Limitations of space and time compel us to give only the main headings and a few bibliographical references. From the historical standpoint we mention that Student (1907) effected one of the very first applications of the negative binomial to the counting of red corpuscles in a "Toma" chamber with 400 small squares. Numerous applications of the negative binomial have been described with and without examples of adjustment, and in comparison with contagious distributions, or by itself (Bliss, 1953).

Just as the negative binomial comes in many different forms, so its field of application is diversified in the extreme. We briefly present a few simple examples in some biological fields.

6.266.1. Counting yeast cells in a "Toma" chamber
(Student, 1907, p. 7)

This was one of the first applications of the negative binomial. We put r for the number of cells per square and f_r for the frequency of the squares with r cells:

r	0	1	2	3	4	5	6	7	8	9	/////
f_r	75	103	121	54	30	13	2	1	0	1	400

$$\bar{r} = \frac{720}{400} = 1.80$$

Poisson adjusted: $\bar{r} = 1.80$

Theoretical probability:

$$\varphi_r = e^{-1.8} \frac{(1.8)^r}{r!}$$

$$\chi^2 \text{ with 6 d.f.} = 9.03 \quad P = 0.25.$$

Negative binomial adjusted: $\varphi_r = (q'-p')^{-k}$

$$P = 0.37$$
$$q' = 1.0889$$
$$p' = 0.0889$$
$$k = 20.2473$$

As required, $\bar{r} = kp' = 1.7998$.

6.266.2. Deaths from a contagious disease (smallpox)

In view of its historical interest, we present the application proposed by Eggenberger–Polya (1923, p. 285), using their urn model in *a*, *b*, *c* (section 6.281). It concerns fatal cases of smallpox observed in Switzerland during the period 1877–1900, i.e. for 288 months. The data are given in a table (r, f_r) which is not reproduced here.

Notation: r—number of fatal cases during one month; $r = 0, 1, 2, \ldots$; $r = 60$ is the extreme number of deaths observed per month, f_r—frequency of the months with r deaths by smallpox.

One calculates that

$$\sum f_r = 288 \text{ months}$$
$$\sum r \cdot f_r = 1584 \text{ fatal cases}$$
$$\bar{r} = 5.5 \text{ deaths per month } (h \text{ in the original text})$$
$$s_r^2 = 83.59, \text{ the variance is significantly larger then the mean.}$$

The calculated relative excess of variance (d in the original text)

$$d_r = \hat{\delta}_r = \frac{83.59 - 5.5}{5.5} = 14.20$$

is very important and one may predict a bad fit by a Poisson law with parameter $\mu = 5.5$, which would imply that the excess is zero.

Just as one passes from the positive binomial to the Poisson law (section 6.231.1), so Eggenberger–Polya (1923) introduced an analogous passage to the limit in the probability urn model in *a*, *b*, *c*, where

$$P_0 = \frac{a}{N_0}, \quad q_0 = \frac{b}{N_0}, \quad \gamma = \frac{c}{N_0} \quad (\delta \text{ in the original}). \tag{a}$$

For $n \to \infty$, they put

$$\lim nP_0 = \mu \tag{b}$$
$$\begin{cases} n \to \infty \\ P_0 \to 0 \end{cases}$$
$$\lim n\gamma = \delta > 0$$
$$\begin{cases} n \to \infty \\ \gamma \to 0, \end{cases}$$

where δ, called the "parameter of interdependence", is also the theoretical relative excess of variance.

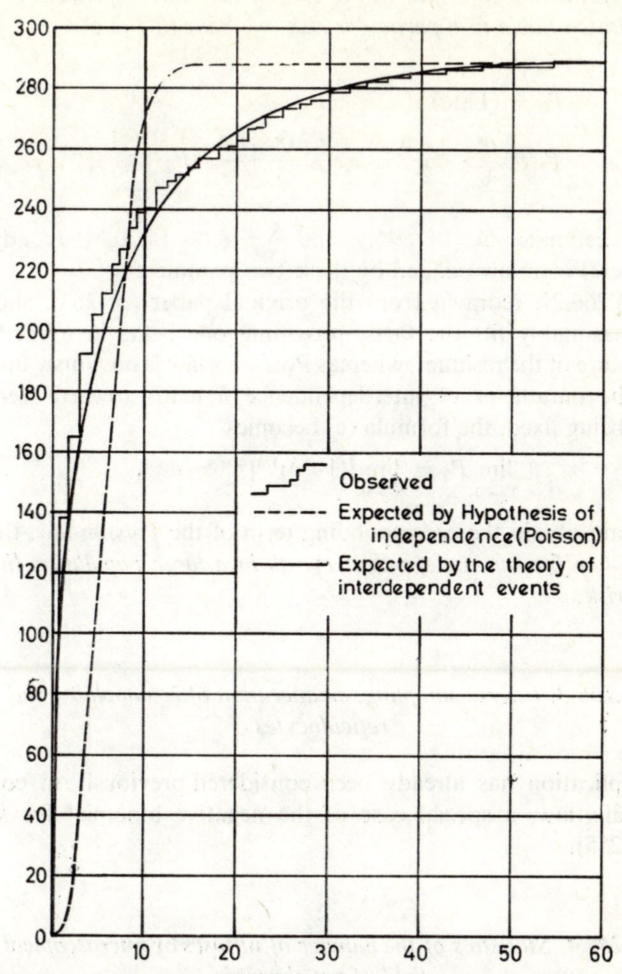

FIG. 6.266.2 I. Comparison between theory and observation in Swiss statistics of smallpox fatality (from Eggenberger–Polya (1923) p. 284). Number of months (ordinate axis) in which the number of fatalities is less than or equal to x (abscissae).

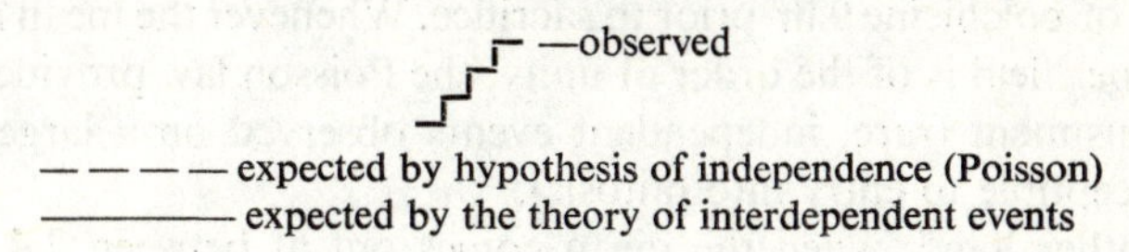

For the *probability limit law in the case of rare interdependent events in a large population liable to a particular risk*, we have (6.262 e)

$$P_0 = (1+\delta)^{-\frac{\mu}{\delta}} \tag{c}$$

$$P_r = \frac{\mu}{1} \cdot \frac{\mu+\delta}{2} \cdots \frac{\mu+(r-1)\delta}{r} \cdot P_0 \cdot \left(\frac{1}{1+\delta}\right)^r . \tag{c'}$$

Using the estimates $\hat{\mu} = \bar{r} = 5.5$ and $\hat{\delta} = \delta_r = 14.20$, they adjust the Eggenberger–Polya law defined by these two parameters.

Figure 6.266.2I, redrawn from the original paper (p. 284), shows that this law reasonably fits the facts, providing one bears in mind the non-random nature of the residues, whereas Poisson's law is obviously unsuitable.

When the parameter of interdependence δ tends towards zero (γ or $c \to 0$), μ being fixed, the formula (c) becomes

$$\lim_{\delta \to 0} P_0 = \lim_{\delta \to 0} [(1+\delta)^{1/\delta}]^{-\mu} = e^{-\mu},$$

and we again obtain the corresponding term of the *Poisson law*, the *probability limit law for rare independent events in a large population liable to a particular risk*.

6.266.3. *Inverse sampling, enumeration of eosinophiles and reticulocytes*

This application has already been considered previously in connection with Pascal's law, a special case of the negative binomial for k integer (section 6.255).

6.266.4. *Statistics of the number of mitoses by microscopical field of rat thyroids*

Bertrand and Thérèse Gayet-Hallion (1952) studied the pattern in the microscopical fields of mitotic figures of rats having received a 0.4 mg/100 g dose of colchicine 9 hr prior to sacrifice. Whenever the mean number $\bar{r}$ of mitoses per field is of the order of unity, the Poisson law provides a satisfactory adjustment (rare, independent events observed on a large number of cells susceptible to entry into mitosis).

On the other hand, when the mean comes out at between 2.4 and 4.9, the authors generally observed a variance which was consistently higher than the mean. A positive relative excess of variance was thus observed which was called the "index of contagion" and expressed as

$$d = \frac{s_r^2 - \bar{r}}{\bar{r}}.$$

The approximation by a negative binomial (μ, δ) (Polya) from the estimates ($\bar{r}, d$) is generally better than with a Poisson law. No physiological explanation has as yet been provided for this "mathematical evidence" of the effect of contagion between the mitoses of the thyroid gland in the colchicinized rat.

6.266.5. Statistics of industrial accidents, "Accident Proneness"

The fundamental paper of Greenwood and Yule (1920) (cited in section 6.265) opened the way to many other publications concerning industrial accidents. Two main concepts are involved, i.e.

(1) a clinical, personal characterization μ specifying the mean (e.g. annual) number of accidents for a definite subject;

(2) a statistical characterization of the population, viz. a gamma law in the parameters (k, γ) or (k, δ) describing the inter-individual variability of μ within this population.

These two concepts are related to one another by a statistical operation of averaging which leads formally to a distribution of contagious type without the agent, or responsible process, being *materially* demonstrable. Hence the reason for the term pseudo-contagion (false contagion) in this context.

Greenwood (1950), in a paper sent to *Biometrika* on the day on which he died, discussed various types of distributions capable of defining "proneness". We shall pass on to one recent example.

Indulski, Rofer and Rolejko (1966) studied absenteeism for sickness by sex in a group of approximately 3000 workers in the period 1961–2 (2 years).

A Poisson law adjustment to the frequency of persons absent due to sickness is, as expected, unacceptable. By contrast, the negative binomial (see 6.261 e′)

$$P_r = (-1)^r \binom{-k}{r} p^k q^r$$

provides an excellent adjustment from the estimates produced as follows.

TABLE 6.266.5A

Estimate	Males	Females
$\hat{p}$	0.44	0.41
$\hat{q}$	0.56	0.59
$\hat{k}$	1.26	1.76

6.266.6. Number of children per family

In a study on gene mutations, Kojima and Thérèse Kelleher (1962) showed that the distribution of families by the number $r = 0, 1, 2, \ldots$ of children can be described by a negative binomial. In particular, U.S. Census data 1950 show a mode of two children, a mean $\bar{r} = 3.0$ and a variance $s_r^2 = 7$.

The corresponding k and q' can be calculated without difficulty:

$$\hat{k} = \frac{\bar{r}^2}{s_r^2 - \bar{r}} = \frac{9}{7-3} = \frac{9}{4} = 2.25,$$

$$\hat{q}' = \frac{s_r^2}{\bar{r}} = \frac{7}{3} = 2.33.$$

Therefore $\hat{p}' = 2.33 - 1 = 1.33$.

The negative binomial is written so:

$$(q' - p')^{-k} = (2.33 - 1.33)^{-2.25}.$$

Upon multiplying $\hat{k}$ by $\hat{p}$ one again finds $2.99 \approx 3.0$.

It is not without interest to look for the probability p of appearing in the statistics (family size, completed fertility):

$$p = \frac{1}{q'} = \frac{1}{2.33} \approx 0.43.$$

The authors in question also show (fig. 1, p. 331) that the Poisson law of mean value 3 (and hence of variance 3) does not correspond to the observed frequencies. The relative excess of variance d is again the same in value as p':

$$d = \frac{s_r^2 - \bar{r}}{\bar{r}} = \frac{7-3}{3} = \frac{4}{3} = 1.33.$$

Analysing the results of the East African Medical Survey 1951–4, Brass (1958) studied four groups of 300 to 400 women aged 35 or more and evaluated them according to the number r of children borne ($r = 0, 1, 2, \ldots 13$ or more). The negative binomial, as adjusted by the maximum likelihood, gives the results shown in Table 6.266.6A.

TABLE 6.266.6A

	Kwinba 40 yr	Bukobu 40–49 yr	Kisii $>$ 35 yr	Msambweni 35 yr
$\bar{r}$	3.52	2.46	6.96	3.37
$\hat{k}$	3.22	3.04	447.7	5.37
$\hat{p}$	0.523	0.447	0.1053	0.386
b of Brass				

In a study of the number of children born alive to Belgian families, based on the Census taken in 1947 in Belgium (General Census VII, 1951), as reduction parameters we have consistently used the mean number $\bar{r}$ of

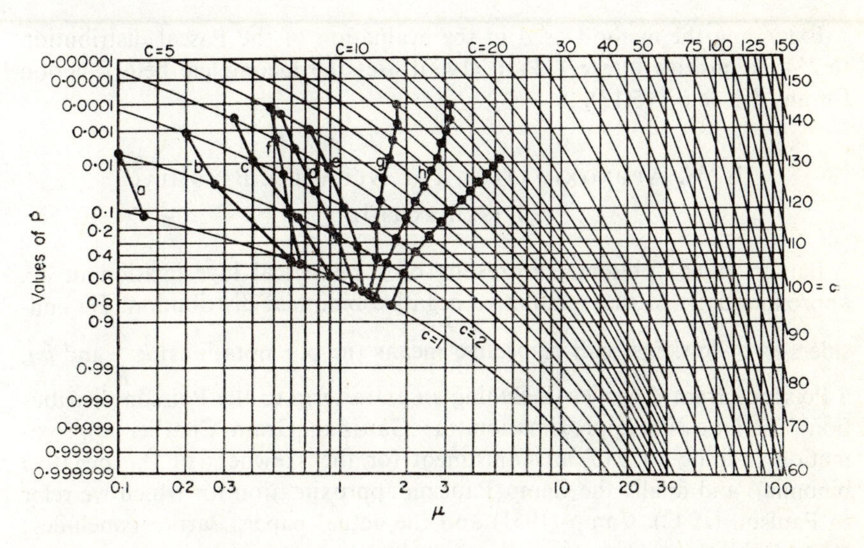

FIG. 6.266.6 I. Disributions of the number of children born alive according to the duration of marriage (in years). a: <1; b: 1 to <2; c: 2 to <3; d: 3 to <4; e: 4 to <5; f: 0 to <5; g: 5 to <10; h: 10 to <15; i: 15 and more.

children by family, the variance s_r^2 and either the relative excess of variance d, or an estimate of Polya's parameter of interdependence (Martin, 1961b, p. 30). The value $d = 1.30$ in relation to Belgium in 1947 is coherent with the $d = 1.33$ stated in 1950 in the U.S. Census.

In the same study, an application of the Poisson-sum probability paper is also shown to illustrate the *development as a whole* of the family of distributions of the number of children born alive according to the duration of marriage (Fig. 6.266.6I).

6.266.7. *Sequential test of hypotheses as to p' and k in the negative binomial*

Grimm and Maly (1962) have proposed a sequential method of using experimental results obeying the negative binomial to test the following hypotheses concerning p' and k:

$$p' = p_1' \text{ versus } p' = p_0' \qquad (k = \text{const}),$$
$$k = k_1 \text{ versus } k = k_0 \qquad (p' = \text{const}).$$

6.267. Evaluation of the Negative Binomial Distribution

Extending the method used in the evaluation of the Pascal distribution (6.254; k positive integer), Patil (1960) uses the incomplete beta function for any k (see 6.315.12).

6.268. Approximating the Negative Binomial Distribution

Bartko (1966) presents the results of a numerical investigation of six approximations to the cumulative negative binomial distribution. He considers two Poissons with respective means (in our notations) $k\frac{q}{p}$ and kq, a Poisson Gram–Charlier (including an extra term to the Poisson distribution), the Gaussian approximation, the Gaussian Gram–Charlier approximation (adding to $\Phi(u)$ an adjustment for the skewness of the negative binomial) and finally the Camp–Paulson approximation for which we refer to Paulson (1942), Camp (1951) and the actual paper. Bartko concludes: "the two best approximations, the two which match the cumulative negative binomial almost exactly, are the Poisson Gram–Charlier and the Camp–Paulson approximations."

6.27. Law in Logarithmic Series

6.271. Point-density Law in Logarithmic Series

6.271.1. Definition

The law in logarithmic series with parameter θ and in the random variable $r = 1, 2, \ldots$ $(0 < \theta < 1)$, is written so:

$$P_r = \frac{-1}{\ln(1-\theta)} \cdot \frac{\theta^r}{r}. \tag{a}$$

Since $\theta < 1$, $\ln(1-\theta)$ is < 0, but the factor (-1) produces a positive P_r

Alternatively,

$$P_r = \alpha \frac{\theta^r}{r}, \tag{a'}$$

where

$$\alpha(\theta) = -\frac{1}{\ln(1-\theta)}. \tag{b}$$

Whence

$$\begin{aligned} P_1 &= \alpha\theta \\ P_2 &= \alpha\frac{\theta^2}{2} \\ P_3 &= \alpha\frac{\theta^3}{3} \\ &\;\;\vdots \end{aligned} \tag{c}$$

Hence the recurrence law

$$P_{r+1} = \theta \cdot \frac{r}{r+1} P_r. \tag{d}$$

In fact we have

$$\sum P_r = 1,$$

for, according to (b),

$$\sum P_r = \frac{-1}{\ln(1-\theta)}\left[\theta+\frac{\theta^2}{2}+\frac{\theta^3}{3}+\ldots\right] = \frac{(-1)[-\ln(1-\theta)]}{\ln(1-\theta)} = +1. \tag{e}$$

The law in logarithmic series was introduced by Sir R. A. Fisher (1943) to describe the relation linking species to the number of individuals belonging to a species.

6.271.2. Reduction parameters

After Kendall and Stuart (1963) and Patil, Kamat and Wani (1964), one calculates the following parameters:

Mean:

$$\mu = E[r] = \alpha\frac{\theta}{1-\theta}$$

$$= -\frac{1}{\ln(1-\theta)}\cdot\frac{\theta}{1-\theta}, \tag{a}$$

it being noted that $P_1 = \alpha\theta$ (see also formulae k′ and l′).

$$\mu = \frac{P_1}{1-\theta}. \tag{a'}$$

Variance:

$$\sigma^2 = V(r) = \frac{\alpha\theta}{(1-\theta)^2}(1-\alpha\theta)$$

$$= E[r]\cdot\frac{1-\alpha\theta}{1-\theta}$$

$$= \frac{\mu}{1-\theta}-\left(\frac{\alpha\theta}{1-\theta}\right)^2 = \frac{\mu}{1-\theta}-\mu^2. \tag{b}$$

Standard deviation:

$$\sigma_r = \sqrt{V(r)} = \sqrt{E[r]\cdot\frac{1-\alpha\theta}{1-\theta}} = \sqrt{\frac{\alpha\theta(1-\alpha\theta)}{(1-\theta)^2}} = \frac{\sqrt{\alpha\theta(1-\alpha\theta)}}{1-\theta}$$

$$= \frac{\sqrt{P_1(1-P_1)}}{1-\theta}. \tag{c}$$

The probability of meeting with $r = 1$ element plays an important part in the formulation of the reduction parameters.

Coefficient of variation:

$$\gamma_r = \frac{\sigma_r}{\mu} = \frac{\sqrt{P_1(1-P_1)}}{1-\theta} \cdot \frac{1-\theta}{P_1} = \sqrt{\frac{1-P_1}{P_1}}. \tag{d}$$

Hence the coefficient of variation of the logarithmic series is less than, equal to or greater than unity according as P_1 is greater, equal or less than $\frac{1}{2}$.

Coefficient of dispersion, relative variance and relative excess of variance. The relative excess of variance δ_r is related to the *coefficient of dispersion* or to the variance/mean quotient of Blackman (1942), by the relation

$$\delta_r = \frac{\sigma^2-\mu}{\mu} = \frac{\sigma^2}{\mu}-1.$$

The quotient σ^2/μ is called the *relative variance* by Clapham (1936).

We express μ and σ^2 as a function of P_1:

$$\mu = \frac{P_1}{1-\theta}, \quad \sigma^2 = \frac{\mu}{1-\theta}-\mu^2. \tag{e}$$

Since $1/(1-\theta) = \mu/P_1$, therefore

$$\sigma^2 = \frac{\mu^2}{P_1}-\mu^2 = \mu^2\frac{1-P_1}{P_1}. \tag{f}$$

Hence the *coefficient of dispersion* is written so:

$$\frac{\sigma^2}{\mu} = \frac{\mu^2}{\mu} \cdot \frac{1-P_1}{P_1} = \frac{\mu(1-P_1)}{P_1} = \frac{1-\alpha\theta}{1-\theta}. \tag{g}$$

This coefficient of dispersion is less than, equal to or greater than unity according as the parameter θ of the geometric series law is less than, equal to or greater than $1-e^{-1} = 1-0.368 = 0.632$.

Consider the case of equality: $\sigma^2/\mu = 1$. Here

$$\frac{1-\alpha\theta}{1-\theta} = 1,$$

$$1-\alpha\theta = 1-\theta.$$

So

$$\alpha = 1.$$

In this case

$$\frac{-1}{\ln(1-\theta)} = 1,$$

or

$$1-\theta = e^{-1},$$

in which case

$$\theta = 1-e^{-1}. \tag{h}$$

The *relative excess of variance* (Martin, 1961) is written:

$$\delta_r = \frac{\sigma^2}{\mu}-1$$

$$= \frac{1-\alpha\theta}{1-\theta}-1 = \frac{1-\alpha\theta-1+\theta}{1-\theta} = \frac{\theta}{1-\theta}(1-\alpha); \tag{i}$$

it is compared with zero and has the sign of $1-\alpha$. In fact we have

$$\delta_r = 0 \quad \text{for} \quad \alpha = 1, \quad \text{or for} \quad \theta = 1-e^{-1}.$$

By virtue of the result obtained previously in respect of the coefficient of dispersion, the relative excess of variance of a logarithmic series law is said to be negative, zero or positive according as α is greater than, equal to, or less than unity, or according as θ is less than, equal to, or greater than $1-e^{-1} = 0.632$.

Shape parameters. The skewness γ_1 and kurtosis γ_2 are based on μ_3 and μ_4, i.e.

$$\mu_3 = \frac{\alpha\theta}{(1-\theta)^3}[1+\theta-3\alpha\theta+2\alpha^2\theta^2],$$

$$\mu_4 = \frac{\alpha\theta}{(1-\theta)^4}[(1+4\theta+\theta^2)-4\alpha\theta(1+\theta)+6\alpha^2\theta^2-3\alpha^3\theta^3], \tag{j}$$

according to Kendall and Stuart (1963, I, p. 113).

The *moment-generating function* is

$$M(t) = -\alpha \ln(1-\theta e^t), \tag{k}$$

hence the mean

$$\mu = M'(t)\Big|_{t=0} = \frac{(-\alpha)(-\theta e^t)}{1-\theta e^t}\Big|_{t=0} = \frac{\alpha\theta}{1-\theta} \tag{k'}$$

etc.

The *probability-generating function* is

$$G(t) = -\alpha \ln(1-\theta t), \tag{l}$$

hence the mean

$$\mu = G'(t)_{t=1} = -\frac{\alpha}{1-\theta t}(-\theta)_{t=1} = \frac{\alpha\theta}{1-\theta}. \tag{l'}$$

6.271.3. Derivation of the logarithmic series law

The logarithmic series law has been obtained as the limit of a negative binomial (written here for k non-integer):

$$P_r = \frac{\Gamma(k+r)}{r!\,\Gamma(k)} \cdot \frac{p'^r}{(1+p')^{k+r}} \tag{a}$$

whenever $k \to 0$, as suggested by the adjustments of the examples studied by Fisher, Corbet and Williams (1943) and Williams (1947). In this process one isolates the factor $1/\Gamma(k)$, the limit of which is denoted by α:

$$\lim_{k \to 0} P_{r_{\text{Bin.}<0}} = \alpha \frac{(r-1)!}{r!} \left(\frac{p'}{1+p'}\right)^r. \tag{b}$$

Setting

$$\frac{p'}{1+p'} = \theta, \tag{c}$$

we get

$$P_{r_{\text{log. ser.}}} = \alpha \frac{\theta^r}{r}. \tag{c'}$$

We have to recall formula (6.271.1 b) which shows that α is really a function of θ.

For an excellent critical presentation of the reasoning of Sir R. A. Fisher in the ecological problem of the species and the individual, as well as the approaches of D. G. Kendall (1948) and Anscombe (1950), reference should be made to chapter II (p. 95) of the work of Patil, Kammat and Wani (1964). In hospital statistics, Herdan (1957) has used the logarithmic series law to describe the association between the number of diseases and the frequency of patients in a community.

6.271.4. Tables of the point-density law in logarithmic series

In depending, in fact, on only a single parameter θ, Fisher's logarithmic law can be tabulated from θ, or from μ which depends only on θ. Tables of P_r have been compiled to five decimal places for a limited number of values of μ (Williams and Bretherton, 1964). Patil, Kammat and Wani (1964, table 16, pp. 116–218) give P_r to six decimal places according to the index of dispersion θ over the interval 0.010 (0.010) 700 (0.005) 900 (0.001) 999.

6.271.5. Connection with the negative binomial via Poisson law

Quenouille (1949) showed that the logarithmic series law with parameter θ, where

$$\alpha = -\frac{1}{\ln(1-\theta)},$$

can serve in the capacity of a compounding function to a Poisson law with parameter μ, resulting in a negative binomial with parameters θ and $\alpha\mu$. Elston and Pickrel (1962) apply Quenouille's reasoning to a statistical approach to ordering and usage policies for a hospital blood bank. Furthermore they describe a method of simulation processed on a Univac 1105 computer.

6.272. Distribution in Logarithmic Series

By definition the distribution in logarithmic series is written:

$$\text{Prob}\,(r \leqslant c) = \sum_{r=1}^{c} P_r(\theta)$$

$$(c = 1, 2, \ldots, \infty).$$

This distribution is tabulated to six decimal places for θ in the interval 0.010 (0.010) 700 (0.005) 900 (0.001) 999 (table 17 of Patil, Kamat and Wani, 1964, pp. 219–389). It should be noted that the spread to the right increases as θ tends to unity.

6.273. Logarithmic Series Law with One or Two Parameters

Written as in 6.271.1 (a) and (a′) the logarithmic series law depends on one parameter θ and has been presented on its own. But, in the analysis of ecological or medical data from statistics presenting the number of groups $n_1, n_r, \ldots$ (total S species) each of which contains a number of individuals (total N), the very nature of the problem leads to a logarithmic series law with *two* parameters α' and R (Bliss, 1965, p. 386).

$$\alpha' R, \alpha' \frac{R^2}{2}, \alpha' \frac{R^3}{3}, \ldots \tag{a}$$

The second parameter

$$\alpha' = \frac{E[n_1]}{R} \tag{b}$$

is not a function of R (or θ) alone (as seen in 6.271.1b), but also of the expected value of the number of groups (species) present with one invidual (see also Martin, 1969).

6.28. Law of Eggenberger–Polya

6.281. Urn of Interdependent Events and Probability Law of Successes

Consider a probabilistic urn which contains a finite number N_0 of balls, of which a are white and b are black:

$$a+b = N_0. \tag{a}$$

After mixing the balls, begin to draw and note the color of the ball, then put back a constant number c of balls of the same color as the drawn ball. We are thus no longer concerned with a fundamental set F_0 of fixed power N_0 as in (a), but with a sequence of fundamental sets F_i of power

$$N_{F_i} = N_0+(i-1)\,c, \tag{b}$$

where i is the ordinal number of the draw, and N_0 is the power of the first set before the first draw. The power of the set of favorable cases thus changes according to the rank of draw *and* the result of the foregoing draw. In Polya's terminology, we thus meet with a case of *inter-dependent events* (1931, p. 134).

Putting A for the issue of a white ball and B for a black one, the probability of the sequence (A, A) is given by the expression

$$P(A, A) = \frac{a}{N_0}\cdot\frac{a+c}{N_0+c}.$$

For the other three types of sequence, we have

$$\begin{aligned} P(A, B) &= \frac{a}{N_0}\cdot\frac{b}{N_0+c} \\ P(B, A) &= \frac{b}{N_0}\cdot\frac{a}{N_0+c} \\ P(B, B) &= \frac{b}{N_0}\cdot\frac{b+c}{N_0+c}. \end{aligned} \tag{c}$$

These four states exhaust the possibilities in two tests. The sum total of the probabilities is therefore equal to unity, which can be easily verified.

Consider now a sequence of successes of type 3, i.e. (A, A, A). Here we have

$$P(A, A, A) = \frac{a}{N_0} \cdot \frac{a+c}{N_0+c} \cdot \frac{a+2c}{N_0+2c}. \tag{d}$$

For a sequence of r successes A, viz. $\begin{pmatrix} A & A & \cdots & A \\ 1 & 2 & \cdots & r \end{pmatrix}$, we have

$$P(A, A, \ldots, A) = \frac{a}{N_0} \cdot \frac{a+c}{N_0+c} \cdot \ldots \frac{a+(r-1)\,c}{N_0+(r-1)\,c}. \tag{e}$$

Finally, we carry out a series of n draws where we observe r successes A, and therefore $(n-r)$ failures. In effect these events realize a random sequence:

$$\begin{matrix} A & B & B & A & A & B & \ldots & B \\ 1 & 2 & 3 & 4 & i & i+1 & & n \end{matrix}. \tag{f}$$

We recall that (formula b)

$$N_{F_i} = N_0+(i-1)c \tag{f'}$$

is the power of the fundamental set before the ith draw, and put

$$n_{A_i} = a+(i-1)c$$

for the power of the sub-set of favorable cases before the ith draw. In this case

$$P(A_i) = \frac{a+(i-1)c}{N_0+(i-1)c}. \tag{f''}$$

For the probability of a sequence A_i, A_{i+1}, we have

$$P(A_i, A_{i+1}) = \frac{a+(i-1)c}{N_0+(i-1)c} \cdot \frac{a+ic}{N_0+ic}, \tag{g}$$

whilst for a sequence A_i, B_{i+1}

$$P(A_i, B_{i+1}) = \frac{a+(i-1)c}{N_0+(i-1)c} \cdot \frac{b+(i-1)c}{N_0+ic}. \tag{h}$$

It is seen that whereas the denominator of the probabilities (g) and (h) depends only on the rank of the test, the numerator depends on the result of the previous test as well as on the rank, by virtue of the rule of play.

One may therefore re-group the random series (f) as two sub-series of type (e), which yields

$$\begin{array}{ccc|c|c|cc} A, & A & \ldots & A & B, & \ldots, & B \\ 1 & 2 & & r & r+1 & & r+n-r \\ & & & & & & \| \\ & & & & & & n \end{array} \quad \text{(i)}$$

The sequence $\begin{pmatrix} A & & A \\ 1, & \ldots, & r \end{pmatrix}$ has its probability given by (e). If we know $P(B_{r+1})$, we know how to write the probability of the sequence $\begin{pmatrix} B & & B \\ r+1, & \ldots, & n \end{pmatrix}$. We therefore calculate the probability of the sequence (A_r, B_{r+1}). By virtue of the rule of play, we have

$$P(A_r, B_{r+1}) = \frac{a+(r-1)c}{N_0+(r-1)c} \cdot \frac{b}{N_0+rc}. \quad \text{(j)}$$

Hence the probability of the sequence $(B_{r+1} \ldots B_n)$, where $n = r+(n-r)$, is

$$P(B_{r+1} \ldots B_{r+(n-r)}) = \frac{b}{N_0+rc} \cdot \frac{b+c}{N_0+(r+1)c} \cdots \frac{b+(n-r-1)c}{N_0+(n-1)c}. \quad \text{(k)}$$

The probability of a particular series of r successes out of n tests, after regrouping in two sub-series $(A \ldots, B \ldots)$, where A goes from rank 1 to rank r and B from rank $r+1$ to n, is written so:

$$\frac{a}{N_0} \cdot \frac{a+c}{N_0+c} \cdots \frac{a+(r-1)c}{N_0+(r-1)c} \cdot \frac{b}{N_0+rc} \cdot \frac{b+c}{N_0+(r+1)c} \cdots \times \frac{b+(n-r-1)c}{N_0+(n-1)c}. \quad \text{(l)}$$

As pointed out by Feller (1958, I), all the particular series of r successes out of n tests can be converted into this type of two sub-series. Each one of these particular series is therefore equal to any other as regards its probability.

Since one can form $\binom{n}{r}$ series with r successes out of n tests, the theorem of total probabilities therefore implies that

$$P_{r,n} = \binom{n}{r} \frac{a}{N_0} \cdots \frac{a+(r-1)c}{N_0+(r-1)c} \frac{b}{N_0+rc} \cdots \frac{b+(n-r-1)c}{N_0+(n-1)c}. \quad \text{(m)}$$

Such is the probability of r successes out of n interdependent tests conducted according to the scheme of Eggenberger–Polya.

Another form of the probability of r successes out of n tests. We introduce the initial elementary probabilities

$$p_0 = \frac{a}{N_0}, \quad q_0 = \frac{b}{N_0}, \quad p_0+q_0 = 1, \tag{n}$$

where the subscript 0 is a reminder that the *correct* situation of the fundamental set is realized only before the first draw. Similarly, we put

$$\gamma = \frac{c}{N_0}, \quad |c| < N_0. \tag{n'}$$

It should be noted that p_0, q_0 and γ are multiples of $1/N_0$.

We divide top and bottom of each factor of expression (m) by N_0. Having regard to (n) and (n′), and specifying the value n of the sample, we get

$$P_{r,n} = \binom{n}{r} \frac{p_0}{1} \cdot \frac{p_0+\gamma}{1+\gamma} \cdots \frac{p_0+(r-1)\gamma}{1+(r-1)\gamma} \cdot \frac{q_0}{1+r\gamma} \times \frac{q_0+\gamma}{1+(r+1)\gamma} \cdots \frac{q_0+(n-r-1)\gamma}{1+(n-1)\gamma}. \tag{o}$$

It should be noted that in this Eggenberger–Polya law, n is finite and fixed, whilst the random variable $r = 0, 1, 2, \ldots, n$ is at most equal to n.

It can be shown that

$$\sum_{r=0}^{n} P_{r,n} = 1. \tag{p}$$

"*Global influence.*" The term "global influence" was introduced by Polya (1931) in order to describe the past history of the probability urn in n draws and the global influence of this upon the result of the $(n+1)$-th draw.

In a Eggenberger–Polya scheme, consider the past situation in which r white balls have been obtained in n draws. *Whatever the configuration $AB \ldots A \ldots B$* with rA's and $(n-r)B$'s, *the probability of drawing* a white of rank $n+1$ in the next draw (either of $r+1$ whites or $n+1$ draws) is always

$$p_{n+1}^{r+1}\Big|_{n}^{r} = \frac{p_0+r\gamma}{1+n\gamma}, \tag{q}$$

since there have been r replacements in favor of whites.

This "global" (overall) number r replacements of whites fixes the numerator, whilst the *total number n* of past draws fixes the denominator of the probability studied. This latter, written in (o), remains unchanged in whatever order the r whites and the $(n-r)$ blacks are presented.

"Contagious" draws. In the original model of Eggenberger and Polya (1923), the added number $c > 0$ is a measure of the risk of success of a "contagious" agent having an initial probability p_0 of infection. Hence, according to the rule of play, if c (Δ of Polya) is positive, then "the number of balls increases after each draw, each success favors the chances of further successes, each non-success worsens the chances of following tests, the success as well as the non-success are contagious".

In their classical article Eggenberger and Polya (1923) give a statistical article concerning the monthly occurrence of smallpox cases in Switzerland during the period 1877–1900 (288 months). The results have already been discussed in section 6.266.2.

6.282. Reduction Parameters

The details of the calculation of the reduction parameters, mean and variance from 6.281 (m), are given by Fisz (1958, pp. 122–123).

Mean:

$$E[r] = n\frac{a}{a+b} = n\cdot\frac{a}{N_0}. \tag{a}$$

In the initial probability p_0, we have

$$E[r] = np_0. \tag{a'}$$

The mean number of successes out of n interdependent trials is equal to the product of n with the *initial probability of success.*

Variance as a function of a, b, c:

$$V(r) = \frac{nab}{(a+b)^2}\cdot\frac{a+b+nc}{a+b+c} = n\cdot\frac{ab}{N_0^2}\cdot\frac{N_0+nc}{N_0+c}. \tag{b}$$

As a function of probabilities p_0, q_0 and γ,

$$V(r) = np_0q_0\frac{1+n\gamma}{1+\gamma}. \tag{b'}$$

Relative excess of variance:

$$\delta_r = \frac{V(r)-E[r]}{E[r]} = q_0 \cdot \frac{1+n\gamma}{1+\gamma} - 1. \qquad \text{(b'')}$$

If c or $\gamma \to 0$, one again finds the relative excess of variance of the positive binomial.

Johnson and Kotz (I, 1969, p. 231) give μ_3 and μ_4.

6.29. Positive or Classical Hypergeometric Law

$$c = -1$$

6.291. Positive Hypergeometric Law—Derivation

The hypergeometric law corresponds to direct sampling in a non-Bernoulli procedure, *without* replacement of the drawn ball. The procedure represents a special case of Eggenberger–Polya draws where the constant c is *negative* and equal to -1 (section 6.281).

From an urn containing initially N_0 balls, of which a are white and b are black, i.e.

$$a+b = N_0,$$

we extract one ball at random. Before the first draw is made, the probability of this being a white ball is

$$p_0 = \frac{a}{N_0}, \qquad \text{(a)}$$

and the probability of it being black is

$$q_0 = \frac{b}{N_0}. \qquad \text{(a')}$$

In this respect

$$p_0+q_0 = 1.$$

If n balls are sampled in succession without replacing the drawn balls, and r white balls are observed, the general scheme can be described as follows in Table 6.291A.

We will now evaluate the probability P_r of observing r whites in n draws without replacing the drawn ball. The number of equally possible combinations is the number of simple combinations of n elements from N_0 elements:

$$_{N_0}C_n = \binom{N_0}{n} = \frac{N_0(N_0-1)\dots(N_0-n+1)}{1\cdot 2\cdot \dots \cdot n} = \frac{N_0!}{n!\,(N_0-n)!}.$$

This is the denominator of P_r.

TABLE 6.291A. BEFORE THE FIRST DRAW

	Balls		
	White	Black	
Sampled	r	$n-r$	n
Non-sampled	$a-r$	$n-r-b$	N_0-n
Initial urn or initial population	a	b N_0-a	N_0
Initial probabilities	p_0	q_0	1

The number of combinations of r whites which can be formed from a whites is

$$\binom{a}{r} = \frac{a(a-1)\ldots(a-r+1)}{1\cdot 2\cdot \ldots \cdot r}.$$

With each of these groupings is associated some one of the $\binom{b}{n-r}$ combinations which can be formed with $(n-r)$ black balls out of a total of b black balls. The numerator of P_r is therefore

$$\binom{a}{r}\cdot\binom{b}{n-r} \quad \text{or} \quad \binom{a}{r}\cdot\binom{N_0-a}{n-r}.$$

Whence

$$P_r = \frac{\binom{a}{r}\cdot\binom{N_0-a}{n-r}}{\binom{N_0}{n}} \qquad (a < N_0)\ (r = 0, 1, \ldots, a). \tag{b}$$

In this scheme with three parameters, N_0 and a (therefore $b = N_0-a$) are *structural parameters of the initial* population. By contrast, n is a *prefixed operational parameter* which defines the effective size of the sample. Finally, by implication, the parameter $(c = -1)$ defines the particular case of the (non-Bernoulli) process of Eggenberger–Polya used. To $c = -1$ we have correspondingly

$$\gamma = -\frac{1}{N_0}. \tag{b'}$$

Upon making this substitution in 6.281 (o), we get

$$P_r = \binom{n}{r}\frac{p_0}{1}\cdot\frac{p_0-\dfrac{1}{N_0}}{1-\dfrac{1}{N_0}}\cdots\frac{p_0-\dfrac{r-1}{N_0}}{1-\dfrac{r-1}{N_0}}\cdot\frac{q_0}{1-\dfrac{r}{N_0}}\cdots\frac{q_0-\dfrac{n-r-1}{N_0}}{1-\dfrac{n-1}{N_0}}. \quad \text{(c)}$$

Canceling the denominators N_0, we have

$$P_r = \binom{n}{r}\frac{N_0p_0(N_0p_0-1)\ldots[N_0p_0-(r-1)]N_0q_0\ldots[N_0q_0-(n-r-1)]}{N_0(N_0-1)\ldots[N_0-(r-1)](N_0-r)\ldots[N_0-(n-1)]}. \quad \text{(d)}$$

We expand $\binom{n}{r}$ and multiply top and bottom $(n-r)!$. Reducing the notations to a minimum, we get

$$P_r = \frac{n\ldots[n-(r-1)]N_0p_0\ldots[N_0p_0-(r-1)]N_0q_0\ldots[N_0q_0-(n-r-1)]\cdot(n-r)\ldots 1}{1\ldots rN_0\ldots[N_0-(r-1)](N_0-r)\ldots[N_0-(n-1)]\cdot(n-r)\ldots 1}.$$

By suitably regrouping the factors, it is seen that formula (b) is again obtained

$$P_r = \frac{\binom{N_0p_0}{r}\binom{N_0q_0}{n-r}}{\binom{N_0}{n}}, \quad \text{(e)}$$

where $a = N_0p_0$ and $b = N_0q_0$.

It can be verified that $$\sum_{r=0}^{a} P_r = 1. \quad \text{(f)}$$

6.292. Reduction Parameters

Mean: As in 6.282. (a), we have

$$E[r] = np_0 = n\frac{a}{N_0}. \quad \text{(a)}$$

The mean number of successes in a sample with no ball replacement is np_0, where p_0 is the initial probability of success and N_0 is the *initial* population.

Variance: We set $\gamma = -\dfrac{1}{N_0}$ in 6.282(b′) and then

$$V(r) = np_0q_0 \frac{1-\dfrac{n}{N_0}}{1-\dfrac{1}{N_0}} = np_0q_0 \frac{N_0-n}{N_0-1}. \tag{b}$$

In a direct sample where the drawn ball is not replaced, if the mean of the numbers of successes is equal to the initial binomial mean, the variance is less than the initial binomial variance np_0q_0.

For large N_0, we have $N_0-1 \approx N_0$. Setting

$$\frac{n}{N_0} = f \tag{c}$$

for the sampled fraction of the population *(fraction decided before drawing)*, the theoretical variance is

$$\sigma^2_{\text{hyperg.}} \approx \sigma^2_{\text{bin.}}(1-f), \tag{c′}$$

and this is equal to the initial binomial variance multiplied by the non-sampled (and therefore unknown) fraction of the initial population.

The variance of the positive hypergeometric law is less than the binomial variance of the initial population.

Standard deviation:

$$\sigma_r = \sqrt{np_0q_0}\sqrt{\frac{N_0-n}{N_0-1}} \approx \sqrt{np_0q_0}\sqrt{1-f}, \tag{d}$$

where $1-f \approx 1-\dfrac{n_0}{N}$.

Coefficient of variation:

$$\gamma_r = \frac{\sigma_r}{\mu} = \frac{\sqrt{np_0q_0(1-f)}}{np_0} \approx \sqrt{\frac{q_0(1-f)}{np_0}}. \tag{e}$$

Relative excess of variance:

$$\delta_r = \frac{\sigma_r^2-\mu}{\mu} = \frac{np_0q_0(1-f)-np_0}{np_0} = q_0(1-f)-1, \tag{f}$$

which is always a negative quantity, as for the positive binomial.

Parameters of shape (Kendall and Stuart, 1963, I, p. 134):
The *skewness* is derived from

$$\mu_{3_{\text{hyperg.}}} = np_0q_0(q_0-p_0)\cdot\frac{(N_0-n)(N_0-2n)}{(N_0-1)(N_0-2)}$$
$$\approx \mu_{3_{(\text{Bin.}>0)}}(1-f)(1-2f), \tag{g}$$

whence, calculation of γ_1 or β_1.

For $N_0 \to \infty$ and n fixed f, $\to 0$, one again finds the third moment of the positive binomial.

Kurtosis. By averaging the approximations $N_0 \approx (N_0-1) \approx (N_0-2) \approx (N_0+1)$, we get

$$\mu_{4_{\text{hyperg.}}} = np_0q_0(1-f)\{1-6f(1-f)+3p_0q_0[(n-2)-nf+6f(1-f)]\}. \tag{h}$$

In the same conditions, ($f \to 0$), by use of the expressions which are independent of f, one again finds the fourth central moment of the positive binomial.

6.293. Origin of the Term "Hypergeometric Law"

The hypergeometric law is so named because its probability-generating function is proportional to the hypergeometric function (M. G. Kendall and Stuart, 1963, I, p. 134):

$$F(\alpha, \beta, \gamma, t) = 1+\frac{\alpha\beta}{\gamma}\frac{t}{1!}+\frac{\alpha(\alpha+1)\beta(\beta+1)}{\gamma(\gamma+1)}\frac{t^2}{2!}+\ldots, \tag{a}$$

where $\alpha = -n$, $\beta = -Np_0$, $\gamma = N_0q_0-n+1$.

6.294. Limit Forms of the Hypergeometric Law

In the form of 6.291(c) it is readily seen that the positive hypergeometric law tends to the positive binomial law in parameters $p = p_0$ and n as $N_0 \to \infty$. Of course, one should then consider a proportion p_0 compatible with a positive binomial which is approximately valid for $f < n/N_0 = 0.1$. To give an example, such a comparison is given in Table 6.294 A.

The case of sampling *with* replacement in a finite population (N_0 and p_0 constants) is governed by the same binomial law as that of sampling *with or without* replacement in a *theoretically infinite* population.

TABLE 6.294A. COMPARISON BETWEEN PROBABILITIES FROM POSITIVE BINOMIAL LAW (b_r) AND POSITIVE HYPERGEOMETRIC LAW (h_r) WITH SAME MEAN 2

Positive binomial urn	Positive hypergeometric urn	r	$\mu = 2$	
			b_r	h_r
$N = \infty$	$N = 1000$	0	0.1326	0.1190
$n = 100$	$n = 100$	1	0.2707	0.2701
$p = 0.02$	$p = 0.02$	2	0.2734	0.2881
		3	0.1823	0.1918
		4	0.0902	0.0895
		5	0.0353	0.0311
		6	0.0114	0.0083
		7	0.0031	0.0018
		8	0.0007	0.0003
		9	0.0002	

When $N_0 \to \infty$ and as $p_0 \to 0$, with n all the time large enough for np_0 to tend to μ, the limit of the hypergeometric law is a Poisson law in μ ($f < 0.1$ and $p_0 < 0.1$).

Finally, just like the positive binomial, the positive hypergeometric law can be approximated by a Gaussian law in the parameters

$$\mu = np_0 \quad \text{and} \quad \sigma^2 = np_0q_0\left(1-\frac{n}{N_0}\right),$$

provided that $\sigma^2 > 9$ (Hald, 1952, p. 691).

6.295. USE OF THE HYPERGEOMETRIC LAW

In connection with this law, Knudsen (1949) has given the probability of acceptance of a lot of pharmaceutical vials in a destructive sampling procedure. The hypergeometric law is widely used in *Quality Control* (see, for example, Dodge and Romig, 1967).

In the *analysis of 2×2 contingency tables* with fixed marginal totals, where the expected value of one cell is less than five, the exact method of Irwin–Fisher–Yates has to be used. Historically, this exact test seems to have emerged in a restricted circle of statisticians round about 1935. This test is based on the hypergeometric law and, from this time, may be found in *Statistical Methods for Research Workers* (Fisher, 1925 and subsequent editions).

Recently, Armitage (1971, pp. 135–138), under the title "The exact test for fourfold tables", has given a numerical recurrent way of calculating the successive terms of the involved hypergeometric law. Moreover, he gives the *correct method for the two-tailed test* which does not correspond to the automatic doubling of the one-tailed probability.

6.210.† Negative Hypergeometric Law

($c = +1$)

6.210.1. Negative hypergeometric law, derivation

If one puts $c = +1$ (addition of a ball of the color drawn) in the general formula for the Eggenberger–Polya urn (6.281), then for $N_0 = a+b$ and $r = 0, 1, 2, \ldots, n$, we get

$$P_r = \binom{n}{r}\frac{a}{N_0}\cdot\frac{a+1}{N_0+1}\cdots\frac{a+(r-1)}{N_0+r-1}\cdot\frac{b}{N_0+r}\cdots\frac{b+(n-r-1)}{N_0+(n-1)}. \quad \text{(a)}$$

On expanding $\binom{n}{r}$,

$$P_r = \frac{\dfrac{a(a+1)\ldots[a+(r-1)]}{r!}\cdot\dfrac{b(b+1)\ldots[b+(n-r-1)]}{(n-r)!}}{\dfrac{N_0(N_0+1)\ldots[N_0+(n-1)]}{n!}}.$$

By definition of $\binom{-a}{r}, \binom{-b}{n-r}, \binom{-N_0}{n}$, and noting that $-b = -(N_0-a) = -N_0+a$, we get

$$P_r = \frac{\binom{-a}{r}\binom{-N_0+a}{n-r}}{\binom{-N_0}{n}}. \quad \text{(b)}$$

Such is the negative hypergeometric law which yields the probability of drawing $r \leqslant n$ white balls in a series of draws where one ball is added ($c = +1$) of the color which has just been extracted.

† 6.210 corresponds in fact to 6.2 X (3 figures) and has not been used before (see p. 206).

The changeover from formula (b) to the corresponding formula of the positive or classical hypergeometric law is made by changing $-a$ into $+a$ and changing $-N_0$ into $+N_0$ (6.291 b).

By the use of symbolism of the type

$$\binom{-a}{r} = (-1)^r \binom{a+r-1}{r},$$

the negative hypergeometric law is also written (Sibuya, Yoshimura and Shimizu, 1964):

$$P_r = \frac{\binom{a+r-1}{r}\binom{N_0-a+n-r-1}{n-r}}{\binom{N_0+n-1}{n}}.$$

This form was also obtained by Skellam (1948) as the result of the composition of a positive binomial ($0 < p < 1$), with a beta law in p with parameters α and β (6.315.11).

The author proposes a method of estimating α and β from factorial moments and possible improvement by the method of maximum likelihood. Moran (1968, chap. 2.16) has also considered the relation between the negative hypergeometric and the distribution responding to the Polya–Eggenberger urn with c non-negative integer. Recently, coming back to Skellam's procedure, Chatfield and Goodhardt (1970) have used the *beta-binomial model* for consumer purchasing behavior. They propose a method of fitting by mean and zeros.

Their study may be applied to drug purchasing habits.

6.210.2. *Reduction parameters*

By analogy with the formulae for the positive or classical hypergeometric law, the reduction parameters of the negative hypergeometric law are:

Mean:

$$E[r] = n\frac{a}{N_0} = np_0; \tag{a}$$

$p_0 = a/N_0$ and $q_0 = b/N_0$ are the *initial* proportions of white and black balls in the urn before adding in $c = +1$.

Variance:

$$V(r) = n \cdot \frac{a}{N_0} \cdot \frac{N_0 - a}{N_0} \cdot \frac{N_0 + n}{N_0 + 1}. \tag{b}$$

For $N_0 \approx (N_0+1)$ and $f = n/N_0$, we have

$$\sigma^2_{\text{hyperg.}<0} \approx np_0q_0(1+f) \approx \sigma^2_{\text{bin.}>0}(1+f). \tag{c}$$

The variance of the negative hypergeometric law is greater than that of the positive binomial corresponding to the initial population. If $N_0 \to \infty$ and $a \to \infty$ such that $a/N_0 \to p$, then the negative hypergeometric law tends towards the positive binomial in (n, p).

The moment generating function is written in Moran (1968, p. 77).

Remark. Compound and generalized discrete distributions. This contribution is mainly devoted to individual discrete and continuous probability laws. We have just mentioned a few cases of compound discrete distributions in sections 6.265, 6.271.5, 6.210.1.

6.3. PROBABILITY LAWS GOVERNING CONTINUOUS RANDOM VARIABLES

6.30. General Considerations

6.301. Continuous Random Variables associated with Probability Density and Distribution Functions

(1) *Dimensional continuous variables* such as the waiting time for an all-or-nothing reaction, length, the weight of some organ, etc.

(2) *Dimensionless continuous variables*, a pure number such as:

(1) A reduced random variable of the type X/σ_X or αX where the dimension of α is the inverse of the dimension of X, which situation is written

$$[\alpha] = \frac{1}{[X]} = [X]^{-1}.$$

Another type of reduced variable is $U/U_{0.5}$, i.e. the ratio of an actual weight U to the median weight of the group; this *reduced weight* has been used by several authors and recently by Walker and Duncan (1967, p. 176) under the name of "relative weight" of sex–height group, as a percentage (see section 6.335.112.1 e).

(2) The logarithm of a magnitude divided by the unit of this magnitude such as $\log t/t_0$ or $\log X/X_0$.

A random variable can traverse a finite or infinite domain or range D. The probability laws of continuous random variables divide into two large classes according as the range of the random variable is finite or infinite. The laws in respect of a finite range are real in the sense that the observable physical or biological quantities are necessarily finite. However, they have fewer physical, biological and mathematical applications than the densities with an infinite range in one or both directions.

Nevertheless, the continuous laws for a finite range deserve to be studied no less for the *pedagogic* advantages which they offer, as for their application to *numerical calculation* (approximation of a given number) and for their *biological* advantages (phenomena observable and finite at the threshold or maximum).

By definition, each value X of the continuous random variable is associated with a probability density function $f(X)$ which is subject to the condition that it is never negative. Furthermore, each type of density function is associated with a closely corresponding domain D, or range, theoretically conceived to be such that the "measure" of this domain is unity. We thus have the formula

$$\int_D f(\mathrm{X})\, d\mathrm{X} = 1 \tag{a}$$

Clearly, the continuous random variable $\boldsymbol{X}$ is contained with certainty within its own range which plays the rôle of fundamental probability set.

Convention: The notation X signifies the (here continuous) current variable $\boldsymbol{X}$ in its instantaneous form, whereas $f(X)$ signifies the numerical value of the probability density at the *fixed* point X. The local probability of finding $\boldsymbol{X}$ between the fixed points $X-\Delta X/2$ and $X+\Delta X/2$ is written so:

$$\Delta F(X) \cong f(X)\, \Delta X.$$

$F(X)$ is the distribution function of $\boldsymbol{X}$ or, emphatically, the cumulative distribution of X. For $\boldsymbol{X}$ the latter is written so:

$$F(X) = \int_a^X f(\mathrm{X})\, d\mathrm{X}, \tag{b}$$

where X is the *fixed* upper limit and a is the origin of the domain.

Since the integral is the limit of a sum of terms such as $\Delta F(X)$ if $\Delta X \to 0$, it may therefore be said that the distribution $F(X)$ is the *total probability* of finding $\boldsymbol{X}$ in some one of the intervals ΔX not overlapping and reconstituting the part of the domain contained between the origin of this domain and the fixed point X. In other words, $F(X)$ is the probability of meeting with one (random) X equal to fixed X or less:

$$\text{Prob.}\,(X \leqslant X) = F(X). \tag{c}$$

In all generality, we have

F (lower limit of the theoretical domain) $= 0$
$F(X)$ is contained between 0 and 1
F (upper limit of the theoretical domain) $= 1$

We shall now distinguish three types of random variables according to their respective domains.

(1) With a random variable X one can associate a "triangular" probability density function:

$$f(X) = \frac{1}{a}\left[1 - \frac{|X-\mu|}{a}\right],$$

where $|X|$ is the value of X in absolute magnitude (or modulus of X) and μ is the mean value which varies in the *finite domain D* of range $2a$.

(2) On the other hand, the *negative exponential probability density function*

$$f(X) = \frac{1}{\lambda} e^{-X/\lambda}$$

refers to a positive random variable X moving in the infinite domain $(0, \infty)$. An example is the life duration of a radioisotope before complete decay.

(3) Finally, in the classical Gaussian curve of mean μ and of variance σ_X^2, the random variable X possesses the probability density

$$g(X) = \frac{1}{\sigma_X\sqrt{2\pi}} \exp\left\{-\frac{(X-\mu)^2}{2\sigma_X^2}\right\},$$

where X traverses the theoretical domain $(-\infty, +\infty)$.

EXAMPLES: The height of a man, birth weight of an infant and salt concentration in tests on the resistance of red corpuscles to increasingly hypotonic solutions. For the theoretically infinite range of the gaussian law there is, of course, the corresponding finite and practical range of the gaussian curve adjusted to an observed histogram.

Huzurbazar (1965, p. 232) also adopts a division of the continuous, distributions into three types based on the "standard ranges" $(-\infty, +\infty)$, $(0, +\infty)$ and $(-1, +1)$, the latter here being the standard case for the laws with finite range.

Terminology. The probability density function $f(X)$, which is an idealization of the histogram (X_i, f_i), is also called the *frequency function*, but we shall use the term "probability density", often shortened to "density". The *function* $F(X)$, or the *distribution of* X, is the idealization of the curve for cumulative relative frequencies in the case of grouped observations. This may explain why many English-speaking statisticians use the term "cumulative distribution function", whereas "distribution" is a sufficient description in itself. Feller (1958, I, p. 168) has also stated that the adjective "cumulative" is superfluous. This point of view is adopted here more particularly because biologists may also come across the term "cumulative function" used in the sense of a "cumulant-generating-function", which in the present paper is designated $K_X(t)$ (6.302.5). In the French $F(X)$ is the *fonction de répartition* which in English is called "distribution".

6.302. Moments, Cumulants and Generating Functions associated with a Continuous Distribution

We consider here the formulae for the moments and cumulants associated with a continuous random variable of positive density $f(X)$ and of distribution $F(X)$, all linked by the relation $f(X)\, d(X) = dF(X)$, on condition that

$$\int_D f(\mathrm{X})\, d\mathrm{X} = 1.$$

The integrals in dX are of Riemann or Lebesgue type, whilst those in dF are Stieltjes integrals (M. G. Kendall and Stuart, 1963, I, p. 33).

This is noted here in order to obtain a complete set of fundamental concepts for the continuous as well as the discrete random variables. For pedagogic reasons the definitions and properties are treated in the particular case of each distribution being studied.

6.302.1. *Moments of a crude variable X*

The initial moment of order i, noted μ_i' is the *expected value* of the ith power of X:

$$\mu_i' = E[X^i] = \int_D \mathrm{X}^i f(\mathrm{X})\, d\mathrm{X} = \int_D \mathrm{X}^i\, d\mathrm{F}. \qquad \text{(a)}$$

Whence $E[X] = \mu_1' = \mu$ is the *expected value* (or true value) of X, i.e.

$$E[X] = \int_D \mathrm{X} f(\mathrm{X})\, d\mathrm{X}.$$

The second moment from the origin is

$$E[X^2] = \mu_2' = \int_D X^2 f(X)\, dX \qquad \text{(b)}$$

and so on.

The central moments μ_i are

$$E[(X-\mu)^i] = \mu_i = \int_D (X-\mu)^i f(X)\, dX, \qquad \text{(c)}$$

whence

$$\mu_1 = \int_D (X-E[X]) f(X)\, dX \equiv 0. \qquad \text{(c')}$$

The second central moment

$$\mu_2 = \int_D (X-\mu)^2 f(X)\, dX \qquad \text{(d)}$$

is the true variance

$$E[(X-\mu)^2] = V(X) = \sigma_X^2.$$

Relations between the set of μ_i' and the set of μ_i. Using the expansion of the binomials $(X-\mu)^i$, within μ_i, there is no difficulty in obtaining the following relations (M. G. Kendall and Stuart, 1963, I, p. 56):

$$\mu_1 = \mu_1' - \mu_1' = 0,$$

$$\mu_2 = \mu_2' - \mu_1'^2,$$

or

$$V(X) = E[X^2] - (E[X])^2 \qquad \text{(e)}$$

$$\mu_3 = \mu_3' - 3\mu_1'\mu_2' + 2\mu_1'^3 .$$

For a symmetrical probability density about the mean, $\mu_3 = 0$,

$$\mu_4 = \mu_4' - 4\mu_1'\mu_3' + 6(\mu_1')^2 \mu_2' - 3\mu_1'^4 . \qquad \text{(e')}$$

The moments μ_i and μ_i' are homogeneous expressions of ith degree in X. The dimension of these moments is related to the dimension of X by the relation

$$[\mu_i] = [\mu_i'] = [X]^i. \qquad \text{(f)}$$

Störmer (1970, p. 23) gives the μ_i' in terms of $F(X)$ exclusively.

6.302.2. *Initial moment generating function*

The set of the μ_i' of the random variable X can be generated by an initial moment generating function:

$$M_X(t) = E[e^{tX}]. \qquad \text{(a)}$$

It is seen that $M_X(t)$ is the expected value of an increasing exponential function of X, where t is an auxiliary variable which is of interest through its increasing powers t, t^2, t^3, etc. The exponents of t correspond to the order of the respective moments:

$$
\begin{aligned}
E[e^{tX}] &= \int_D e^{tX} f(X)\, dX = \int_D e^{tX}\, dF \\
&= \int_D \left(1+X\frac{t}{1}+X^2\frac{t^2}{2!}+\ldots\right) f(X)\, dX \\
&= \int_D dF\cdot 1+\int_D X\, dF\cdot\frac{t}{1}+\int_D X^2\, dF\cdot\frac{t^2}{2!}+\ldots. \qquad \text{(b)}
\end{aligned}
$$

Hence

$$M_X(t) = E[e^{tX}] = 1+\mu_1'\frac{t}{1}+\mu_2'\frac{t^2}{2!}+\ldots, \qquad \text{(c)}$$

on condition that the integral exists.

In any case, we always have the identity

$$M_X(0) = \int_D dF = 1. \qquad \text{(d)}$$

Providing again that the integrals described in (b) do exist, one can successively isolate $\mu_1', \mu_2', \ldots$ by taking the first, second, etc. derivatives of $M_X(t)$ at the point $t = 0$. Thus, for example,

$$M_X'(t) = 0+\mu_1'+\mu_2't+\ldots \qquad \text{(e)}$$

for $t = 0$, and there remains

$$\mu_1' = M_X'(0) \quad \text{or} \quad [DM_X(t)]_{t=0}. \qquad \text{(f)}$$

In the following pages several other examples are given.

6.302.3. *Central moment generating functions*

The generating function of the central moments refers to the deviation about the mean

$$x = X-\mu \qquad \text{(a)}$$

and is written as $M_x(t)$ or $m_x(t)$ in analogy with $m(t)$ or $m(z)$ of section 6.204 where the subscript r is omitted

$$m_x(t) = E[e^{tx}] = E[e^{t(X-\mu)}] = E[e^{tX}]\cdot e^{-\mu t}. \qquad \text{(a')}$$

Therefore since μ is constant in the evaluation of $E[e^{tX}]$ we have

$$m_x(t) = e^{-\mu t} \cdot M_X(t). \tag{b}$$

The central moment generating function can therefore be calculated without difficulty from the initial moment generating function.

The notation $m_x(t)$ is redundant since it is enough to write $M_x(t)$ for the central moment generating function while the initial moment generating function is written $M_X(t)$.

Just as in expression (b) in section 6.302.2, we expand (a):

$$\begin{aligned} m_x(t) &= \int_D e^{t(X-\mu)}\, dF \\ &= \int_D \left[1+(X-\mu)\frac{t}{1}+(X-\mu)^2\frac{t^2}{2!}+\dots\right] dF, \\ &= \int_D dF\cdot 1+\int_D (X-\mu)\, dF\cdot\frac{t}{1}+\int_D (X-\mu)^2\, dF\cdot\frac{t^2}{2!}+\dots. \end{aligned} \tag{c}$$

Hence

$$m_x(t) = 1+\mu_1\frac{t}{1}+\mu_2\frac{t^2}{2!}+\dots. \tag{d}$$

We have

$$m_x(0) = \int_D dF = 1 \tag{e}$$

as previously in 6.302.2(d).

But by virtue of the formulae 6.302.1(e), we have $\mu_1 = 0$. Therefore

$$m_x(t) = 1+\mu_2\frac{t^2}{2!}+\mu_3\frac{t^3}{3!}+\dots. \tag{f}$$

Upon applying differentiation with respect to $t = 0$, we get

$$\begin{aligned} m'_x(0) &= 0 \\ m''_x(0) &= \mu_2 = \sigma_x^2 \\ m'''_x(0) &= \mu_3 \quad \text{etc.} \end{aligned} \tag{g}$$

6.302.4. Characteristic functions

From Poincaré (1912, p. 126), whenever the integral 6.302.2 (b) does not exist for real t, the real t is replaced by the purely imaginary it $\left(i = \sqrt{-1}\right)$, and the characteristic function of an arbitrary distribution F is then written so:

$$E[e^{itX}] = \int_D e^{itX}\,dF = \varphi_X(t). \tag{a}$$

For example, $M_X(t)$ does not exist for the Cauchy law, whereas $\varphi_X(t)$ does have meaning.

Under certain general conditions, if the characteristic function is known, this is equivalent to knowing the distribution, and *vice versa* (M. G. Kendall and Stuart, 1963, I, chapter 4). This function comes from studies in mathematical statistics and probability analysis, so it is not used in the present context where the purposes of biologists are mainly concerned.

Owing to the formal analogy between the formulae 6.302.2 (b) and 6.302.4 (a), the characteristic function is also the moment generating function.

Furthermore, the characteristic function is the *Fourier transform* of the particular probability density corresponding to the distribution $F(X)$.

We again take formula (a), but now for a domain $D = (-\infty, +\infty)$ and after having introduced the probability density $f(X)$:

$$\varphi_X(t) = \int_{-\infty}^{+\infty} e^{itX} f(X)\,dX. \tag{b}$$

Under certain conditions, by applying the *theorem of inversion*, one can isolate $f(X)$ by entering $\varphi_X(t)$ within the integral:

$$f(X) = \frac{1}{2\pi} \int_{-\infty}^{+\infty} \varphi_X(t) e^{-iXt}\,dt. \tag{c}$$

Such is the "dual role" of the characteristic function and of the probability density.

In the present context $\varphi_X(t)$ is the first characteristic function, whilst

$$\psi_X(t) = \ln \varphi_X(t)$$

is the second characteristic function (Roy, 1965, pp. 38 and 42). The latter is referred to by some English-speaking writers as the "cumulative function" (Cornish and Fisher, 1937; and *inter alia* Kendall and Stuart, 1963, I, p. 68).

6.302.5. *Cumulant generating functions*

In mathematical physics one requires quantities which are invariant to changes of coordinates of different complexity. In statistics, consider the relations 6.302.2 (d) and 6.302.3 (e) which yield the zero-th initial moment $M_X(0) = 1$ or central moment $m_x(0) = 1$. These remain invariant in the transformation of origin by μ in passing from the original variable X to the deviation $x = X-\mu$. In fact it is certain that X is to be found in D and that x is in the corresponding domain. On the other hand, the higher moments are not invariant in this transformation process (see the formulae 6.302.1 e). However, Thiele (1903) introduced a system of ***semi-invariants*** or ***cumulants*** which were designated as $\varkappa_i$ by R. A. Fisher and have been studied in Cornish–Fisher (1937).

The cumulants $\varkappa_i$ of a distribution $F(X)$, or of a variable X, are found from the cumulant generating function introduced by *Laplace* (Cornish–Fisher, 1937, p. 2):

$$K_X(t) = \ln M_X(t). \tag{a}$$

Upon expanding $K_X(t)$, we obtain the cumulants

$$K_X(t) = \varkappa_1' \frac{t}{1} + \varkappa_2 \frac{t^2}{2!} + \varkappa_3 \frac{t^3}{3!} + \ldots, \tag{b}$$

where $K_X(0) = \ln 1 = 0$.

The relation between the moments μ_i' and the cumulants $\varkappa_i$ is given by the identity

$$\varkappa_1' \frac{t}{1} + \varkappa_2 \frac{t^2}{2!} + \varkappa_3 \frac{t^3}{3!} + \ldots \equiv \ln \left(1 + \mu_1' \frac{t}{1} + \mu_2' \frac{t^2}{2!} + \ldots\right)$$

$$\equiv \ln M_X(t). \tag{c}$$

In order to isolate $\varkappa_1'$, we differentiate this identity with respect to t

$$\varkappa_1' + \varkappa_2 t + \ldots = \frac{1}{M_X(t)} \frac{\partial M_X(t)}{\partial t}. \tag{d}$$

At the point $t = 0$, there remains

$$\varkappa_1' = \frac{1}{M_X(0)} \cdot \mu_1' = \mu_1'. \tag{d'}$$

Taking the first derivative

$$\varkappa_1' + \varkappa_2 t + \varkappa_3 \frac{t^2}{2!} + \varkappa_4 \frac{t^3}{3!} \ldots = \frac{M_X'(t)}{M_X(t)},$$

we pass on to the second derivative

$$K_X''(t) = \varkappa_2+\varkappa_3 t+\varkappa_4 \frac{t^2}{2} \dots = \frac{M_X(t)\, M_X''(t)-[M_X'(t)]^2}{[M_X(t)]^2}\,. \qquad \text{(e)}$$

At the point $t = 0$, there remains

$$K_X''(0) = \varkappa_2 = M_X(0)\, M_X''(0)-[M_X'(0)]^2, \qquad \text{(f)}$$

for the denominator is equal to unity since $M_X(0) = 1$.

By definition, $M'(0) = \mu_1'$ and $M''(0) = \mu_2'$; the second cumulant of X is therefore written as

$$\varkappa_2 = M_X''(0)-[M_X'(0)]^2 = \mu_2'-\mu_1'^2 = \mu_2\,. \qquad \text{(f')}$$

It is thus seen that the second-order cumulant is no other than the second central moment of X or the variance of X. Indeed the relation of invariance as regards the transformation $\pm a$ is

$$\varkappa_2(X+a) = \varkappa_2(X). \qquad \text{(f'')}$$

In particular, for the deviation $x = X-\mu$, we have

$$\varkappa_2(x) = \varkappa_2(X-\mu) = \varkappa_2(X). \qquad \text{(f''')}$$

We could likewise obtain $\varkappa_3$ and $\varkappa_4$, etc.

It has just been shown that the cumulants $\varkappa_i$ can be obtained by simple partial differentiations on $K_X(t)$ (in order to obtain the desired cumulant), and others on $\ln M_X(t)$ (to obtain the ith order cumulant as a function of the μ_j' $(j \leqslant i)$). In this section we use the classical method of identification in order to outline the calculation. M. G. Kendall and Stuart (1963, I, p. 70) have formed the relations for the cumulants $\varkappa_1, \varkappa_2, \dots, \varkappa_{10}$ from a process of identification. We write here the first four cumulants of the original variable X as a function of the μ_i' and also of the μ_i from 6.302.1:

$$\begin{aligned} \varkappa_1' &= \mu_1' \\ \varkappa_2 &= \mu_2'-\mu_1'^2 = \mu_2 \\ \varkappa_3 &= \mu_3'-3\mu_2'\mu_1'+2\mu_1'^3 = \mu_3 \\ \varkappa_4 &= \mu_4'-4\mu_3'\mu_1'-3\mu_2'^2+12\mu_2'\mu_1'^2-6\mu_1'^4 = \mu_4-3\mu_2^2. \end{aligned} \qquad \text{(g)}$$

6.302.6. Cumulant generating function of the central variable
$x = X - \mu$

Consider once more the generating function of the central moments μ_i:

$$M_x(t) = e^{-\mu t} M_X(t). \tag{a}$$

The generating function of the cumulants of x (or centred for X) is written as follows

$$K_x(t) = \ln M_x(t) = -\mu t + \ln M_X(t). \tag{b}$$

It should be noted that $K(0) = 0$ since $\ln 1 = 0$.

The subscript x of the cumulants $\varkappa_i$ denotes the random variable to which these cumulants relate.

Upon expanding $K_x(t)$, we get

$$\varkappa_{1x}\frac{t}{1} + \varkappa_{2x}\frac{t^2}{2!} + \varkappa_{3x}\frac{t^3}{3!} + \ldots = -\mu t + \ln M_X(t). \tag{c}$$

We derive this relation

$$\varkappa_{1x} + \varkappa_{2x}t + \ldots = -\mu + \frac{M_X'(t)}{M_X(t)}. \tag{d}$$

Therefore the derivative $K_x'(0)$ is

$$\varkappa_{1x} = -\mu + \frac{M_X'(0)}{M_X(0)} = -\mu + \mu_1' = -\mu + \varkappa_{1X}' = 0, \tag{e}$$

since

$$\mu = \mu_1' = \varkappa_{1X}'. \tag{e'}$$

The *non-invariance* of the first cumulant is thus shown in relation to a transformation of amplitude μ.

Concisely, for $x = X - \mu$,

$$\varkappa_{1X}' = \mu_1' = \mu$$
$$\varkappa_{1x} = 0.$$

Successive differentiations of (d) show simply that

$$\begin{aligned} \varkappa_{2x} &= \varkappa_{2X} \\ \varkappa_{3x} &= \varkappa_{3X} \end{aligned} \tag{f}$$

etc.

In fact, the constant μ disappears as from the second derivative of (b). These equalities are justifiably called *semi-invariants* (in relation to translations) of the quantities $\varkappa_i$.

6.302.7. Effect of a change of scale on moments, cumulants and their generating functions

We change over from the crude random variable X to the reduced variable αX ($\alpha > 0$, constant) and $[\alpha] = [X]^{-1}$ where [ ] means "the dimension of" as used in physics (6.301). The following relations hold good (the relevant variable appears in round brackets):

1. *Reduced variable*: αX

$$\mu_i'(\alpha X) = \alpha^i \mu_i'(X)$$

$$\mu_i(\alpha X) = \alpha^i \mu_i(X) \tag{a}$$

$$\varkappa_i(\alpha X) = \alpha^i \varkappa_i(X)$$

$$M_{\alpha X}(t) = E[e^{t\alpha X}] = M_X(\alpha t) \tag{b}$$

$$K_{\alpha X}(t) = \ln M_{\alpha X}(t) = \ln M_X(\alpha t) \tag{c}$$

2. *Reduced deviate*: $z = \alpha x = \alpha(X-\mu)$; $[\alpha] = [X]^{-1}$ From

$$\frac{z}{\alpha} = X - \mu \tag{d}$$

we see that the ratio z/α is equal to the deviate $X-\mu$. We know that

$$M_{X-\mu}(t) = e^{-\mu t} M_X(t). \tag{e}$$

For the ratio z/α, we have

$$M_{z/\alpha}(t) = E[e^{t(z/\alpha)}] = E[e^{(t/\alpha)z}] = M_z\left(\frac{t}{\alpha}\right). \tag{f}$$

Whence the formula (Gumbel, 1960, p. 14) giving the moment generating function of z.

$$M_z\left(\frac{t}{\alpha}\right) = e^{-\mu t} M_X(t). \tag{g}$$

3. *Standard deviate.*

For the standard deviate

$$\xi = \frac{X-\mu}{\sigma_X}, \tag{h}$$

formulae (f) and (g) are written with the replacement of $1/\alpha$ by σ_X.

6.302.8. Parameters of variability and shape for a probability law (K. Pearson diagram)

With a distribution for which the moments or cumulants are known, one generally uses $\varkappa_2 = \mu_2 = \sigma^2$ as the absolute measure of the variability.

For a variable X having the deviation $x = X-\mu$, the variance is unchanged by any translation of X. Hence

$$\sigma_X^2 = \sigma_x^2 .$$

The parameters of shape are based on the moments or cumulants of order 3 and 4 divided by a convenient power of $\mu_2 = \sigma^2$, in order to obtain a dimensionless number. For K. Pearson the parameters of shape are written from consideration of the central moments:

$$\begin{aligned} &\text{Skewness} \quad \beta_1 = \frac{\mu_3^2}{\mu_2^3}, \\ &\text{Kurtosis} \quad \beta_2 = \frac{\mu_4}{\mu_2^2}. \end{aligned} \tag{a}$$

But R. A. Fisher proceeds from the cumulants 3 and 4:

$$\begin{aligned} &\text{Skewness} \quad \gamma_1 = \frac{\varkappa_3}{\varkappa_2^{3/2}} = \sqrt{\beta_1}, \\ &\text{Kurtosis} \quad \gamma_2 = \frac{\varkappa_4}{\varkappa_2^2} = \frac{\mu_4}{\mu_2^2}-3 = \beta_2-3. \end{aligned} \tag{b}$$

From the point of view of symmetry, a density of reference is any symmetrical density for which $\mu_3 = 0$; therefore $\beta_1 = 0$ and $\gamma_1 = 0$. From the kurtosis point of view, the density of reference is the Gaussian density for which $\mu_4 = 3\sigma^4$; therefore $\beta_2 = 3$ and $\gamma_2 = 0$.

Of course, when one compares the kurtosis of two symmetrical curves, the latter are superposed with the same variance. The reader is referred to R. A. Fisher (1922, p. 322) for the superposition of two "symmetrical error curves of equal intrinsic accuracy" with the equations, respectively,

$$df = \frac{1}{\pi}\,\frac{dx}{1+x^2} \qquad \text{(Cauchy)},$$

$$df_1 = \frac{1}{2\sqrt{\pi}}\,e^{-x^2/4}\,dx \qquad \text{(Gauss)}.$$

In this exposition of general concepts valid for the majority of continuous distributions, it is now the appropriate time to draw special attention to the existence of the Pearson diagram (β_1, β_2) (Pearson–Hartley, 1954, p. 210). In this particular representation *all* the Gauss curves at triple infinity ∞^3 (n, μ, σ) are represented by a unique point $\beta_1 = 0, \beta_2 = 3$. The diagram is divided into separate zones by lines where β_2 is a known function of β_1; the zones and lines all correspond to types of probability laws of Pearson's system (see e.g. Elderton, 1953; M. G. Kendall and Stuart, 1963, I, pp. 148–154 and 167–173 for the system introduced by Johnson (1949).

When a biologist finds a considerable amount of new data and he has made a serious reduction of these data using statistics $\bar{X}$, s_X^2, g_1 and g_2 or b_1 and b_2, the localization of the point (b_1, b_2) in the Pearson diagram is always useful. However, this point (b_1, b_2) does not have the value of the theoretical point $(\beta_1; \beta_2)$ for it is subject to sampling variation (Martin, 1972).

6.302.9. Generating functions of a sum of independent random variables

The following theorems can be stated:

(1) The generating function of the moments μ_i' of a sum of independent variables

$$X = X_1 + X_2 + \ldots + X_n \tag{a}$$

is equal to the product of the generating functions of the analogous moments of the constituent variables:

$$\begin{aligned} M_X(t) &= M_{X_1+\ldots+X_n}(t) = E[\exp\{X_1+X_2+\ldots+X_n\}t] \\ &= M_{X_1}(t)\cdot M_{X_2}(t)\cdot\ \ldots\ \cdot M_{X_n}(t). \end{aligned} \tag{b}$$

Hence

$$M_{\sum_i X_i}(t) = \prod_{i=1}^{n} M_{X_i}(t). \tag{c}$$

Special case: if X is the sum of n independent random variables with the same distribution $F(X)$, we get

$$M_{nX} = \underbrace{M_{X+X+\ldots+X}}_{n \text{ terms}}(t) = [M_X(t)]^n \tag{c'}$$

(2) The generating function of the cumulants of the sum of n independent variables $X = X_1 + \ldots X_n$ is equal to the sum of the cumulant generating

functions $K_{X_i}(t)$ in the constituent variables. In fact, we have

$$K_{X_1+X_2 \dots +X_n}(t) = \ln (M_{X_1+X_2+ \dots X_n}(t))$$
$$= \ln M_{X_1}(t) + \ln M_{X_2}(t) + \dots \ln M_{X_n}(t)$$
$$= K_{X_1}(t) + K_{X_2}(t) + \dots + K_{X_n}(t). \quad \text{(d)}$$

Hence

$$K_{\sum_i X_i}(t) = \sum_i K_{X_i}(t). \quad \text{(e)}$$

Special case: if X is the sum of n independent variables with the same distribution $F(X)$, then

$$K_{nX}(t) = \underbrace{K_X(t) + \dots K_X(t)}_{n \text{ terms}} = nK_X(t). \quad \text{(e')}$$

6.31. Continuous Random Variables in a Finite Domain

6.311. Rectangular or Uniform Law

6.311.1. Uniform law in the original variable X

6.311.11. Uniform probability density in X

6.311.111. Definition. The rectangular or uniform law associates with the continuous random variable moving in the finite domain or range (a, b), the probability density

$$f(X) = K \text{ const.} \quad \text{(a)}$$

The dimension of K is the inverse of the dimension of X. Thus, if X is a dimension expressed in cm, the constant K is denoted in cm^{-1} as has been said in 6.302.7:

$$[K] = [X]^{-1}. \quad \text{(a')}$$

The condition

$$\int_a^b f(X)\, dX = 1 \quad \text{(b)}$$

expresses the certainty of finding a random value for X between a and b.

Hence the rectangular or uniform probability density function in the domain (a, b) of measure $R = b-a$, is

$$f(X) = \frac{1}{b-a} = \frac{1}{R}. \tag{c}$$

If X is given in cm, the probability density $f(X)$ is in cm^{-1}.

The *elementary probability* of finding (random) $\boldsymbol{X}$ between $X-(dX/2)$ and $X+(dX/2)$ (X fixed) is written as $f(X)\,dX = dX/R$. This probability is a dimensionless, pure number.

The discrete probability P_r which was associated with the discrete random variable r, is replaced here by the infinitesimal probability

$$dF(X) = f(X)\,dX \tag{d}$$

associated with X.

The discrete summation $\sum_r$, defined over the domain of r, is replaced by the limit of a sum of infinitesimals, or the integral over the domain D of X.

The probability of finding the random value $\boldsymbol{X}$ in the infinitesimal interval $(X-dX/2, X+dX/2)$, centred on fixed X, is written so:

$$dF(X) = \text{Prob}\left\{X-\frac{dX}{2} X \leqslant \boldsymbol{X} \leqslant X+\frac{dX}{2}\right\} = \frac{dX}{R}. \tag{e}$$

Depending on the nature of X, this *infinitesimal probability* is of *local type* (X = length for instance), or *instantaneous* (X = time).

6.311.112. *Reduction parameters*

The *range*, or domain of variation, is

$$R = b-a.$$

Moments. As in the discrete case, the reduction parameters of the probability density functions of continuous random variables are based on the moments. In the present case the initial rth order moment is no longer written as a sum $\left(\sum_r\right)$, but as the integral

$$\mu_i' = E[\boldsymbol{X}^i] = \int_D X^i f(\mathrm{X})\,d\mathrm{X}, \tag{a}$$

the definite integral here extending over the fundamental domain $\boldsymbol{D}$ of measure R.

The central *r*th order moment (about the mean) is

$$\mu_i = E[(X-\mu)^i] = \int_D (X-\mu)^i f(X)\, dX. \tag{b}$$

Introducing the absolute deviation $x = X-\mu$, we have

$$\mu_i = \int_D x^i f(x+\mu)\, dx.$$

Mean. The mean value μ is the initial first-order moment or the expected value of X:

$$\mu = \mu_1' = E[X] = \int_a^b X f(X)\, dX$$

$$= \frac{1}{R}\int_a^b X\, dX = \frac{1}{R}\left[\frac{X^2}{2}\right]_a^b = \frac{1}{2R}(b^2-a^2) \tag{b'}$$

$$= \frac{1}{2R}(a+b)\cdot(b-a)$$

$$\mu = E[X] = \frac{a+b}{2}.$$

Here the letter E stands for "expected value of" as seen in 6.302.1.

The mean of the rectangular or uniform distribution occurs at the mid-point between the extremities of the domain (see Fig. 6.311.112I).

In the following calculations the origin is transferred to μ, hence the *absolute deviation* is $x = X-\mu$.

Clearly,

$$dx = dX.$$

When X varies from a to b, x traverses the interval $(-R/2, +R/2)$.

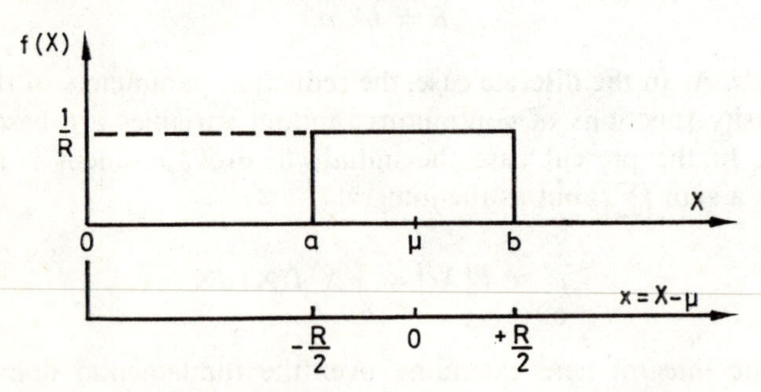

FIG. 6.311.112 I. Rectangular (uniform) probability density.

Variance:

$$\sigma^2 = V(X) = \mu_2 = \int_a^b (X-\mu)^2 f(X)\, dX. \tag{c}$$

By virtue of the symmetry about μ, we have

$$V(X) = \frac{2}{R}\int_0^{R/2} x^2\, dx = \frac{2[x^3]_0^{R/2}}{3R} = \frac{2R^3}{24R} = \frac{R^2}{12}. \tag{d}$$

The variance of the rectangular or uniform density of range R is

$$\frac{R^2}{12} = 0.0833R^2. \tag{e}$$

This result is used to introduce Sheppard's correction, applied to the original variance in the case of bell-shaped histograms where R is the class interval.

Standard deviation:

$$\sigma = \frac{R}{\sqrt{12}} = \frac{R}{2\sqrt{3}} = \frac{R}{3.46}. \tag{f}$$

Skewness:

$$\gamma_1 = \frac{\mu_3}{\sigma^3}. \tag{g}$$

A priori, since the distribution is symmetrical about μ, the skewness should therefore be zero. It is of interest to verify this by calculating γ_1 and find

$$\gamma_1 = 0. \tag{g'}$$

Kurtosis:

$$\gamma_2 = \frac{\mu_4}{\sigma^4} - 3. \tag{h}$$

By definition and owing to the symmetry of $f(X)$, we have

$$\mu_4 = \frac{2}{R}\int_0^{R/2} x^4\, dx = \frac{2}{5R}[x^5]_0^{R/2} = \frac{2}{5\times 32R}\times R^5 = \frac{R^4}{80}$$

$$\sigma^4 = \frac{R^4}{144}.$$

Therefore,

$$\gamma_2 = \frac{R^4}{80} \times \frac{144}{R^4} - 3 = 1.8 - 3 = -1.2. \qquad \text{(h')}$$

The uniform or rectangular distribution is platykurtic, as expected.

6.311.12. Rectangular or uniform distribution

6.311.121. Definition. The definition of the distribution function given in the form of a *sum* for discrete random variables, applies also to continuous variables with finite (or infinite) range. The probability of seeing an instantaneous value **X** at most equal to a fixed value *X*, or a cumulative probability up to *X*, is now written by means of a definite *integral*;

$$F(X) = \text{Prob.}\,(\mathbf{X} \leqslant X) = \int_a^X f(\mathrm{X})\,d\mathrm{X}, \qquad \text{(a)}$$

where *a* is the lower limit of the interval of variation of *X*.

For the uniform or rectangular distribution of range *R*, we have

$$F(X) = \text{Prob.}\,(\mathbf{X} \leqslant X) = \frac{1}{R}\int_a^X d\mathrm{X} = \frac{1}{R}\,[\mathrm{X}]_a^X = \frac{\mathrm{X}-a}{R}. \qquad \text{(b)}$$

For $X = b$, the value 1 is again found. Figure 6.311.121I shows the rectangular or uniform density function and the corresponding distribution function.

The uniform distribution function increases linearly with the upper limit *X* assigned to the instantaneous value **X**. It is represented by the shaded area under the density $f(X)$ and by the corresponding ordinate of the distribution function $F(X)$. The slope of the distribution straight-line of *X* is $1/R$, since in the right triangle *a, b, c*

$$\tan\theta = \frac{1}{R}.$$

Thus, in conclusion, the uniform or rectangular distribution function of *X* is defined by the following relations:

$$\begin{array}{ll} X \leqslant a & F(X) = 0 \\ a < X < b & F(X) = \dfrac{X-a}{R} \\ X \geqslant b & F(X) = 1 \end{array} \qquad \text{(c)}$$

6.311.122. Reduction parameters. To the classical parameters of reduction which have already been discussed (6.311.112), it is convenient to add the case where distributions are finite in range. The quantile 0 is designated as a, and for the quantile 1 we put b. In order to be specific, but perhaps unrealistic, we shall assume a rectangular distribution of the "sensitivity" of a biological material to toxic effects; as it is not certain that this model is applicable, we shall make this assumption provisionally.

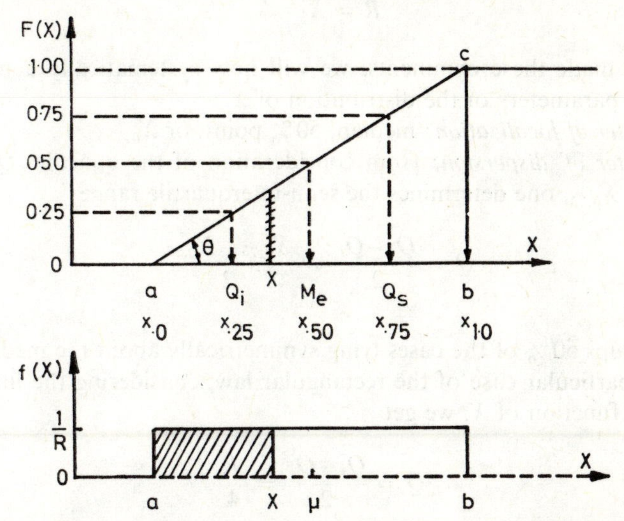

FIG. 6.311.121 I. Rectangular (uniform) density $f(X)$ and distribution $F(X)$.

Threshold point a is the point at which the biological material should commence to respond; the quantile $X_{0.0} = a$ corresponds to 0% response; but there is a margin of variability.

Median point $X_{0.5}$ (50% point): the point at which half of the most sensitive part of the population has already reacted while the least sensitive half has yet to react. This median point localizes the distribution on the physical or toxic scale OX. It corresponds to the root of the equation

$$F(X_{0.5}) = 0.5. \quad \text{(d)}$$

Point X_1 *of integral response* (100% point): the point at which the biological material as a whole has responded and beyond which 100% of the population would continue to respond; there is also a margin of variability at this terminal point.

It should be mentioned that these three points are not independent, for

$$X_{0.5} = \frac{X_{0.0}+X_1}{2}. \tag{e}$$

The median is thus equal to the mean since $X_0 = a$ and $X_1 = b$. Moreover, the domain of "sensitivity" extends from $X_0 = a$ to $X_1 = b$, i.e. within the range

$$R = X_1 - X_0. \tag{f}$$

Having made these comments, we will now systematically consider the reduction parameters of the distribution of X.

Parameter of localization: median, 50% point, or $X_{0.5}$

Parameter of dispersion: from consideration of the quartiles $Q_i = X_{0.25}$ and $Q_s = X_{0.75}$, one determines the semi-interquartile range

$$Q = \frac{Q_s - Q_i}{2} = \frac{X_{0.75} - X_{0.25}}{2} \tag{g}$$

which groups 50% of the cases lying symmetrically about the median.

In the particular case of the rectangular law, considering the linearity of $F(X)$ as a function of X, we get

$$Q = \frac{Q_s - Q_i}{2} = \frac{R}{4}. \tag{h}$$

The interval of variation or range R and the standard deviation $\sigma = R:2\sqrt{3}$ are determined from the density $f(X)$. Likewise, the semi-interquartile range is a parameter of dispersion determined from the distribution $F(X)$. For the uniform law, we have the relation

$$\sigma = \frac{2Q}{\sqrt{3}} = \frac{Q_s - Q_i}{\sqrt{3}}. \tag{i}$$

Skewness. After R. A. Fisher the skewness parameter is denoted by γ_1 when it is determined from the probability density $f(X)$. We shall put γ_1' when the skewness is determined from the distribution $F(X)$. In this respect we adopt the definition given by Kelley (1924, p. 77), and also used by Peatman (1947, p. 392), for the Gauss curve:

$$\gamma_1' = \frac{X_{0.10} + X_{0.90}}{2} - X_{0.5} \tag{j}$$

Since the rectangular or uniform law is symmetrical, therefore $\gamma_1' = 0$, just the same as γ_1 (formula 6.311.112 g).

Kurtosis. After Kelley and using the proposed notation, the parameter of kurtosis is

$$\gamma_2' = \frac{X_{0.75} - X_{0.25}}{2} \cdot \frac{1}{X_{0.90} - X_{0.10}}. \tag{k}$$

The kurtosis parameter is thus defined for the distribution curve as the quotient of the semi-interquartile range by the interdecile range $X_{0.90} - X_{0.10}$.

6.311.2. Standard rectangular probability law

Consider a continuous random variable X which represents a distance measured in cm. The rectangular density for distances X measured in cm is defined by the mean μ and by the range R cm. From R we calculate σ cm $= R : 2\sqrt{3}$ cm and $f(X) = 1/R$ cm^{-1}.

6.311.21. Standard rectangular probability density

If a rectangular density independent of the unit of measurement is desired, the following operations must be performed:

(1) *Translation* of the physical origin $X = 0$, which is *identical* for all cases, into $X = \mu$, which then acts as *specific* origin to each particular case; hence the original values X_{cm} are replaced by the absolute deviation

$$x_{cm} = X_{cm} - \mu_{cm}; \tag{a}$$

(2) *Change of scale* to be obtained by expressing the absolute deviation x_{cm} in units σ_{cm}; the standard deviation thus justifies its name for it serves as a *specific standard* to each case for measuring the absolute deviation. The *rectangular standard deviate* ξ (a dimensionless number) is the quotient of

$$\xi = \frac{X_{cm} - \mu_{cm}}{\sigma_{cm}} = \frac{x_{cm}}{\sigma_{cm}} \tag{b}$$

These transformations are illustrated in Figs. 6.311.21I and II.

In the standard deviate ξ, the standardized range becomes the pure number

$$\frac{R_{cm}}{\sigma_{cm}} = \frac{R_{cm}}{R_{cm}/2\sqrt{3}} = 2\sqrt{3}. \tag{c}$$

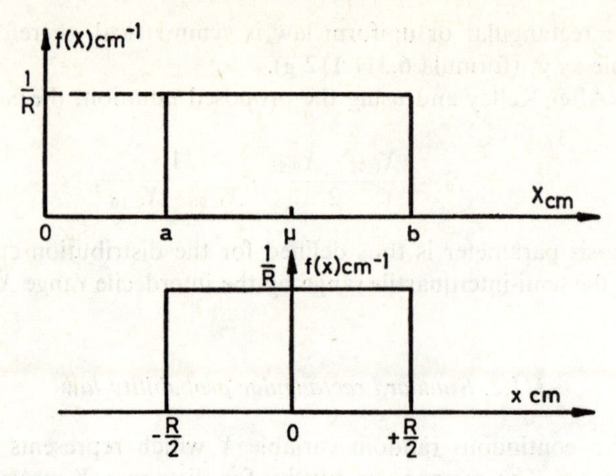

FIG. 6.311.21 I. Change of origin in the rectangular (uniform) probability density: $x = X - \mu$.

The domain of the standard rectangular deviate is therefore contained between $-\sqrt{3}$ and $+\sqrt{3}$.

In order to find the probability density $\varphi(\xi)$ of the standard deviate, we avail ourselves of the property of invariance of the total differential. The corresponding conservation of the local probability about X and about ξ

$$\frac{X-\mu}{\sigma_X} = \xi, \tag{d}$$

is given by the relation

$$f(X)\,dX = \varphi(\xi)\,d\xi. \tag{e}$$

But since μ and σ_X are constants,

$$d\xi = \frac{dX}{\sigma_X}. \tag{f}$$

Therefore

$$\varphi(\xi) = f(X)\frac{dX}{d\xi} = f(X)\cdot\frac{dX}{dX/\sigma_X} = \sigma_X f(X). \tag{g}$$

This general formula is utilized in the *transformation of random variables*.

The probability density of the standard deviate (pure number) ξ is equal to the product of the σ_X cm of the original variable with the probability

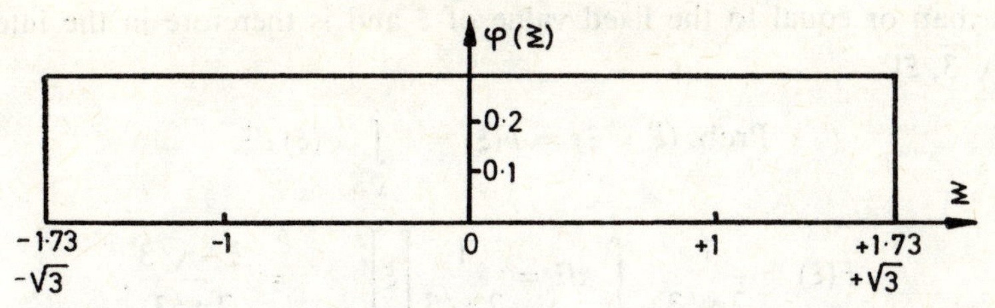

FIG. 6.311.21 II. Change of scale in the centered rectangular (uniform) density: $\xi = x/\sigma_x$.

$$\varphi(\xi) = \frac{1}{2\sqrt{3}} \approx 0.29.$$

$$E\xi = 0. \quad V\xi = 1. \quad \gamma_1 = 0. \quad \gamma_2 = -1, 2.$$

density $f(X)_{\mathrm{cm}^{-1}}$ of this same original variable. This product, as expected, is a pure number.

Consider now the special case of the rectangular law:

$$f(X) = \frac{1}{R}, \tag{h}$$

$$\sigma_X = \frac{R}{2\sqrt{3}}. \tag{i}$$

By virtue of (g), one obtains the standard rectangular law which is unique

$$\varphi(\xi) = \frac{1}{R} \times \frac{R}{2\sqrt{3}} = \frac{1}{2\sqrt{3}}, \tag{j}$$

and is represented in Fig. 6.311.21II.

We in fact again get the standardized range, the inverse of which is the ordinate of $\varphi(\xi)$. This transformed law still remains a probability law in ξ. In fact the probability of finding the random ξ in this domain is equal to unity:

$$\int_D \varphi(\xi)\, d\xi = \frac{1}{2\sqrt{3}} \int_{-\sqrt{3}}^{+\sqrt{3}} d\xi = \frac{1}{2\sqrt{3}} \Big[\xi\Big]_{-\sqrt{3}}^{+\sqrt{3}} = 1.$$

6.311.22. Standard rectangular distribution. The distribution of ξ is, by definition, the probability of meeting by chance with a value ξ which is less than or equal to the fixed value of ξ and is therefore in the interval $(-\sqrt{3}, \xi)$:

$$\text{Prob. } (\boldsymbol{\xi} \leqslant \xi) = F(\xi) = \int_{-\sqrt{3}}^{\xi} \varphi(\xi)\, d\xi, \tag{a}$$

$$F(\xi) = \frac{1}{2\sqrt{3}} \int_{-\sqrt{3}}^{\xi} d\xi = \frac{1}{2\sqrt{3}} \Big[\xi\Big]_{-\sqrt{3}}^{\xi} = \frac{\xi+\sqrt{3}}{2\sqrt{3}}. \tag{b}$$

This distribution law of ξ is represented by a segment of slope $1/2\sqrt{3}$ in the domain $(-\sqrt{3}, +\sqrt{3})$. But we are really concerned with a distribution. In fact,

$$F(-\sqrt{3}) = \frac{-\sqrt{3}+\sqrt{3}}{2\sqrt{3}} = 0,$$

$$F(+\sqrt{3}) = \frac{\sqrt{3}+\sqrt{3}}{2\sqrt{3}} = 1. \tag{c}$$

$$F(0) = \frac{\sqrt{3}}{2\sqrt{3}} = 0.5.$$

The relations between the (local) density $\varphi(\xi)$ and the (total) probability $F(\xi)$ are shown in Fig. 6.311.22I.

The shaded area adjoining the density function $\varphi(\xi)$ is contained between the origin and ξ and has the same measure as the ordinate $F(\xi)$ represented by a line segment with oblique crossings. In particular, the area adjoining $\varphi(\xi)$ and situated to the left of $\xi = 0$ has the measure of 0.5 which is equal to the measure of the segment $F(0)$.

Since the *standard distribution* $F(\xi)$ is *unique*, it can be tabulated once and for all, because it depends only on ξ. So for the rectangular distribution one may outline the Table 6.311.22A.

TABLE 6.311.22A. STANDARD RECTANGULAR DISTRIBUTION $F(\xi)$

ξ	$-\sqrt{3}$ -1.732	-1	$-\frac{\sqrt{3}}{2}$ -0.866	0	$+\frac{\sqrt{3}}{2}$ $+0.866$	$+1$	$+\sqrt{3}$ $+1.732$
$F(\xi)$	0.00	0.21	0.25	0.50	0.75	0.79	1.00

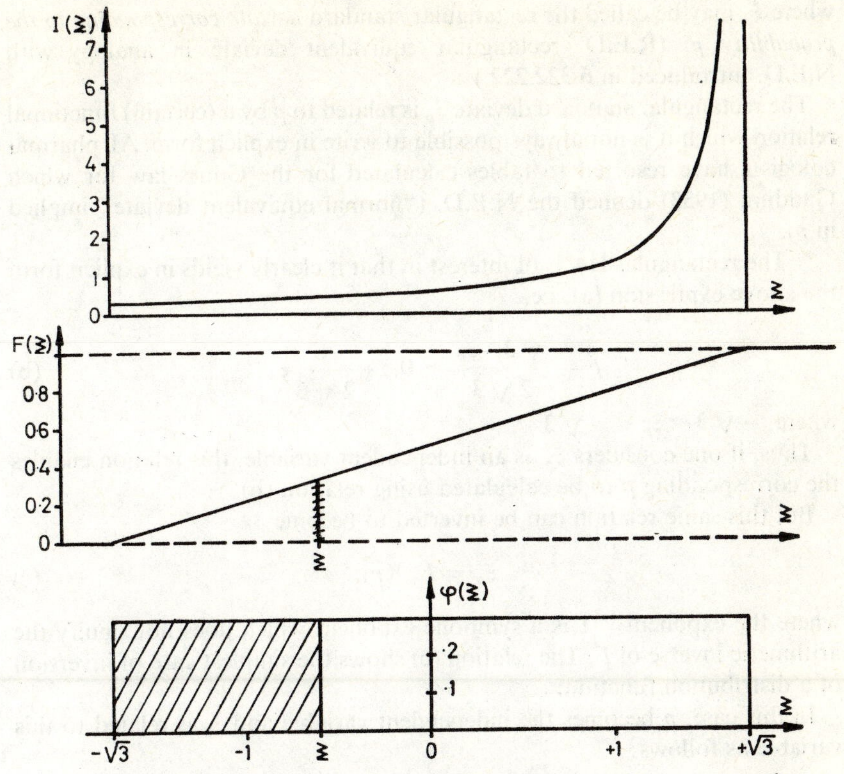

FIG. 6.311.22 I. Standard rectangular or uniform law. Probability density $\varphi(\xi)$. Distribution $F(\xi)$. Imminence rate $I(\xi)$.

6.311.3. Inversion of the standard uniform distribution

We will now introduce a general form of notation which will be found indispensable in the following pages. Since $F(\xi)$ is contained between 0 and 1, we can put

$$p = F(\xi) \tag{a}$$

for the (total) probability of finding a random $\boldsymbol{\xi}$ less than or equal to the given ξ.

We can go further and apply the subscript p to ξ and then write the identity

$$p \equiv F(\xi_p), \tag{a'}$$

where ξ_p may be called the rectangular standard *deviate corresponding to the probability* p. (R.E.D. rectangular equivalent deviate in analogy with N.E.D. introduced in 6.322.222.)

The rectangular standard deviate ξ_p is related to p by a (certain) functional relation which it is not always possible to write in explicit form. All pharmacologists have resorted to tables calculated for the Gauss law for which Gaddum (1933) defined the N.E.D. ("normal equivalent deviate" implied in p).

The rectangular law is of interest in that it clearly yields in explicit form the above expression (a), i.e.

$$p = \frac{\sqrt{3}+\xi_p}{2\sqrt{3}} = 0.5+\frac{1}{2\sqrt{3}}\xi_p, \tag{b}$$

where $-\sqrt{3} \leq \xi_p \leq +\sqrt{3}$.

Thus, if one considers ξ_p as an independent variable, this relation enables the corresponding p to be calculated using relation (b).

But this same relation can be inverted to become

$$\xi_p = F^{-1}(p), \tag{c}$$

where the exponent -1 is a symbolic exponent which does not signify the arithmetic inverse of F. The relation (c) shows the simplest case of inversion of a distribution function.

In this case, p becomes the independent variable and ξ_p is related to this variable as follows:

$$\xi_p = (p-0.5)2\sqrt{3}. \tag{d}$$

Clearly, the standard rectangular deviate corresponding to the probability p is equal to the product of the difference $p-0.5$ with the measure of the domain of ξ_p.

The relation (d) can be written more concisely as

$$\xi_p = (2p-1)\sqrt{3}. \tag{d'}$$

Some values of the couple (p, ξ_p) are tabulated below.

TABLE 6.311.3A. INVERSION OF THE RECTANGULAR DISTRIBUTION

p	0.00	0.25	0.50	0.75	1.00
ξ_p	$-\sqrt{3}$	$-\frac{\sqrt{3}}{2}$	0.00	$+\frac{\sqrt{3}}{2}$	$+\sqrt{3}$

According to the relation (d), the ξ_p are negative in 50% of the cases, that is to say, for any p less than 0.5. The situation can, however, be improved by adding to ξ_p a constant such that ξ_p is either zero or positive.

For $p = 0$, we have

$$\xi_0 = -\tfrac{1}{2}\cdot 2\sqrt{3} = -\sqrt{3}.$$

If one adds $\sqrt{3}$ to ξ_p defined by (d), one defines a rectangular probability unit (see 6.322.225.1 e) which, by analogy with the classical probit of C. I. Bliss (1935), is denoted by y_p (without introducing here an index for specifying the rectangular law):

$$y_p = \sqrt{3}+\xi_p, \tag{e}$$

$$y_p = \sqrt{3}+(p-0.5)2\sqrt{3} = 2\sqrt{3}\cdot p. \tag{f}$$

Since p varies from 0 to 1, the y_p traverses the domain of ξ from 0 to $2\sqrt{3}$, following a straight-line segment, the slope of which is the reduced range $2\sqrt{3}$.

If one compares the relation (e) with the relation (b) of section 6.311.22, it is seen that y_p is the numerator of (b).

The inversion of (f) leads to

$$p = \frac{y_p}{2\sqrt{3}}. \tag{g}$$

The unit of probability which we defined in order to avoid, with absolute certainty, negative values of ξ_p, is the fraction of the domain ξ which permits the probability p to be found; the domain is here counted as from the origin $-\sqrt{3}$.

However, the probability-unit contracted in probit was introduced by Bliss (1935) and R. A. Fisher (1935) in the unique context of u_p, the Gaussian standard deviate, by adding 5 to u_p so that there would be a practical certainty of meeting with no negative value. Hence the classical (Gaussian) probit is

$$Y_p = 5+u_p,$$

where u_p is the N.E.D. of Gaddum (1933) (the normal equivalent deviate). Hence the reason why to achieve agreement with the addition of 5 to u_p after Bliss (1935) for the Gaussian distribution, we add the number 5 instead of $\sqrt{3}$ to ξ_p.

The suffix R indicates that it is defined from a *rectangular* (uniform) law.

In this system, when p varies from 0 to 1, the rectangular probability-unit ranges between $5-\sqrt{3} = 3.268$ and $5+\sqrt{3} = 6.732$. The median point $X_{0.5}$

and the probability-unit 5 correspond, by definition, to the percentage $100p\% = 50\%$.

Upon replacing ξ_p by its value from (d′), one obtains explicitly the rectangular probability-unit of p:

$$Y_{R,p} = 5+\sqrt{3}(2p-1). \tag{h}$$

It is possible to compile a table of $Y_{R,p}$ in the finite interval $(5-\sqrt{3};\ 5+\sqrt{3})$ and centered on 5 for $100p\% = 50\%$. This leads us on to performing the transformation of the percentages into probability-units for the special rectangular case, thus enabling the reader to realize for himself all the steps involved whenever one has to be content with calculated tables for the Gaussian curve.

TABLE 6.311.3B TABLE OF RECTANGULAR PROBABILITY-UNITS $Y_{R,p}$

100%	p	Y_{Rp}
0	0	3.268
10	0.10	3.614
20	0.20	3.691
25	0.25	4.134
30	0.30	4.307
40	0.40	4.654
50	0.50	5.000
60	0.60	5.346
70	0.70	5.693
75	0.75	5.866
80	0.80	6.039
90	0.90	6.386
100	1.00	6.732

However, to get the same center 5 of a rectangular probability-unit when $p = 0.5$, we proceed as Bliss (1935) (who worked in the context of the Gaussian standard deviate) by adding 5 to ξ_p. The reason why we add 5 to the Gaussian standard deviate or Gaddum's (1933) N.E.D. is explained in section 6.322.225.1 (e).

In these conditions, the rectangular probability-unit $Y_{R,p}$ corresponding to the fraction p of cases is written

$$Y_{R,p} = 5+\xi_p. \tag{i}$$

From the standpoint of rectifying the curve $p = F(X)$ which one meets with so often in biological tests in all or nothing, there is clearly no need or rectification in the perhaps theoretical case of a rectangular density of sensitivity since the relation $p = F_R(X)$ is linear from the outset. On the other hand, for a gaussian distribution with distribution function $p = F_G(X_p)$ having a sigmoid form, the probit transformation is indispensable for rectifying the dose–response or log dose–response curve as will be seen in 6.322.225.1. Here the subscripts R and G stand for *rectangular* and *Gaussian* respectively.

Probability-unit in the general case. In formulae (h) and (i), we have introduced a rectangular probability-unit as the rectangular standard deviate added to the centering constant 5. For other symmetrical distributions, we use the same definition when the standard deviate is specific to the distribution considered (see Subject Index).

6.311.4. Moment generating function of the rectangular law

For the sake of convenience, we consider firstly the uniform continuous law defined on the domain (0, 1):

$$f(X) = 1. \tag{a}$$

The initial moment generating function (in relation to the origin) is written as

$$M_X(t) = E[e^{tX}] = \int_0^1 e^{tX}\, dX$$

$$= \frac{1}{t}\left[e^{tX}\right]_0^1 = \frac{1}{t}(e^t - 1). \tag{b}$$

After the expansion of e^t, we get

$$M_X(t) = \frac{1}{t}\left(1 + \frac{t}{1!} + \frac{t^2}{2!} + \ldots - 1\right)$$

$$= 1 + \frac{t}{2!} + \frac{t^2}{3!} + \ldots, \tag{c}$$

where $M_X(0) = 1$.

Hence successively differentiating $M_X(t)$ and making $t = 0$, we get

(a) the mean value

$$M'_X(0) = \mu'_1 = \tfrac{1}{2} \tag{d}$$

as expected,

(b) the second-order initial moment

$$M''_X(0) = \mu'_2 = \tfrac{1}{3}; \tag{e}$$

(c) therefore the variance (section 6.311.112 (d) for $R = 1$) is written as

$$\sigma^2 = \mu'_2 - \mu'^2_1 = \frac{1}{3} - \left(\frac{1}{2}\right)^2 = \frac{1}{3} - \frac{1}{4} = \frac{1}{12}, \tag{e'}$$

or Sheppard's correction for a unit interval;

(d) in general, the tth order initial moment is

$$\mu'_i = \frac{1}{i+1}. \tag{f}$$

Uniform central density. The rectangular density in x about the origin and of range R, is written as

$$f(x) = \frac{1}{R}; \qquad -\frac{R}{2} \leqslant x \leqslant +\frac{R}{2}. \tag{a}$$

The (central) moment generating function is

$$m_x(t) = \frac{1}{R} \int_{-R/2}^{+R/2} e^{tx}\, dx = \frac{1}{Rt} \left[e^{tx} \right]_{-R/2}^{+R/2}$$

$$= \frac{e^{(R/2)t} - e^{-(R/2)t}}{Rt}. \tag{b}$$

Upon introducing the relation $\sinh y = \dfrac{e^y - e^{-y}}{2}$, this relation becomes

$$m_x(t) = \frac{\sinh (R/2)t}{(R/2)t}. \tag{c}$$

Since $\sinh x$ and x are two odd functions of x, the quotient (c) is an even function of $(R/2)t$ which is written:

$$m_x(t) = 1 + \left(\frac{R}{2}\right)^2 \frac{t^2}{3!} + \left(\frac{R}{2}\right)^4 \frac{t^4}{5!} + \ldots. \tag{d}$$

Therefore all the odd order central moments are zero; $\mu_1 = \mu_3 = \mu_5 = \ldots = 0$; this is expected by virtue of the symmetry of $f(x)$ about $x = 0$.

The moment of even order $2i$ is written as

$$\mu_{2i} = \frac{(R/2)^{2i}}{(2i+1)} \qquad (i = 1, 2, \ldots) \tag{e}$$

6.311.5. *Use of the rectangular law, or distribution*

The rectangular law, although very simple, is interesting in several respects:

(1) It is an elementary example of a law with finite range about which biologists are satisfied in their own mind.

(2) From the teaching point of view, the developments are readily comprehensible (e.g. Finney, 1964, p. 453).

(3) It describes the random aspect of the numerical approximation (rounding error) and is the basis of Sheppard's correction for the variance.

(4) By summation of rectangular variables of given range, one successively generates the triangular law of double range to arrive with four variables at a close image of a Gaussian law with finite range equaling 4 times the "range of departure" (van der Waerden, 1967, pp. 105–106).

(5) The rectangular law has, as an important special case, the above uniform transformation defined on the domain (0, 1):

$$dp = dF.$$

E. S. Pearson (1938) has shown its value in goodness of fit tests and in combining independent tests of significance.

6.311.6. *Random standard rectangular deviates*

From section 6.322.4 devoted to standard Gaussian (normal) deviates, we refer to Quenouille's (1959) table of standard rectangular deviates ξ_p (6.311.2 b) linked to the standard normal deviate u_p (6.322.222 b) corresponding to the same crude observation X_p.

This *relation of equi-probability* p of the standard rectangular deviate ξ_p and the standard normal deviate u_p is written

$$p = F(\xi_p) = \Phi(u_p) \tag{a}$$

Quenouille (1959, p. 178) subtracts 0.5 from each term of (a) so that from

6.311.3 b and 6.322.222 c, we may write

$$p-0.5=\frac{\xi_p+\sqrt{3}}{2\sqrt{3}}-0.5=\frac{1}{\sqrt{2\pi}}\int_{-\infty}^{u_p} e^{-\frac{1}{2}u^2}\,du-0.5. \tag{b}$$

This simplifies into

$$p-0.5=\frac{\xi_p}{2\sqrt{3}}=\Phi(u_p)-0.5. \tag{c}$$

Whence, starting from u_p or u_p+5 tables (6.322.222) we get the ξ_p.

Here is the correspondence between Quenouille's notation in his tables and ours in this work:

$$\begin{aligned} x_1 &\to u_p \\ x_2 &\to \xi_p \end{aligned} \tag{d}$$

Such tables of ξ are useful for *simulation or Monte-Carlo* methods where the purpose is to investigate the sampling effects of departure from the normal (Gaussian) or any other standard rectangular distribution. For other standard distributions, Quenouille's tables will just be referred by relations such as (a) or similar to (b). The simple case of rectangular standard random deviates has been taken as pedagogic example.

6.311.7. Imminence rate (hazard function) for the standard rectangular law

The conditional probability for a random $\boldsymbol{\xi}$ to be in the interval ξ, $\xi+d\xi$, knowing that $\boldsymbol{\xi}$ is greater than the fixed value ξ is written (6.122)

$$\text{Prob}\,(\xi \leq \boldsymbol{\xi} \leq \xi+d\xi \,|\, \boldsymbol{\xi} \geq \xi) = I(\xi)\,d\xi \tag{a}$$

where the imminence rate or hazard function is written

$$I(\xi)=\frac{\varphi(\xi)}{1-F(\xi)}. \tag{a'}$$

Owing to 6.311.21 formula (j) and 6.311.22 formula (b), we write

$$I(\xi)=\frac{1}{\sqrt{3}-\xi}. \tag{b}$$

The vertical $\xi=\sqrt{3}$ is an asymptote to $I(\xi)$ which is represented in Fig. 6.311.22I, upper part.

6.312. TRIANGULAR LAWS

6.312.1. *Unilateral triangular law*

6.312.11. *Decreasing triangular density*

The unilateral triangular probability density is defined in the domain $(0, a)$ by the relation

$$f(X) = \frac{2}{a}\left(1 - \frac{X}{a}\right) \qquad (0 \leqslant X \leqslant a),$$

and represented in Fig. 6.312.11I.

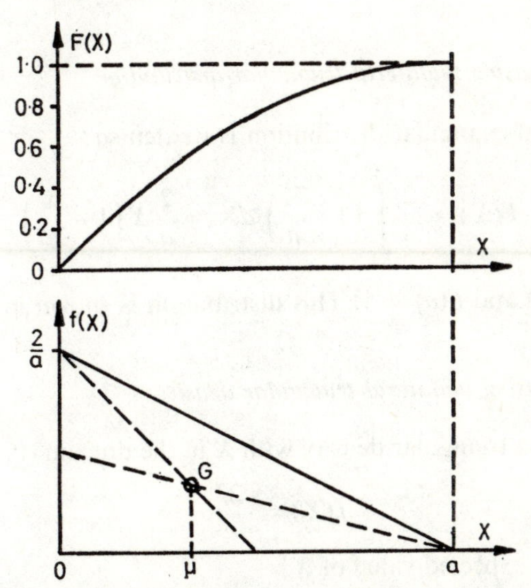

FIG. 6.312.11 I. Decreasing triangular law. Probability density $f(X)$. Distribution $F(X)$.

Using the formula for the area of a right triangle, it is verified that the area subjoining $f(X)$ is unity; we give the result of a number of calculations performed by applying general formulae:

(1) The mean $E[X] = \mu = \dfrac{a}{3}$ is the abscissa of the centre of gravity G of the triangle.

(2) The variance is $V(X) = \sigma_X^2 = \frac{a^2}{18}$.

(3) The standard deviation is $\sigma_X = \frac{1}{3}\frac{a}{\sqrt{2}}$.

(4) As to be expected, the coefficient of variation is considerable:

$$\text{C.V.\%} = \frac{100}{\sqrt{2}} = 71\%.$$

A consideration which is of interest as regards the unilateral triangular law within the finite domain (0, a), is the generation of the negative exponential law within the domain (0, ∞) by the process of exponentiation (Martin, 1955). It also sets in play a random variable having an upper limit a, and, permits either the quotient $X/a < 1$ or the difference $a-X > 0$ to be considered.

6.312.12. Decreasing unilateral triangular distribution

The unilateral triangular distribution is written so:

$$F(X) = \frac{2}{a}\int_0^X \left(1-\frac{X}{a}\right) dX = \frac{2}{a}X\left(1-\frac{X}{2a}\right) \qquad \text{(a)}$$

where $F(0) = 0$ and $F(a) = 1$. This distribution is shown in Fig. 6.312.11I.

6.312.13. Increasing unilateral triangular density

The increasing triangular density with X in the domain (0, a) is written as

$$f(X) = \frac{2X}{a^2}. \qquad \text{(a)}$$

The mean or expected value of X

$$E[X] = \frac{2a}{3} \qquad \text{(b)}$$

is found without difficulty as the abscissa of the centre of gravity of the area subjacent to $f(X)$.

The variance is

$$V(X) = \frac{a^2}{18}, \qquad \text{(c)}$$

as obtained for the decreasing triangular density with the same range.

A. Weber (1967) uses a graphical approximation of an asymmetrical density in X by two triangular laws framing one uniform law. This makes it possible to estimate the variances of a covariate X appearing in ratio estimates of the type $\bar{Y}/\bar{X}$.

The increasing unilateral triangular distribution is written as

$$F(X) = \frac{2}{a^2}\int_0^X X\,dX = \frac{2}{a^2}\,\frac{X^2}{2}.$$

Hence

$$F(X) = \frac{X^2}{a^2} \tag{d}$$

where $F(0) = 0$ and $F(a) = 1$.

6.312.2. *Bilateral triangular law*

6.312.21. *Bilateral triangular law in original variable X*

6.312.211. *Bilateral triangular probability density*

6.312.211.1. Definition. The triangular density associates the continuous random variable X moving in the finite range $R = 2a$, with the triangular law:

$$f(X) = \frac{1}{a}\left[1 - \frac{|X-\mu|}{b}\right] \tag{a}$$

the symbol $|\ |$ here signifying "absolute value of".

As the absolute deviation, $x = X-\mu$, we have

$$f(x) = \frac{1}{a}\left[1 - \frac{|x|}{b}\right]. \tag{b}$$

The condition that the area subjacent to $f(X)$ should be unity, leads to the relation

$$b = a. \tag{b'}$$

Therefore the definitive equation of the triangular density is

$$f(x) = \frac{1}{a}\left[1 - \frac{|x|}{a}\right]. \tag{c}$$

It depends only on the single parameter a. The abscissae for $f(X) = 0$ are $\pm a$ and $R = 2a$; the ordinate at the origin is $f(0) = 1/a$ (see Fig. 6.312.211.1I)

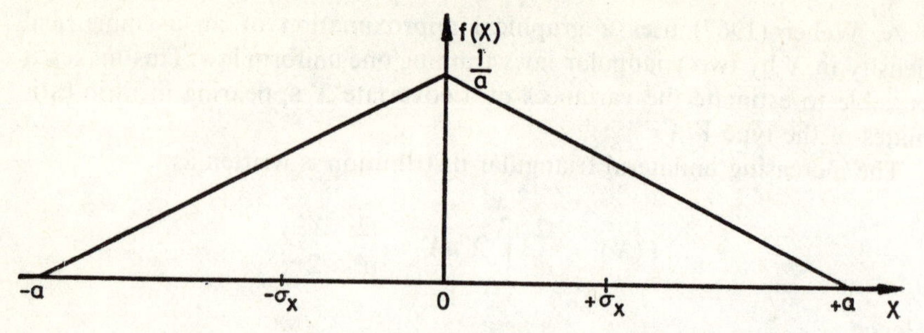

FIG. 6.312.211.1 I. Bilateral triangular probability density.

As the original variable we have

$$f(X) = \frac{1}{a}\left[1-\frac{|X-\mu|}{a}\right], \tag{d}$$

X here varying in the domain from $X_0 = \mu-a$ to $X_1 = \mu+a$.

6.312.211.2. Reduction parameters. Since the calculations have already been developed for the rectangular law and in the present case they are equally simple, we shall only give the results.

Range: $R = X_1 - X_0$

$$R = 2a, \quad \text{therefore} \quad a = \frac{R}{2},$$

Mean: $$E[X] = \mu = \frac{X_0+X_1}{2}, \tag{a}$$

Variance: $$V(X) = \sigma_X^2 = \frac{R^2}{24}, \tag{b}$$

Standard deviation: $$\sigma_X = \frac{R}{2\sqrt{6}} = 0.2041\ R, \tag{c}$$

$$\sigma_X \approx 0.20\ R. \tag{c'}$$

It will be recalled that $\sigma_X = \sigma_x$.

$\gamma_1 = 0$ symmetrical distribution, (d)

$\gamma_2 = 0.25$ slightly leptokurtic distribution. (e)

6.312.212. Distribution function. This will be given for the standard form.

6.312.22. Standard bilateral triangular law

6.312.221. Standard triangular density. We express x by means of the standard deviation taken as unity, or in the standard triangular deviate

$$\xi = \frac{x}{\sigma_x} = \frac{x}{R} 2\sqrt{6}. \qquad \text{(a)}$$

The reduced triangular law is (see 6.311.21g):

$$\varphi(\xi) = \sigma_x f(\xi\sigma) = \frac{R}{2\sqrt{6}} \frac{2}{R}\left[1 - \frac{|\xi|\cdot R}{\frac{R}{2}\cdot 2\sqrt{6}}\right]. \qquad \text{(b)}$$

After reduction, the standard bilateral triangular density is

$$\varphi(\xi) = \frac{1}{\sqrt{6}}\left[1 - \frac{1}{\sqrt{6}}|\xi|\right], \qquad \text{(c)}$$

where ξ traverses the domain from $-\sqrt{6}$ to $+\sqrt{6}$; in fact the range R expressed in units $\sigma_x = \dfrac{R}{2\sqrt{6}}$ is

$$\frac{R}{R/2\sqrt{6}} = 2\sqrt{6}.$$

For the sake of convenience and clarity, we replace the exact density (c) by the *approximate triangular density* which corresponds to the approximate value (c') i.e.:

$$\sigma_x \approx 0.2R. \qquad \text{(c')}$$

We thus obtain the following simple expression (Martin, 1952)

$$\varphi(\xi) = 0.4[1 - 0.4|\xi|], \qquad \text{(d)}$$

where ξ traverses the domain $(-2.5, +2.5)$ and is represented in Fig. 6.312.221I, part a.

6.312.222. Approximate standard triangular distribution. The approximate reduced triangular distribution function is defined by

$$p = \text{Prob}(\boldsymbol{\xi} \leqslant \xi) = F(\xi) = 0.4 \int_{-2.5}^{\xi} [1 - 0.4|\xi|]\cdot d\xi, \qquad \text{(a)}$$

and represented in Fig. 6.312.222I, part b.

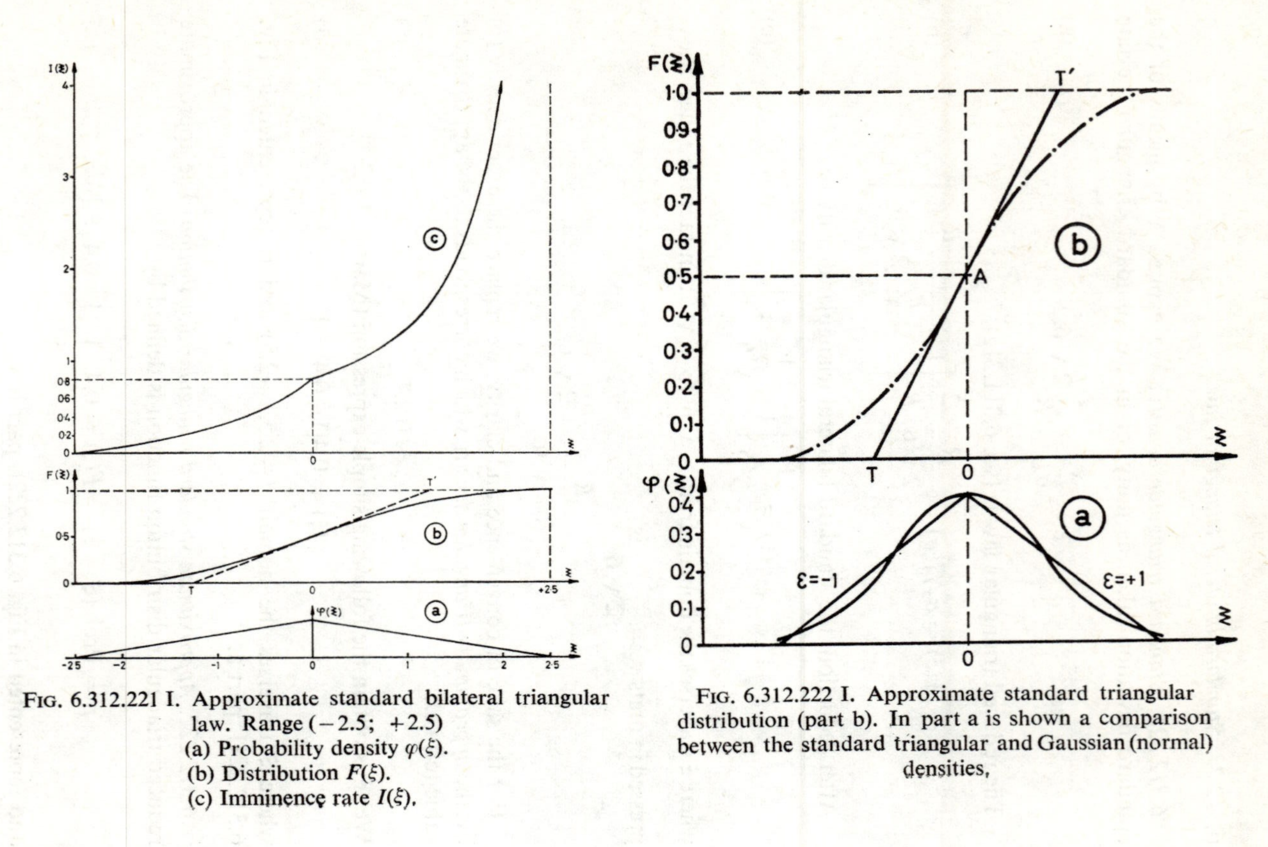

FIG. 6.312.221 I. Approximate standard bilateral triangular law. Range (− 2.5; + 2.5)
(a) Probability density $\varphi(\xi)$.
(b) Distribution $F(\xi)$.
(c) Imminence rate $I(\xi)$.

FIG. 6.312.222 I. Approximate standard triangular distribution (part b). In part a is shown a comparison between the standard triangular and Gaussian (normal) densities,

This integral consists of two consecutive parabolic arcs joining at the point $\xi = 0$, following the common tangent TT'. In this respect the equations are

$$F(\xi) = 0.5+\frac{\xi^2}{12.5}+\frac{\xi}{2.5} \quad 0 \leqslant p < 0.5 \quad -2.5 \leqslant \xi < 0 \quad \varepsilon = -1$$

$$F(\xi) = 0.5 \quad p = 0.5 \quad \xi = 0 \quad \varepsilon = 0 \qquad \text{(b)}$$

$$F(\xi) = 0.5-\frac{\xi^2}{12.5}+\frac{\xi}{2.5} \quad 0.5 < p \leqslant 1 \quad 0 < \xi \leqslant 2.5 \quad \varepsilon = +1.$$

The symbol ε of the relations (c) and (d) is -1 for the first arc and $+1$ on the second arc.

The bi-parabolic sigmoid corresponding to the approximate standard triangular distribution can be written in either form (c) or (d):

$$p = F(\xi) = 0.5-\varepsilon\frac{\xi^2}{12.5}+\frac{\xi}{2.5}, \qquad \text{(c)}$$

ε being defined in the relations (b) according to the value of p: $\varepsilon = -1$ for $p < 0.5$, $\varepsilon = 0$ for $p = 0.5$ and $\varepsilon = +1$ for $p > 0.5$. In this regard

$$p = 0.5+0.4\xi(1-0.2\varepsilon\xi). \qquad \text{(d)}$$

This relation is tabulated below.

TABLE 6.312.222A. APPROXIMATE TRIANGULAR STANDARD DISTRIBUTION

ξ	−2.5	−2	−1.5	−1.0	−0.73	−0.5	0.0	0.5	0.73	1.0	1.5	2.0	2.5
$F(\xi)$	0.00	0.02	0.08	0.18	0.25	0.32	0.50	0.68	0.75	0.82	0.92	0.98	1

N.B. In part a of Fig. 6.312.222I is shown a comparison between the standard triangular density and the standard Gaussian (normal) density (see 6.322.212).

6.312.223. Inversion of the standard triangular distribution. The *approximate triangular probability unit* is $Y_{T,p}$. In order to obtain in explicit form the reduced deviation ξ_P corresponding to p, it is necessary to invert the definite integral, which entails solving the second order equation in ξ appearing as (c) in the foregoing section.

We write it in the following form in ξ_P:

$$0.08\varepsilon\xi_p^2 - 0.4\xi_p + (p - 0.5) = 0 \quad \text{(a)}$$

and, retaining only the sign – before the radical, we get

$$\xi_p = \varepsilon 2.5[1 - \sqrt{1 - 2\varepsilon(p - 0.5)}]. \quad \text{(b)}$$

The *triangular probability-unit* corresponds to the fraction p, is noted $Y_{T,\,p}$ and is defined by adding 5 to ξ_p in the relation (b) to obtain the same centering 5 for $p = 50\%$ as for the Gaussian probit

$$Y_{T,\,p} = 5 + 2.5\varepsilon[1 - \sqrt{1 - 2\varepsilon(p - 0.5)}]. \quad \text{(c)}$$

Table 6.312.223A shows the values of the relation (c).

TABLE 6.312.223A. APPROXIMATE TRIANGULAR PROBABILITY UNIT

General formula: $Y_{T,\,p} = 5 + \varepsilon\, 2.5[1 - \sqrt{1 - 2\varepsilon(p - 0.5)}]$

$p < 0.5,\ \varepsilon = -1 \qquad p = 0.5,\ \varepsilon = 0 \qquad p > 0.5,\ \varepsilon = +1$

$Y_{T,\,p} = 5 - 2.5[1 - \sqrt{1 + 2(p - 0.5)}] \qquad Y_{T,\,0.5} = 5$

$Y_{T,\,p} = 5 + 2.5[1 - \sqrt{1 - 2(p - 0.5)}]$

p	$Y_{T,p}$	p	$Y_{T,p}$	p	$Y_{T,p}$
0.000	2.5000	0.50	5.00	0.55	5.1285
0.001	2.6117			0.60	5.2640
0.005	2.7500			0.65	5.4085
0.01	2.8535			0.70	5.6292
0.02	3.0000			0.75	5.7232
0.03	3.1122			0.80	5.9190
0.04	3.2070			0.85	6.1307
0.05	3.2905			0.90	6.3820
0.10	3.6180			0.95	6.7095
0.15	3.8692			0.96	6.7930
0.20	4.0810			0.97	6.8877
0.25	4.2677			0.98	7.0000
0.30	4.4362			0.990	7.1465
0.35	4.5915			0.995	7.2500
0.40	4.7360			0.999	7.3882
0.45	4.8715			1.000	7.5000

6.312.224. Applications of the triangular law. (1) The table of triangular probability which is of double interest (a) *pedagogic*, for the reader can tabulate it for himself, and (b) *statistical*, for it provides a good approximation to the normal or Gaussian probits, as will be seen in section 6.314.7.

(2) Rapoport (1952) has used the triangular law as reference from the point of view of concentration in the central zone. This author calculated the output of a set of simple neuron counters by making the hypothesis that the refractory period of the independent units is governed by the triangular law.

(3) Approximation to the Gaussian law. Van Deuren (1934, I, p. 224) gives the triangular relation

$$y = -0.645\,|x| + 1.07 - 0.6$$

as approximation to $y = e^{-x^2}$. The same author (*ibid.*, pp. 288 and 289) discusses the triangular laws of roof form $y = 1 - |x|$, and in the form of a trench with sides inclined at 45°, i.e. $y = |x|$.

6.312.225. Imminence rate (hazard function) of the standard bilateral triangular law. As explained in 6.122 and in 6.311.7 for the standard rectangular law, the imminence rate $I(\xi)$ (or hazard function) of the standard bilateral triangular law is easily calculated when one knows $\varphi(\xi)$ and $F(\xi)$, according to the formula

$$I(\xi) = \frac{\varphi(\xi)}{1 - F(\xi)}. \qquad \text{(a)}$$

This function of ξ in the range $(-2.5, +2.5)$ is represented in Fig. 6.312.221I, part c.

As for the rectangular law, the end of the range is an asymptote for $I(\xi)$. As explained in Martin (1966), the curve $I(\xi)$ shows a break in slope at the point $\xi = 0$ $I(\xi) = 0.8$.

6.313. Parabolic Law

6.313.1. Parabolic probability law

6.313.11. Parabolic probability density function

After the rectangular and triangular laws the parabolic law follows quite naturally; it is the simplest case of a curve of the type Pearson II (Elderton, 1953, p. 86).

Moreover in the range (0, 1) the parabolic density appears as a symmetrical Beta–1 probability law (section 6.315.11).

Definition. The parabolic density function associates with the continuous random variable X moving in the finite range $R = 2\sqrt{a:c}$, the parabolic density:

$$f(X) = a - c(X - \mu)^2. \qquad \text{(a)}$$

In absolute deviations $x = X-\mu$ with symmetry about $x = 0$, we have

$$f(x) = a-cx^2 \tag{b}$$

where a is the ordinate of the peak on the axis of symmetry $X = \mu$, and $f(x)$ vanishes at the points $x = \pm\sqrt{a/c}$, hence the range is

$$R = 2\sqrt{a/c}. \tag{c}$$

If X and x are lengths measured in cm, then the dimensions of $f(X)$ and of the coefficients a and c denoted [ ] are given in the following units:

$[f(x)] = L^{-1}$, given in cm^{-1} for $f(x)\,dx$ is a pure number,

$[a] = L^{-1}$, given in cm^{-1} for a is a term of $f(X)$,

$[c] = L^{-3}$, given in cm^{-3} for cx^2 should have the dimension L^{-1}.

The relation (c) enables the coefficient c to be expressed as a function of a and R:

$$a = c\frac{R^2}{4}. \tag{d}$$

The second condition which relates a to c via R, is given by requiring that the area subjacent to $f(x)$ should be equal to unity:

$$\int_0^{R/2} (a-cx^2)\,dx = \tfrac{1}{2}. \tag{e}$$

We will use here the symbol [ ] which indicates the result of integration taken to the upper and lower limits respectively (no risk of confusion!):

$$\left[ax - c\frac{x^3}{3}\right]_0^{R/2} = \frac{1}{2},$$

$$a\cdot\frac{R}{2} - c\cdot\frac{R^3}{24} = \frac{1}{2},$$

$$aR - c\frac{R^3}{12} = 1.$$

We replace a by its value given in (d). Thus

$$c\frac{R^3}{4} - c\frac{R^3}{12} = 1.$$

Therefore

$$c = \frac{6}{R^3}. \tag{f}$$

Substituting the values of a and c as in the relation (b), we obtain the parabolic law in absolute deviation x of range R:

$$f(x) = \frac{3}{2R} - \frac{6}{R}\frac{x^2}{R^2} = \frac{3}{2R}\left[1-4\left(\frac{x}{R}\right)^2\right]. \tag{g}$$

Figure 6.313.11I objectifies the relations between the rectangular (uniform) law and the parabolic law of the same range R. In particular, the ordinate

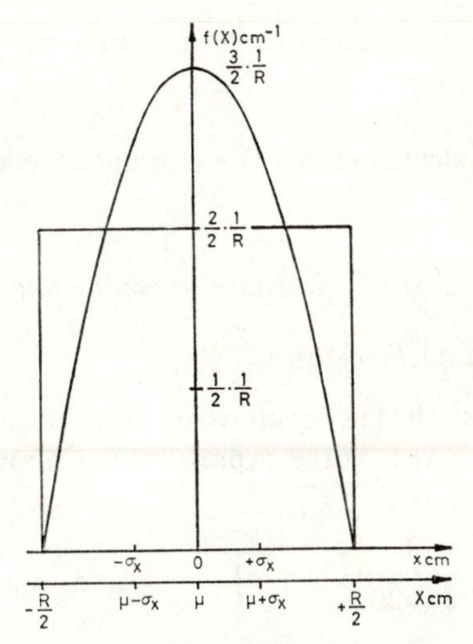

FIG. 6.313.11 I. Parabolic and uniform (rectangular) probability densities with the same range R.

of the rectangular density of same range is situated at $\frac{2}{3}$ of the ordinate to the origin of the parabolic density which is $\frac{3}{2R}$.

Reduction parameters. It is supposed that the parabolic variable varies between X_0 and X_1.

Range: $R = X_1 - X_0$, (h)

Mean: $E[X] = \mu = \dfrac{X_1+X_0}{2}$, (i)

Variance: $V(X) = \sigma_X^2 = \dfrac{R^2}{20}$. (j)

Standard deviation: $\frac{R}{\sqrt{20}} = 0.2236R.$ (k)

We also have:

$\gamma_1 = 0$ symmetrical distribution, (l)

$\gamma_2 = -0.86$ (platykurtic distribution). (l′)

The constant c is inversely proportional to σ_x^3.

$$c = 0.067 : \sigma^3. \quad \text{(m)}$$

Thus c is a parameter of concentration inversely proportional to the cube of the standard deviation.

6.313.12. Parabolic distribution function (see under reduced form below).

6.313.2. *Standard parabolic law*

6.313.21. Standard parabolic density

We express the parabolic law (g) above in terms of standardized deviations $y = x : \sigma$, where σ is given by the expression (k) above. Having regard to (6.311.21 g), we have

$$h(y) = \frac{3}{2\sqrt{20}}\left(1-\frac{y^2}{5}\right) \qquad -\sqrt{5} \leqslant y \leqslant +\sqrt{5} \quad \text{(a)}$$

$$= 0.3354\left(1-\frac{y^2}{5}\right).$$

To get a parabolic density defined in the range (−2.5, +2.5) we now introduce the variable

$$\xi = \frac{5}{2\sqrt{5}}y = 1.118y. \quad \text{(a′)}$$

After reduction, we get a standard parabolic density

$$\varphi(\xi) = 0.3[1-(0.4\xi)^2], \quad \text{(b)}$$

where ξ traverses the range between −2.5 and +2.5 (see Fig. 6.313.21I).

This standard parabolic density (P) is usefully compared with the standard triangular density (T) and standard rectangular (uniform) density (R), all defined in the same range in Table 6.314.7A.

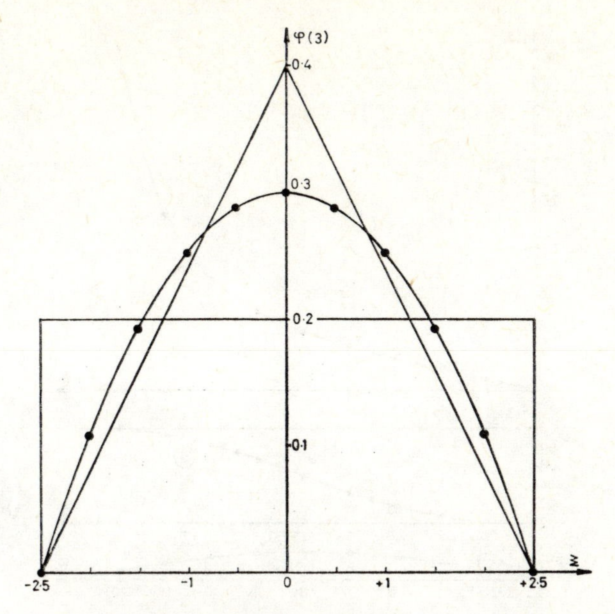

FIG. 6.313.21 I. Standard parabolic, triangular and rectangular (uniform) probability densities with range $(-2.5; +2.5)$.

6.313.22. Standard parabolic distribution

The standard parabolic distribution function is written so:

$$\text{Prob}\,(\boldsymbol{\xi} \leqslant \xi) = F(\xi) = 0.3 \int_{-2.5}^{\xi} (1-0.16\xi^2)\, d\xi, \tag{a}$$

and is represented in Fig. 6.313.22, curve b.

This integral is a cubic curve with a kink at $\xi = 0$, the equation of which is

$$F(\xi) = 0.5 + 0.3\xi \left(1 - \frac{0.16\xi^2}{3}\right). \tag{b}$$

No difficulty arises in calculating the values (Table 6.313.22A) with reference to Fig. 6.313.22I (curve b).

TABLE 6.313.22A. APPROXIMATE STANDARD $F(\xi)$ PARABOLIC DISTRIBUTION

ξ	-2.5	-2.0	-1.5	-1.0	-0.5	0.0	$+0.5$	$+1.0$	$+1.5$	$+2.0$	$+2.5$
$F(\xi)$	0.000	0.030	0.104	0.216	0.352	0.500	0.648	0.784	0.896	0.970	1.0

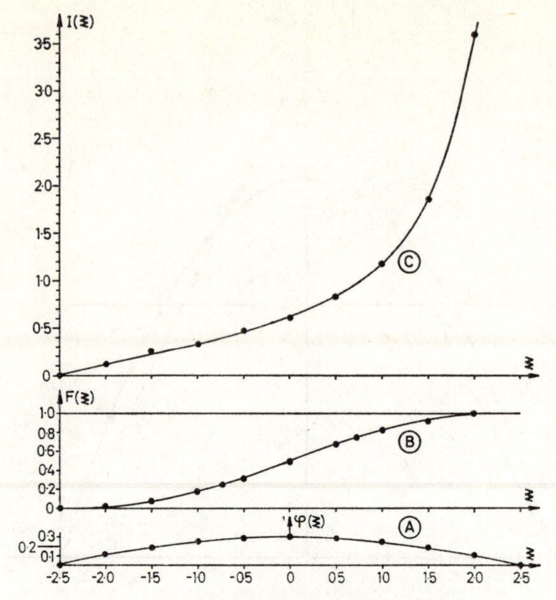

FIG. 6.313.22 I. Standard parabolic law.
(a) Probability density.
(b) Distribution.
(c) Imminence rate.

6.313.23. Inversion of the standard parabolic distribution

The parabolic probability unit $Y_p(\xi)$ corresponds to the fraction p of cases. The formula of the above sub-section shows that the cumulative probability as from the value $\xi = -2.5$ is given by a cubic relation in ξ, i.e.

$$p = 0.5+0.3\xi_p\left(1-\frac{0.16\xi_p^2}{3}\right). \tag{a}$$

The classical method of solving the equation

$$x^3+px+q = 0, \tag{b}$$

involves difficulties in an inversion of (a). Therefore, unlike the triangular law which gave p as a second degree function in ξ, there is no simple way of inverting the cubic relation in ξ. However, the construction of a table with a step-length or interval less than 0.1 for ξ would permit a numerical or graphical inversion. For example, it is noticed that the reduced deviations $\xi = -2.0$ and $\xi = +2.0$ correspond to $p = 0.03$ and 0.97 respectively. We write the parabolic probability unit corresponding to p in the accepted form for the Gaussian probit and centering on 5 (C. I. Bliss, 1935):

$$Y_{P,\,p} = \xi_p + 5, \qquad \text{(c)}$$

and we then have

$$Y_{P,\,(0.03)} = 5-2 = +3,$$

$$Y_{P,\,(0.97)} = 5+2 = +7.$$

These values correspond to fractions $p = 0.02$ and 0.98 in the case of the triangular law, this being a difference of about 1% from $p = 0.03$ and $p = 0.97$. For fractions closer to 0.5, the correspondence of the results is less distinct.

It is interesting to compare the percentages corresponding to the probability units given for laws having a finite range, with the classical Gaussian probits. This comparison will be made in section 6.314.7 for the rectangular, triangular, parabolic and angular laws.

6.313.24. Imminence rate (hazard function) of the standard parabolic density

From the definition of the imminence rate or hazard function (6.122), it is easy to calculate for the standard parabolic law

$$I(\xi) = \frac{\varphi(\xi)}{1-F(\xi)} \qquad (-2.5 \leqslant \xi \leqslant +2.5). \qquad \text{(a)}$$

The result shown in 6.313.22 I, curve c, is a monotonic function increasing to infinity for $\xi \to +2.5$.

6.314. ANGULAR LAW

The angular law plays an important part in mathematical statistics (Feller, 1958; Parzen, 1960) and is used in the statistical treatment of the response in all-or-nothing as a function of the dose or of a suitable metamer or transform of the dose such as log dose (Knudsen and Curtiss, 1947).

Unlike most statistical laws, the angular law $p = \sin^2 \theta$ is introduced more naturally as a distribution than in the form of a probability density.

6.314.1. *Angular law in θ*

6.314.11. *Angular distribution*

In section 6.221.8 the representation of the binomial point (r independent reactors out of n exposed) in the binomial ruling (probability paper) with coordinates $(\sqrt{n-r}, \sqrt{r})$ defines a split which makes an angle θ linked with the probability p of response by the relation

$$p = \sin^2 \theta. \tag{a}$$

The domain of variation of θ extends from 0 to $\pi/2$. As a first approximation the probability p is estimated by the fraction r/n of reactors

$$\hat{p} = \frac{r}{n}. \tag{a'}$$

Reverting to the usual notation, the distribution of θ is written as

$$p = F(\theta) = \sin^2 \theta, \tag{b}$$

where

$$F(0) = \sin^2 0 = 0 \tag{b'}$$

and

$$F\left(\frac{\pi}{2}\right) = \sin^2 \frac{\pi}{2} = 1.$$

$F(0)$ evolves monotonically from 0 to 1.

We recall that in the binomial ruling

$$\theta = \arctan \sqrt{p/q} = \tan^{-1} \sqrt{p/q}, \tag{c}$$

according to the two types of accepted notations.

We invert (c), and have

$$\tan \theta = \frac{\sin \theta}{\cos \theta} = \frac{\sqrt{p}}{\sqrt{q}}. \tag{c'}$$

This relation is satisfied by setting

$$\begin{aligned} \sin \theta &= \sqrt{p} \\ \cos \theta &= \sqrt{q}. \end{aligned} \tag{d}$$

Squaring we obtain the two relations which added term by term equal unity:

$$\sin^2\theta = p \qquad \text{(d')}$$
$$\cos^2\theta = q$$

$$\sin^2\theta+\cos^2\theta = p+q = 1. \qquad \text{(d'')}$$

One thus again finds the relation (a); but by virtue of symmetry, the relation in $\cos^2\theta = q$ could be and elsewhere has been used.

We have thus recalled the intimate relation which exists between the split θ of the representation used in the binomial ruling, and the angular distribution $p = \sin^2\theta$ for which the median value of θ is $\pi/4$ radians or 45°. This is also the mean value by virtue of the symmetry of the distribution.

The tangent at the point of inflection has the value of the derivative in $\theta = \pi/4$ for its slope. The derivative of p with relation to θ (probability density studied in section 314.12) is written:

$$\frac{dp}{d\theta} = 2\sin\theta\cos\theta = \sin 2\theta. \qquad \text{(e)}$$

For $\theta = \pi/4$, the derivative is

$$\sin 2\cdot\frac{\pi}{4} = \sin\frac{\pi}{2} = 1.$$

The second derivative of p

$$\frac{d^2p}{d\theta^2} = 2\cos 2\theta \qquad \text{(f)}$$

vanishes for $\theta = \pi/4$. Hence the median point $\theta = \pi/4$ is a point of inflection for the distribution. In Fig. 6.314.11I (part b) the inflectional tangent may be traced to the point ($\pi/4$; 0.5).

Furthermore, the first derivative vanishes for $\theta = 0$ and $\theta = \pi/2$; hence the distribution p starts with a zero slope at $\theta = 0$, exhibits a point of inflection at $\theta = \pi/4$ and then attains its maximum for $\theta = \pi/2$ with a zero slope.

Having established the general shape of the distribution, and also the localization parameter (median $= \pi/4$), we calculate a *measure of dispersion* in the form of the *semi-interquartile range*.

The lower quartile Q_1 is the value of θ which corresponds to the equation

$$0.25 = \sin^2 Q_1 = \sin^2\theta_{0.25}\,. \qquad \text{(g)}$$

Hence

$$\sin Q_1 = \sqrt{0.25} = 0.50$$

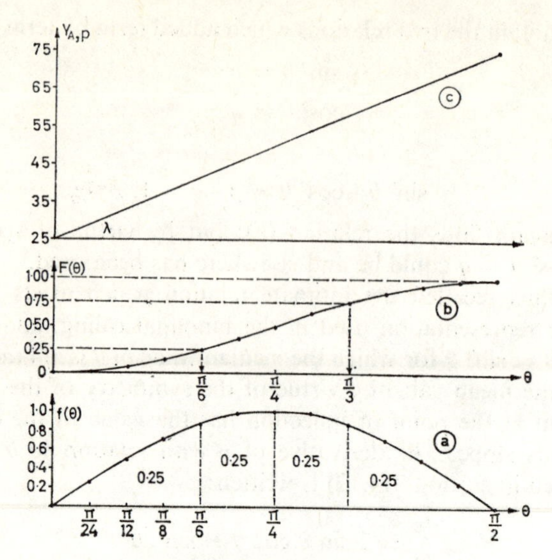

FIG. 6.314.11 I. Angular law.
(a) Probability density.
(b) Distribution.
(c) Angular probability unit (6.314.3).

and

$$Q_1 = \frac{\pi}{6} \quad \text{or} \quad 30°. \tag{g'}$$

By symmetry, the upper quartile is $Q_s = \pi/3$ or 60°. For this symmetrical distribution, the semi-interquartile range is therefore

$$\frac{\theta_{0.75} - \theta_{0.25}}{2} = \frac{Q_s - Q_i}{2} = \frac{1}{2}\left(\frac{\pi}{3} - \frac{\pi}{6}\right) = \frac{\pi}{12} \quad \text{or} \quad 15°. \tag{h}$$

The angular distribution may be cut into four equal parts by three cardinal points of the distribution which are submultiples of π, i.e.

$$\text{in radians:} \quad 0, \frac{\pi}{6}, \frac{\pi}{4}, \frac{\pi}{3}, \frac{\pi}{2};$$

$$\text{in degrees:} \quad 0°, 30°, 45°, 60°, 90°.$$

The area subjacent to the probability density is cut into four parts of measure 0.25 in Fig. 6.314.11I, part a.

6.314.12. Angular probability density

We shall again take the relation (6.314.11a) which defines the angular distribution

$$p = \sin^2 \theta. \tag{a}$$

The angular probability density function is then written as

$$f(\theta) = \frac{dp}{d\theta} = 2 \sin \theta \cos \theta = \sin 2\theta. \tag{b}$$

It is represented in Fig. 6.314.11I, part b.

This curve is symmetrical about

$$\theta = \frac{\pi}{4}.$$

At this point the ordinate is

$$f(0) = \sin 2 \cdot \frac{\pi}{4} = \sin \frac{\pi}{2} = 1.$$

This ordinate is maximal, for the first derivative

$$f'(\theta) = 2 \cos 2\theta \tag{c}$$

vanishes at $\theta = \pi/4$, whilst the second derivative

$$f''(\theta) = -4 \sin 2\theta \tag{d}$$

is negative at the same point

$$f''\left(\frac{\pi}{4}\right) = -4. \tag{d'}$$

It is interesting to verify directly the value of the mean and important to calculate the variance of the angular variable θ.

Mean:

$$E[\boldsymbol{\theta}] = \int_0^{\pi/2} \theta \sin 2\theta \, d\theta = \tfrac{1}{4} \int_0^{\pi} 2\theta \sin 2\theta \, d2\theta. \tag{e}$$

We put $x = 2\theta$ and integrate by parts

$$\int x \sin x \, dx = -x \cos x + \sin x.$$

Therefore at the limits 0 and π,

$$E[\boldsymbol{\theta}] = \frac{1}{4}\Big[-x \cos x + \sin x\Big]_0^{\pi} = \frac{1}{4}[-\pi(-1)] = \frac{\pi}{4}, \tag{e'}$$

and one again finds the center of symmetry.

Variance: We calculate firstly μ'_2

$$\mu'_2 = \int_0^{\pi/2} \theta^2 \sin 2\theta \, d\theta = \tfrac{1}{8}\int_0^{\pi} (2\theta)^2 \sin 2\theta \, d(2\theta), \tag{f}$$

where $x = 2\theta$, and one writes

$$8\mu'_2 = \int_0^{\pi} x^2 \sin x \, dx.$$

Integrating by parts, we have

$$8\mu'_2 = \Big[[-x^2 \cos x + 2x \sin x + 2 \cos x\Big]_0^{\pi}$$
$$= -\pi^2(-1) + 2(-1) - 2 = \pi^2 - 4.$$

Hence

$$\mu'_2 = \frac{\pi^2}{8} - \frac{1}{2}. \tag{f'}$$

The variance is written as

$$V(\boldsymbol{\theta}) = \sigma_\theta^2 = \mu_2 = \mu'_2 - \mu_1'^2 = \frac{\pi^2 - 8}{16}. \tag{g}$$

In radians

$$\sigma_\theta^2 = \frac{9.8696 - 8.0000}{16} = 0.11685. \tag{g'}$$

The standard deviation is

$$\sigma_\theta = \frac{\sqrt{\pi^2 - 8}}{4} = \sqrt{0.11685} = 0.342. \tag{h}$$

The coefficient of variation of the angular variable is written as

$$\gamma_\theta = \frac{\sigma_\theta}{E\theta} = \frac{\sqrt{\pi^2 - 8}}{4} \cdot \frac{4}{\pi} = \sqrt{1 - \frac{8}{\pi^2}} = \sqrt{1 - 0.8106}$$
$$= \sqrt{0.1894} = 0.4352. \tag{i}$$

6.314.13. Moment generating function associated with the angular law

The initial moment generating function of the angular law in the original random variable is written as

$$M_\theta(t) = \int_0^{\pi/2} e^{\theta t} \sin 2\theta \, d\theta. \tag{a}$$

Taking advantage of the relation (12) appearing in Defares and Sneddon (1960, p. 288), we get

$$M_\theta(t) = \left[\frac{e^{\theta t}}{4+t^2}(t \sin 2\theta - 2\cos 2\theta)\right]_0^{\pi/2}$$

$$= \frac{e^{(\pi/2)t}}{4+t^2}(0+2) - \frac{1}{4+t^2}(0-2).$$

Hence

$$M_\theta(t) = \frac{2}{4+t^2}(1+e^{(\pi/2)t}), \tag{b}$$

with

$$M_\theta(0) = \frac{2\times 2}{4} = 1. \tag{b'}$$

The first derivative

$$M'_\theta(t) = \frac{-4t}{(4+t^2)^2}(1+e^{(\pi/2)t}) + \frac{\pi}{4+t^2}e^{(\pi/2)t}, \tag{c}$$

taken at $t = 0$, yields

$$\mu'_1 = M'_\theta(0) = 0 + \frac{\pi}{4} = \frac{\pi}{4}, \tag{c'}$$

as required.

The second derivative, not reproduced here, evaluated at $t = 0$, yields

$$\mu'_2 = M''(0) = \frac{\pi^2}{8} - \frac{1}{2}. \tag{d}$$

The variance of θ is therefore written by means of

$$\sigma_\theta^2 = \mu_2 = \mu'_2 - \mu_1'^2 = \frac{\pi}{8} - \frac{1}{2} - \left(\frac{\pi}{4}\right)^2 = \frac{\pi^2}{16} - \frac{1}{2}, \tag{e}$$

or

$$\sigma_\theta^2 = \frac{\pi^2 - 8}{16} = \frac{9.8696 - 8}{16} = 0.11685 \text{ radians} \tag{e'}$$

as obtained in 6.314.12 (g').

6.314.2. *Angular law in* ξ

6.314.21. *Angular distribution in* ξ

In the study of the laws with finite range up to the present point (triangular and parabolic), we have been considering the reduced range $R/\sigma_X = 5$.

For the angular law a variable ξ of the standardized type can be defined in this finite range $R = 5$ with $-2.5 \leqslant \xi \leqslant +2.5$. Thus ξ is introduced as a linear function of the angle θ:

$$\theta = \frac{\pi}{10}(2.5+\xi), \tag{a}$$

for ξ ranging from -2.5 ($\theta = 0$) to 2.5 ($\theta = \pi/2$).

The new variable ξ is written as

$$\xi = \frac{10}{\pi}\theta - 2.5. \tag{a'}$$

In this case the expected value $E[\xi] = 0$ corresponds to the median (and mean) $\theta_{0.5} = \pi/4$.

TABLE 6.314.21A. ANGULAR DISTRIBUTION IN ξ
(range -2.5; $+2.5$)

ξ	θ		$p = \sin^2 \theta$
	radians	degrees	
$-2.5 = -\frac{10}{4}$	0	0	0.0
$-0.8333 = -\frac{10}{12}$	$\frac{\pi}{6}$	30°	$\sin^2 \frac{\pi}{6} = \left(\frac{1}{2}\right)^2 = 0.25$
0	$\frac{\pi}{4}$	45°	$\sin^2 \frac{\pi}{4} = \left(\frac{\sqrt{2}}{2}\right)^2 = 0.5$
$+0.8333 = +\frac{10}{12}$	$\frac{\pi}{3}$	60°	$\sin^2 \frac{\pi}{4} = \left(\frac{\sqrt{3}}{2}\right)^2 = 0.75$
$+2.5 = +\frac{10}{4}$	$\frac{\pi}{2}$	90°	1.0

The *angular distribution in* ξ is written so:

$$p = \sin^2 \left\{ \frac{\pi}{10} (2.5+\xi) \right\}. \tag{b}$$

Since the angle θ in (a) is given in radians, we have the following correspondences with the values of ξ (see Table 6.314.21A).

6.314.22. *Angular density in* ξ

In order to write the probability density in ξ, we firstly differentiate the relation (a) in section 6.314.21; we have

$$d\theta = \frac{\pi}{10}\, d\xi. \tag{a}$$

TABLE 6.314.22A

	Angle θ		Angular probability density $\varphi(\xi) = \frac{\pi}{10} \sin \frac{2\pi}{10} (\xi+2.5)$	Angular distribution $p = \sin^2 \frac{\pi}{10} (\xi+2.5)$
1	2	3	4	5
ξ	$\frac{2\pi}{10}(\xi+2.5)$ radians	degrees	$\varphi(\xi)$	p
−2.5	0	0	0	0.000
−2.0	0.1π	18	0.097	0.024
−1.5	0.2π	36	0.185	0.095
−1.0	0.3π	54	0.809	0.219
−0.5	0.4π	72	0.299	0.346
0	0.5π	90	0.314	0.500
0.5	0.6π	108	0.299	0.654
1.0	0.7π	126	0.254	0.794
1.5	0.8π	144	0.185	0.905
2.0	0.9π	172	0.017	0.995
2.5	π	180	0	1.000

The reduced angular probability density $\varphi(\xi)$ is written from cosideration of $f(\theta) = \sin 2\theta$

$$\varphi(\xi) = \frac{dp}{d\xi} = \frac{dp}{d\theta} \cdot \frac{d\theta}{d\xi} = \frac{d\theta}{d\xi} f(\theta). \tag{b}$$

Having regard to the relations (a) and (b) above and to (a) in section 6.314.21, we have

$$\varphi(\xi) = \frac{\pi}{10} \sin \frac{2\pi}{10}(2.5+\xi). \tag{c}$$

The (maximal) ordinate to the origin is

$$\varphi(0) = \frac{\pi}{10} \sin \frac{\pi}{2} = \frac{\pi}{10} = 0.314. \tag{d}$$

Table 6.314.22A shows the ordinates of the angular density (column 4) and of the angular distribution (column 5) in ξ in the finite range $(-2.5, +2.5)$ for ξ and the corresponding angles θ.

6.314.3. Inversion of the angular distribution. Angular probability unit

The inversion of the angular distribution

$$F(\theta) = p = \sin^2 \theta \tag{a}$$

is immediate and is written so:

$$\begin{aligned} \theta_p &= \text{arc sin} \sqrt{p} \\ &= \sin^{-1} \sqrt{p} \end{aligned} \tag{b}$$

according to the two types of notations accepted. The symbol θ_p stands for "angle θ corresponding to p".

We transfer into (b) the value θ taken from the relation 6.314.21 (b). We thus have a linear relation between the co-graduated values θ_p and ξ_p:

$$\theta_p = \frac{\pi}{10}(2.5+\xi_p). \tag{c}$$

With regard to ξ_p, the reduced angular deviate corresponding to p, we get:

$$\xi_p = \frac{10}{\pi}\theta_p - 2.5. \tag{d}$$

As with Bliss (1935), so likewise one can add 5 to the reduced angular deviate to define an angular probability unit.

$$\begin{aligned} Y_{A,p} &= 5+\xi_p \\ &= 2.5+\frac{10}{\pi}\,\theta_p, \end{aligned} \tag{e}$$

which is centered on $Y_{A(0.5)} = 5$.

Upon substituting into (e) the value of θ_p defined in (b), the angular probability unit is obtained as an explicit function of p:

$$Y_{A,p} = 2.5+\frac{10}{\pi}\text{ arc sin }\sqrt{p}. \tag{e$'$}$$

There is therefore no difficulty in compiling a Table of $Y_{A(p)}$ using the methods described in section 6.314.4. In Table 6.314.7A are given the $100\,p\%$ which correspond to values of angular probability units contained between 2.5 and 7.5, with a view to comparing them with the classical Gaussian probits and with probability units defined from consideration of other distributions with finite range.

Figure 6.314.11I represents the following three elements in the domain $(0, \pi/2)$ of θ:

(a) angular probability density

$$f(\theta) = \sin 2\theta$$

(b) angular distribution

$$p = F(\theta) = \sin^2\theta$$

(c) angular probability unit

$$Y_{A,p} = 2.5+\frac{10}{\pi}\,\theta_p = 2.5+\frac{10}{\pi}\text{ arc sin }\sqrt{p}$$

$$\text{slope:}\quad \tan\lambda = \frac{5}{\pi/2} = \frac{10}{\pi} = 3.183.$$

6.314.4. Angular transformation tables

The relative frequency $\mathcal{P} = r/n$ follows asymptotically a Gaussian distribution (see 6.322.11) with the parameters

$$E[\mathcal{P}] = p$$

$$V(\mathcal{P}) = \frac{p(1-p)}{n}. \tag{a}$$

The approximation is valid if $np(1-p) > 9$. The transform in arc sin or $\sin^{-1}$ of $\mathcal{P}$, where $\mathcal{P} = r/n$ is an estimate of the probability p, can take either of two forms.

(1) *Transformation* θ. This transformation

$$\theta = \text{arc sin } \sqrt{p} = \sin^{-1} \sqrt{p} \tag{b}$$

was introduced in 6.314.3 (b).

Given that $\mathcal{P} = r/n$ is an estimate of p and that $\hat{\theta} = \sin^{-1} \mathcal{P}$, we have, for θ *expressed in radians*,

$$E[\hat{\theta}] = \sin^{-1} \sqrt{p}$$

$$V(\hat{\theta}) = \frac{1}{4n}. \tag{c}$$

This variance is independent of the true value of θ, but it does depend on the number n of the exposed individuals. Therefore the angular transformation of p into θ only stabilizes the variance on the condition that the observed proportions of reactors

$$\mathcal{P}_i = \frac{r_i}{n}$$

are all based on the same n.

If θ is expressed in degrees, we have

$$\theta \text{ degrees} = \frac{180}{\pi} \cdot \theta \text{ radians.} \tag{d}$$

Hence on passing to the variance

$$V(\hat{\theta}) \text{ degrees} = \left(\frac{180}{\pi}\right)^2 V(\theta) \text{ radians} = \frac{180^2}{\pi^2} \cdot \frac{1}{4n} = \frac{820 \cdot 7}{n}. \tag{e}$$

Table X of Fisher–Yates (1957) is published with the heading *Angular Transformation*. It transforms the value of $100p\%$ by steps of 0.1%, into θ given in degrees according to the relation

$$\theta \text{ degrees} = \text{arc sin } \sqrt{100p\%} = \sin^{-1} \sqrt{100p\%}.$$

For an entry $100p = 50.0\%$, one reads $\theta = 45°$.

Table XI of Fisher–Yates (1957) gives the transformation of the fractions of reactors into degrees. Example: 1 reactor out of 30 exposed, leads to $\theta = 10.5°$.

(2) *Transformation* $y = 2\theta$. Several authors have also considered the transformation

$$y = 2\theta$$
$$y = 2 \sin^{-1} \sqrt{\mathcal{P}}. \tag{f}$$

The distribution of y is approximately Gaussian with

$$E[y] = 2\sin^{-1}\sqrt{p}$$
$$V(y) = \frac{1}{n}, \qquad \text{(g)}$$

for $np(1-p) > 9$ (Hald, 1952b, p. 18). We thus have a second *numerical* system for the same angular transformation, i.e. $y = 2\theta$.

Table $y = 2$ arc sin $\sqrt{p}$ for $0 \leq p \leq 1$ by steps of 0.001; y in radians varies from 0 to $\pi = 3.1416$ (Hald, 1952b, pp. 70–71).

Example: entry $p = 0.5$; one reads $y = 2$ arc sin $\sqrt{0.5} = 1.5708$.

In accordance with (c) the variance of y is $1/n$. This relation is consistent with $V(\theta) = 1/4n$ since $y = 2\theta$. There are other useful tables in the context of the angular transformation.

In the scientific tables (Geigy, 1963) there is a table arc sin $p = \sin^{-1}p$ (p. 69), where our p is replaced by x:

$$y \text{ radians} = \text{arc sin } p,\ p = 0.000, (0.001), 1.$$

It should be noted that the entry is here p and not $\sqrt{p}$.

Finally, it is mentioned that the construction of $\varphi(\theta) = \sin 2\theta$ and $p = \sin^2\theta$ is facilitated if the table of natural sines is used (Table XXXI of Fisher–Yates, 1957), where the entry is in degrees and minutes after conversion of θ radians.

(3) *Redefinition of the angular transform.* For r reactors on n exposed, i.e. the proportion $\mathcal{P} = r/n$ instead of (b), Anscombe (1956, p. 464) proposes the transformation we denote by

$$\theta_A = \sin^{-1}\sqrt{\frac{r+\frac{1}{2}}{n+\frac{1}{2}}} \qquad \text{(h)}$$

and which is asymptotically unbiased.

6.314.5. *Arc sine transformation ruling of J. Berkson*

J. Berkson has proposed an arc sine transformation ruling with arithmetically equidistant abscissae.† There are two scales or ranges of ordinates:

(1) *left-hand ordinates*

$$p = \sin^2\left(\frac{k}{2}+\frac{\pi}{4}\right). \qquad \text{(a)}$$

† Paper No. 32.452, Arc sine transformation ruling, Codex Co., Norwood, Mass., U.S.A.

Berkson uses $p\%$ in multiplying the second term by 100; k is in radians and centered on zero. The above relation corresponds to 6.314.21 (b).

$$p = \sin^2\left(\frac{\pi\xi}{10} + \frac{\pi}{4}\right),$$

where ξ is the "reduced angular deviation" (pure number) introduced in 6.314.3 (c).

(2) *right-hand ordinates*, centered on zero

$$k_{\text{radians}} = 2\sin^{-1}\sqrt{p} - \frac{\pi}{2}. \qquad \text{(b)}$$

The linear relation between k and ξ is written so:

$$\frac{k}{2} = \frac{\pi}{10}\xi,$$
$$k = \frac{\pi}{5}\xi. \qquad \text{(c)}$$

6.314.6. *Use of the angular law*

(1) In its reduced form 6.314.1 (a):

$$p = F(\theta) = \sin^2\theta, \qquad \text{(a)}$$

the $\sin^2\theta$ law plays a useful rôle in the representation of a *dose–response sigmoid* or *metamer* of the dose–response sigmoid for events in all-or-nothing over a *finite* domain (θ in $0, \pi/2$).

(2) But in its inverse form

$$\theta = \sin^{-1}\sqrt{p} \qquad \text{(b)}$$

it is currently used to *rectify the dose–response sigmoid* or metamer of the dose (see Knudsen and Curtiss, 1947, p. 4 and Miller, 1950, pp. 910–911).

(3) In order that the range of θ should become equal to unity, θ is expressed as a fraction of $\pi/2$; then one writes the expression

$$\frac{2}{\pi}\theta = \frac{2}{\pi}\sin^{-1}\sqrt{p}. \qquad \text{(c)}$$

This form is one of the vital nodes of the Calculus of Probabilities (Feller, 1958, I, p. 80). For example, in an honest game of heads or tails, after a large number n of tosses, the random fraction $\boldsymbol{p}$ of time during which one of the two players is winning, is written so:

$$\text{Prob}\,(\boldsymbol{p} \leqslant p) = \frac{2}{\pi}\,\theta_p = \frac{2}{\pi}\,\text{arc sin}\,\sqrt{p}. \tag{d}$$

This law in p is considered further in section 6.315.13.

6.314.7. Comparison of three reduced distributions having finite range, and of the angular law, with range (−2.5; 2.5) with the standard Gaussian distribution

Having by now defined probability units centered on 5 for $p = 0.5$ for four distributions with finite range, it is of interest to compare these scales of probability with the Gaussian (normal) probability scale (see 6.322.225.1 and 2).

From Table 6.314.7A it will be seen that the parabolic law is not satisfactory. But the triangular distribution is generally intermediate for values of the reduced deviate between 3 and 7 for the Gaussian distribution and the angular distribution.

TABLE 6.314.7A. COMPARISON OF THE PERCENTAGES 100p% CORRESPONDING TO VARIOUS PROBABILITY UNITS FOR THE REDUCED RECTANGULAR, TRIANGULAR, PARABOLIC, GAUSSIAN LAWS AND THE ANGULAR LAW WITH THE SAME RANGE (−2.5; +2.5) AS THE TRIANGULAR AND PARABOLIC

Probability density	Probability units								
	2.5	3	3.27	4	5	6	6.73	7	7.5
Rectangular R			0	20	50	80	100		
Triangular T	0	2	4.74	18	50	82	95.26	98	100
Parabolic P	0	3	6.4	21.6	50	78.4	93.6	97	100
Angular A	0	2.45	5.76	20.6	50	79.4	94.24	97.55	100
Gaussian	0.62	2.25	4.18	16	50	84	95.82	97.75	99.38

6.314.8. Angular imminence rate (hazard function)

The imminence rate, or hazard function (section 6.122c, d), for the angular law is written as follows:

$$I(\theta) = \frac{f(\theta)}{1-F(\theta)} = \frac{\sin 2\theta}{1-\sin^2\theta} = \frac{2\sin\theta\cos\theta}{\cos^2\theta} = 2\tan\theta. \tag{a}$$

This function increases monotonically from 0, for $\theta = 0$, to ∞, for $\theta = \pi/2$, and passing through the value 2 at $\theta = \pi/4$.

6.314.9. Standard angular law

We have considered the crude dimensionless angular variable θ radians (6.314.11 and 6.314.12) centered on $\pi/4$ and the centered angular deviate (which is also of the reduced type) ξ (6.314.3c).

Berkson's k (6.314.5a and c) is also centered on 0. There is still room for a standard angular deviate.

$$\eta = \frac{\theta - \pi/4}{\sigma_\theta}, \tag{a}$$

$\sigma_\theta \approx 0.342$ (6.314.12h).

The density is written

$$\psi(\eta) = \sigma_\theta \sin 2\left(\frac{\pi}{4} + \sigma_\theta \eta\right). \tag{b}$$

The finite range of η is $\left(-\dfrac{\pi}{4\sigma_\theta},\ +\dfrac{\pi}{4\sigma_\theta}\right)$, or $(-2.2965, +2.2965)$.

The ordinate at the origin is

$$\psi(0) = 0.342. \tag{b'}$$

6.315. Beta Laws of Probability

6.315.1. Beta law of the first kind

6.315.11. Reduced beta density of the first kind

Definition. Since the beta distribution is required in the calculation of the moments of the beta density, our exposition does not follow the scheme adopted for the other laws.

The *reduced beta density of the first kind* depends on the two parameters l, m; the dimensionless variable x evolves in the finite domain (0, 1). This variable is said to be a β_1 or beta -1 variate, designated as $\beta_1(l, m)$ (Weatherburn, 1949, p. 153).

$$b_1(x; l, m) = \frac{1}{\beta(l, m)} x^{l-1}(1-x)^{m-1}, \tag{a}$$

where $0 \leq x \leq 1$, l and $m > 0$, and $\beta(l, m)$ is defined mathematically from the function $\Gamma(l)$ of section 6.334.10. Here

$$\beta(l, m) = \int_0^1 x^{l-1}(1-x)^{m-1}\, dx \tag{b}$$

is the integral of $x^{l-1}(1-x)^{m-1}$ over the range of definition of x which is finite.

$\beta(l, m)$ is evaluated from the gamma function (see 6.334.10):

$$\beta(l, m) = \frac{\Gamma(l)\,\Gamma(m)}{\Gamma(l+m)}. \tag{c}$$

For integers l and m, we have

$$\beta(l, m) = \frac{(l-1)!\,(m-1)!}{(l+m-1)!}. \tag{d}$$

The notation $B(l, m)$ is also used.

Parameters of definition. By virtue of the structure of $\beta(l, m)$, the initial moments of order i can be written directly as follows:

$$\mu_i' = \frac{1}{\beta(l, m)} \int_0^1 x^i \cdot x^{l-1}(1-x)^{m-1}\, dx$$

$$= \frac{\beta(l+i, m)}{\beta(l, m)} \tag{e}$$

$$\frac{\beta(l+i, m)}{\beta(l, m)} = \frac{l}{l+m} \cdot \frac{l+1}{l+m+1} \cdots \frac{l+i-1}{l+m+i-1}.$$

Hence the mean

$$E[x] = \mu_1' = \mu = \frac{l}{l+m} \tag{f}$$

and

$$\mu_2' = \frac{l}{l+m} \cdot \frac{l+1}{l+m+1}. \tag{f'}$$

The variance is

$$V(x) = \sigma_x^2 = \mu_2' - \mu_1'^2 = \frac{l}{l+m} \cdot \frac{l+1}{l+m+1} - \frac{l}{l+m} \cdot \frac{l}{l+m}$$

$$= \frac{lm}{(l+m)^2\,(l+m+1)}. \tag{g}$$

Standard deviation:

$$\sigma_X = \frac{1}{l+m}\sqrt{\frac{lm}{l+m+1}}. \qquad (g')$$

Coefficient of variation:

$$\gamma_x = \frac{\sigma_x}{E[x]} = \frac{1}{l}\sqrt{\frac{lm}{l+m+1}}. \qquad (g'')$$

Parameters of shape. β_1 and β_2 are tabulated for the couples (l, m), or rather (p, q) varying by steps 0.5 in Table II of K. Pearson (1934). For $l = m$ or $p = q$, we have a symmetrical curve and $\beta_1 = 0$.
For example: (1) $l = m = 2 \mid p = q = 2$ (p. 437); we have ($Mo = \tilde{x} =$ mode):

$$E[x] = 0.5 \qquad Mo = \tilde{x} = 0.5 \qquad \sigma = 0.2236$$

$$\beta_1 = 0 \text{ (predicted value)} \qquad \beta_2 = 2.2338;$$

(2) $l = 2,\ m = 3 \mid p = 2,\ q = 3$:

$$E[x] = \tfrac{2}{5} = 0.4 \qquad \sigma = 0.2$$

$$\frac{1}{Mo} = 1 + \frac{3-1}{2-1} = 3 \text{ (see j \textit{infra})}$$

$$Mo = \tfrac{1}{3} = 0.3333.$$

These modes appear in Fig. 6.315.11I and are indicated by arrows on the x axis: the ordinates b_1 according to (a) are on the left scale

$$\beta_1 = 0.0816 \qquad \beta_2 = 2.3571$$

$$\gamma_1 = 0.2857 \qquad \gamma_2 = 0.6429$$

Shape of the beta density of the first kind. Taking the derivative of (a), we get

$$\begin{aligned} b_1'(x; l, m) &= \frac{1}{\beta(l, m)}\left[(l-1)\, x^{l-2}(1-x)^{m-1} - x^{l-1}(m-1)\,(1-x)^{m-2}\right] \\ &= \frac{1}{\beta(l, m)}\, x^{l-2}(1-x)^{m-2}\left[(l-1)\,(1-x)-(m-1)x\right]. \end{aligned} \qquad (h)$$

This derivative vanishes at three points:

(a) *origin:* $x = 0$ if $l > 2$,

(b) *extremity of the range* $x = 1$ if $m > 2$,

(c) *mode: Mo* is the root of the equation written on canceling the factor between brackets in (h), which yields

$$(l-1)\,(1-x) = (m-1)x$$

$$x[m-1+l-1] = l-1.$$

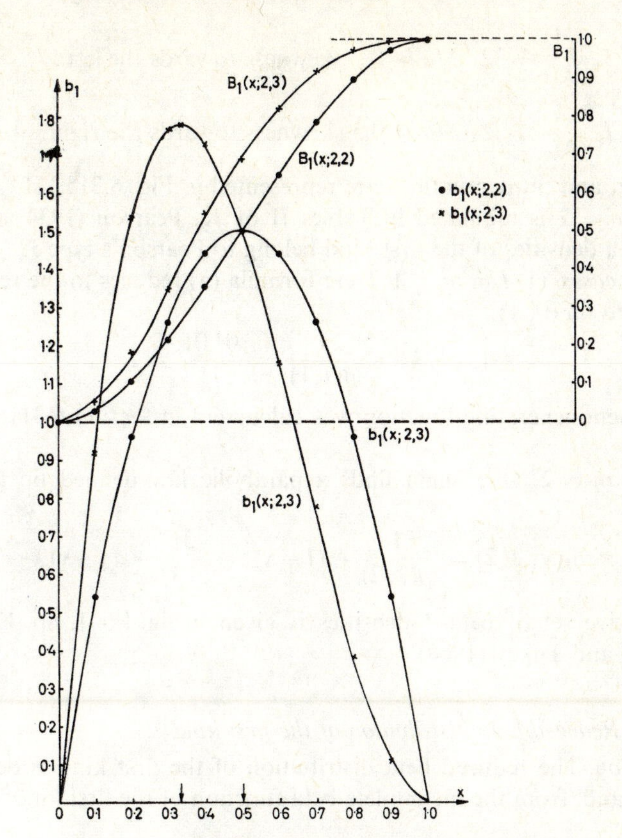

FIG. 6.315.11 I. Reduced beta-laws of the first kind.
$l = m = 2$; $l = 2$, $m = 3$.
Probability densities (ordinates on the left scale).
Distributions (ordinates on the right scale).

Hence the mode is

$$Mo = \frac{l-1}{l+m-2}. \tag{i}$$

We write this relation in the reciprocal of the mode

$$\frac{1}{Mo} = \frac{(l-1)+(m-1)}{l-1} = 1+\frac{m-1}{l-1}. \tag{j}$$

For $m = l$, $\frac{1}{Mo} = 2$; $Mo = 0.5$ (symmetry).

For $m < l$, $\dfrac{1}{Mo} < 2$; $Mo > 0.5$ skewness towards the left.

For $m > l$, $\dfrac{1}{Mo} > 2$; $Mo < 0.5$ skewness towards the right.

The first and third situations are represented in Fig. 6.315.11I.

The $Mo = \tilde{x}$ is tabulated in Tables II of K. Pearson (1934 and 1948).

The beta densities of the first kind belong to Pearson's type I.

Special cases. (1) $l = m = 1$. Here formula (a) reduces to the rectangular law defined on (0, 1):

$$b_1(x; 1, 1) = \frac{1}{\beta(1,1)} = \frac{0!\,0!}{1!} = 1. \tag{k}$$

The moment-generating-function was calculated in section 6.311.4 for this law.

(2) $l = m = 2$. One again finds a parabolic law defined on the range (0, 1).

$$b_1(x; 2, 2) = \frac{1}{\beta(2, 2)} \cdot x(1-x) = \frac{3!}{1!\,1!} \cdot x(1-x). \tag{l}$$

An extensive set of beta–1 densities is given in fig. F6–1, pp. 176–177 of Mosteller and Tukey (1968).

6.315.12. *Reduced beta distribution of the first kind*

Definition. The reduced beta distribution of the first kind is defined, on the one hand, from the incomplete beta function of the first kind

$$\beta_x(l, m) = \int_0^x \mathrm{x}^{l-1}(1-\mathrm{x})^{m-1}\, d\mathrm{x}, \tag{a}$$

and, on the other hand, from the complete beta function of the first kind.

The classical form of the beta distribution of the first kind is

$$I_x(l, m) = \frac{\beta_x(l, m)}{\beta(l, m)} = \frac{1}{\beta(l, m)} \int_0^x \mathrm{x}^{l-1}(1-\mathrm{x})^{m-1}\, d\mathrm{x}, \tag{b}$$

with the obvious relations

$$I_0(l, m) = 0$$
$$I_1(l, m) = \frac{\beta(l, m)}{\beta(l, m)} = 1. \tag{b'}$$

K. Pearson (1934 and 1948) and a team of co-workers have produced tables of the beta distribution of the first kind which is denoted as $\beta(p, q)$ for $\beta(l, m)$ and as $\beta_x(p, q)$ for $\beta_x(l, m)$.

By analogy with the beta density of the first kind $b_1(x; l, m)$, one logical notation for the beta distribution of the first kind is

$$B_1(x; l, m) = I_x(l, m). \tag{c}$$

Two examples are given in the upper part of Fig. 6.315.11I (previous section).

Tables of the beta distribution of the first kind. The tables edited by K. Pearson (1934 and 1948) show the values of the beta distribution of the first kind $I_x(p, q)$, which we write as $I_x(l, m)$,

$$I_x(l, m) = \frac{\beta_x(l, m)}{\beta(l, m)} \tag{d}$$

using the recurrence formula

$$I_x(l, m) = xI_x(l-1, m)+(1-x)\,I_x(l, m-1). \tag{e}$$

Moreover

$$I_X(l, m) = 1-I_{1-X}(m, l). \tag{f}$$

This complementary relation saves about the half of the surface of the tables.

In Pearson's notation, $I_x(l, m) = I_x(p, q)$ is tabulated for x varying by steps of 0.01 from $p = 0.5$, $q = 0.5$ to $p = 50$, $q = 50$, and at several values of $p \geqslant q$ corresponding to q fixed, e.g. for $q = 2$ one finds $I_x(p, 2)$ for p evolving from 2 to 50 by steps of $+1$. By way of illustration the reader can verify the values of the distributions represented in Fig. 6.315.11 I (ordinates on the right).

For $q = 2$, $p = 2$ (pp. 44–45 of Tables I of Pearson) the calculated distribution is

$$B_1(x; 2, 2) = I_x(2, 2) = \tfrac{6}{6}x^2(3-2x),$$

the factor $\frac{1}{6} = 0.6666667$ appearing here through the column ($q = 2, p = 2$); furthermore,

$$\beta(2, 2) = \frac{\Gamma(2)\cdot\Gamma(2)}{\Gamma(2+2)} = \frac{1}{2.3} = \frac{1}{6}.$$

Likewise, for $p = 2$, $q = 3$, $\beta(2, 3) = \frac{1}{12} = 0.83333333\cdot\frac{1}{10}$, the calculated distribution is written

$$B_1(x; 2, 3) = I_x(2, 3) = \frac{1}{\beta(2, 3)}\int_0^x \mathrm{x}(1-\mathrm{x})^2\,dx = \frac{12}{12}(6x^2-8x^3+3x^4)$$

$$= x^2(6-8x+3x^2).$$

This distribution is also easily obtained using relation (d), K. Pearson's Table I (pp. 43–44) and reading $I_x(3, 2)$ since $p \geqslant q$.

6.315.13. First arc sine law (Feller)

6.315.131. Definition. In the study of the probability where one of the two players at heads or tails wins for a large number n of tosses or maintains the winning position on balance (probability of long leads), Feller (1958, I, p. 77) presents the *first arc sine law* in the following form: for α fixed $(0 < \alpha < 1)$ and $n \to \infty$, the probability that the fraction r/n of time in the lead should be less than α tends to a limit which is written as $A(\alpha)$:

$$A(\alpha) = \frac{1}{\pi}\int_0^\alpha \frac{dx}{\sqrt{x(1-x)}} = \frac{2}{\pi} \text{ arc sin } \sqrt{\alpha} = \frac{\text{arc sin } \sqrt{\alpha}}{\pi/2}. \tag{a}$$

The letter A is here the first character of arc sine.

The probability density of the first arc sine law is written so:

$$a(x) = \frac{dA(x)}{dx} = \frac{1}{\pi}\frac{1}{\sqrt{x(1-x)}}. \tag{b}$$

It is represented in part a of Fig. 6.315.131I by a curve in U, the ordinate of which tends to infinity as $x \to 0$ or $x \to 1$. The minimum of $f(x)$ occurs at $x = \frac{1}{2}$ (antimode) and here we have

$$a\left(\frac{1}{2}\right) = \frac{2}{\pi} \approx 0.637.$$

The first arc sine distribution

$$A(x) = \frac{2}{\pi} \text{ arc sin } \sqrt{x} \tag{c}$$

is represented in part b of Fig. 6.315.131I.

The point of inflection at $x = 0.5$ corresponds to the minimum $a\,(0.5) = 0.637$; the inflectional tangent has the slope 0.637.

From fig. 4 of Feller (1958, I, p. 80), we form the table of deciles x_A (which would appear in Feller as t_p).

TABLE 6.315.131A. DECILES OF THE FIRST ARC SINE LAW
(from fig. 4 of Feller 1958, I. p. 80)

$A(x)$	0	0.1	0.2	0.3	0.4	0.5	0.6	0.7	0.8	0.9	1.0
x_A	0	0.024	0.095	0.206	0.345	0.500	0.655	0.794	0.905	0.976	1.000

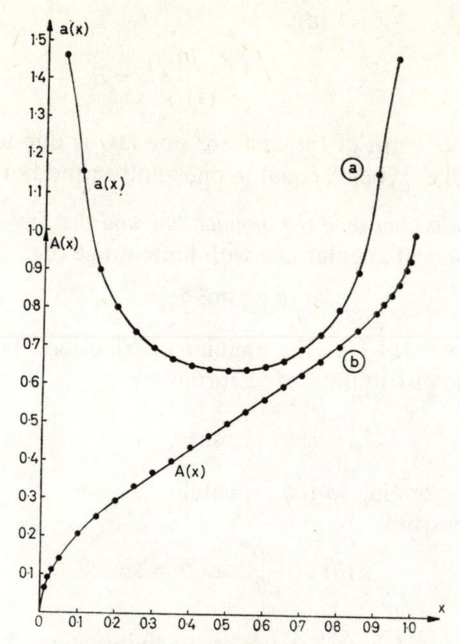

FIG. 6.315.131 I. First arc sine law.
(a) Probability density.
(b) Distribution.

The formula which relates x_A to A $(0 \leqslant A \leqslant 1)$ is

$$x_A = \sin^2 \frac{\pi A}{2}. \tag{d}$$

Newell (1965) has given another derivation of (b) and (c) by considering the situation where a variable which oscillates sinusoidally is sampled at random or systematically with a sampling interval unequal to a rational fraction of the period of oscillation (see also section 6.315.134).

6.315.132. Relation between the first arc sine law and the law β_1. The density (*b*) of the preceding of section can be written in the form of a beta density of first kind with the parameters $l = \frac{1}{2}$ and $m = \frac{1}{2}$:

$$a(x) = \frac{1}{\pi} x^{-\frac{1}{2}}(1-x)^{-\frac{1}{2}}. \tag{a}$$

The variable x is presented as a $\beta_1(\frac{1}{2}, \frac{1}{2})$ variate

$$a(x) = \frac{1}{\beta(\frac{1}{2}, \frac{1}{2})} x^{-\frac{1}{2}}(1-x)^{-\frac{1}{2}}, \tag{a'}$$

where, according to 6.315.11 (c),

$$\beta(\tfrac{1}{2}, \tfrac{1}{2}) = \frac{\Gamma(\frac{1}{2})\cdot\Gamma(\frac{1}{2})}{\Gamma(1)} = \pi. \tag{b}$$

The symmetric U form of the first arc sine law is due to the parameters of the variable β_1, i.e. $\frac{1}{2}$, being equal to one another and both less than unity.

6.315.133. Relation between the angular law and the first arc sine law. We again take the classical angular law with finite range $(0, \pi/2)$,

$$p = \sin^2 \theta \tag{a}$$

defined in section 6.314.1 as the angular distribution. The angle θ is the argument; p is the distribution of θ; furthermore,

$$\theta_p = \text{arc sin } \sqrt{p} \tag{b}$$

is the angle corresponding to the probability p.

The angular density

$$f(\theta) = \frac{d}{d\theta} \sin^2 \theta = \sin 2\theta \tag{c}$$

is a bell-shaped curve where θ traverses the finite range $(0, \pi/2)$.

In the first arc sine law of Feller which we have just studied, θ is no longer the independent variable but p. The relation (b) links θ_p to p. In this regard the argument p of the angular law corresponds to the notation α of Feller (6.315.131a), whilst the present p corresponds to x in the relation 6.315.131 (b).

For p varying from 0 to 1, if we multiply θ_p by $2/\pi$, or if we divide θ_p by $\pi/2$, we get

$$\frac{2}{\pi}\,\theta_p = \frac{\theta_p}{\pi/2} = \frac{2}{\pi} \text{ arc sin } \sqrt{p}. \tag{d}$$

This relation defines a monotonically increasing function which is no other than the distribution of p.

$$A(p) = \frac{2}{\pi} \text{ arc sin } \sqrt{p}, \tag{e}$$

where

$$A(0) = 0$$

$$A(1) = \frac{2}{\pi} \text{ arc sin } \sqrt{1} = \frac{2}{\pi}\cdot\frac{\pi}{2} = 1. \tag{f}$$

This distribution $A(p)$, a function of p (which is itself the distribution of θ), is a linear function of θ:

$$A(p) = \frac{2}{\pi}\,\theta_p. \tag{g}$$

The probability density p is written as

$$\begin{aligned} a(p) &= \frac{dA(p)}{dp} = \frac{2}{\pi}\frac{1}{\sqrt{1-(\sqrt{p})^2}}\cdot\frac{d}{dp}\sqrt{p} \\ &= \frac{2}{\pi}\frac{1}{\sqrt{1-p}}\,\frac{1}{2\sqrt{p}}, \end{aligned} \tag{h}$$

and we again find the arc sine density

$$a(p) = \frac{1}{\pi}\frac{1}{\sqrt{p(1-p)}}. \tag{h'}$$

The function $a(p)$ and $A(p)$ are represented in the same coordinate system with x instead of p in Fig. 6.315.131I.

6.315.134. Applications. Stating that "simple mathematical models may give rise to unusual frequency distributions", Newell (1965, p. 165) considers a U-shaped density which arose in the paper industry with the conclusion that "a punched card of average thickness is the most uncommon type of card, since the mean of this distribution is antimodal". In addition to the classical field of sky cloudiness the author suggests other fields of application of U-shaped density such as instantaneous pulse pressures, gas concentrations in the lung, variables related to menstruation or to daily or annual climatic variation.

In the Bayesian analysis of contingency tables by the logit (6.323.222) method of Lindley (1964), Gart (1966) notes that the first arc sine density $a(p)$ (formula 6.315.133h') serves as a prior density for the positive binomial; this prior assumption is equivalent to assuming that the corresponding distribution in $\sin^{-1}\sqrt{p}$ (6.315.133e) is uniformly distributed.

6.315.14. Distribution of order statistics

We consider a population X of probability density $f(X)$ and of distribution $F(X)$. From this population we take a random sample of n independent values which we arrange in a progressively increasing order:

$$\underbrace{X_{(1)}\,X_{(2)}\,\ldots\,X_{(r-1)}}_{r-1}\;\underbrace{X_{(r)}}_{1}\;\underbrace{X_{(r+1)}\,X_{(n)}}_{n-r}, \tag{a}$$

distinguishing the rank value (r).

The probability of observing the particular sequence observed with the n values is the composite probability

$$[F(X_{(r)})]^{r-1}\, dF(X_{(r)})\cdot[1-F(X_{(r)})]^{n-r}, \tag{b}$$

i.e.

(a) the $(r-1)$ values less than $X_{(r)}$ each have a probability $F(X_{(r)})$ of being less than $X_{(r)}$,

(b) the value $X_{(r)}$ has a probability

$$dF\,(X_{(r)}) = f(X_{(r)})\, dX_{(r)}$$

of finding itself in the interval $X_{(r)}-\dfrac{dX_{(r)}}{2},\ X_{(r)}+\dfrac{dX_{(r)}}{2}$,

(c) the $(n-r)$ values greater than $X_{(r)}$ each have a probability $1-F(X_{(r)})$ of exceeding $X_{(r)}$.

The number of series such as (a) is the number of tri-partitions of n into $(r-1)$, 1 and $(n-r)$, i.e.

$$\frac{n!}{(r-1)!\,1!\,(n-r)!} = \frac{\Gamma(n+1)}{\Gamma(r)\cdot\Gamma(n-r+1)} = \frac{1}{\beta(r, n-r+1)}. \tag{c}$$

according to 6.315.11 (c).

The probability $dG(X_{(r)})$ of finding the order statistic $X_{(r)}$ in the neighborhood of $X_{(r)}$ is a total probability obtained by multiplying (b) with (c):

$$dG(X_{(r)}) = \frac{1}{\beta(r, n-r+1)}\cdot F_r^{r-1}(1-F_r)^{n-r}\, dF_r, \tag{d}$$

on writing as M. G. Kendall and Stuart (I, 1963, p. 325):

$$F_r = F(X_{(r)}). \tag{d'}$$

The range of F_r is clearly (0, 1).

One thus sees in (d) an example of a beta distribution of the first kind. The value $X_{(r)}$ in the ordered sequence (a) is the order statistic r (rth order statistic) over the n values. The full notation should give n explicitly, but this is not customary. It will be seen, however, that n is in fact specified in the tables of order statistics. In particular, if $n = 2r+1$, then there are r values which are superior to it. Hence $X_{(r+1)}$ is the *median* of the random sample of n values with $F(X_{(r+1)}) = q = p = 0.5$. The *quartiles* and the *deciles* are also order statistics.

The formula (d) shows that there are as many distributions $G(X_{(r)})$ of the rth order statistic in a random sample of n, as there are distribution $F(X)$ with continuous derivatives.

A monographical exposition would doubtless involve studying, for each distribution $F(X)$, the associated distribution of the order statistics $G(X_{(r)})$.

In particular, for the rectangular (uniform) law defined over the domain (0, 1),

$$F(X) = X, \tag{e}$$

and then $G(X_{(r)})$ reduces to a beta law of the first kind in X.

The order statistics associated with the Gaussian case will be treated later in section 6.322.226. It is believed that these concepts should penetrate further into the biological field and in this connection the reader is referred to the important monograph of Sarhan and Greenberg (1962).

6.315.15. Transformed beta-variables (Blom, 1954)

Stating from the inversion of the normal distribution

$$u_p = \phi^{-1}(p) \qquad 0 \leqslant p \leqslant 1$$

and the problem of plotting points on a normal probability paper, Blom (1954) studied the properties of order statistics and of end corrections. He took advantage of the work by Chernoff and Lieberman (1954) who showed the close relationship between the plotting problem on one hand and the general problem of estimating parameters by means of linear functions of order statistics, on the other hand.

Blom calls transformed variates in the argument $p(0 \leqslant p \leqslant 1)$ *transformed beta-variates* which are particularly useful for the purpose of normalizing positive binomial, negative binomial, Poisson and χ^2 variables (Blom, 1954, pp. 33 and 37–39).

6.315.2. Beta law of second kind

6.315.20. Transformation

In

$$\beta(l, m) = \int_0^1 x^{l-1}(1-x)^{m-1}\, dx, \tag{a}$$

we perform the change of variable

$$x = \frac{1}{1+y} \qquad 1-x = \frac{y}{1+y}$$
$$dx = -\frac{dy}{(1+y)^2}. \tag{b}$$

As x varies from 0 to 1, y varies from $+\infty$ to 0. We obtain a second form of the mathematical function $\beta(l, m)$:

$$\beta(l, m) = +\int_0^\infty \frac{1}{(1+y)^{l-1}} \cdot \frac{y^{m-1}}{(1+y)^{m-1}} \frac{dy}{(1+y)^2} = \int_0^\infty \frac{y^{m-1}}{(1+y)^{l+m}} dy. \quad \text{(c)}$$

6.315.21. Beta density of second kind

6.315.211. Definition. One can thus define a reduced beta density of second kind in y (β prime variate or β_2 (or $\beta-2$) variate) whose density is written

$$b_2(y; l; m) = \frac{1}{\beta(l, m)} \frac{y^{m-1}}{(1+y)^{l+m}} \quad \text{(a)}$$

$$0 \leqslant y < \infty.$$

Contrary to the β_1 variate x which varies from 0 to 1, the β_2 variate y varies between 0 and ∞.

Logically, this section should appear under general heading 6.3 since the range of the beta-2 variate is $(0, \infty)$. We leave it here because of the close connection between beta-1 and beta-2 variates.

The family of the β_2 variates belongs to the type IV of K. Pearson. These curves are tangential to Oy at $y = 0$ for $l > 2$.

The derivative of $b_2(y; l; m)$ is written as

$$b_2' = y^{m-2}(1+y)^{-(l+m+1)}[(m-1)(1+y)-(l+m)y]. \quad \text{(b)}$$

Here the sign (′) has the classical significance of a derivative. The derivative vanishes at the origin for $m > 2$.

The mode, given by the expression

$$Mo = \frac{m-1}{l+1}, \quad \text{(c)}$$

exists for $m > 1$.

If one performs on Mo_y the inverse transformation of expression 315.20 (b), i.e.

$$y = \frac{1-x}{x}, \quad \text{(d)}$$

we again find the mode of x given in the expression 6.315.11 (j).

6.315.212. Reduction parameters. From 6.315.20 (c), the initial moments of order i are as follows:

$$\mu_i' = \frac{\beta(l+i, m-i)}{\beta(l, m)} = \frac{l}{m-1} \cdot \frac{l+1}{m-2} \cdot \ldots \cdot \frac{(l+i-1)}{m-i}. \quad \text{(a)}$$

Hence the *mean* is

$$E[y] = \mu_1' = \frac{l}{m-1}, \quad \text{(b)}$$

and the *variance* is calculated to be

$$V(y) = \sigma_y^2 = \frac{l(l+m-1)}{(m-1)^2(m-2)}. \quad \text{(c)}$$

6.315.3. Relations between the variables $\beta_1(l, m)$, $\beta_2(l, m)$ and $\gamma(l)$

The beta variables of first kind (beta-1) and second kind (beta-2) are written as $\beta_1(l, m)$ and $\beta_2(l, m)$. There is no risk of confusion with the parameters of shape β_1 and β_2.

Of the important theorems presented in Weatherburn (1949, pp. 153–159), we shall state there of them:

(1) If x is a beta-2 variate of the type $\beta_2(l, m)$, then its inverse

$$y = \frac{1}{x}$$

is a beta-2 variate with interchanged parameters, i.e it is of the type $\beta_2(m, l)$.

This important property intervenes in the test $F(\nu_1, \nu_2)$, where ν_1 and ν_2 are degrees of freedom.

(2) If x and y are two independent gamma variates with the parameters l and m respectively, then the quotient

$$z = \frac{x}{x+y}$$

is a variate of the type $\beta_1(l, m)$.

(3) If x and y are two independent gamma variates with the parameters l and m respectively, then the quotient

$$w = \frac{x}{y}$$

is a $\beta_2(l, m)$ variate.

Application. By virtue of the foregoing theorems, the quotient $z^2 = u^2/v^2$ of the squares of two independent Gaussian standard deviates (each governed by a gamma law with parameter $\frac{1}{2}$) is a $\beta_2(\frac{1}{2}, \frac{1}{2})$ variate whose density is written so

$$df(z^2) = \frac{dz^2}{\pi(1+z)\sqrt{z^2}}$$

over the domain $(0, \infty)$.

The distribution of $z = u/$ over the domain $(-\infty, +\infty)$ is written as

$$df_1(z) = \frac{1}{\pi}\frac{dz}{1+z^2},$$

and is no other than the Cauchy density (see 6.324.11a).

6.32. Continuous Laws with Infinite Domain in the Two Directions

6.321. Law of Laplace

6.321.1. Laplacian law in the original variable

6.321.11. Laplacian density of probability

The negative bi-exponential density. This law is an immediate extension of the bilateral triangular law treated in section 6.312.2. It is known as the first law of Laplace and will be simply named Laplacian law.† In fact this density associates with the random variable X moving in the domain $(-\infty, +\infty)$, the density function

$$f(X)\,\text{cm}^{-1} = \frac{1}{2a}\exp\{-|X-\mu|/a\}, \qquad \text{(a)}$$

where $|X-\mu|$ is the absolute value of the deviation $X-\mu$.

† This avoids confusion with the "double negative exponential" considered in section 6.326.

If X is measured in cm, then μ and a are also in cm, for the argument of the exponential is a pure number. In this case $f(X)$ is given in cm^{-1}, this being imposed by the scale-factor a_{cm} in the denominator of the coefficient of the exponential.

In the absolute deviation

$$x = X - \mu,$$

the relation (a) becomes the central bi-exponential density in x:

$$f(x) = \frac{1}{2a} e^{-|x|/a}, \tag{a'}$$

which is symmetrical about the origin $x = 0$. The letter $|x|$ is here the absolute value of x. The peak at the origin corresponds to a turning point (Fig. 6.321.11I).

It is varified without difficulty that the integral of $f(x)$ over the domain of x is equal to unity:

$$\frac{1}{2a} \int_{-\infty}^{+\infty} e^{-|x|/a}\, dx = \frac{2}{2a} \int_{0}^{\infty} e^{-|x|/a}\, dx = 1. \tag{b}$$

Parameters of description. In view of the symmetry, the *mean* of x is zero and that of X is μ:

$$E[x] = 0,$$

$$E[X] = \mu. \tag{c}$$

The *variance* $\mu_2 = V(x) = \sigma^2$ is calculated from $E[x^2]$:

$$E[x^2] = 2\frac{1}{2a} \int_{0}^{\infty} x^2 e^{-x/a}\, dx = 2a^2 \tag{d}$$

Hence

$$V(x) = \sigma_x^2 = 2a^2. \tag{e}$$

Therefore

$$\sigma_x = \sigma_X = a\sqrt{2} \tag{f}$$

and

$$a = \frac{\sigma_x}{\sqrt{2}}. \tag{g}$$

The parameter a is proportional to the standard deviation and provides a characterization of the dispersion. If a increases, the ordinate to the origin decreases and, conversely, the curve spreads out (Fig. 6.321.11I).

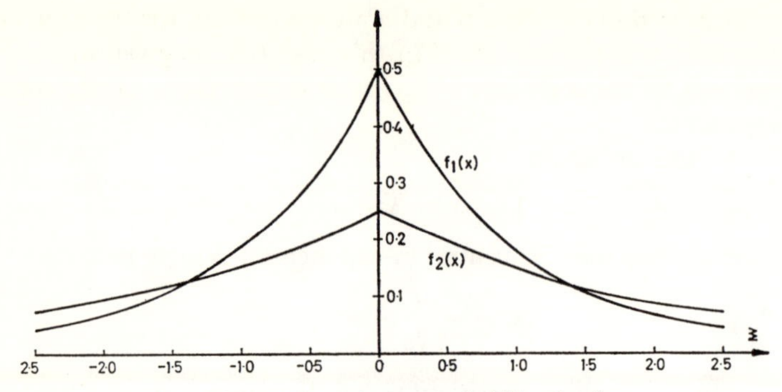

FIG. 6.321.11 I. Laplacian probability density

$$f(x) = \frac{1}{2a} e^{-|x|/a}$$

$$f_1(x) \quad a_1 = 1$$
$$f_2(x) \quad a_2 = 2.$$

Substituting into the relation (a′) above the value of $a = \dfrac{\sigma_x}{\sqrt{2}}$ given by (g), the Laplacian density in the parameter σ becomes

$$f(x) = \frac{1}{\sqrt{2}\sigma_x} \exp\left\{-\sqrt{2}\,|x|/\sigma_x\right\}. \tag{h}$$

The parameters of shape are the following:

$$\gamma_1 = 0, \quad \text{for owing to symmetry} \quad \mu_3 = 0. \tag{i}$$

One then calculates

$$\mu_4 = 4!\, a^4,$$

whence

$$\gamma_2 = \frac{\mu_4}{\sigma^4} - 3 = \frac{1\cdot 2\cdot 3\cdot 4\cdot a^4}{4a^4} - 3 = 6-3 = 3. \tag{j}$$

This result shows that the bi-exponential distribution is strongly leptokurtic; it is even the leptokurtic curve *par excellence* with no asymmetry, compared with the Gaussian curve.

6.321.12. Laplacian distribution

The negative bi-exponential distribution gives the probability of $\boldsymbol{x}$ being at most equal to x:

$$\text{Prob}\,(\boldsymbol{x} \leqslant x) = F(x) = \frac{1}{2a} \int_{-\infty}^{x} e^{-|x|/a}\, dx. \tag{a}$$

Just the same as for the triangular law, it is a convenience to distinguish three zones by a coefficient ε, for the derivative is discontinuous at point $x = 0$ corresponding to $p = 0.5$. We then have

$$F(x) = 0.5+\varepsilon 0.5(1-e^{-|x|/a}), \tag{b}$$

for $\varepsilon = -1$, 0 and $+1$.

The expressions for $F(x)$ are shown for each of the three zones in reference to $p = 0.5$ and $x = 0$ in Table 6.321.12A.

TABLE 6.321.12A. LAPLACIAN DISTRIBUTION

Zones	ε	p	$F(x)$
1. $-\infty \leqslant x < 0$	-1	< 0.5	$0.5-0.5(1-e^{-\|x\|/a})=0.5e^{-\|x\|/a}$
2. $x = 0$	0	$= 0.5$	0.5
3. $0 < x \leqslant +\infty$	$+1$	> 0.5	$0.5+0.5(1-e^{-\|x\|/a})$

The distribution $F(x)$ of sigmoid shape consists of two exponential arcs ($\varepsilon = +1$) and ($\varepsilon = -1$) meeting at the point of inflection $x = 0$ following a common tangent with the slope

$$F'(0) = f(0) = \frac{1}{2a}. \tag{c}$$

Hence, as indicated in the diagram (Fig. 6.321.12I),

$$\tan\theta = \frac{1}{2a} = \frac{1}{\sigma\sqrt{2}}. \tag{d}$$

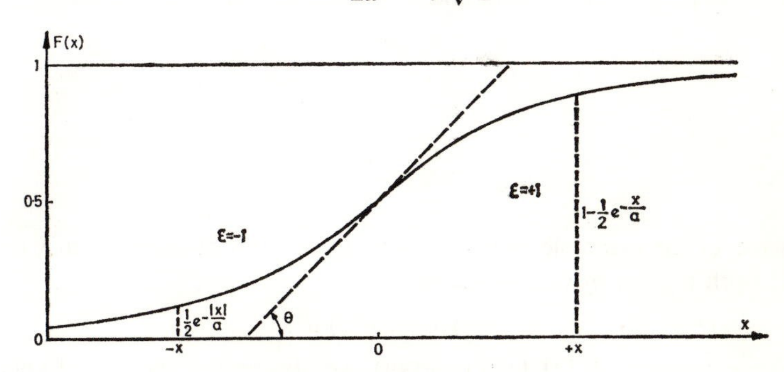

FIG. 6.321.12 I. Laplacian distribution

For $x < 0$, $F(x) = \frac{1}{2}e^{-|x|/a}$ $\varepsilon = -1$
$x = 0$, $F(0) = \frac{1}{2}$
$x > 0$, $F(x) = 1-\frac{1}{2}e^{-|x|/a}$ $\varepsilon = +1$.

At best the standard deviation σ (or the parameter a) is low, and the slope of the bi-exponential sigmoid is steep. One can also define reduced bi-exponential laws and distributions and rectify the bi-exponential sigmoid.

6.321.13. *Some applications of the Laplacian distribution*

Faced with the lack of an explicit formula for the Gaussian distribution, one has recourse to the negative bi-exponential distribution. Such was the solution urged by Landahl (1939) and Rashevsky (1940, p. 161) in the quantitative study of psychological distrimination.

In another order of ideas, Greenwood (1946) considers whether the statistician may not perhaps assist the epidemiologist studies on the intervals between change events. Taking up this problem further, Pyke (1965, p. 425) notes that the negative bi-exponential or Laplacian distribution leads to convenient expressions to handle.

6.321.2. *Standard Laplacian law*

6.321.21. *Standard Laplacian probability density*

The bi-exponential law in X depends on two parameters μ and a, one of which (μ) was taken as the origin in considering the absolute deviation $x = X - \mu$.

We will now go further by introducing a change of scale and taking σ as a new unit of x. We thus define the standard Laplacian deviate

$$\xi = \frac{X-\mu}{\sigma} = \frac{x}{\sigma_x} \tag{a}$$

Hence

$$d\xi = \frac{dx}{\sigma_x}. \tag{a'}$$

By virtue of the principle of invariance of the local probability in x (or X) and ξ, both related by (a), we write

$$f(x)\,dx = \varphi(\xi)\,d\xi.$$

Having regard to 6.321.11 (f) and (a), we ob tain the standard Laplacian density

$$\varphi(\xi) = \frac{1}{\sqrt{2}} e^{-\sqrt{2}\,|\xi|}, \tag{b}$$

from which it is seen that the subjacent area is equal to 1. In fact, we have

$$\frac{1}{2}\int_{-\infty}^{+\infty} e^{-\sqrt{2}|\xi|}\,d\xi = \frac{2}{\sqrt{2}}\int_{0}^{\infty} e^{-\sqrt{2}\xi}\,d\xi = \int_{0}^{\infty} e^{-\sqrt{2}\xi}\,d(\sqrt{2}\xi) = 1. \quad \text{(b')}$$

The graphical representation is given in Fig. 6.321.21I (curve a).

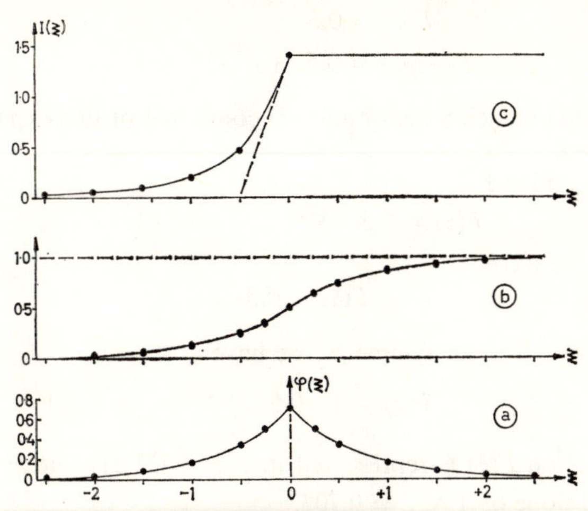

FIG. 6.321.21 I. Standard Laplacian law.
(a) Probability density.
(b) Distribution.
(c) Imminence rate.

The parameters of the standard Laplacian density are:

$$\xi = \frac{x}{\sigma}$$

$$E[\boldsymbol{\xi}] = 0 \qquad \gamma_1 = 0 \qquad \text{(c)}$$
$$V(\boldsymbol{\xi}) = 1 \qquad \gamma_2 = +3$$

6.321.22. Standard Laplacian distribution

The probability of meeting with a reduced deviation $\boldsymbol{\xi}$ at most equal to ξ fixed, is written so:

$$\text{Prob}\,(\boldsymbol{\xi} < \xi) = F(\xi) = \frac{1}{\sqrt{2}}\int_{-\infty}^{\xi} e^{-\sqrt{2}|\xi|}\,d\xi. \qquad \text{(a)}$$

Since the origin $\xi = 0$ is a discontinuity of $\varphi(\xi)$, we mark out zones of ξ according to the values of $\varepsilon = -1, 0, +1$ (see previous section).

We then have

$$F(\xi) = 0.5 + \varepsilon 0.5(1 - e^{-\varepsilon\sqrt{2}\xi}), \qquad \text{(b)}$$

and it is found that

$$F(-\infty) = 0.5 + (-1)0.5 = 0,$$
$$F(0) = 0.5$$
$$F(+\infty) = 0.5 + (+1)0.5 = 1$$

The standard Laplacian distribution is composed of two exponential arcs joining at $\xi = 0$.

For $\xi < 0$, we have

$$F(\xi) = 0.5e^{+\sqrt{2}\xi} \qquad \varepsilon = -1. \qquad \text{(c)}$$

For $\xi = 0$, we have

$$F(\xi) = 0.5. \qquad \text{(d)}$$

For $\xi > 0$, by virtue of symmetry, we have

$$F(\xi) = 1 - 0.5e^{-\sqrt{2}\xi} \qquad \varepsilon = +1. \qquad \text{(e)}$$

The distribution $F(\xi)$ is represented in Fig. 6.321.21I (curve b) and has an inflectional slope of $1/\sqrt{2} = 0.707$.

Practically 50% of all cases are concentrated between the standard Laplacian deviates -0.5 and $+0.5$. This concentration is greater than for the Gaussian curve of the same standard deviation 1 where half the cases are between $-\frac{2}{3}$ and $+\frac{2}{3}$, or more exactly between -0.6745 and $+0.6745$.

6.321.23. Inversion of the standard Laplacian distribution. Laplacian probability unit

The relation (b) in the foregoing sub-section can be written with the introduction of $p = F(\xi)$. Moreover, ξ_p, the standard deviate corresponding to p, is defined by inverting the distribution

$$\xi_p = F^{-1}(p). \qquad \text{(a)}$$

From

$$F(\xi) = p = 0.5 + \varepsilon 0.5(1 - e^{-\varepsilon\sqrt{2}\xi_p}), \qquad \text{(b)}$$

we get

$$\varepsilon 0.5(1 - e^{-\varepsilon\sqrt{2}\xi_p}) = p - 0.5. \qquad \text{(b')}$$

In this form it is seen that ε plays the part of zone indicator since it is of the same sign as $p-0.5$. It can be left indeterminate or equal to 0 for $p = 0.5$, for the parenthesis cancels out automatically since $\xi_{0.5} = 0$.

We isolate the exponential term leaving ε for $1/\varepsilon$, because these two terms are equal:

$$1-e^{-\varepsilon\sqrt{2}\xi_p} = \varepsilon \frac{p-0.5}{0.5}, \tag{c}$$

$$e^{-\varepsilon\sqrt{2}\xi_p} = 1-\varepsilon(2p-1).$$

Hence the inversion of the distribution (a) is realized *explicitly* in the form

$$\xi_p = -\frac{\varepsilon}{\sqrt{2}} \ln [1-\varepsilon(2p-1)], \tag{d}$$

where ξ_p is the standard Laplacian deviate corresponding to the probability p alternatively called the standard Laplacian equivalent deviate.

In the zone where $p < 0.5$, $\varepsilon = -1$ and a more compact expression is

$$\xi_p = \frac{1}{\sqrt{2}} \ln 2p, \tag{e}$$

and ξ_p is indeed negative for $-\varepsilon = +1$, and $\ln 2p$ is negative since $2p < 1$. Finally, we have

$$\xi_p = +\frac{1}{\sqrt{2}} \ln 2p, \quad p < 0.5. \tag{e'}$$

Conversely, in the zone where $p' > 0.5$, $\varepsilon = +1$ and we have

$$\xi_{p'} = -\frac{1}{\sqrt{2}} \ln (1-2p'+1),$$

$$\xi_{p'} = -\frac{1}{\sqrt{2}} \ln [2(1-p)]. \tag{f}$$

As $p' > 0.5$, $1-p'$ is < 0.5 and $2(1-p')$ is <1. In these conditions $\ln [2(1-p')]$ is < 0 and $\xi_{p'}$ is > 0.

If $p' = 1-p$, we have

$$\xi_{1-p} = -\frac{1}{\sqrt{2}} \ln \{2[1-(1-p)]\}.$$

$$\xi_{1-p} = -\frac{1}{\sqrt{2}} \ln (2p); \tag{g}$$

ξ_{1-p} is positive since if $1-p$ is >0.5, we have $p < 0.5$ and $2p < 1$, hence $\ln 2p < 0$.

Upon comparing (e) and (g), it is seen that (Fig. 6.321.23 I)

$$\xi_{1-p} = -\xi_p. \tag{h}$$

The standard deviates corresponding to two symmetrical probabilities about 0.5 are values which are equal in absolute magnitude and opposite in sign. This property holds good for any symmetrical distribution.

The values of ξ_p are negative for $p < 0.5$, or in half of all cases. This is the reason why, taking up the idea of C. I. Bliss (1935), we introduce a *Laplacian probability unit centered* on $p = 0.5$ by adding 5 to ξ_p:

$$Y_{L,p} = 5+\xi_p. \tag{i}$$

For $p < 0.5$, we have

$$Y_{L,p} = 5+\frac{1}{\sqrt{2}} \ln 2p.$$

For $p = 0.5$, we have

$$Y_{L,p} = 5. \tag{j}$$

For $p > 0.5$, we have

$$Y_{L,p}= 5-\frac{1}{\sqrt{2}} \ln 2(1-p).$$

In Table 6.321.23A we show values of the Laplacian probability and the corresponding values of the Gaussian probit extracted from the Tables of Fisher and Yates (1957).

TABLE 6.321.23A. LAPLACIAN PROBABILITY UNIT $Y_{L,p}$ AND THE GAUSSIAN PROBIT Y_P

p	$\frac{1}{2} \ln 2p$	$Y_{L,p}$	Y_P
0.025	−2.1183	2.8817	3.0400
0.05	−1.6282	3.3718	3.3551
0.10	−1.1380	3.8620	3.7184
0.20	−0.6479	4.3521	4.1584
0.25	−0.4901	4.5099	4.3255
0.30	−0.3612	4.6388	4.4756
0.40	−0.1578	4.8422	4.7467
0.50	0.0000	5.0000	5.0000
0.60	+0.1578	5.1578	5.2533
0.70	+0.3612	5.3612	5.5244
0.75	+0.4901	5.4901	5.6745
0.80	+0.6479	5.6479	5.8416
0.90	+1.1380	6.1380	6.2816
0.95	+1.6282	6.6282	6.6449
0.975	+2.1183	7.1183	6.9600

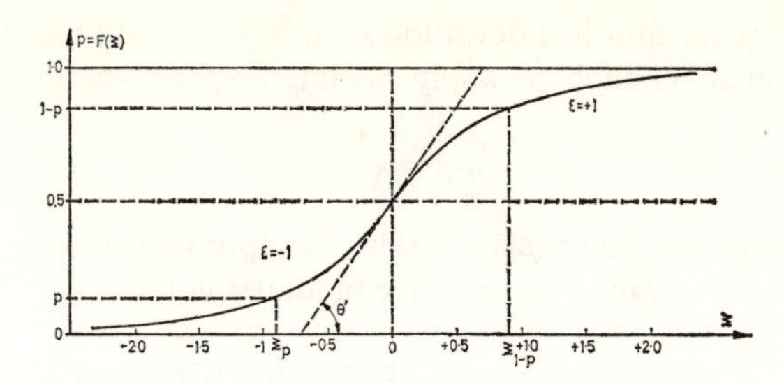

FIG. 6.321.23 I. Inversion of the standard Laplacian distribution. From p to ξ_p;

$$\xi_p = -_{1-p}.$$

$$\tan \theta' = \frac{1}{\sqrt{2}}.$$

6.321.3. Laplacian imminence rate or hazard function

The hazard function (imminance rate) for the Laplacian law is a nice application of relation 6.122 (e″) to the distribution of section 6.321.22.

For $\xi < 0$ (6.321.22 c), the imminence rate

$$I(\xi) = -D \ln\left[1-0.5e^{+\sqrt{2}\xi}\right] \tag{a}$$

is represented in the curved section of (c) in Fig. 6.321.21I. For $\xi > 0$ (6.321.22 e), we have

$$I(\xi) = -D \ln 0.5e^{-\sqrt{2}\xi} = +\sqrt{2} = 1.414. \tag{b}$$

This horizontal is represented in (c) of the same figure which also shows the tangent at the point (0−) of the curve $I(\xi)$. So for $\xi > 0$, the hazard function of the Laplacian or negative bi-exponential law is similar to the hazard function of a (single) negative exponential law (6.331.15).

6.322. GAUSS LAW

6.322.0. Introduction

We outline here the transition from the binomial law to the Gauss or Normal distribution in the simplest case ($p \approx \frac{1}{2}$).

We consider the binomial

$$P_r = \binom{n}{r} p^r q^{n-r} = \frac{n!}{r!\,(n-r)!}\, p^r q^{n-r} \tag{a}$$

of mean value $\mu = np$ and the variance $\sigma^2 = npq$.

We introduce the absolute deviation $r-\mu = r-np$ and make believe that n and r are great enough to apply Stirling's approximation (see section 6.334.10):

$$n! \approx n^n e^{-n} \sqrt{2\pi n}. \tag{b}$$

After replacing the factorials of (a) by the approximation (b), we get the Gaussian *approximation* of the positive binomial in the form:

$$\underset{\substack{n \text{ large} \\ r \text{ large}}}{P_r} \approx \frac{1}{\sqrt{2\pi}\sqrt{npq}} \exp\{-(r-np)^2/2npq\} \tag{c}$$

and where the ordinate of the mean is

$$\frac{1}{\sqrt{2\pi}\sqrt{npq}} \approx \frac{0.4}{\sqrt{npq}}. \tag{c'}$$

If now n is made to tend to infinity *without* p or q tending to zero (otherwise we have Poisson's case, see section 6.231.1), it is found that:

(1) the mean value np tends to infinity and becomes a transient mean like the variable n;

(2) the standard deviation $\sqrt{npq}$ also tends to infinity; the standard of elastic or extensible measure increases as $\sqrt{n}$;

(3) the ordinate of the mean tends to zero as $1/\sqrt{n}$.

To sum up, when $n \to \infty$, the binomial law and also the Gaussian which approximates it, both experience a *translation* to infinity jointly with a corresponding *dilatation* and *depression* leading at the limit to a complete collapse.

In order to remedy this regrettable state of affairs, mathematical statisticians perform the following operations:

(1) the fleeting mean $\mu = np$ is taken as a new origin of the deviation $r-\mu$;

(2) the extensible standard deviation $\sqrt{npq}$ is taken as the new standard of measure for the step of the deviation $r-np$.

These operations stabilize the curve as a whole and, in particular, the ordinate at the origin becomes fixed and equal to

$$1:\sqrt{2\pi} \approx 0.4.$$

In fact, the replacement of the step 1 of r by the new step $\sqrt{npq}$ leads to multiplying the ordinate P_r by $\sqrt{npq}$ in order to assure conservation of the areas.

TABLE 6.322.0A
$\lim_{n \to \infty} \sigma P_\mu$ given in column 5

1	2	3	4	5
n	$\mu = np$	$\sigma = \sqrt{nqp}$	$P_r = \binom{n}{r} p^r q^{n-r}$	$\sqrt{npq} \cdot P_\mu$
		for $p = q = \frac{1}{2}$ and even n		
n	$\mu = \frac{n}{2}$	$\sigma = \frac{\sqrt{n}}{2}$	$P_r = \binom{n}{n/2} \cdot \frac{1}{2^n}$	$\frac{\sqrt{n}}{2} \binom{n}{n/2} \frac{1}{2^n}$
2	1	0.707	$\frac{2}{4} = 0.500$	0.354
4	2	1.000	$\frac{6}{16} = 0.375$	0.375
6	3	1.225	$\frac{20}{64} = 0.312$	0.383
8	4	1.414	$\frac{70}{256} = 0.273$	0.387
10	5	1.581	$\frac{252}{1024} = 0.246$	0.389
20	10	2.236	$\frac{184{,}756}{1024 \times 1024} = 0.176$	0.394
↓	↓	↓	↓	↓
∞	∞	∞	0	0.399

To sum up, the limit curve has been stabilized as a whole by taking the mean as the origin, by expressing the deviation $r-\mu$ in terms of the unit $\sqrt{npq}$, and correspondingly by multiplying each P_r written in (c) by $\sqrt{npq}$, which neutralizes the factor of collapse.

Hence, we have

$$\lim_{\substack{n \to \infty \\ r \to \infty}} \sqrt{npq} \binom{n}{r} p^r q^{n-r} = \frac{1}{\sqrt{2\pi}} e^{-u^2/2}, \tag{d}$$

providing r varies about $\mu = np$ in such a way that

$$\lim_{n \to \infty} \frac{r-np}{\sqrt{npq}} = u. \tag{d'}$$

This pair of relations (d) and (d′) shows that the sequence of standardized binomials in the paremeters $p = q = \frac{1}{2}$ tends to the unique and universal standard Gaussian curve (see section 6.322.2). By way of illustration we give the evolution of the relations (c) and (d) when n increases in even values and confining ourselves to the value of P_r or mode obtained for $r = n/2$.

The number of combinations $\binom{n}{n/2}$ of column 4 is taken from Geigy Tables (1963, p. 25).

It is seen that as n passes from 2 to 20, the modal ordinate decreases from 0.500 to 0.176 (column 4), whereas in column 5 the product of this modal ordinate for the stabilizing factor $\sqrt{npq} = \frac{1}{2}\sqrt{n}$ passes from 0.354 to 0.394, a value close to 0.399 which corresponds to the limit for $n \to \infty$.

Position with regard to the Gauss law. The *Gauss* law is the prototype probability law for a continuous variable. It plays a central rôle in the Calculus of Probabilities and Mathematical Statistics. In its standard form, and only in this form, it derives from the standardized binomial law expressed, not in the random variable r, but in the reduced deviate

$$u = \frac{r - np}{\sqrt{npq}}.$$

Astronomers, geodeticians, physicists and chemists have widely applied the Gauss law to the errors of measures.

In his *Physique Sociale* (1835–1869) and *Anthropométrie* Adolphe Quetelet has declared as a *leit-motiv* that the distribution of sizes follows the *symmetric* binomial distribution or its classical limit, the Gaussian distribution.

Following the "mystique of Quetelet", biologists have in turn considered this law to apply frequently in biology.

Nobody contests the central position of the Gauss law in mathematical statistics. However, when it is desired to apply the law to experimental results, one ought to bear in mind the famous gibe of the French physicist Lippmann, reported by Henri Poincaré in his *Calculus of Probabilities* (1912, pp. 170–171). The law of errors (or the Gauss law) "is not obtained by rigorous deductions; no demonstration of it by anyone is anything but crude.... Everyone believes in it, however, said to me one day M. Lippmann, for the experimentalists believe that it is a theorem of mathematics and the mathematicians believe that it is an experimental fact."

As further papers were accumulated, so in proportion the doubt grew as to the universal field application of the Gaussian law. K. Pearson, cited by Yule–Kendall (1950, p. 188), wrote that the reader may enquire whether it is possible to find matter which obeys the normal law in the probable limits. He replied in the affirmative, but added that this law is not a universal law of Nature: "We must hunt for cases."

For this reason it is appropriate to look for such cases in biological or clinical material, and so much the more because Geary (1947, p. 241) writes: "Normality is a myth, there never has been a normal distribution and there will never be one." However, this author adds the qualification: "this statement is necessary from the practical point of view, but it represents a surer mental attitude than all those which have been fashionable in the last two decades".

We will now consider the Gauss law in a continuous variable in its own right. The so-called Gaussian probability law has been obtained as an approximation to the positive binomial by de Montmort and studied by Laplace (second Laplace law).

Owing to the great use made by Gauss in the theory and treatment of errors of observations, several authors, mainly French, use the denomination law of Laplace–Gauss. We have fully credited Laplace with his first law (6.321.1) so we use the shorter expression Gaussian law for the "second Laplace law".

According to M. G. Kendall and Stuart (1963, I, p. 135, note) "the description of the distribution as the 'normal' due to K. Pearson is now almost universal among English writers".

However, not as a continental writer, but as being mainly concerned with biological and medical problems, we would leave this privilege of using the "normal law" to the mathematical statisticians in their own field and prefer to use the "gaussian law" for describing the variability of normal or pathologic biological material or subjects.

6.322.1. Gauss law in the original variable X

6.322.11. Gaussian probability density

6.322.111. Definition. The Gauss law corresponds to the probability density fixed by the parameters μ and σ_X:

$$g(X)\,\mathrm{cm}^{-1} = \frac{1}{\sigma_X\sqrt{2\pi}} \exp\left\{-\frac{(X-\mu)^2}{2\sigma_X^2}\right\}. \tag{a}$$

Such is the Gauss law of mean μ and standard deviation σx, where the random variable X traverses theoretically the range $(-\infty, +\infty)$. It is represented in Fig. 6.322.111I. The theoretical domain of definition of the random variable X extends from $-\infty$ to $+\infty$.

The three cardinal points μ, $\mu-\sigma_X$ and $\mu+\sigma_X$ partition the subjacent area into four parts which are roughly equal to 0.16, 0.34, 0.34 and 0.16 respectively, the sum total of these values being unity. Accordingly, we anticipate the standard Gaussian curve (section 6.322.2) which will be used for demonstration purposes and numerical values.

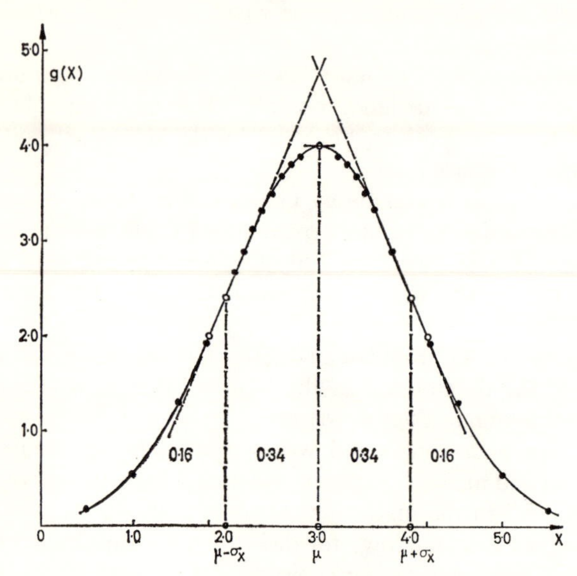

FIG. 6.322.111 I. Gaussian (normal) probability density. Ordinates are $10g(X)$.

Likewise, in the histogram the sum of the relative frequencies is unity, and the area subjacent to the Gauss curve is unity.

The curve $g(X)$ is symmetrical about a vertical line erected from the mean value μ, since $X-\mu$ appears squared in $g(X)$.

The density $g(X)$ shows a maximum at the point $X = \mu$. At this point the maximal ordinate is

$$g(\mu) = \frac{1}{\sigma_X\sqrt{2\pi}}. \tag{b}$$

Since

$$\frac{1}{\sqrt{2\pi}} = 0.3989 \approx 0.4,$$

we have

$$g(\mu) = \frac{0.4}{\sigma_X}. \tag{c}$$

The ordinate in μ is inversely proportional to the standard deviation. For a smaller σ_X, the ordinate to the mean is higher, the curve is sharper and the values X are more concentrated about the mean μ. That is why engineers and mathematicians use the *index of precision of measures:*

$$h = \frac{1}{\sigma_X}\,\text{cm}^{-1}. \tag{d}$$

For a smaller σ_X, the index of accuracy is greater.

The concept of precision is connected with the *reproducibility* of measures or with the concentration of results. Since $g(\mu)$ is maximum, the mode of X is equal to the mean. Moreover, owing to the symmetry, the median is equal to the mean. Hence,

$$\text{Mean} = \text{Median} = \text{Mode}.$$

The curve $g(X)$ shows two symmetric points of inflection at two of the cardinal points: $\mu-\sigma_X$ and $\mu+\sigma_X$. At the points of inflection the ordinate is equal to 60% of the ordinate maxima:

$$g(\mu-\sigma_X) = g(\mu+\sigma_X) = 0.6g(\mu). \tag{e}$$

This property is used to indicate the skeletal framework of a plausible Gauss curve. The inflectional slopes are respectively given by $g'(\mu-\sigma_X)$ and $g'(\mu+\sigma_X)$.

The maximal ordinate is reduced to half at the points X satisfying the relation

$$g(X) = \tfrac{1}{2}g(\mu)$$

$$\frac{1}{\sigma_X\sqrt{2\pi}}\exp\left\{-\frac{(X-\mu)^2}{2\sigma_X^2}\right\} = \frac{1}{2}\cdot\frac{1}{\sqrt{2\pi}}\cdot\frac{1}{\sigma_X}. \tag{f}$$

We simplify and take natural logarithms

$$-\frac{(X-\mu)^2}{2\sigma_X^2} = -\ln 2$$

$$(X-\mu)^2 = \sigma_X^2 2\ln 2.$$

The required points have as abscissas

$$X = \mu\pm\sqrt{2\ln 2}\,\sigma_X, \tag{g}$$

or, approximately,

$$X = \mu\pm 1.18\sigma_X. \tag{g'}$$

Finally, the curve $g(X)$ tends asymptotically to zero as X tends to $-\infty$ or to $+\infty$.

6.322.112. Parameters of definition. Theoretically the *interval of variation* or range of X extends from $-\infty$ to $+\infty$.

Over this interval the area subjacent to $g(X)$ is equal to unity:

$$\int_{-\infty}^{+\infty} g(X)\, dX = 1. \qquad \text{(a)}$$

Clearly, there is an *absolute certainty* ($p = 1$) of finding the Gaussian variable between $-\infty$ and $+\infty$.

Localization parameter. The mean or expected value of X is μ:

$$E[X] = \mu,$$

where E is an operator signifying "expectation of X",

$$E[X] = \int_{-\infty}^{+\infty} Xg(X)\, dX = \mu. \qquad \text{(b)}$$

In the continuous variable X the true mean μ of a Gaussian population, the probability density of which is $g(X)$, is no other than the *weighted mean* of X. The local weight is $g(X)\, dX$ and the sum of the weights is unity by virtue of (a).

Parameters of variability.

Absolute variability: The range R theoretically extends from $-\infty$ to $+\infty$, but *in practice* R is finite. A good example is the case where the height X of 1237 Belgian army recruits remained within the interval 154–194 cm (6.322.225.1).

Variance: By definition the application of the operator V to the continuous variable X will realize the *weighted* mean of the squares of the deviations about the mean μ. The weight for X is $g(X)\, dX$ and, as we have seen, the sum of these weights is equal to unity. Thus

$$V(X) = E[(X-\mu)^2], \qquad \text{(c)}$$

or, explicitly for the Gaussian weights,

$$V(X) = \int_{-\infty}^{+\infty} (X-\mu)^2 g(X)\, dX. \qquad \text{(d)}$$

Such is the theoretical variance σ_X^2 of a Gaussian population. This theoretical variance is also the theoretical moment μ_2 in relation to the mean, or central second moment. The theoretical variance of a Gaussian population is *not* of the form

$$\frac{\Sigma(X_i-\mu)^2}{n}.$$

This expression is valid only for an equi-weighted variable such as one of the equi-probable decimals 0, 1, 2, . . ., 9 with a constant weight

$$w_i = \tfrac{1}{10}.$$

Standard deviation σ_X: In a Gaussian law the mean μ and the variance σ_X^2 must *both* be given in order to define a particular Gaussian curve. In representative samples extracted from a Gaussian population, the calculated mean $\bar{X}$ and the observed variance S_X^2 show a zero correlation:

$$\text{Cov}(\bar{X}, S_X^2) = 0.$$

Relative variability.

Theoretical coefficient of variation:

$$\gamma_X = \frac{\sigma_X}{\mu} \tag{e}$$

$$\text{C.V.\%} = 100 \frac{\sigma_X}{\mu} \% \quad \text{or hundredths.}$$

In view of the absence of relation between μ and σ_X^2, the coefficient of variation of a Gaussian random variable can assume ∞^2 values.

Signal/noise ratio:

This parameter is used in information theory and is the inverse of the coefficient of variation:

$$\frac{\mu}{\sigma_X} = \frac{1}{\gamma_X}. \tag{e'}$$

Parameters of shape.

These are very important parameters for they serve as a reference for appraising the Gaussian or non-Gaussian nature of observed histograms (6.302.8). The parameters defining the shape or form of a distribution depend on the central moments μ_3 and μ_4 in Pearson's notation, and on the cumulants $\varkappa_3$ and $\varkappa_4$ in the notation of R. A. Fisher.

To obtain pure numbers independent of the units in which the random variable is expressed, the moments or cumulants are divided by an appropriate power of σ.

Parameter of skewness:

$$\beta_1 = \frac{\mu_3^2}{\mu_2^3} = \frac{\mu_3^2}{\sigma^6} \quad \text{(K. Pearson)}, \tag{f}$$

$$y_1 = \sqrt{\beta_1} = \frac{\varkappa_3}{\varkappa_3^{3/2}} = \frac{\varkappa_3}{\sigma^3} \quad \text{(R. A. Fisher)}, \tag{f'}$$

where $\mu_3 = \varkappa_3$.

It should be noted that β_1 and γ_1 are zero in cases of symmetric density and, in particular, for the Gaussian density.

Kurtosis:

$$\beta_2 = \frac{\mu_4}{\mu_2^2} = \frac{\mu_4}{\sigma^4} \quad \text{(K. Pearson);} \tag{g}$$

for the Gaussian (mesokurtic) density, $\mu_4 = 3\sigma^4$. Therefore $\beta_2 = 3$.

$$\gamma_2 = \frac{\mu_4 - 3\sigma^4}{\sigma^4} = \beta_2 - 3 \quad \text{(R. A. Fisher);} \tag{g'}$$

for the Gaussian density which serves as reference for the kurtosis, $\gamma_2 = 0$.

6.322.113. Gaussian density for a sample of N subjects. For a random sample of N subjects, the ordinates of the Gauss curve are

$$Ng(X) = \frac{N}{\sigma_X\sqrt{2\pi}} \exp\left\{\frac{-(X-\mu)^2}{2\sigma_X^2}\right\}. \tag{a}$$

The area subjacent to this curve is equivalent to N. When N, μ and σ_X vary according to special cases, the curves belong to a family containing virtually a *triple* infinity (∞^3) of curves.

The frequencies f_i of the histogram (X_i, f_i) of class I interval are homologable to

$$f_i \approx Ng(X_i)\cdot I = \frac{N\cdot I}{\sigma_X\sqrt{2\pi}} \exp\left\{-\frac{(X_i-\mu)^2}{2\sigma_X^2}\right\}. \tag{b}$$

The calculation of $g(X_i)$ is practicable when the unique, universal standard Gauss law is known, for which tables have been compiled (see 6.322.2).

6.322.12. Gaussian distribution

6.322.121. Definitions. The *Gaussian distribution* of X gives the probability of observing a random value $\boldsymbol{X}$ less than or equal to a fixed upper limit X:

$$\text{Prob}\,(\boldsymbol{X} \leqslant X) = F(X) = \int_{-\infty}^{X} g(X)\, dX. \tag{a}$$

It is seen that $F(X)$ expresses the *total probability* of finding a random value $\boldsymbol{X}$ between $-\infty$ and X. The *distribution of X* is the integral of the *probability density in X*. For X varying from $-\infty$ to $+\infty$, we have

$$\begin{aligned} &(F-\infty) = 0 \\ 0 < F(X) < 1 \qquad &-\infty < X < +\infty \\ &F(+\infty) = 1. \end{aligned} \tag{b}$$

For the Gauss law the explicit equation of $F(X)$ is

$$F(X) = \frac{1}{\sigma_X\sqrt{2\pi}} \int_{-\infty}^{X} \exp\left\{-\frac{(X-\mu)^2}{2\sigma_X^2}\right\} dX. \tag{c}$$

Such is the equation of the *Gaussian sigmoid* which has a point of inflection for $X = \mu$, a point which also marks off the maximum of the derived curve $g(X)$:

$$\frac{dF(X)}{dX} = g(X). \tag{d}$$

At the point μ the slope of the Gaussian sigmoid is

$$F'(\mu) = g(\mu) = \frac{0.4}{\sigma_X}. \tag{e}$$

The slope of the sigmoid at point μ is correspondingly steeper for a smaller σ_X.

The *Gaussian* sigmoid for a continuous random variable corresponds to the discrete sigmoid defined for the symmetrical binomial law.

6.322.122. Inversion of the Gaussian distribution. In a distribution the probability p of finding a random value at most equal to X is given by the relation

$$p = F(X_p) \qquad 0 \leqslant p \leqslant 1. \tag{a}$$

We apply the operator F^{-1} to this relation; in other words, we in theory realize the inversion of $F(X_p)$. By writing

$$\begin{aligned} F^{-1}(p) &= F^{-1}F(X_p), \\ X_p &= F^{-1}(p), \end{aligned} \tag{b}$$

we obtain the *quantile X_p equivalent to the probability p* of observing a random $\boldsymbol{X}$ at most equal to X_p.

Whenever the analytic relation (a) between p and X_p can be inverted, the *quantile X_p* is obtained in *explicit form as a function of p*. This operation is not possible for the Gauss law and we shall see below (6.322.2) that the result appears in tabular form.

For a Gaussian distribution of heights of mean value μ, one may write:

$$\text{median height} = X_{0.50} = F^{-1}(0.50) = \mu.$$

The *percentile* $X_{100p\%}$ is defined as

$$F(X_{100p\%}) = 100p\%. \tag{c}$$

For example, the percentile 25 (first quartile Q_1 or lower quartile Q_i) corresponds to the value $100p = 25\%$. Therefore,

$$Q_1 = Q_i = X_{25\%}.$$

The percentilage of a distribution is the operation which should fix the values X_{100p} for a series of percentages $100p\%$. In this respect, *decilage* is the particular operation of fixing the deciles $X_{0.10}$, $X_{0.20}$, etc.

We shall see that the parameters describing the Gaussian sigmoid are all written in terms of five quantiles, viz.

(1) median $X_{0.50}$

(2) to quartiles: $X_{0.25}$ and $X_{0.75}$

(3) two deciles: $X_{0.10}$ and $X_{0.90}$

6.322.123. Parameters of definition for the Gaussian distribution.

Parameter of localization: The *median*

$$Me(X) = X_{0.50} \tag{a}$$

is defined by the relation

$$F(Me(X)) = 0.50,$$

or

$$F(X_{0.50}) = 0.50. \tag{a'}$$

If the distribution is symmetrical and, in particular, Gaussian, then the mean μ, the median *Me* and the mode *Mo* will all coincide, as mentioned in 6.322.111.

Parameter of dispersion: By virtue of the symmetry of the Gaussian law, 25% of all cases occur between the lower quantile $Q_i = X_{0.20}$ and the median, whilst another 25% of cases occur between the median and the upper quantile $Q_s = X_{0.75}$. Hence the distances $X_{0.50} - X_{0.25}$ and $X_{0.75} - X_{0.50}$ are equal to the *semi-inter-quartile range:*

$$Q = \frac{X_{0.75} - X_{0.25}}{2}. \tag{b}$$

This is the parameter of dispersion or scatter which is proper to symmetrical distributions, and therefore to the Gaussian. The semi-inter-quartile range, or median deviation, groups half of the cases symmetrically about the mean, e.g. for artillery gunners. It is also the deviation which has one chance in two of being or not being exceeded.

In reference to the table for $\Theta(u)$ (see 6.322.212A and 6.322.224.1) for the Gaussian standard deviate, the standard deviation σ_x of the Gaussian probability density is related to the *semi-inter-quartile range Q* for the Gaussian sigmoid by the relation

$$Q = 0.6745\sigma_X. \tag{c}$$

N.B. The semi-inter-quartile range is not suitable for estimating the dispersion of an asymmetrical distribution. In this case, use is made of the *inter-quartile range*

$$K = X_{0.75} - X_{0.25},$$

which in non-symmetrical fashion covers 50% of the cases about the median. One dimensionless expression is written as K/σ_X.

Parameters of shape

Skewness. The Gauss curve possesses the axis of symmetry $X = \mu$. The Gaussian sigmoid is symmetrical about the point (μ; 0.50). To measure the degree of skewness, the (dimensional) coefficient

$$\frac{X_{0.10}+X_{0.90}}{2} - X_{0.50} \tag{d}$$

has been proposed. This is zero for a symmetrical distribution. In order to test the symmetry of a cumulative relative frequency curve, however, it is expedient to use the dimensionless coefficient

$$y_1' = \frac{X_{0.10}+X_{0.90}-2X_{0.5}}{\sigma_X}. \tag{d'}$$

We suggest the notation γ_1' for measuring the degree of skewness of a distribution, whilst γ_1 plays the same role for the probability density.

Kurtosis. For a symmetrical sigmoid we have the coefficient (Kelley, 1924, p. 77; Peatman, 1947, p. 390):

$$y_2' = \frac{1}{2}\,\frac{X_{0.75}-X_{0.25}}{X_{0.90}-X_{0.10}}. \tag{e}$$

From consideration of the reduced Gauss curve (6.322.224), the value of this coefficient is found without difficulty for a Gaussian distribution:

$$y_2' = \frac{0.6745\sigma_X}{2.5632\sigma_X} = 0.2632. \tag{e'}$$

This coefficient is used to test the kurtosis of a cumulative relative frequency curve in relation to the kurtosis of the Gaussian sigmoid. Of course, the parameter γ_2' in (e) will only give the degree of pure kurtosis for a symmetrical distribution. If the distribution is asymmetrical, a symmetrifying transformation is required for the random variable before calculating γ_2' which will then represent the pure kurtosis of the transformed variable.

6.322.124. Plotting a Gaussian distribution. Suppose that it is required to plot a Gaussian distribution corresponding to the population $n = 100$ and to the parameters $\mu = 5$ and $\sigma = 2$. In accordance with the formula (6.322.113a), we have

$$100g(X) = \frac{100}{\sigma\sqrt{2\pi}} \exp\left\{-\frac{1}{2}\left(\frac{X-\mu}{\sigma}\right)^2\right\}. \qquad \text{(a)}$$

This density is calculated in section 6.322.224.6. The corresponding distribution, owing to 6.322.211 (e), is

$$100F(X) = 100 \int_{-\infty}^{X} g(X)\, dX = 100 \int_{-\infty}^{u} \varphi(u)\, du = 100\, \Phi u, \qquad \text{(b)}$$

where (see 6.322.211 c)

$$u(X) = \frac{X-\mu}{\sigma} = \frac{X-5}{2}. \qquad \text{(c)}$$

The five cited points are obtained without difficulty from $X = \mu$ by calculating the u corresponding to the other four X.

TABLE 6.322.124A. CALCULATING FIVE POINTS OF A GAUSSIAN DISTRIBUTION WITH GIVEN PARAMETERS

1	2	3	4	5
X	$X-\mu$	$u = \frac{X-\mu}{\sigma}$	$\Phi(u)$	$100F(X)$
$\mu = 5$	0	0	0.500	50
$\mu-\sigma = 3$	$-\sigma$	-1	0.158	15.8
$\mu-2\sigma = 1$	-2σ	-2	0.023	2.3
$\mu+\sigma = 7$			0.842	84.2
$\mu+2\sigma = 9$			0.977	97.7

In column 1 one has to *write* the points X which are symmetrical about those for which the ordinate $100F(X)$ has been calculated.

6.322.13. Gauss law in biology and clinical work

In the Introduction 6.322.0, under the sub-heading "Position with regard to the Gauss law", the importance of the Gauss law was stressed in the Calculus of Probabilities and Mathematical Statistics where from another quarter it has received the name of "normal law". If this term satisfies the mathematician, it risks being ambiguous in meaning for the biologist; in fact, many biological quantities sampled from a *clinically normal* group exhibit a fundamental variability between now-pathologic individual entities which is described by *other* laws of probability than the "normal law" of the theoreticians. Hence the caustic opinion of H. Poincaré, the reservations of K. Pearson and Geary's extreme position in principle. The authoritative voice of Fry, a statistician confronted with engineering problems, should also be mentioned (1928, p. 205): "... experience has taught that very few sets of experimental data appear to follow the (Gaussian) law—in fact, very few of them are even symmetrical. Hence, though it can do us no harm to know how far men have been able to go in putting a foundation under an ancient monument, we shall be wise not to overlook the fact that it is, in a sense at least, venerable principally for its age. It has its uses, but it is not divinely ordained for the cure of all statistical woes."

A critical study of the consequences of non-normality in Quality Control has been done by Moore (1956). And still today E. S. Pearson (1968) heavily contests the uncritical use of Gaussian hypothesis which is so often and too slavishly set up. In our teaching at the undergraduate and graduate levels we also feel that it is a duty to call attention to this fundamental point.

It is imperative that young doctors should escape from the mental conditioning that equates a unimodal frequency density with a Gauss curve; indeed, when the asymmetry has been too flagrant, some have been seen to write "asymmetrical Gauss curve"! But let us now consider a few examples of biological quantities for which the model of Gaussian variability is not inconsistent with the observations in homogeneous groups of individuals which are clinically in good health. This condition is crucial for the definition of norms and in the realization of biopharmaceutical studies (Wagner, 1966, p. 61).

In *human biometry* one may mention:

(1) height at birth and of adults,
(2) characteristics of elongation (arm-span, height etc.),
(3) weight at birth.

In reference to *clinical biology*, the reader should consult the pioneer work of King and Wootton (1956, pp. 1–2), who produced centiles of 1, 10, 90 and 99% (see Fig. 6.322.225.7II) for the following elements:

Total blood—uric acid (mg per 100 ml); phosphate (inorganic expressed in P) (mg per 100 ml); glucose (mg per 100 ml); chlorides (in NaCl) (mg per 100 ml).

Serum and plasma—sodium (mEq per l), calcium (mEq per l), chlorides (mEq per l), CO_2—combining power (mEq per l), inorganic phosphates (mEq per l), total proteins (g per 100 ml) (see also 6.322.224.7), albumins (g per 100 ml), fibrin (g per 100 ml), inorganic phosphate (mg per 100 ml), phosphate (total acid-soluble). From a group of 100 blood donors, the Gaussian character of serum chloride is confirmed by Healy (1968).

We also mention the *resistance of the red corpuscles to hypotonia* in NaCl (Gallo, 1961) (see also Fig. 6.322.225.7I).

In *cardiac physiology:* the period $T = RR$ of the electrocardiogram; a good example is given by Hald (1952a, pp. 177–8). Hence, for clinicians, it is preferable to use the period, rather than the frequency $f = 1/T$.

The reader is requested to refer to section 6.335.17 where, under the log-normal sub-heading, we have collected biological data, the histograms of which are positively skew.

6.322.2. Standard Gauss law

6.322.21. Gaussian standard deviate u and standard Gaussian probability density $\varphi(u)$

6.322.211. Obtaining $\varphi(u)$. As we have seen above (6.322.113), a triple-infinity ∞^3 (n, μ, σ_X) of Gauss curves exists. In passing from $ng(X)$ to $g(X)$, the quantity n is eliminated by an operation homologous to the change from the absolute frequencies to the relative frequencies. This sub-family ∞^2 (μ, σ_X) then only depends on μ and σ_X. If the physical origin is transported to the biological origin, the original value X is transformed into the absolute deviate x

$$x = X - \mu, \tag{a}$$

the σ_X here remaining unchanged and equal to σ_x, whilst the mean of x becomes equal to zero:

$$\begin{cases} E[x] = E[(X-\mu)] = E[X] - \mu = 0, \\ \sigma_x = \sigma_X. \end{cases} \tag{b}$$

There thus remains a simple infinity of variables x depending only on σ_x, i.e. ∞^1 (σ_x).

It is reasonable to express the deviate x in its "natural" unit which is the *standard* deviation σ_x, thereby making a change of scale which leads to the *standard* Gaussian or *normal* deviate.

$$u = \frac{x}{\sigma_x} = \frac{X-\mu}{\sigma_X}, \tag{c}$$

which is unique and dimensionless.

This universal random variable is defined by the following parameters:

$$\begin{cases} E[\boldsymbol{u}] = \dfrac{1}{\sigma_x} E[\boldsymbol{x}] = 0, \\ \sigma_u = \dfrac{1}{\sigma_x} \sigma_x = 1. \end{cases} \tag{d}$$

These parameters are calculated by applying simple general rules governing random variables.

It now remains for us to obtain the probability densities of x and u from $g(X)$. Accordingly, we use the principle of conservation of probabilities which corresponds to the principle of invariance of the differential in the changes of variable (a) and (c):

$$g(X)\,dX = g(x) = \varphi(u)\,du, \tag{e}$$

where $\varphi(u)$ is the probability density of the standard deviate, and

$$x = X - \mu; \quad dx = dX$$

$$u = \frac{X-\mu}{\sigma_X} = \frac{x}{\sigma_X}; \qquad du = \frac{dX}{\sigma_X} = \frac{dx}{\sigma_x}. \tag{f}$$

We proceed from

$$g(X)\,dX = \frac{1}{\sigma_X\sqrt{2\pi}}\exp\left\{-\frac{(X-\mu)^2}{2\sigma_X^2}\right\}dX, \tag{g}$$

and replace

$$\begin{cases} X-\mu & \text{by} \quad x, \\ dX & \text{by} \quad dx, \\ \sigma_X & \text{by} \quad \sigma_x. \end{cases}$$

We then have

$$g(x)\,dx = \frac{1}{\sqrt{2\pi}}\exp\left\{-\frac{x^2}{2\sigma_x^2}\right\}\frac{dx}{\sigma_x}. \tag{h}$$

Substituting into this relation u for x/σ_x and du for dx/σ_x, we get

$$g(x)\,dx = \frac{1}{\sqrt{2\pi}}e^{-u^2/2}\,du = \varphi(u)\,du,$$

where

$$\varphi(u) = \frac{1}{\sqrt{2\pi}}e^{-u^2/2}. \tag{i}$$

The probability density of the normal standard deviate u is a unique, universal curve depending on no other parameter. The law $\varphi(u)$ is certainly a probability density since it can be shown that

$$\int_{-\infty}^{+\infty}\varphi(\mathrm{u})\,d\mathrm{u} = 1. \tag{j}$$

For the evaluation of this integral, the reader is referred e.g. to Weatherburn (1949, p. 148).

6.322.212. Study of $\varphi(u)$. Shape of the curve. Since we are concerned with a probability density, we will study its form as in Fig. 6.322.224.5I, where the unit on the axis of the abscissae is the same as that on the axis of the ordinates $\varphi(u)$. The columns refer to Table 6.322.212A.

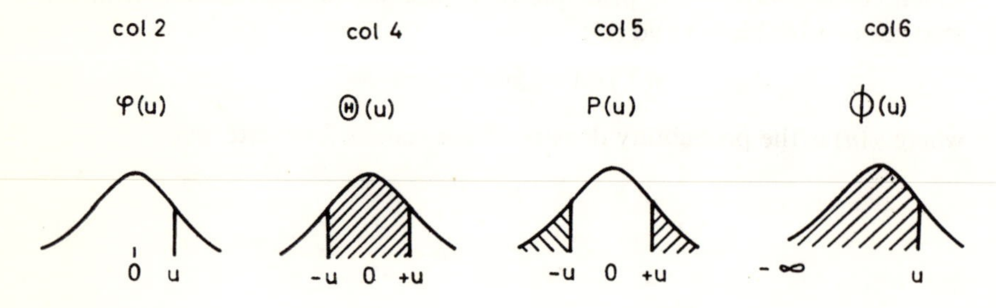

The standard density is symmetrical about the mean $E[\boldsymbol{u}] = 0$

$$\varphi(u) = \varphi(-u). \tag{a}$$

It presents a maximum for $u = 0$. Let us in fact differentiate $\varphi(u)$:

$$\varphi'(u) = D\frac{1}{\sqrt{2\pi}}e^{-u^2/2} = \frac{1}{\sqrt{2\pi}}(-u)e^{-u^2/2}. \tag{b}$$

This derivative vanishes at $u = 0$.

The derivative is positive as u varies from $-\infty$ to 0, which is an interval where $\varphi(u)$ is increasing. Then the derivative becomes negative in the interval $(0, +\infty)$ in which $\varphi(u)$ decreases.

The origin u is therefore a maximum for $\varphi(u)$, a conclusion which is confirmed by the sign of the second derivative $\varphi''(0)$.

In the neighborhood of the origin, the expansion of $\varphi(u)$, limited to the first two terms (u^2 small), is written as

$$\varphi_1(u) = \frac{1}{\sqrt{2\pi}}\left(1-\frac{1}{2}u^2+\ \dots\right), \tag{c}$$

thus representing an arc of a parabola osculating to the Gaussian density at $u = 0$. This arc of parabola, when extended beyond the area of validity (u^2 no longer small), cuts the axis of the abscissae at the points $u = \pm\sqrt{2} = \pm 1.4$. In the neighborhood of the origin, the density $\varphi(u)$ is therefore concave towards the base. But, however, for u large, the negative exponential resumes its rights and $\varphi(u)$ is then concave upwards. Hence the existence of two points of inflection symmetrical about the origin and corresponding to the roots of $\varphi''(u) = 0$.

We calculate the second derivative

$$\varphi''(u) = \frac{1}{\sqrt{2\pi}}u^2e^{-u^2/2} - \frac{1}{\sqrt{2\pi}}e^{-u^2/2} = \frac{1}{\sqrt{2\pi}}e^{-u^2/2}(u^2-1). \tag{d}$$

This second derivative vanishes at the points where (u^2-1) vanishes, i.e. for the roots of $u^2-1 = 0$,

$$u = \pm 1. \tag{d'}$$

Therefore the points of inflection of the standard normal deviate $\boldsymbol{u}$ are situated at -1 and $+1$ respectively. For $x = X-\mu$, the points are situated at $-\sigma_x$ and $+\sigma_x$. Finally, for the X's, the points of inflection have as their abscissae $\mu-\sigma_X$ and $\mu+\sigma_X$, a result given in formula 6.322.111 (e).

The negative sign of
$$\varphi''(0) = -\frac{1}{\sqrt{2\pi}}$$
confirms the maximality of $\varphi(u)$ at the origin $u = 0$. This maximum has the value $\varphi(0) = 1/\sqrt{2\pi} = 0.3989 \approx 0.4$. At the point of inflection $u = -1$, the curve $\varphi(u)$ crosses the tangent of positive slope
$$\varphi'(-1) = +\frac{1}{\sqrt{2\pi}} e^{-1/2} \approx 0.4 \times 0.6 \approx 0.24.$$

At $u = +1$, the inflectional tangent possesses a negative slope
$$\varphi'(+1) = -\frac{1}{\sqrt{2\pi}} e^{-1/2} \approx -0.24.$$
As X tends to $-\infty$ or $+\infty$, $\varphi(u)$ tends asymptotically towards zero.

Tabulation of $\varphi(u)$. By the use of a negative exponential e^{-x} table, no difficulty arises in compiling a table showing $\varphi(u) = \varphi(-u)$ as a function of the entry u (Table 6.322.212A below; see also Fisher–Yates, 1957, table II, where u is designated as x; Pearson–Hartley, 1954, p. 104; and Geigy Tables, 1963, where the Table of Ordinates of $\varphi(u)$ appears as $f(c)$, p. 31).

In particular, $\varphi(-1) = \varphi(+1) = 0.24$.

The following notation is adopted (columns 1–6).

(1) Standard deviate u.

(2) Absolute ordinates: $\varphi(u) = \frac{1}{\sqrt{2\pi}} e^{-u^2/2}$ (probability density of u).

(3) Relative ordinates: $\varphi(u)/\varphi(0)$ (Table of Davenport). In this quotient the ordinate at $u = \pm 1$ equals about 60% of the ordinate at the origin (0); in fact, $0.24 = 0.4 \times 0.6$.

(4) Function $\Theta(u)$: the probability of observing a random reduced deviate between $-u$ and $+u$
$$\Theta(u) = \text{Prob}\{-u < \boldsymbol{u} < +u\} = \frac{1}{\sqrt{2\pi}} \int_{-u}^{+u} e^{-u^2/2}\, du.$$

(5) Function $P(u) = 1 - \Theta(u)$: the probability of observing a random reduced deviate $\boldsymbol{u}$ exceeding in absolute magnitude a fixed value of u:
$$P(u) = \text{Prob}\{|\boldsymbol{u}| > u\}.$$

(6) $\Phi(u)$ is the distribution of u:
$$\Phi(u) = \text{Prob}\{\boldsymbol{u} \leqslant u\}: \quad \text{if} \quad u < 0, \quad \Phi(-u) = \tfrac{1}{2}P(u)$$
$$\text{if} \quad u = 0, \quad \Phi(0) = \tfrac{1}{2}$$
$$\text{if} \quad u > 0, \quad \Phi(+u) = \frac{1}{2} + \frac{1}{2}\,\Theta(u) = 1 - \frac{P(u)}{2}.$$

TABLE 6.322.212A. STANDARD GAUSS CURVE AND ASSOCIATED INTEGRALS (COPYRIGHT L. MARTIN, 1966)

1	2	3	4	5	6	
u	$\varphi(u)$	$\frac{\varphi(u)}{\varphi(0)}$	$\Theta(u)$	$P(u)$	$\Phi(u)$ $u-$	$u+$
0.00	0.399	1.000	0.000	1.000	0.5	0.5
0.05	0.398	0.999	0.040	0.960	0.480	0.520
0.10	0.397	0.995	0.080	0.920	0.460	0.540
0.15	0.391	0.989	0.119	0.881	0.440	0.560
0.20	0.391	0.980	0.159	0.841	0.420	0.580
0.25	0.387	0.969	0.197	0.803	0.401	0.599
0.30	0.381	0.956	0.236	0.764	0.382	0.618
0.35	0.375	0.941	0.274	0.726	0.363	0.637
0.40	0.368	0.923	0.311	0.689	0.344	0.666
0.45	0.360	0.904	0.347	0.653	0.326	0.674
0.50	0.352	0.882	0.383	0.617	0.308	0.692
0.55	0.343	0.860	0.418	0.582	0.291	0.709
0.60	0.333	0.835	0.452	0.548	0.274	0.726
0.65	0.323	0.810	0.484	0.516	0.258	0.742
0.6745	0.318	0.797	0.500	0.500	0.250	0.750
0.70	0.312	0.783	0.516	0.484	0.242	0.758
0.75	0.301	0.755	0.547	0.453	0.226	0.774
0.80	0.290	0.726	0.576	0.424	0.212	0.788
0.85	0.278	0.697	0.605	0.395	0.197	0.803
0.90	0.266	0.667	0.632	0.368	0.184	0.816
0.95	0.254	0.637	0.658	0.342	0.171	0.829
1.00	0.242	0.607	0.683	0.317	0.158	0.842
1.05	0.230	0.576	0.706	0.294	0.147	0.843
1.10	0.218	0.546	0.729	0.271	0.135	0.865
1.15	0.206	0.516	0.750	0.250	0.125	0.875
1.18	0.200	0.500	0.762	0.238	0.119	0.881
1.20	0.194	0.487	0.770	0.230	0.115	0.885
1.25	0.183	0.458	0.789	0.211	0.105	0.895
1.30	0.171	0.430	0.806	0.194	0.097	0.903
1.35	0.160	0.402	0.823	0.176	0.088	0.912
1.40	0.150	0.375	0.838	0.162	0.081	0.919
1.45	0.139	0.349	0.853	0.147	0.0735	0.926
1.50	0.130	0.325	0 846	0.134	0.067	0.933

Table 6.322.212A. Cont'd

1	2	3	4	5	6	
1.55	0.120	0.301	0.879	0.121	0.0605	0.9395
1.60	0.111	0.278	0.890	0.110	0.055	0.945
1.645	0.103	0.258	0.900	0.100 10^{-1}	0.050	0.950
1.65	0.102	0.256	0.901	0.099	0.0495	0.9505
1.70	0.094	0.236	0.911	0.089	0.0445	0.9555
1.75	0.086	0.216	0.920	0.080	0.0400	0.9600
1.80	0.079	0.198	0.928	0.072	0.0360	0.9640
1.85	0.072	0.181	0.935	0.065	0.0325	0.9675
1.90	0.066	0.164	0.942	0.058	0.0290	0.9710
1.95	0.060	0.149	0.949	0.051	0.0255	0.9745
1.96	0.0584	0.146	0.950	0.050	0.025	0.975
2.00	0.0540	0.135	0.9545	0.0455	0.0228	0.9772
2.05	0.0488	0.122	0.9596	0.0404	0.0202	0.9798
2.10	0.0440	0.110	0.9643	0.0357	0.0178	0.9822
2.15	0.0396	0.099	0.9684	0.0316	0.0158	0.9842
2.20	0.0355	0.089	0.9722	0.0278	0.0139	0.9861
2.25	0.0317	0.080	0.9755	0.0245	0.0123	0.9877
2.30	0.0283	0.071	0.9785	0.0215	0.0108	0.9892
2.35	0.0252	0.063	0.9811	0.0189	0.0095	0.9905
2.40	0.0224	0.056	0.9836	0.0164	0.0082	0.9918
2.45	0.0198	0.050	0.9857	0.0143	0.0072	0.9928
2.50	0.0175	0.044	0.9876	0.0124	0.0062	0.9938
2.55	0.0155	0.039	0.9892	0.0108	0.0054	0.9946
2.576	0.0145	0.036	0.9900	0.010 10^{-2}	0.005	0.995
2.60	0.0136	0.034	0.9907	0.0093	0.0047	0.9953
2.65	0.0119	0.030	0.9920	0.0080	0.0040	0.9960
2.70	0.0104	0.026	0.9931	0.0069	0.0035	0.9965
2.75	0.0091	0.023	0.9940	0.0060	0.0030	0.9970
2.80	0.0079	0.020	0.9949	0.0051	0.0025	0.9975
2.85	0.0069	0.017	0.9956	0.0044	0.0022	0.9978
2.90	0.0060	0.015	0.9963	0.0037	0.0019	0.9981
2.95	0.0051	0.013	0.9968	0.0032	0.0016	0.9984
3.00	0.0044	0.011	0.9973	0.00270	0.00135	0.99865
3.05	0.0038	0.0095	0.9977	0.00229	0.00115	0.99885
3.10	0.0033	0.0083	0.9981	0.00194	0.00097	0.99903
3.15	0.0028	0.0070	0.9984	0.00163	0.00082	0.99918
3.20	0.0024	0.0060	0.9986	0.00137	0.00069	0.99431
3.25	0.0020	0.0050	0.9989	0.00116	0.00058	0.99442
3.29	0.0018	0.0045	0.9990	0.001 10^{-3}	0.00050	0.99950

Table 6.322.212.A. Cont'd

1	2	3	4	5	6	
3.30	0.0017	0.0043	0.9990	0.00097	0.00048	0.99952
3.35	0.0015	0.0038	0.9991	0.00081	0.00041	0.99959
3.40	0.0012	0.0030	0.9993	0.00067	0.00034	0.99966
3.45	0.0010	0.0025	0.9994	0.00060	0.00030	0.99970
3.50	0.00087	0.0023	0.99953	0.00047	0.00024	0.99976
3.60	0.00061	0.0015	0.99968	0.00032	0.00016	0.99984
3.70	0.00043	0.0011	0.99978	0.00022	0.00011	0.99989
3.80	0.00029	0.0007	0.99986	0.00014	0.00007	0.99993
3.89	0.00021	0.0005	0.99990	0.00010 10^{-4}	0.00005	0.99995
4.00	0.00013	0.0003	0.99994	0.00006	0.00003	0.99997
4.42	0.00011		0.99999	0.00001 10^{-5}	0.00005	0.99995
4.89				0.000001 10^{-6}		
5.32				10^{-7}		
5.73				10^{-8}		
6.11				10^{-9}		

Differential equation governing $\varphi(u)$. An interesting property of the standard normal density is stated here: the absolute value of the slope of the inflectional tangent is equal to the ordinate of $\varphi(u)$ at the points of inflection. Having regard to the signs, we write

$$\varphi'(-1) = (-1)\varphi(-1), \quad \varphi'(+1) = (+1)\varphi(+1).$$

These two relations are both special cases of a general relation written from consideration of (b):

$$\varphi'(u) = -u\varphi(u). \tag{e}$$

Clearly, the slope at some point u is equal to minus the measure of the rectangle of base u and height $\varphi(u)$.

We will write this latter relation as a logarithmic derivative

$$\frac{\varphi'(u)}{\varphi(u)} = -u, \tag{f}$$

or

$$D \ln \varphi(u) = -u. \tag{g}$$

Clearly, therefore, the relative increment of the ordinate at point u is equal, but opposite in sign to the abscissa u. Such is the simplest of the equations of the *system of curves of K. Pearson* introduced in 6.302.8. For the present, we are content to show that the relation (g) certainly generates a Gaussian probability density.

In differential notation the relation (g) becomes

$$\frac{d \ln \varphi(u)}{du} = -u, \tag{h}$$

$$d \ln \varphi(u) = -u\, du = -\tfrac{1}{2}\, du^2 . \tag{i}$$

Taking the indefinite integral and introducing the constant C, we have

$$\ln \varphi(u) = C - \tfrac{1}{2}u^2 . \tag{j}$$

Hence

$$\varphi(u) = e^{C-\frac{1}{2}u^2} = e^C \cdot e^{-\frac{1}{2}u^2} = Ke^{-\frac{1}{2}u^2} . \tag{k}$$

Determining K by the normalization condition,

$$K \int_{-\infty}^{+\infty} e^{-\frac{1}{2}u^2}\, du = 1, \tag{l}$$

we get

$$K = \frac{1}{\sqrt{2\pi}} \qquad \text{[see 6.322.211 (j)].}$$

The function $\varphi(u)$ which is positive, and the integral of which is equal to unity over the range of definition of u, is therefore a probability density.

6.322.213. Parameters of definition for $\varphi(u)$. The expressions (a), (b), (d) and (e) are evaluated by means of integration by parts. We shall return elsewhere to the calculation of μ_2, μ_3 and μ_4 (see 6.322.3).

Parameters of localization.

Mean value. It has already been seen that $E[u] = 0$. This can be shown from consideration of $\varphi(u)$:

$$E[\boldsymbol{u}] = \int_{\infty-}^{+\infty} u\, \varphi(u)\, du$$

$$= \frac{1}{\sqrt{2\pi}} \int_{-\infty}^{+\infty} ue^{-\frac{1}{2}u^2}\, du = \frac{1}{\sqrt{2\pi}} \int_{-\infty}^{+\infty} e^{-\frac{1}{2}u^2}\, d\frac{1}{2}u^2 .$$

Hence

$$E[u] = -\frac{1}{\sqrt{2\pi}}\left[e^{-\frac{1}{2}u^2}\right]_{-\infty}^{+\infty} = 0. \tag{a}$$

This result could have been foreseen because the integrand is an odd function of u.

Parameters of dispersion:
Range $(-\infty, +\infty)$.
Variance:

$$\mu_2 = V(u) = E[(u-0)^2] = \int_{-\infty}^{+\infty} u^2 \varphi(u)\, du. \tag{b}$$

Standard deviation: $\sigma_u = 1$. (b′)

From (a) and (b′), u is a standard (here Gaussian) deviate.

Coefficient of variation: Biologists will be surprised to see that the coefficient of variation of the standard Gaussian variable is equal to infinity:

$$\gamma_u = \frac{\sigma_u}{E[u]} = \frac{1}{0} = \infty. \tag{c}$$

Conversely, the signal/noise ratio is zero.

Parameters of shape:

Skewness. The degree of asymmetry is given from the expression

$$\mu_3 = \frac{1}{\sqrt{2\pi}} \int_{-\infty}^{+\infty} (u-0)^3 e^{-\frac{1}{2}u^2}\, du. \tag{d}$$

Since the integrand is an odd function of u, therefore $\mu_3 = 0$, hence

$$\beta_1 = 0, \quad \gamma_1 = 0. \tag{d′}$$

The density $\varphi(u)$ is symmetrical about the origin $u = 0$.

Kurtosis. The parameters of kurtosis are obtained by applying integration by parts to the formula

$$\mu_4 = \frac{1}{\sqrt{2\pi}} \int_{-\infty}^{+\infty} (u-0)^4 e^{-\frac{1}{2}u^2}\, du. \tag{e}$$

As the result of the calculation,

$$\mu_4 = 3.$$

Hence, since $\sigma = 1$, we have

$$\beta_2 = 3, \quad \gamma_2 = \beta_2 - 3 = 0 \qquad \text{(e')}$$

The normal or Gaussian density $\varphi(u)$ is mesokurtic and serves as a reference for measuring the kurtosis of a particular probability density.

Notation: The *Gaussian* standard deviate (or standard normal deviate) u of mean value 0 and standard deviation 1 may also be written as

$$N(0, 1),$$

the capital letter N here being the first letter of the word "normal" in its Gaussian statistical sense. This notation and the wording "standard normal deviate" are in usage amongst mathematical statisticians and no biological normality is necessarily implied (see 6.322.0, final paragraph). This is why we deliberately use Gaussian "instead of normal" for the sake of unambiguity for the biologists.

6.322.22. Standardized Gaussian distribution

6.322.221. Definition. The standard Gaussian distribution expresses the (total) probability of finding a random reduced deviate $\mathbf{u}$ at most equal to a fixed upper limit u. Thus

$$\text{Prob}\,(\mathbf{u} \leqslant u) = \Phi(u) \quad [-\infty < u < +\infty], \qquad \text{(a)}$$

where $\Phi(u)$ is the integral of the standard Gaussian probability density $\varphi(u)$:

$$\Phi(u) = \int_{-\infty}^{u} \varphi(\mathrm{u})\, d\mathrm{u} = \frac{1}{\sqrt{2\pi}} \int_{-\infty}^{u} e^{-\mathrm{u}^2/2}\, d\mathrm{u}, \qquad \text{(b)}$$

where

$$\begin{aligned} \Phi(-\infty) &= 0, \\ 0 < \Phi(u) &< 1, \\ \Phi(+\infty) &= 1, \\ \Phi(0) &= \tfrac{1}{2}. \end{aligned} \qquad \text{(b')}$$

The relation (b) does not exist in explicit form. It is tabulated in Pearson–Hartley (1954, table 1) for $u > 0$, in Geigy Tables (1963, table on right of p. 28) and in the present chapter (Table 6.322.212A, col. 6) for $u < 0$ as well as $u > 0$. Here, for $u = 1.00$ (col. 1), column 6 on the left-hand side headed $u(-)$ we have $\Phi(-1) = 0.158 \approx 0.16$, and on the right-hand side headed $u(+)$, it is seen that $\Phi(+1) = 0.842 \approx 0.84$.

The Gaussian distribution appears as a symmetrical sigmoid represented by the *same unit* for the axis of the abscissae u as for the axis of the ordinates $p=\Phi(u)$ (see Fig. 6.322.224.5I, curve b). We recall that the ordinates $\varphi(u)$ of the density are also given in the same unit as u.

The slope of the inflectional tangent at $u = 0$ is equal to

$$\Phi'(0) = \varphi(0) = \frac{1}{\sqrt{2\pi}} \approx 0.4. \tag{c}$$

The term "inflectional slope" will be retained (section 6.323.25).

It is useful to show biologists at least one procedure for compiling a table of $\Phi(u)$.

For $u > 0$, one writes

$$\Phi(u) = \frac{1}{2} + \frac{1}{\sqrt{2\pi}} \int_0^u e^{-\frac{1}{2}u^2}\, du.$$

The function within the integral is expanded as a MacLaurin series:

$$e^{-\frac{1}{2}u^2} = 1 - \frac{1}{2}u^2 + \frac{1}{2!}\frac{u^4}{2^2} - \frac{1}{3!}\frac{u^6}{2^3} + \dots . \tag{d}$$

Integrating termwise between 0 and u and putting in evidence the u, we have

$$\begin{aligned} \Phi(u) &= \frac{1}{2} + \frac{u}{\sqrt{2\pi}}\left(1 - \frac{1}{2!}\frac{u^2}{3} + \frac{1}{2!}\frac{1}{2^2}\frac{u^4}{5} \dots\right) \\ &\approx 0.5 + 0.4u\left(1 - \frac{u^2}{6} + \dots\right). \end{aligned} \tag{e}$$

Consider a few examples of the *approximation* ($0.3989 \approx 0.4$) to clarify the method for the biologist:

Approximation (e)	*Exact*
$\Phi(0.10) \approx 0.50 + 0.4 \times 0.10 \approx 0.54$	(0.540)
$\Phi(0.25) \approx 0.50 + 0.4 \times 0.25 \approx 0.60$	(0.599)
$\Phi(0.50) \approx 0.50 + 0.4 \times 0.5\left(1 - \frac{0.25}{6}\right) \approx 0.69$	(0.692)

$$\Phi(0.6745) \approx 0.50 + 0.4 \times 0.6745\left(1 - \frac{0.6745^2}{6}\right) \approx 0.7493 \quad (0.750) \tag{f}$$

On comparing the approximate values calculated by the first or first two terms of the parenthesis, good agreement is found with the values appearing in column 6 of the Table 6.322.212A in the case where u varies from 0 to 0.6745, the abscissa of the right-hand point of inflection.

This result could be expected for it is in the region $(-0.6745 \leq u \leq +0.6745)$ that the parabola

$$y(u) = \frac{1}{\sqrt{2\pi}}\left(1-\frac{u^2}{2}\right), \tag{g}$$

which corresponds to the first two terms of $\varphi(u)$, is osculatory to $\varphi(u)$.

When u is large, a mathematical expansion treated in M. G. Kendall and Stuart (1963, I, pp. 136–138) leads to Mill's ratio (see section 6.322.224.3).

6.322.222. Inversion of the standard Gaussian distribution. In regard to the relation (b) of the sub-section immediately above, we explicitly set

$$p = \Phi(u_p). \tag{a}$$

The inversion, performed symbolically,

$$u_p = \Phi^{-1}(p) \tag{b}$$

yields the "normal equivalent deviate" (N.E.D.) of Gaddum (1935), or the "normit" of Berkson (1955) corresponding to p. In the form

$$p = \frac{1}{\sqrt{2\pi}} \int_{-\infty}^{u_p} e^{-u^2/2}\, du, \tag{c}$$

u_p is seen to be the upper limit of the area subjacent to the reduced Gauss density and the measure of which is p.

Tabulation of N.E.D.'s or normits. The standardized deviates corresponding to a probability p are tabulated in Fisher–Yates (1957, table I), in Pearson–Hartley (1954, table 4 in conjunction with table 3 for small p), and in Geigy Tables (1963, left-hand table on p. 28). Finally, Table 6.322.212A above will show the inversion of $\Phi(u)$ by entering with p in column 6 and then reading the N.E.D. or normit u_p in column 1. For example, by setting $p = 0.05$ in column 6 on the left-hand side designated $u(-)$, the value $u_{0.05} = -1.645$ is read in column 1. If one enters with $p = 0.95$ on the right-hand side, designated $u(+)$, one comes away with $u_{0.95} = +1.645$.

6.322.223. Parameters of definition for the standard Gaussian distribution. The median of the standard Gaussian distribution is zero:

$$u_{0.5} = 0. \tag{a}$$

The upper and lower quartiles are respectively

$$u_{0.25} = -0.6745 \approx -\tfrac{2}{3}; \qquad u_{0.75} = +0.6745 \approx +\tfrac{2}{3}. \tag{b}$$

As the measure of dispersion or scatter, the semi-inter-quartile range is

$$\frac{u_{0.75}-u_{0.25}}{2} = 0.6745 \approx \frac{2}{3}, \tag{c}$$

whilst

$$\sigma_u = 1. \tag{c'}$$

As in section 6.322.123, the skewness is expressed by the coefficient

$$\gamma_1' = u_{0.1}+u_{0.9}-2u_{0.5} = 0, \tag{d}$$

which bears witness to the symmetry.

Finally, the kurtosis is defined in the standard Gaussian distribution by

$$\gamma_2' = \frac{1}{2}\,\frac{u_{0.75}-u_{0.25}}{u_{0.9}-u_{0.1}} = \frac{0.6745}{2\times 1.2816} = \frac{0.6745}{2.5632} = 0.2632. \tag{e}$$

6.322.224. Other integrals associated with the standard Gauss law

6.322.224.1. The function $\Theta(u)$. The probability (total and centred on $u = 0$) of meeting with a random standard deviate $\boldsymbol{u}$ within the symmetrical limits $(-u, +u)$ is written as $\Theta(u)$. This also represents the probability of meeting with a standard deviate u which is less in absolute magnitude than some fixed upper limit u. Thus

$$\text{Prob}\,(|\boldsymbol{u}| \leqslant u) = \text{Prob}\,(-u \leqslant \boldsymbol{u} \leqslant +u) = \Theta(u) = \int_{-u}^{+u} \varphi(\mathrm{u})\, d\mathrm{u}. \tag{a}$$

Hence

$$\Theta(u) = \frac{1}{\sqrt{2\pi}} \int_{-u}^{+u} e^{-\frac{1}{2}\mathrm{u}^2}\, d\mathrm{u}, \tag{a'}$$

or, by virtue of the symmetry of $\varphi(\mathrm{u})$,

$$\Theta(u) = \frac{2}{\sqrt{2\pi}} \int_{0}^{u} e^{-\mathrm{u}^2/2}\, d\mathrm{u}. \tag{a''}$$

Taking the (calculated) risk of appearing old-fashioned, we do not omit the equation

$$\Theta(u) = 0.5, \tag{b}$$

which defines the median deviation[†] (artillery), i.e. the deviation which has

[†] In French: *écart médian (des artilleurs)*.

a 0.5 probability of not being exceeded in absolute value (or of being exceeded in absolute probability). By setting $\Theta(u) = 0.5$ in column 4 of Table 6.322.212A, in column 1 we have $u = 0.6745$, hence

$$\Theta(0.6745) = 0.5. \qquad \text{(b')}$$

Therefore the median deviation is 0.6745. This is also the semi-inter-quartile range or quartile deviation.

The equation

$$\Theta(u) = 0.68 \qquad \text{(c)}$$

defines the limits -1 and $+1$ which cover practically 68% of the cases. Thus

$$\Theta(1) = 0.68. \qquad \text{(c')}$$

There are two main applications of the function $\Theta(u)$: (a) the proposition of inclusion; (b) the proposition of estimation.

(a) *Proposition of inclusion of a random individual value* within a centered interval with probability $\Theta(u)$.

A statistical concept which is very important for biologists, namely, the *interval of inclusion of measure 0.95*.†

The equation

$$\Theta(u) = 0.95 \qquad \text{(d)}$$

defines the limits -1.96 and $+1.96$ of the centered *interval of 0.95 measure* for the inclusion of a Gaussian random deviate $\boldsymbol{u}$. The fact that u is a universal variable $N(0, 1)$ enables all *problems of inclusion* to be solved *for, and only for, a Gaussian variable* $N(\mu, \sigma_X)$.

Consider, for example, the human height X with $\mu = 173$ cm and $\sigma_X = 6$ cm, and let us write the *proposition of inclusion with a probability 0.95* for a random Gaussian individual value X. The lower and upper limits of this interval of inclusion centered on μ are respectively $\mu - 1.96\sigma_X$ and $\mu + 1.96\sigma_X$. The proposition is therefore written as

$$\text{Prob}\,\{\mu - 1.96\sigma_X \leqslant X \leqslant \mu + 1.96\sigma_X\} = 0.95. \qquad \text{(e)}$$

Using the common approximation $1.96 \approx 2$, the expression becomes

$$\text{Prob}\,\{\underset{161\text{ cm}}{173 - 2\sigma_X} \leqslant X \leqslant \underset{185\text{ cm}}{173 + 2\sigma_X}\} = 0.95. \qquad \text{(e')}$$

If, therefore, an individual belonging to a Gaussian population ($\mu = 173$; $\sigma_X = 6$) is extracted at random, one has a 95% chance of finding the height of this individual to be between 161 cm and 185 cm. We say that the measure of the *interval of inclusion* for a random *individual value* is 0.95, and the

† In French: *intervalle d'appartenance de mesure 0.95.*

interval thus defined has an amplitude of two sigma. Such is the *rule of two sigma* which is often used to fix on an *interval of biological normality*, although from further afield this is seen to be arbitrary. Persons engaged in experimental research are reminded that *this rule is only applicable if the observed biological data are Gaussian.*

If it is desired to extend this statistical range of "biological normality" and increase its measure from 0.95 to 0.99, the value 1.96 should be replaced by 2.576 (see Table 6.322.212A, column 4) because $\Theta(2.576) = 0.99$.

For any unimodal probability density which is symmetrical about the mean, a rule of the $k\times$*sigma type* is definable for the half-amplitude of the interval of inclusion of fixed measure 0.95 or 0.99. Here the constant k depends on the type of symmetrical density of probability.

Finally, for any *unimodal density* of an observed variable X, the upper and lower limits of the inclusion interval of 0.95 or 0.99 measure are defined in an *entirely general* way by the following quantiles:

$$\begin{aligned} &X_{0.975} \text{ and } X_{0.025} \text{ for the interfal of measure } 0.95; \\ &X_{0.995} \text{ and } X_{0.005} \text{ for the interval of measure } 0.99. \end{aligned} \tag{f}$$

This interval is not symmetrical about the mean nor the median for the skew probability densities.

As an example for the special case of the Gauss frequency curve, the limits of the symmetrical interval of inclusion of measure 0.95 are

$$\begin{aligned} X_{0.025} &= \mu - 1.96\sigma_X, \\ X_{0.975} &= \mu + 1.96\sigma_X. \end{aligned} \tag{f'}$$

The relation (e) is thus refound.

(b) *Proposition of estimation of an unknown mean.* The estimation of the unknown mean μ of a unimodal population of variance σ_X^2 using the calculated mean $\bar{X}$ of a large sample of n values, is based on the *standard error of the observed mean*, i.e.

$$\sigma_{\bar{X}} = \frac{\sigma_X}{\sqrt{n}}. \tag{g}$$

For a 0.95 probability of confidence (or fiducial probability), we have

$$\text{Prob}\,\{\bar{X} - 1.96\sigma_{\bar{X}} > \mu < \bar{X} + 1.96\sigma_{\bar{X}}\} = 0.95. \tag{h}$$

For a 0.99 probability of confidence, 1.96 is replaced by 2.576 in accordance with the table $\Theta(u)$. In general, for a fiducial or confidence probability $\Theta(u)$, we put u_Θ for the value of u which imparts the value Θ to $\Theta(u)$; this realizes the inversion of $\Theta(u)$. The statement then becomes, *without ambiguity*,

$$\text{Prob}\,\{\bar{X} - u_\Theta \sigma_{\bar{X}} < \mu < \bar{X} + \mu_\Theta \sigma_{\bar{X}}\} = \Theta(u). \tag{i}$$

Such is one of the three systems of notation for intervals. Some authors, polarized by the notion of (significance) level, speak of a confidence interval of measure $\alpha = 0.05$, for instance, but this would be rather a question of an interval of distrust.

In the present notes, the probability of an even variable such as u and t_ν (Student's t with ν degrees of freedom) *exceeding in absolute value* some fixed value u or t, is written as $P(u)$ or $P(t)$, and specifying the argument. There is no risk of confusion with P_r in the case of discrete random variables, or with the notation P or p (without argument) used to denote the degree of significance in the tests u, t, χ^2 and F.

6.322.224.2. Function $P(u)$. The test of significance for the difference between two true means μ_1 and μ_2 from populations of variance σ_1^2 and σ_2^2 using large samples (n_1 and n_2) is based on the standard Gaussian deviate

$$u = \frac{(\bar{X}_1 - \bar{X}_2) - (\mu_1 - \mu_2)}{\sqrt{\dfrac{\sigma_1^2}{n_1} + \dfrac{\sigma_2^2}{n_2}}}. \tag{a}$$

The null hypothesis (H_0) that $\mu_1 = \mu_2$ is rejected at the significance level $\alpha = 0.05$ if u is *greater than* 1.96 in absolute value, or if the function $P(u)$ is *less than* 0.05.

It is convenient to introduce along with Fisher–Yates (1957, table 1, commentary) a table of values where u "is the deviation such that the probability of an observation falling outside the range from $-u$ to $+u$ is P". (The x of Fisher–Yates appears here as u.)

Since these tables are in general use amongst experimenters, we believe it to be essential than the function

$$P(u) = 1 - \Theta(u) \tag{b}$$

should be considered in its own right.

For the convenience of the user, the function $P(u)$ is tabulated in column 5 of Table 6.322.212A.

EXAMPLE: If $u = 1.96$, it is immediately seen that $P(1.96) = 0.05$.

And if, for instance, it is required to know the standard deviate u_P corresponding to $P = 0.01$, one sets $P = 0.01$ in column 5 to read $u_P = 2.576$ in column 1.

Table 1 of Fisher–Yates (1957) gives the standard deviate u_P at the place where the entry of P to two decimal places is found by adding the entry of the line (0.0, 0.1, ..., 0.9) to the entry of the column (0.00, ..., 0.09). The u_P corresponding to $P = 0.10$ is thus equal to 1.645.

In 10% of the cases the Gaussian standard deviate is greater than 1.645

in absolute magnitude. By reason of the symmetry, u will exceed $+1.645$ *(in only the positive sense)* in 5% of the cases; and also u will be less than -1.645 *(in only the negative sense)* in 5% of the cases. For this reason, in a *unilateral or one-tailed test at the level 0.05*, the null hypothesis H_0 is rejected if u is greater than $+1.645$; whereas in a bilateral or two-tailed test at the same threshold 0.05, the H_0 is rejected if u is *either* greater than $+1.96$, *or* less than -1.96, or concisely if $|u|$ is greater than 1.96.

Biologists will thus realize that prior knowledge of the obligatory direction of a comparison makes for a more sensitive test of significance between two means. Anticoagulants which tend *solely* to *prolong coagulation time*, are a well-known example.

6.322.224.3. Function R(u), Mill's ratio. Mill's ratio (1926) is the ratio of the complementary distribution $1-\Phi(u)$ to the ordinate $\varphi(u)$ which bounds the above cited area from the fixed point u:

$$R(u) = \frac{1-\Phi(u)}{\varphi(u)} = e^{\frac{1}{2}u^2} \int_u^\infty e^{-u^2/2}\,du. \tag{a}$$

This ratio is of great value in series expansion of $R(u)$ or $\Phi(u)$ when u is large (Kendall and Stuart, 1962, I, pp. 136–138).

For biologists it should be mentioned that Mill's ratio is the inverse of the Gaussian imminence rate considered in section 6.322.228:

$$R(u) = \frac{1}{I(u)}. \tag{b}$$

6.322.224.4. Probability relations between the three expressions for Gaussian random variables. Between the original variable X, the absolute deviate x and the standard deviate u, all connected by the relations

$$u = \frac{X-\mu}{\sigma_X} = \frac{x}{\sigma_x} \qquad (\sigma_X = \sigma_x), \tag{a}$$

there are *propositions of inclusion* in which σ is written without the subscript:

$$\text{Prob}\,(\mu-\sigma < X < \mu+\sigma) = 0.68, \tag{b}$$

$$\text{Prob}\,(-\sigma < x < +\sigma) = 0.68, \tag{c}$$

$$\text{Prob}\,(-1 < u < +1) = 0.68. \tag{d}$$

Whence, expressing X in terms of u and σ

$$X = \mu+u\sigma, \tag{e}$$

one can revert from (d) in u to (b) in X now written as

$$\text{Prob}\left[\mu-\sigma < \left\{\begin{matrix} X \\ \text{or} \\ \mu+u\sigma \end{matrix}\right\} < \mu+\sigma\right] = 0.68. \tag{f}$$

Between the original variable X, the absolute deviation X and the standard deviate u, connected by the relation (a), there are *relations of the differential type:*

$$du = \frac{dX}{\sigma} = \frac{dx}{\sigma}. \tag{g}$$

Hence the following relations exist between the *local probabilities:*

$$g(X)\,dX = \frac{1}{\sigma\sqrt{2\pi}} \exp\left\{-\frac{(X-\mu)^2}{2\sigma^2}\right\} dX, \tag{h}$$

$$g(x)\,dx = \frac{1}{\sigma\sqrt{2\pi}} \exp\left\{-\frac{x^2}{2\sigma^2}\right\} dx, \tag{i}$$

$$\varphi(u)\,du = \frac{1}{\sqrt{2\pi}} e^{-u^2/2}\,du. \tag{j}$$

Finally, the *probability densities* are

$$g(x) = g(u\sigma) = \frac{1}{\sigma\sqrt{2\pi}} \exp\left\{-\frac{(u\sigma)^2}{2\sigma^2}\right\} = \frac{1}{\sigma}\varphi(u), \tag{k}$$

$$g(X) = g(\mu+u\sigma) = \frac{1}{\sigma\sqrt{2\pi}} \exp\left\{\frac{[(\mu+u\sigma)-\mu]^2}{2\sigma^2}\right\} = \frac{1}{\sigma}\varphi(u). \tag{l}$$

The factor $1/\sigma$ confers on $\varphi(u)$ the dimension of $g(X)$.

6.322.224.5. Graph integrating the functions of point or interval associated with the standard Gaussian law. The standard Gauss curve and the integrals which are associated with it, are presented in two forms, viz. as a point function such as $\varphi(u)$ which is also symmetrical

$$\varphi(-u) = \varphi(+u), \tag{a}$$

or as interval functions considered as definite integrals. Three types of intervals are to be distinguished.

The distribution $\Phi(u)$ is an additive function of interval defined on the interval $(-\infty, u)$:

$$\begin{aligned} \Phi(-u) &< 0.5, \\ \Phi(0) &= 0.5, \\ \Phi(+u) &> 0.5. \end{aligned} \tag{b}$$

The complementary distribution is defined on $(u, +\infty)$:

$$\Psi(u) = 1-\Phi(u). \tag{c}$$

The difference $\Phi(+u)-\Phi(-u)$ generates another additive function of interval $\Theta(u)$ which, itself, is defined on the symmetrical interval $(-u, +u)$, or even on the intervals $(-u, -0)$ and $(+0, +u)$ joined by the point $u = 0$:

$$\Phi(+u)-\Phi(-u) = \Theta(u). \tag{d}$$

Finally, the function

$$P(u) = 1-\Theta(u) \tag{e}$$

is an additive function of interval defined on the union of the two symmetrical disjoint intervals $(-\infty, -u)$ and $(+u, +\infty)$.

From the numerical point of view, the two symmetrical functions of interval $\Theta(u)$ and $P(u)$ which have just been considered, are presented as double the functions thus defined:

$$\begin{aligned}\Theta(u) &= 2[\tfrac{1}{2}-\Phi(u)]\\ &= 2[\Phi(+u)-\tfrac{1}{2}],\end{aligned} \tag{f}$$

$$\begin{aligned}P(u) &= 2\Phi(-u)\\ &= 2[1-\Phi(+u)].\end{aligned} \tag{g}$$

Each of these couples in $(-u)$ or in $(+u)$ checks with the relation $\Theta+P = 1$ of the complementarities.

Plotted without difficulty from the Table 6.322.212A, Fig. 6.322.224.5I graphically represents the functions which have just been discussed. The same scale is used for the variable u and the various functions. In fact, each function expresses the *measure* of the interval over which it is defined, this measure here being taken in reference to the probability density $\Phi(u)$.

N.B. For the sake of being comprehensive, we also show the imminence rate or hazard function $I(u)$ (6.122 and 6.322.228).

$$I(u) = \frac{\varphi(u)}{1-\Phi(u)} \tag{h}$$

Like the *unconditional* local probability density $\varphi(u)$, the *conditional* local probability density $I(u)$ is a function of the point u. In particular

$$I(0) = \frac{\varphi(0)}{1-\Phi(0)} \approx \frac{0.4}{0.5} = 0.8 \tag{h'}$$

6.322.224.6. Construction of a Gaussian density of given μ and σ. Suppose it is required to form a Gaussian density corresponding to the population $n = 100$ and with the parameters $\mu = 5$ and $\sigma = 2$. In accordance with the

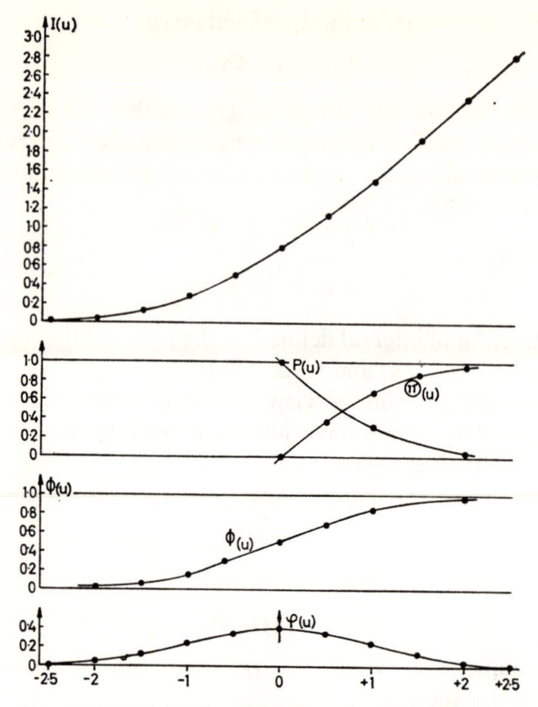

FIG. 6.322.224.5 I. Standard Gaussian (normal) law.
Probability density $\varphi(u)$.
Distribution $\Phi(u)$.
Integrals $\Theta(u)$ and $P(u)$ (see text).
Imminence rate $I(u)$.

formula 6.322.113 (a),

$$100\,g(X) = \frac{100}{\sigma\sqrt{2\pi}} \exp\left\{-\frac{1}{2}\left(\frac{X-\mu}{\sigma}\right)^2\right\}. \tag{a}$$

Introducing the standard deviate u as some function of X, say $u(X)$, we get

$$100\,g(X) = \frac{100}{\sigma\sqrt{2\pi}} e^{-u(X)^2/2} = \frac{100}{\sigma}\varphi(u), \tag{b}$$

where

$$u(X) = \frac{X-\mu}{\sigma} = \frac{X-5}{2}. \tag{c}$$

We calculate the ordinates of the three cardinal points and of two other framing points, averaging the approximation $0.3989 \approx 0.4$. In this respect the ordinate at the mean is written as

$$g(\mu) = \frac{100}{\sigma}\varphi(0) = \frac{100\times 0.4}{2} = \frac{40}{2} = 20. \quad \text{(d)}$$

The five points in question are readily obtained from $X = \mu$ by calculating the u corresponding to the 4 other X and by applying to $g(\mu)$ the reduction $\varphi(u)$: $\varphi(0)$ of Davenport's table (Table 6.322.224.6A).

TABLE 6.322.224.6A. CALCULATING FIVE POINTS OF A GAUSSIAN DENSITY WITH GIVEN PARAMETERS
$\mu = 5$, $\sigma = 2$

1	2	3	4	5
X	$X-\mu$	$u = \frac{X-\mu}{\sigma}$	$\frac{\varphi(u)}{\varphi(0)}$	$100\, g(X)$
$\mu = 5$	0	0	1.000	20
$\mu-\sigma = 3$	$-\sigma$	-1	0.607	12.14
$\mu-2\sigma = 1$	-2σ	-2	0.135	2.70
$\mu+\sigma = 7$	$+\sigma$	$+1$		12.14
$\mu+2\sigma = 9$	$+2\sigma$	$+2$		2.70

By virtue of symmetry, it is unnecessary to calculate $g(X)$ for $X > \mu$ when the value for $X < \mu$ is known. But the points X which are symmetrical about those for which the ordinate 100 $g(X)$ has been calculated, should be *written* in column 1. For ease of reference, the ordinates corresponding to the abscissae of column 1 are also written in column 5 under the horizontal broken line.

6.322.224.7. Framing of a histogram by a Gaussian density. Without repeating here the method of adjustment for a Gaussian law, we will briefly show the "framing" of a histogram by a Gaussian density.

Consider the Table 6.322.224.7 which gives the frequencies of the total protein level X, as sampled on 56 normal subjects (Gatez, 1966).

Statistics of reduction:

Mean	$\bar{X} = 69.29$	
Variance	$S_X^2 = 21.58$	
Standard deviation	$S_X = 4.65$	(a)

Coefficient of variation C.V.% = 6.7%

Statistics of shape with their standard error $g_1 = -0.15 \pm 0.33$

$g_2 = -0.8 \pm 0.65$

TABLE 6.322.224.7A. TOTAL PROTEINS IN g/l

1	2	3	4	5
Class	Frequencies	Cumulative frequencies	Relative cumulative frequencies	Probit (2 decimal places)
58–61.99	4	4	7.14	3.53
62–65.99	9	13	23.21	4.27
66–69.99	17	30	53.57	5.09
70–73.99	18	48	85.71	6.07
74–77.99	7	55	98.21	7.10
78–81.99	1	56	100.00	↑

The Gaussian hypothesis ($\gamma_1 = 0$; $\gamma_2 = 0$) may therefore be not rejected.

In accordance with the formula 6.322.113(a), we will calculate the ordinate at the mean with the approximation $0.3989 \approx 0.4$:

$$56\, g(\bar{X}) = \frac{56}{4.65\sqrt{2\pi}} = \frac{0.4 \times 56}{4.65} = \frac{22.4}{4.65} = 4.82. \tag{b}$$

At the points $\bar{X} - 1\sigma_x$: $= 64.64$ and $\bar{X} + 1\sigma_x = 73.94$, Davenport's table (column 3 in Table 6.322.212A) shows that the ordinate at $\bar{X}$ is reduced to 60.7% of its value. Hence

$$g(\bar{X} - 1\cdot\sigma_x) = g(\bar{X} + 1\cdot\sigma_X) = 4.82 \times 0.607 = 2.93. \tag{c}$$

Thus the ordinates at the three cardinal points are obtained.

To complete the frame, we further calculate the ordinates at the points at a distance $2\sigma_X$ from $\bar{X}$, for instance,

$$\bar{X} - 2\sigma_X = 59.99$$

and

$$\bar{X} + 2\sigma_X = 78.59.$$

We write

$$g(\bar{X} - 2\cdot\sigma_X) = g(\bar{X} + 2\cdot\sigma_X) = 4.82 \times 0.135 = 0.65 \tag{d}$$

and the general frame is thus complete. The reader is invited to put on the same graph the histogram (Table A) and the sketch of the Gaussian density from points calculated in (b), (c) and (d).

The calculation can obviously be performed also for other points; in particular one can calculate the ordinates $56g(X_i)$ at the centers of classes and then pass to the theoretical frequencies $\varphi_i = 56 \times Ig(X_i)$, where I is the class interval. The χ^2 test for goodness of fit is (pooling the tails up to 5)

$$\chi^2 = \sum_{i=1}^{g} \frac{(f_i - \varphi_i)^2}{\varphi_i},$$

where φ_i is a theoretical frequency and g is the number of groups. The number of degrees of freedom is $\nu = g - 3$, for three linear relations between the f_i are used to adjust the constants $n = 56$, $\bar{X}$ and s_X^2.

6.322.224.8. Tukey's hanging rootogram. A clever way of detaching single frequencies showing a departure from a fitted Gaussian (normal) density is given by J. Tukey (unpublished work) diffused and illustrated by Healy (1968, p. 211). This method of "hanging rootogram" is operated so that rectangular elements of the histogram are replaced by their respective squared roots and hung from the fitted curve. "The bases of the rectangles now fluctuate about the base line, and the square-root transformation results in this fluctuation being associated with a constant standard deviate ± 0.5. The eye can thus be guided by the addition of limit lines one unit above and below the base line—we expect all but one or two of the rectangles' bases to fall between these limits." In his Figs. 3 and 3', Healy (1968) shows in (a) the classical standing histogram and the hanging rootogram describing the variability of log-serum urea (6.335.17) from 1000 blood donors.

6.322.225. The concept of probit

6.322.225.1. Rectification of the Gaussian sigmoid by an anamorphosis. Hazen's or Henry's straight line. The normit-X or probit-X line. As starting from the 19th century, *Fechner*, a psychophysiologist, investigated methods of rectification of the curve of the *cumulative relative frequencies* $100\,p\%_0(X)$, or alternatively, for the Gaussian sigmoid. This sigmoid is shown at (d) in Fig. 6.322.225.1I illustrating the fundamentals of the rectification process (Martin, 1956 and 1960).

At (A) the normal curve is shown adjusted to the histogram for the height X of Belgian recruits of the 1953 intake born in 1933 and on the French roll. Here the descriptive parameters (taken as theoretical) are: $\mu = 172.40$ cm and $\sigma_X = 6.16$ cm. The multiples marked for the standard deviation

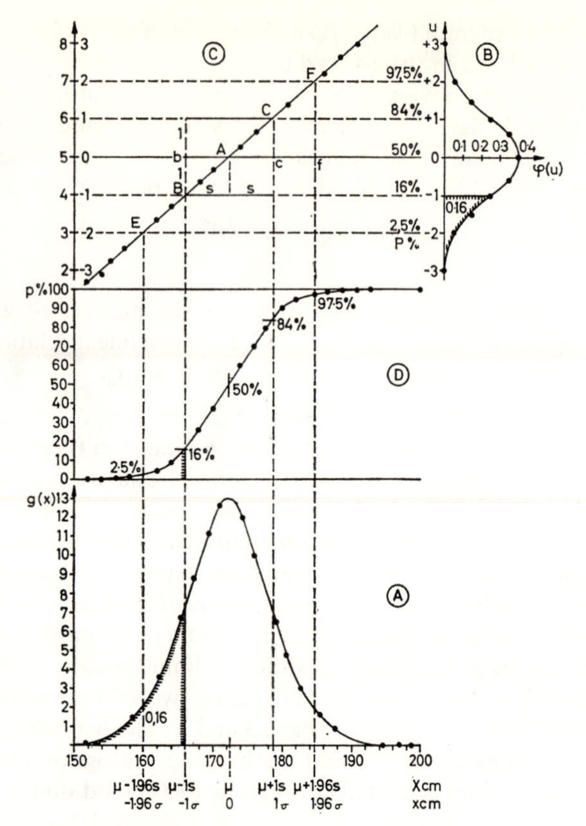

FIG. 6.322.225.1 I. Graph probit $-X$ and normit $-X$.

(A) Gaussian density probability of the stature X.
Abscissae: upper scale: X cm
lower scale: $x = (X-\mu)$ cm
Ordinates: $g(X)\ \text{cm}^{-1}$.

(B) Standard gaussian probability density in u.

$$\text{Abscissae: (vertical-right) } u = \frac{X-\mu}{\sigma_X}$$

Ordinates: (horizontal) $\varphi(u)$

(C) Probit $-X$ and normit $-X$ straight line
Abscissae: X_{cm}
Ordinates: (left) normit (N.E.D.): u
probit: $Y = 5+u$

(D) Gaussian distribution of the stature X.

$\sigma_X = 6.16$ cm are superposed on the abscissae X cm or the absolute deviations $x = X-\mu$. For each point, the abscissa of which is measure in σ_X, we have noted the theoretical percentage of cases falling below the absolute deviation x. For instance, under the point $\mu - 1.\sigma_X$, or $X = 166.25$ cm, we have set $100p = 16\%$, which corresponds to the shaded zone, the measure of which is 0.16 as a fraction of unity.

At (B) thes tandard Gaussian curve is shown as a function of the standard deviate u which is associated as follows with x or X:

$$u = \frac{x}{\sigma_x} = \frac{X-\mu}{\sigma_X} \quad (\sigma_x = \sigma_X). \tag{a}$$

It will be recalled that the standard deviation of u is equal to unity,

$$\sigma_u = 1.$$

Please note that the abscissae axis for u is perpendicular to that of X. The percentage $100p\%$ of the cases below this standard deviation is set under each of the standardized deviates. For example, 16% appears under the standard deviate $u = -1$, for there is a fraction 0.16 of unity of cases to the left of $u = -1$. After Gaddum (1933) $u_{0.16} = -1$ is the standard normal deviate corresponding to the percentage $100p\% = 16\%$. More generally u_p is said to be the standard normal deviate corresponding to p, or called the normal equivalent deviate N.E.D. (see Fig. 6.322.225.1II).

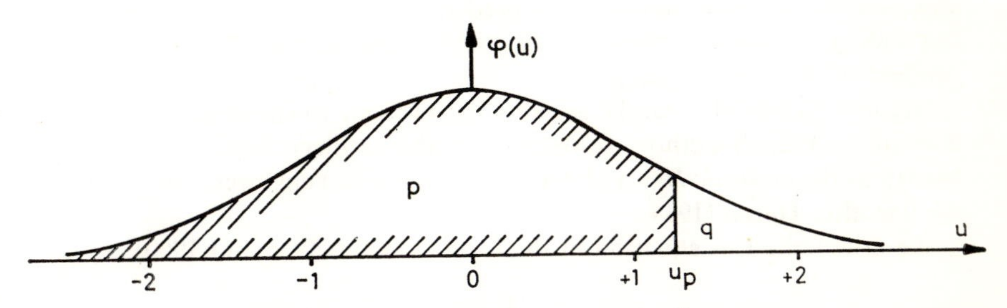

FIG. 6.322.225.1 II. N.E.D. u_p corresponding to p.

One can now eliminate graphically the cumulative probability p between x_p and u_p and plot the graph showing u as a function of x. Thus, in (A) the value $x = 0$, and in (B) the value $u = 0$, correspond to the percentage $100p = 50\%$; these two latter coordinates define the point A (0,0) of the graph in coordinates (x, u), scale u to the left of (C). The point $X = 166.24$ cm marks a height such that 16% of the individuals are less than 166.24 cm tall. More concisely, we will say that $X = 166.24$ cm is the height corresponding to the percentage $100\,p = 16\%$. Generally the height corresponding to p or quantile p is denoted as X_p.

Likewise, in (A) the value $x = \mu - 1 \cdot \sigma_X$, and in (B) the reduced deviation $u = -1$, correspond to the percentage $100p = 16\%$; these two coordinates define the point B $(-\sigma_X, -1)$ of segment C. Then the points of the curves (A) and (B) corresponding to $100p = 84\%$ lead to the point C $(+\sigma_X, +1)$. One can proceed in the same way for the percentages 2.5 and 97.5% which define the points $E(-1.96\sigma_x;\ -1.96)$ and $F(+1.96\sigma_x;\ +1.96)$ respectively.

From consideration of the equal or similar triangles ABb, ACc and AFf, for instance, the points E, B, A, C and F lie along a straight line having the equation

$$\begin{aligned} u &= \alpha + \beta x \\ &= \alpha + \beta(X - \mu), \end{aligned} \tag{b}$$

where $\alpha = 0$ since $u = 0$ for $x = 0$ or $X = \mu$, and

$$\beta = \frac{1}{\sigma_X} = \frac{1}{\sigma_x}. \tag{c}$$

Any point $M(X, u)$, the coordinates u and X of which satisfy the relation (a), is situated on the line (C), the slope of which is equal to the reciprocal of the standard deviation of x or of X.

The *rectification of the normal sigmoid* is thus achieved by taking as abscissae the absolute deviate measured in cm, and as ordinates the corresponding standard normal deviates which are pure numbers and independent of the unit of measure.

Negative values of x can be avoided by reverting to the original variable $X = \mu + x$. French authors (especially in the artillery field) refer to the line (C) as the *droite de Henry* (1894), whilst American engineers have named the line after Hazen (1914).

We now write this equation in the form (Martin, 1956 and 1960)

$$u_p = \frac{X_p - \mu}{\sigma_X}. \tag{d}$$

This expression is none other than the definition of the standard deviate where the fraction p of cases common to the quantiles X_p and u_p is explicitly stated. From this standpoint, the line of Henry or Hazen realizes a co-graduation (Gini, 1916; Martin 1960) between X_p and u_p; the subscript p here is common to X and to u. The standard deviate u_p corresponding to p or to the N.E.D. of Gaddum (1933), was rechristened *normit* by Berkson (1955). The probit-X line (C) thus also represents the *normit-X line*.

As one moves away from the mean by values less than X, the probability of finding values still less than X progressively decreases. On the scale of the standard deviates, the probability of meeting with standard normal values less than -5, or above $+5$, is very weak, of the order of 10^{-6} (Fisher–Yates, 1957, table I). In other words, in a population of mean value 0 and of unit standard deviation, one has in practice a very great degree of certainty that no values will fall outside the interval $(-5, +5)$.

After Bliss (1935), negative values of the standard deviate u are avoided by adding 5 to it. This is of considerable interest in the performance of calculations by desk calculating machines or electronic computers.

The quantity

$$Y_p = u_p + 5 \tag{e}$$

defines a *Gaussian probability unit, abbreviated to probit corresponding to the percentage* 100 p% of cases which occur to the left of the standard deviate u_p in the standard normal curve (B), or to the left of the value $\mu + u_p \sigma_X$ measured in cm on the scale of sizes in curve (A). This scale of ordinates Y is situated to the left of the scale u of line C in Fig. 6.322.225.1 I.

In the original variable X the equation of the *probit-X line* is written as

$$Y_p = \frac{X_p - \mu}{\sigma_X} + 5. \tag{f}$$

Summing up, the Gaussian sigmoid is rectified by conserving the abscissae X, but whilst replacing the ordinates $100p\%(X)$ of (D) [shown as *hatched area* in the curve (α) and as *hatched ordinate* in (α) by the standard normal deviate u_p taken in curve (B).

The ordinates u_p with the addition of 5 yield the required probit-X line.

*6.322.225.2. **Probit tables.*** To facilitate practical applications of the rectification of sigmoids for the % response as a function of the dose (or of the log-dose), tables have been compiled which show the probit Y_p corresponding to the fraction p and these tables render great services daily in biological test laboratories. One of the most frequently used tables (table IX of Fisher–Yates, 1957) gives to four decimal places the probits corresponding to fractions of 100 p% shown to one decimal place, i.e. in steps of 0.1%. For instance, the probit of exactly 2.5% is equal to 3.0400.

To demist the eyes of biologist users of the probit table, we will calculate the probit corresponding to 2.5% (or 0.025) from Table 6.322.212A.

Since 2.5% is less than 0.50%, we begin in the column $\Phi(u)$, for $u < 0$ and stop at 0.025, or in the position

$$0.025 = \Phi(u). \tag{a}$$

The entry for the reduced deviate u_Φ corresponding to 2.5% realizes the inversion of the relation (a)

$$u_{0.025} = \Phi^{-1}(0.025). \tag{b}$$

In column 1 we thus have

$$u_{0.025} = -1.96. \tag{b'}$$

The minus sign is present for 0.025 is less than 0.5. Since the probit corresponding to p is obtained by adding 5 to u_p, the result from the table of Fisher–Yates is found to be

$$Y_{0.025} = Y_{2.5\%} = 5+(-1.96) = 3.04. \tag{b''}$$

Pearson and Hartley (1954) have constructed their table 6 for p in the range 0.500–0.999968 with the theoretical (expected) probits Y_p ranging from 5.00 to 9.00 in steps of 0.01. Entries complementary to 10 for the probits are noted $1-p$, which is easily evaluated.

Finally, Geigy Tables (1963, pp. 54–55) reproduce the table IX of Fisher–Yates.

Relation between the probits of two complementary fractions. Consider now two fractions which are symmetrical about 0.5, for instance, $p = 0.158$ and $1-p = 0.842$. In Table 6.322.212A, we have

$$u_{0.158} = -1 \quad \text{and} \quad u_{0.842} = +1.$$

Hence

$$u_{0.158}+u_{0.842} = 0.$$

Since the probits are obtained by adding 5 to each reduced deviate, therefore

$$Y_{0.158}+Y_{0.842} = 0+5+5 = 10.$$

In general one can write

$$Y_p+Y_{1-p} = 10. \tag{c}$$

Fisher–Yates (1957, p. 65) give the entries $100p\%$ in steps of 0.1% between 0.1 and 97.9%. But in regard to the imprecision which vitiates the tails of the Gaussian distribution, these authors tabulated the $100p\%$ in steps of 0.01% between 98.00 and 99.99%.

Hence the relation (c) becomes

$$Y_p = 10-Y_{1-p}, \tag{c'}$$

and in this form the probit of any percentage between 0.1% and 1.99% inclusively is readily furnished in steps of 0.01%.

EXAMPLE. What is the probit of 0.52%? One calculates

$$100.00-0.52 = 99.48\%.$$

The probit of 99.48% is equal to 7.5622. Therefore the probit of 0.52% is

$$Y_{0.52\%} = 10.0000-7.5622 = 2.4378.$$

As a cross-check, it is seen that the value found appears between the 0.5% and 0.6% probits on the first line of table IX of Fisher–Yates.

6.322.225.3. Weighting of probits. Besides p and Y_p, other items are taken into consideration in biological tests on quantal response. This topic will be treated in the adjustment of a linear weighted regression of Y on X (dose or log-dose). In particular, the "weighting coefficient" column will hold our attention for this is a general statistical concept.

For a working probit corresponding to p, where p is estimated by the proportion of reactors observed

$$\mathcal{P} = \frac{r}{n}, \tag{a}$$

the weighting coefficient per unit exposed to risk is written as a function of quantities referred to the true value p, viz.

$$w_{Y\mathcal{P}} = \frac{\varphi^2(u_p)}{pq}, \tag{b}$$

where u_p is the N.E.D. corresponding to $p(q = 1-p)$,

$Y_p = 5+u_p$,

$\varphi(u_p) =$ the ordinate at u_p of the standard Gaussian density.

For n independent units exposed to the risk of action in all or nothing, the weight of the observed probit is equal to $nw_{Y\mathcal{P}}$. The weight of the working probit corresponding to p is no other than the reciprocal of the variance of $Y_{\mathcal{P}}$:

$$w_{Y\mathcal{P}} = \frac{1}{\sigma^2_{Y\mathcal{P}}}. \tag{c}$$

By virtue of the definition of the probit as the sum of a constant 5 and u_p, we have

$$\sigma^2_{Y\mathcal{P}} = \sigma^2_{u\mathcal{P}}. \tag{d}$$

For a probability density $f(x)$, the variance of the quantile $x_{\mathcal{P}}$, the theoretical value of which is $\sigma^2_{x_p}$, is written in general as (M. G. Kendall and Stuart, 1963, I, p. 237):

$$V(x_{\mathcal{P}}) = \sigma^2_{x\mathcal{P}} = \frac{pq}{n} \cdot \frac{1}{f^2(x_p)}. \tag{e}$$

We recall that the observed proportion $\mathcal{P} = r/n$ of r independent reactors out of n exposed to risk, derives from a binomial variable $r = 0, 1, 2, \ldots, n$ where the probability p of reaction is constant

$$V(x_{\mathcal{P}}) = V(\mathcal{P})\frac{1}{[f(x_p)]^2}. \tag{f}$$

Returning to u_p with $f(x)$ replaced by the symmetrical standard Gaussian density $\varphi(u)$, we have

$$V(u_{\mathcal{P}}) = \sigma^2_{u\mathcal{P}} = V(\mathcal{P})\frac{1}{\varphi^2(u_p)} = \frac{pq}{n}\cdot\frac{1}{\varphi^2(u_p)}, \tag{g}$$

where

$$\varphi(u_p) \approx \varphi(u_{\mathcal{P}}). \tag{g'}$$

This structure of the variance of the quantile $u_{\mathcal{P}}$ shows the separation according to two phenomena:

(1) *quantal aspect* of the local binomial variance

$$V(\mathcal{P}) = \frac{pq}{n}; \tag{h}$$

(2) *continuous aspect* of the contribution in $\varphi^2(u_p)$, where $\varphi(u_p)$ is the probability density of the true continuous distribution

$$p = \Phi(u), \tag{i}$$

relating to the Gaussian (or log-Gaussian) hypothesis the true responses to the doses (or log-doses) written as $X = \mu + u\sigma$.

6.322.225.4. Probit rulings and normit rulings. Numerous probit or normit rulings are produced commercially. Paper No. 32.451 of Codex Book Co., Norwood, Mass., U.S.A. was introduced by J. Berkson, having an arithmetic scale in abscissae. The right-hand ordinates are graduated directly in % (from 1 to 99), whilst the left-hand scales are graduated in normits (or standard normal deviates or N.E.D.) designated as t, although this notation does not seem to us to be a happy one for amongst biologists t with ν degrees of freedom should be reserved to its specific usages in estimation and for testing statistical hypotheses. Another similar paper is the *Sandsynligheds-papier* J.C.P. No. 1000/100 A of J. Chr. Petersens, Copenhagen K.

The *Codex* paper No. 41.453 (normal-normal ruling of J. Berkson) has both scales expressed in normits. We ourselves also use the paper No. 485 *Schäfers Feinpapier*, Plauen (Vogtl), G.D.R. (logarithmic abscissae).

6.322.225.5. Construction of a probit–X diagram, plotted from a frequency table. The probit–seroprotein level diagram corresponding to the data of Table 6.322.224.7A is obtained without difficulty by representing the probits (column 5) corresponding to the relative cumulative frequencies (column 4) as a function of the *end of the class-interval*. These probits have been extracted from table IX of Fisher–Yates (1957). This table gives the probits to four decimal places, but two are sufficient for a graphical illustration. The points (end of interval; probit corresponding to the relative cumulative frequency at the end of the interval) may be plotted just above the basic histogram. The linearity is adjusted visually, enabling a median $Me = 69.15$ and a graphical $S_x = 4.50$ to be taken.

The standard deviation is the length defined by the intersection of the probit line with the horizontal lines drawn, for instance, through $Y = 5$ and $Y = 6$. Hence the slope is seen to be $(6-5)/\sigma_x = 1/\sigma_x$.

These graphically determined data are in good agreement with the mean (69.29) and the standard deviation (4.65) as calculated. This should be so because the values of $g_1 = -0.15$ and $g_2 = -0.80$ are not contradictory to the Gaussian hypothesis.

The simple visual test of the random position of the points probit-X along a linear trend must be completed by stating the limits of the dispersion interval for a given probability; this interval varies with the number n of data in the sample.

For such numbers as 20, 60, 100 and 1000, Jacquet (1962) has produced a diagram of the 0.95 limits for the points probit–X: *abscissae:* universal scale in $\mu+k\sigma_x$ with k as a multiple of 0.2; *ordinates:* classical probit scale.

An example of this diagram is given in J. Martin (1967, p. 367). Finally Jacquet has also given 95 and 99% confidence zones for selected values of n between 20 and 1000.

6.322.225.6. Plotting a probit-X line corresponding to a Gaussian law (μ, σ_X). According to formula 6.322.225.1 (f), the probit line as a function of X corresponding to a Gaussian law with mean μ and standard deviation σ_X, has the equation

$$Y_p = 5+\frac{X-\mu}{\sigma_X} \tag{a}$$

or

$$Y_p-5 = \frac{1}{\sigma_X}(X-\mu). \tag{a'}$$

In a bi-arithmetic millimeter ruling, this line will pass through the point $(X = \mu;\ Y = 5)$ with the slope $1/\sigma_X$. Hence the following graphical construction: from the point (μ, 5) draw a horizontal segment of length σ_X; at the extremity of this segment, raise a perpendicular of unit length, thus obtaining the slope $\beta = 1/\sigma_X$ in the *directional triangle* of base σ_X and of height $+1$. An arbitrary percentile X_p is immediately estimated graphically. Thus, proceeding from the probability p, read off the probit Y_p in the probit table and then, through the ordinate Y_p, draw a line parallel to the OX axis. This segment cuts the probit line at a point which, when dropped onto OX, will directly give the graphical estimate of X_p.

Now plot the line (a′) on a probit ruling with an arithmetical scale X and a functional scale graduated in p or $100\,p\%$. On this scale the vertical segment of length $+1$ is marked out according to the distance between the ordinate $Y_{84.13\%}$ and the ordinate $Y_{50\%}$, which again brings us to the previous case.

6.322.225.7. Value of the probit representation and its limitations. The probit representation renders great services in the following domains:

(1) *Clear representation of statistical data which,* in frequency density form, would be intricate with bell-shaped curves in juxtaposition. A good example is given in A. A. Weber (1960, p. 97) and describes the evolution according to age of the proportion of children with trachoma scars.

(2) *Graphical test for the Gaussian nature of a density.* Any Gaussian density leads to a linear graph in probit–X coordinates. In principle, the (X–probit)-points should be randomly disposed about a linear tendency. This is the case in Fig. 6.322.225.1 I regarding the height of Belgian recruits. The reader should also use the confidence zones for the probit–X line (Jaquet, 1962) as explained in 6.322.225.5.

However, the reciprocal, which is of special interest here, is not always unbiased. Thus, if the probit–X points describe a linear trend, the χ^2-test based on the comparison of observed and expected frequencies according to a Gaussian law, can lead to rejection of this Gaussian hypothesis suggested for the graphical test. If, for instance, a parabolic trend, even only faintly apparent, is superposed on the general linear tendency, it is necessary to verify the conclusion by a χ^2-test, or to calculate g_1 and g_2. This precautionary step is correspondingly more necessary for a larger sample (owing to the numerical scale of χ^2). In Geigy Tables (1963, pp. 165–6) an example is given of the histogram for the diameter of $n = 2000$ erythrocytes in which the probit–diameter graph that seems linear would lead to non-rejection of the hypothesis of a Gaussian population. But the adjustment of a Gaussian density shows a distinct effect of asymmetry. This non-Gaussian nature is confirmed by the result of the χ^2-test for goodness of fit which is significant at the level 0.001. If the probit–U diagram is markedly parabolic with concavity towards the base, a transformation log $(U-K)$ should be tried (see 6.335.21). If, on the other hand, a parabola with concavity upwards is observed, a transformation log $(K-U)$ can be tried (see 6.335.22).

(3) *Use of information contained in the first open class of a histogram.* The class mid-points are used in calculating the mean of classified data. On the other hand, the final class (whether open or not) is always lost in the probit representation. Furthermore, the "center" of this class when open should not be used any more than the "center" of the first open class in calculating the mean.

For reasons of economy in the presentation of data, official statistics often make use of open classes. No objections arise so long as the statistician is able to evaluate the open classes separately.

(4) In all cases where the *data from observation correspond directly to a distribution*, the probit graph is widely used. Such is the well-known case of transformation of the response–dose (or log-dose) curve in the biological tests of all-or-nothing type. But since the log-dose is generally used, this case will be treated in section 6.335 along with the log-Gaussian law. There is, however, another interesting example of distributions recorded directly during experiments on the resistance of the red corpuscles to hypotonia.

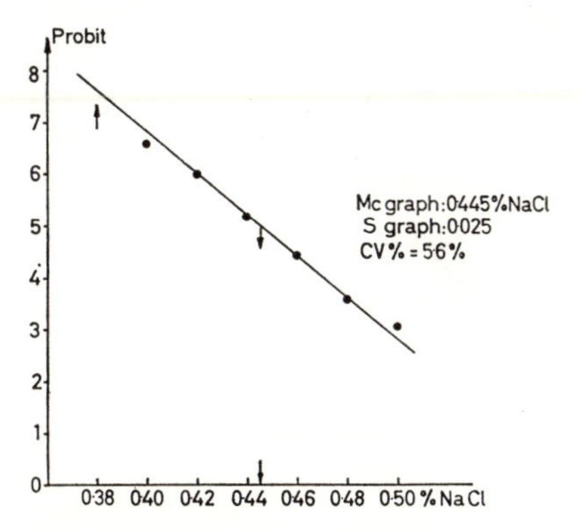

FIG. 6.322.225.7 I. Probit representation of hemolysis by NaCl (data from Gallo, 1961).

Figure 6.322.225.7I is prepared from the data of Gallo (1961). The NaCl concentration is marked off along the axis of the abscissae, and along the ordinates the probit corresponding to the fraction hemolysed at the indicated concentration.

In this probit–salt concentration diagram, the median $Me = 0{\cdot}445\%$ NaCl and the standard deviation $\sigma = 0.025$ are found without difficulty, and hence a coefficient of variation of 5.6%. Besides this probit representation, one can also use a "*logit*" *diagram* (see 6.323.222f).

(5) *Graphical estimation of the median and standard deviation.* A notable example is provided by the frequency distribution table for the height of 1237 Belgian recruits, for whom the statistics of reduction were

$$\bar{X} = 173.35 \text{ cm}, \qquad s_x = 6.28 \text{ cm},$$
$$g_1 = 0.048, \qquad g_2 = 0.07.$$

From the graph, the median is $Me = 173.3$ and the standard deviation from the probit line mentioned under (2) is 6.3 cm (Fig. 6.322.225.1 I).

In this way it is possible to verify all the operations implied in mechanographical mass calculations.

(6) The probit–X representation enables the *part of a distribution which is useful to a biologist* to be seized upon as a whole. Whether the (unimodal) distribution is Gaussian or not, the probit–X curve leads to a graphical estimate not only of the median, but also to any quantile in the range 5–95% where the accuracy is satisfactory.

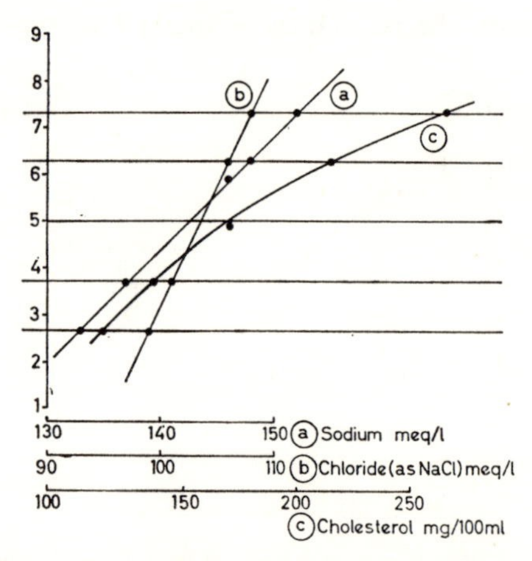

FIG. 6.322.225.7 II. Probit diagram for clinical data (data from King and Wootton, 1956).

One application of the technique has enabled the probit curves or line to be reconstituted approximately from four quantiles $X_{0.01}$, $X_{0.10}$, $X_{0.90}$ and $X_{0.99}$ as produced by King and Wootton (1956, pp. 1–2).

Figure 6.322.225.7II shows the different behavior of cholesterol (lognormal), of Na and chlorides as NaCl (Gaussian), and also yields any percentile desired and, in particular, the median. It will be noted that the measure of the two proposed ranges of biological normality for these authors is 0.80 for the severe domain, and 0.98 for the widened domain. Another range of biological normality with measure 0.95 may also be considered.

A tendency is nowfortunately manifest amongst clinicians to produce frequency curves of biological quantities from reasonably ample series, rather than being content with the single extreme values of the range of "normality".

Epidemiologists also utilize the representation of distributions as a series of percentiles. Thus, for instance, McDowell (1967, pp. 11 and 13) has presented hematocrit distributions by age×sex and race×sex for percentiles selected between 2.5 and 97.5%.

(7) In an analysis of a 2^3 factorial experiment designed against trachoma in Morocco, Weber and Linder (1960) start from the slopes of probit–age straight lines to conduct proper analysis of variance.

6.322.226. Rankits, scores for ordinal or ranked data

6.322.226.1. Definition of rankit. For small series of independent biological or clinical observations, it is important to see whether they derive from a Gaussian population or not. In this respect the important problem is to specify the probability law which would be better decided upon if more numerous and, in that case, grouped data were collected.

Moreover, certain statistical tests, such as Student's test by pairs, are rather *sensitive* to the non-Gaussian nature of analysed differences.

After G. Box (1953) it is said that some tests *lack robustness vis-à-vis the non-Gaussian nature of groups analysed effectively.*

A simple technique, the rankit value diagram, permits a graphical check on whether the Gaussian hypothesis can be rejected; although, clearly, the check is unable to affirm positively that a small series is in fact an extract from a Gaussian population (μ; σ^2).

The term "rankit" is a contraction of *rank* and un*it*, introduced by Ipsen and Jerne (1944, p. 353) in analogy with the term "probit" (*prob*ability un*it*) introduced by Bliss (1935) (see 6.322.225.1e).

The word *rank* implies an arrangement in precedence, a series of increasing values $X_{(r)}$:

$$X_{(1)}, \ldots, X_{(2)}, \ldots, X_{(r)}, \ldots, X_{(n)},$$

where the subscript in brackets signifies an *index of ordination by increasing values.* Clearly, one may also rank the observations by decreasing values.

The rankit–X diagram is closely associated with the method of co-graduation used to define the probit–X diagram (see section 6.322.225.1).

Each random term of the ordinal series observed $X_{(r)}$ $(r = 1, 2, \ldots, n)$ is set in relation to n fixed positions, the rankits $u_{(r|n)}$ being defined in reference to the standard normal curve. The rankit $u_{(r|n)}$ is the rth rankit in the fixed series $u_{(1|n)} \ldots u_{(n|n)}$.

In accordance with the formula (6.315.14d), the distribution of the rth ordinal standard Gaussian deviate $u_{(r|n)}$ in a random series of n, is written as

$$dG(u_{(r|n)}) = \frac{1}{B(r, n-r+1)} \Phi_r^{r-1}(1-\Phi_r)^{n-r}\, d\Phi_r, \tag{a}$$

where

$$\begin{aligned} \Phi_r &= \Phi(u_{(r|n)}) = q, \\ 1-\Phi_r &= p. \end{aligned} \tag{a'}$$

The position of the rth standard Gaussian deviate on the axis $(-\infty, +\infty)$ is a random position depending on the particular representative sample. But if, however, from the standpoint of the theory of estimation, the sampling is repeated, *in the long run*, the position of each order statistic $u_{(r|n)}$ will stabilize itself in the average about some fixed position $E[\boldsymbol{u}_{(r|n)}]$.

The mean value of this position $u_{(r|n)}$ is written as

$$E[\boldsymbol{u}_{(r|n)}] = \int_{-\infty}^{+\infty} \mathrm{u}_{(\mathrm{r})}\, d\mathrm{G}(\mathrm{u}_{(\mathrm{r})}), \tag{b}$$

or, after Fisher–Yates (1957, p. 28),

$$E[\boldsymbol{u}_{(r|n)}] = \int_{-\infty}^{+\infty} \frac{n!}{(r-1)!\,(n-r)!}\, p^{n-r} q^{r-1} \mathrm{u}\varphi(\mathrm{u})\, du.$$

For each set $n = 2, 3, \ldots$ of values, the rankits $u_{(r|n)}$, or scores for ranked or ordinal data, have been tablulated for $r = 1, 2, \ldots, n$.

In view of the symmetry of the parent Gaussian law, tables XX and XXI of Fisher–Yates (1957) are presented in abridged form, eliminating the negative values of the rankits, and omitting the median value

$$u_{(r+1|n)} = 0 \quad \text{for all} \quad n = 2r+1.$$

In Table 6.322.226.1A, we sketch out the beginnings of a complete table according to the effective value n of the sample and the ranks (r) within the sample n.

From this table it will be seen that the range $u_{(n|n)} - u_{(1|n)}$ becomes larger in proportion as n is increased.

TABLE 6.322.226.1A. RANKITS $u_{(r|n)}$ FOR THE FIRST VALUES n OF THE SAMPLE. EXTRACT FROM FISHER–YATES (1957, TABLE XX)
(To be read vertically)

	n – effective value of sample			
(r) rank	2	3	4	5
1	−0.56	−0.85	−1.03	−1.16
2	+0.56	0	−0.30	−0.50
3		+0.85	+0.30	0
4			+1.03	+0.50
5				+1.16

Figure 6.322.226.1 I objectifies this fact and illustrates the symmetry of all the $u_{(r)}$ and the progressive decrease of the linear density of the rankits on moving away from the center towards the periphery.

Tables of rankits or scores for ordinal (*or ranked*) *data*. The positive values of the rankits for n ranging from 2 to 50 have been tabulated and may be

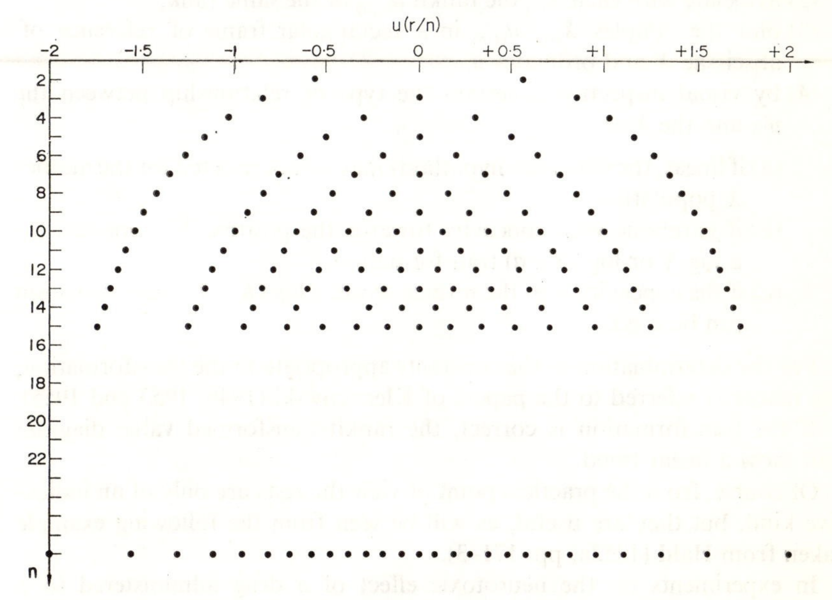

FIG. 6.322.226.1 I. Rankits $u_{(r|n)}$ for $n = 2, 3, \ldots, 27$ (data from Fisher–Yates, 1957, table XX).

found in Fisher–Yates (1957, pp. 86–87) and in Pearson–Hartley (1954, p. 175), Harter (1961) and Bliss (II, 1970, pp. 552–3 and p. 612).

6.322.226.2. Use of rankits. (α) *Rankit transformation.* The analysis of preference test results, or of non-metric, but ordinable experimental data, can be conducted after rankit transformation. Thus, in the analysis of tests of organoleptic characteristics, Bliss (1960) introduced the rankits of preference numbers $n_{ij} = Np_{ij}$. These rankits can be introduced into Analysis of Variance as standard deviates. N.E.D.s or normits, to which they are closely similar. Vessereau (1965, p. 121) emphasizes with good reason the advantage of rankits over normits, for the former will conserve a finite value up to the bounds 0 or N of the range of the number of preferences.

In order to simplify the calculations, Fisher–Yates (1957, p. 86) gives in table XXI the sum of the squares (*SS*) of the deviations (about the mean 0) of the rankits for each n (effective sample size) from 2 to 50.

(β). *Graphical test for the Gaussian nature of small series.* For a graphical test of the Gaussian nature of small series of n observations, the procedure is as follows:

(1) rank the observations in an increasing progression $X_{(1)}, \ldots, X_{(r)}, \ldots, X_{(n)}$,
(2) associate with each $X_{(r)}$ the rankit $u_{(r|n)}$ of the same rank,
(3) plot the couples $X_{(r)}$, $u_{(r|n)}$ in a rectangular frame of reference of abscissae X and ordinates u,
(4) by visual inspection ascertain the type of relationship between the u's and the X's:

(a) if linear, the Gaussian hypothesis may be not rejected for the mother X population.
(b) if parabolic with concavity towards the positive X's, one can try a $\log X$ or $\log (X+a)$ transformation,
(c) if the concavity is in the inverse sense, a $\log (K-X)$ transformation can be tried.

For the determination of the constants appropriate to the transformation, the reader is referred to the papers of Kleczkowski (1949, 1953 and 1955).

If the transformation is correct, the rankit-transformed value diagram will show a linear trend.

Of course, from the practical point of view the tests are only of an indicative kind, but they are useful, as will be seen from the following example taken from Hald (1952a, pp. 171–3).

In experiments on the neurotoxic effect of a drug administered to a group of $n = 16$ flies for 30 sec, the reaction time t_i was recorded as from the instant of application to the moment when the fly fell on its back.

Hald used a probit ruling for which the $T_{\mathcal{P}_i}$ entries corresponding to the $t_{(i)}$ ranked in ascending magnitude were obtained from the estimated total probability of response in the interval $(0, t_i)$:

$$\mathcal{P}_i = \frac{i-0.5}{n}. \qquad \text{(c)}$$

The graph for $Y_{\mathcal{P}_i}$ as a function of $\log_{10} t_i$ yields a satisfactory linear relationship where the median from the graph is approximately 16.5 sec; the graph is closely similar to that plotted for a group of 15 flies treated for 60 sec (median from the graph = 15.0 sec) (see fig. 7.9 of Hald, 1952a, p. 172).

In this way Hald produced an interesting example of a probit–log t diagram with parallel lines for two *discrete* series.

It seems interesting to show, e.g. for the group of flies treated for 30 sec, how the diagrams in $t_{(i)}$ and $\log_{10} t_{(i)}$ compare on the same rankit scale.

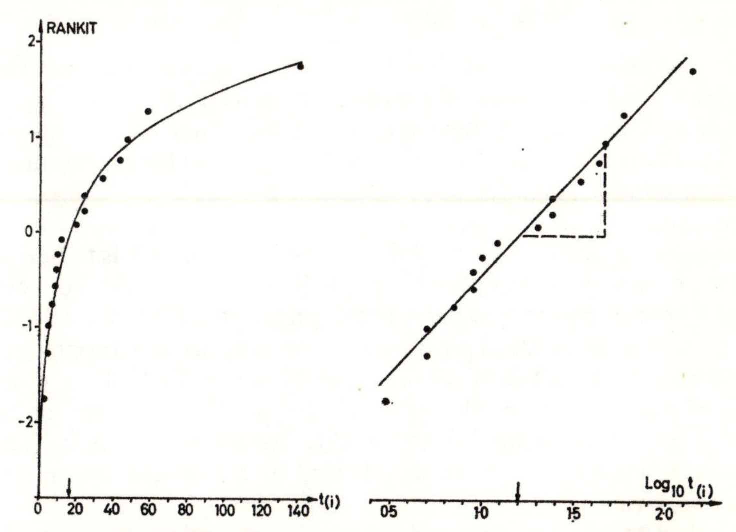

FIG. 6.322.226.2 I Rankits diagrams in $t_{(i)}$ and $\log_{10} t_{(i)}$. t = neurotoxic reaction time of flies. (data from Hald, 1952a, pp. 172–3.)

Figure 6.322.226.2 I shows on the left-hand side these times $t_{(i)}$ $(i = 1, 2, \ldots, 16)$ arranged in an increasing progression in a rankit (.) diagram.

The general tendency, traced by hand, is manifestly nonlinear and suggests that a logarithmic transformation of the abscissae may lead to a linear trend. This is in fact so, as shown by the right-hand part of the graph (x) in $\log_{10} t_i$.

From the graph the median of the log t_i is equal to 1.205, in good correspondence with the mean of the logarithms calculated by Hald and which equals 1.219.

The method of the directional triangle provides a graphical estimate of $\sigma_{\log_{10} t_i} = 0.46$, corresponding excellently to the value 0.456 calculated by Hald.

In section 6.335.112.1 (m) devoted to the log-gaussian law, a formula is given which, transposed into log-decimals, is written as

$$\text{Median of all } t_i = \text{antilog}_{10} \text{ of the mean of all } \log_{10} t_i. \tag{d}$$

Using the values calculated by Hald, we have

$$Me\ t_i = \text{antilog}_{10}\, 1.219 = 16.56,$$

whilst using our graphical procedure

$$Me\ t_i = \text{antilog}_{10}\, 1.205 = 16.07.$$

These two values agree satisfactorily with the graphical median of the t_i ($Me\ t_i$ graph = 16.5) defined on the rankit–t_i curve.

By thus anticipating, for pedagogic reasons, the log-Gaussian law discussed later, the rankit diagram is seen to be useful as a guide to the transformation of random variables, even for small discrete series.

If the rankit–$t_{(i)}$ diagram is concave towards the axis of the rankits, a transformation of the log $(K-t_i)$ type could be tried, where K is to be determined by trial and error, or by referring to Kleczkowki's method. To provide a useful if not necessary complement to the graphical test, Shapiro and Wilk (1965) have given a statistical procedure for checking the visual appreciation of the graphical rankit test of the Gaussian character of data from a small sample of size n (n up to 50). After the graphical analysis, the ordered observations $y_{(i)}$ are multiplied by suitable factors $a_{(i)}$ depending on n, the products are added and the sum divided by the sample numerator of variance s_X^2, giving a ratio W.

This calculated ratio is compared with a table of significance points. This method is used and commented in E. S. Pearson (1968).

(γ) *Graphical test for the Gaussian character of grouped series.* In the foregoing section the data were treated individually as single $X_{(i)}$. However, many biological observations can be presented in the form of grouped data. The latter data are of two types, viz.

(a) data of the continuous type presented in the form of frequency tables, e.g. Table 6.322.224.7A which gives the frequency of the classes of serum-proteins in normal subjects;

(b) essentially discrete counting data $r = 0, 1, 2, \ldots$, e.g. data encountered in the Poisson, binomial and other laws, or scores associated with responses which may be quantified by an appropriate system.

These two types of data can both be treated by the rankit method, providing the total number does not exceed 50, this being the limit of table XX of Fisher–Yates (1956). From Ib. Andersen, a table is given by Ipsen and Jerne (1944, p. 359) for the scores $r = 0, 1, 2, \ldots, 6$ used to describe the

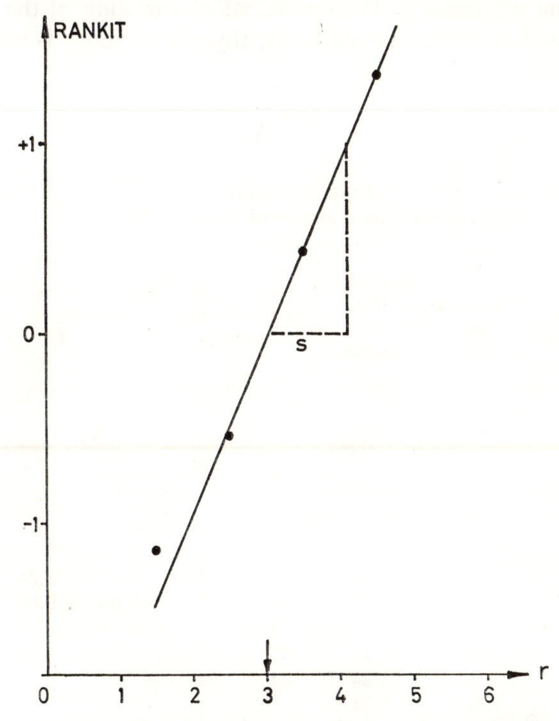

FIG. 6.322.226.2 II. Rankit diagram for the score of thyroid reaction to thyroid hormone. (data from Ipsen and Jerne, 1944, p. 359.)

thyroid reaction of 24 guinea pigs after administration of thyroid hormone. A score $r = 1$ corresponds to continuous limits 0.5 and 1.5. The rankits with the limits $u_{(i|24)}$ are written as a function of the cumulative rank i out of 24. The mean rankit at the end (1.5) of the class of score 1 is the arithmetic mean of the rankits surrounding the point 1.5, say -1.24 at the point $i = 3$ and -1.04 at the point $i = 4$, on the other side of the limits 1.5. In Fig. 6.322.226.2. II, these mean rankits are shown at the end of interval 1.5, 2.5, etc. A line section is adjusted visually with regard to the (psychological) weight of the abscissa points 4.5 and 3.5 which frame the median score, the graphical value of which is *Me* graph = 3.05.

The linearity of the rankit–X relationship makes it possible not to reject the hypothesis as to the Gaussian nature of the scores.

By transposing the directional triangle of the probit–X graph to the rankit–X diagram, a graphical estimate of the standard deviation is obtained ($s_r = 1.10$). The calculated values $\bar{r} = 3.0$ and $s_r = 1.14$ are closely similar to those found graphically. The coefficient of variation of the scores indicating the reaction of the thyroid to the thyroid hormone is important and equals

$$\text{C.V.\%} = 100\frac{1.14}{3.0} = 38\%.$$

TABLE 6.322.226.2A
(from Ipsen and Jerne, 1944, p. 359)

Score	Upper limit of classes	Frequencies	Cumulative frequencies	Frequencies at the limits of classes		Rankits with limits for $n = 24$		Mean rankits at the class limits
I	II	III	IV	V		VI		VII
				start	end			
0	0.5	0	0					
1	1.5	3	3		3		−1.24	−1.14
2	2.5	4	7	4	7	−1.04	−0.60	−0.54
3	3.5	9	16	8	16	−0.48	+0.37	+0.43
4	4.5	6	22	17	22	+0.48	+1.24	+1.37
5	5.5	2	24	23	24	+1.50	—	
6	6.5	0	24	—		—		
		24						

6.322.226.3. Standardized range. Suppose that a random sample of n values X_i is extracted from a Gaussian population (μ, σ). Stating n explicitly as a subscript of R, the range of this sample is

$$R_n = X_{(n)} - X_{(1)}. \tag{a}$$

Upon adding and subtracting the constant μ, the standardized range is

$$W_n = \frac{R_n}{\sigma} = \frac{X_{(n)}-\mu}{\sigma} - \frac{X_{(1)}-\mu}{\sigma}. \tag{b}$$

Each random sample has some corresponding range. The mean value of R_n/σ is written so:

$$E[W_n] = \frac{1}{\sigma}E[R_n] = E\left[\frac{X_{(n)}-\mu}{\sigma}\right]-E\left[\frac{X_{(1)}-\mu}{\sigma}\right]. \tag{c}$$

In accordance with the definition of rankit given in 6.322.226.1,

$$E\left[\frac{X_{(n)}-\mu}{\sigma}\right] = u_{(n|n)},$$
$$E\left[\frac{X_{(1)}-\mu}{\sigma}\right] = u_{(1|n)}. \tag{d}$$

Hence the mean value of the standardized range is no other than the difference between the rankit of rank n and the rankit of rank 1, this difference being denoted as d_n by Pearson–Hartley (1954, table 20, right-hand part):

$$E[W_n] = u_{(n|n)}-u_{(1|n)} = d_n. \tag{e}$$

For the values $n = 2$ (1) 50, the reader can himself obtain these values from table XX of Fisher–Yates (1957).

An exhaustive table has been compiled by Tippett (1925) and this is reproduced in table 27 of Pearson–Hartley (1954) for $n = 2$ (1) 499 and then $n = 500$ (10) 1000. From such tables, the biologist may consider the range of observations which, on average, he may expect to cover by different numbers of *random* realizations in a Gaussian population ont the assumption that σ is known.

For $n = 5$, we have $d_5 = 2.33$; this same result is obtained from Table 6.322.226.1A:

$$d_5 = 1.16-(1.16) = 2\times 1.16 = 2.32.$$

A biologist likes to know that with Gaussian random samples of $n = 10$ elements, on average he only covers 3σ of the range (-1.54 to $+1.54$ in sigma units on either side of the mean); whereas 500 Gaussian random observations are required for this range to be doubled and marked at 6σ.

The distributions of the range $F(W_n)$ have been tabulated by E. S. Pearson (1942) and taken up as table 22 of Pearson and Hartley (1954) for $n = 2$ (1) 20. The quantity $1/d_n$ is also given there (table of peripheral quantiles W_F). Table 23 of these same authors gives for $n = 2$ (1) 20 the distribution function $F(W_n)$, where W is graduated in steps of 0.05 between 0.00 and $W_2 = 5.75$, $W_{20} = 7.25$. The parameters of this law of the range W_n (mean d_n, σ, σ^2, β_1, β_2 and the coefficient of variation squared) are given in table 20 of the same authors for $n = 2$ (1) 20. The unfolding evolution of $\beta_1 = 0.99$

and $\beta_2 = 3.87$ for $n = 2$ towards the values 0.16 and 3.26 respectively, shows that the probability density of the range W_n tends to a Gaussian law as n increases.

The tables 20 and 23 (excepting β_1 and β_2) have been grouped together by Hald (1952b, table VIII). In Table 6.322.226.3A we give an extract from

TABLE 6.322.226.3A. EXTRACT FROM HALD (1952b, TABLE VIII)

n	α_n	β_n	γ_n	Quantiles $W_{n, P}$				
				2.5	5.0	50.0	95.0	97.5
2	1.128	0.853	0.756	0.04	0.09	0.95	0.77	3.17
10	3.078	0.797	0.259	1.67	1.86	3.02	4.47	4.79
20	3.735	0.729	0.195	2.45	2.63	3.69	5.01	5.30

these tables for $n = 2$, 10 and 20, and for the percentage values 2.5, 5.0, 50.0, 95.0 and 97.5 of $F(W_n)$. Hald uses the notation α_n, β_n and γ_n as shown below (Table 6.322.226.3B).

TABLE 6.322.226.3B

	Pearson–Hartley (1954, p. 164)	Hald (1952b, pp. 60–61)
$E[W_n]$	d_n	$\alpha_n = m(W)$
$V(W_n)$	$V_n =$ variance	
σ_{W_n}	S.D.	$\beta_n = \sigma(W)$
$\dfrac{\sigma_{W_n}}{E[W_n]}$	$\left(\sqrt{\dfrac{d_n^2}{V_n}}\right)^{-1}$	$\gamma_n = \dfrac{\sigma}{m}$

That $F(W_n)$ tends to a symmetrical law is objectified by the mean–median difference decreasing with n.

6.322.227. Half-normal plots. In the analysis of factorial tests without replication in exploratory research on four factors or more, Birnbaum (1959) considers the questions of validity of normal models and of statistical inference. Apart form statistical rules, he suggested a graphical method (half-normal plots) which was developed in a companion article by Daniel (1959).

The half-normal probability density of the essentially null or positive variable $x \geqslant 0$ of origin O and of variance σ_X^2, is written as

$$g_1(x) = \sqrt{\frac{2}{\pi\sigma_x^2}} \exp\left\{-\frac{x^2}{2\sigma_x^2}\right\}. \quad (x \geqslant 0) \tag{a}$$

The distribution of $x \geqslant 0$ or of $|x|$ is

$$F_1(x) = 2[\Phi(u) - \tfrac{1}{2}] \quad (x \geqslant 0) \tag{b}$$

where the standard deviate u is

$$u = \frac{x}{\sigma_x}. \tag{b'}$$

Thus

$$F_1(0) = 0 \quad \text{for} \quad \Phi(0) = \tfrac{1}{2},$$
$$F_1(+\infty) = 1 \quad \text{for} \quad \Phi(+\infty) = 1. \tag{b''}$$

Therefore the probability corresponding in $(0, +\infty)$ to the upper quartile $x_{0.75}$ of the integer variable x is

$$F_1(x_{0.75}) = 2[0.75 - 0.50] = 0.50$$

for the essentially positive variable.

The median of $|x|$ is therefore equal to the upper quartile of x:

$$|x|_{0.5} = x_{0.75}. \tag{c}$$

So if one proceeds from the normit–x or probit–x plot, marked off in p, the half-normal plot is obtained by setting

$$p' = 2p - 1 \tag{d}$$

for the points corresponding to $|x|$, i.e. to the primordial part of the ordinate axis.

An example of a *general-purpose half-normal grid* is given by Daniel (1959, p. 313) and is the equivalent of the probit–X plot. On this grid the ordinal quantities $X_{(1)} \ldots X_{(i)} \ldots X_{(n)}$ are marked off as a function of

$$p' = \frac{i - 0.5}{n}. \tag{e}$$

Since 2^{n-1} "contrast-sums" correspond to one factorial test 2^n, the author has constructed special-purpose grids for $n = 15$, 31, 63 and 127 which are simpler to use for it is no longer necessary to pass from rank i through n to p.

Whenever the representative points of factorial effects deviate sufficiently from a straight line through the origin, these effects merit special consideration which may lead to them being recognized as real. But the author does alert the reader to the fact that this graphical method is *no substitute* for analysis of variance, and it should be used with a critical mind.

Tukey (1959) communicated to Daniel (1959, pp. 317–18), to whom we refer the reader, a method of half-normal plotting as a function of the logarithm of the contrasts, following the idea that the relative magnitude of the contrasts has an intuitive value. Sparks (1970) has written in ASA Fortran a program for half normal plotting.

Table of positive rankits for half-normal plotting. In his recent book, Bliss (II, 1970, p. 612) produces a very useful table (table A39) of rankits for plotting half-normal diagrams with $n = 25$ to 31 and 61 degrees of freedom, for samples of size $N = 2n+1$. Due reference is given to a technical report of Aeronautical Res. Labs. written by H. L. Harter in 1960. This table extends the classical tables of rankits (Fisher–Yates (eds.), 1957; Pearson–Hartley (eds.), 1954) as well as the one of Harter (1961) adapted with permission by Bliss (II, 1970, pp. 552–3) in his table A6.

Applications. Daniel gives several examples taken from the published literature. In particular, in the context of the pharmaceutical industry, he treats the results of a factorial design 2^5 on the production of penicillin given in Davies (ed.) (1956, p. 417). He also shows a plot in accordance with the recommendations of Tukey. Bliss (II, 1970, pp. 411 and 474) gives other examples.

6.322.228. Imminence rate or hazard function. We will now consider the concept of imminence rate or hazard function for the standard Gaussian law. In accordance with the definition of section 6.122, the imminence rate of the standard Gaussian deviate u is written as

$$I(u) = \text{Prob}\ [(\boldsymbol{u} in (u,\ u+du) | \boldsymbol{u} \geqslant u)] = \frac{\varphi(u)}{1-\Phi(u)}. \qquad \text{(a)}$$

This conditional probability is thus defined only for standard deviates $\boldsymbol{u}$ greater than fixed u, i.e. over the range $(\boldsymbol{u} > u)$ of measure $1-\Phi(u)$. This is shown in Fig. 6.322.224.5I top curve.

The three reference values are verified without difficulty:

$$I(0) = \frac{\varphi(0)}{1-0.5} = 2\times 0.399 = 0.798 \approx 0.8,$$

$$I(-0.6745) = \frac{0.318}{1-0.25} = \frac{4\times 0.318}{3} = 0.424,$$

$$I(+0.6745) = \frac{0.318}{1-0.75} = 4\times 0.318 = 1.272.$$

The imminence rate monotonically increases between 0 and infinity (see Feller, 1958, I, p. 166):

$$I(-\infty) = \frac{\varphi(-\infty)}{1} = 0,$$

$$I(+\infty) = \lim_{u=\infty} \frac{\varphi(u)}{1-\Phi(u)} = \infty.$$

This function is also given in coordinates $[u, \log_{10} I(u)]$ in Gumbel (1960, p. 22) where $I(u)$ is denoted $\mu(u)$ (intensity).

6.322.229. Gaussian elasticity. To be as comprehensive as possible in this anatomical study of distributions, we mention the economic concept of *elasticity* which is well-known to specialists in this field, and for which a probabilistic interpretation has been given (Aitchison and Brown, 1957, pp. 130–1). Here we shall consider only the formal aspect, for the actual variables (consumption and income) are of the log-Gaussian type which will be considered in section 6.335.

In their table A_5 these authors (1957, pp. 162–3) condense the necessary elements to the adjustment of a probit-X line and add another column $Z(Y)/P(Y)$ to the table under the heading "elasticity" and which we qualify as Gaussian elasticity.

In our notation for elasticity, we have

$Y \rightarrow u$ is the normit or N.E.D. (do not confuse this Y with the probit notation);

$Z(Y)$ becomes $\varphi(u)$ and is the ordinate of the standard Gaussian density;

$P(Y)$ becomes $\Phi(u)$ and is the standard Gaussian distribution.

The expression for Gaussian elasticity now becomes

$$\frac{\varphi(u)}{\Phi(u)}. \tag{a}$$

By virtue of the symmetry of the Gaussian law, we have

$$\begin{aligned} \varphi(-u) &= \varphi(u), \\ \Phi(-u) &= 1-\Phi(u), \\ \Phi(+u) &= 1-\Phi(-u). \end{aligned} \tag{b}$$

Hence the Gaussian elasticity should be represented in the diagram by the mirror image articulated on $u = 0$ of the imminence density $I(u)$.

In section 6.335.16 we refer to a remark on the concept of elasticity in a more realistic context as log-Gaussian elasticity.

6.322.3. Generating functions associated with the Gaussian law

Here we shall apply the formulae of sections 6.302.2 to 6.302.7.

6.322.31. Initial moment generating function

For the Gaussian law (μ, σ^2), the moment generating function in relation to the origin is written as

$$M_X(t) = E[(e^{tX})] = \int_{-\infty}^{+\infty} g(X)e^{tX}\, dX$$

$$= \frac{1}{\sigma_X\sqrt{2\pi}} \int_{-\infty}^{+\infty} \exp\left\{ -\frac{1}{2}\left(\frac{X-\mu}{\sigma_X}\right)^2 + tX \right\} dX. \qquad \text{(a)}$$

We introduce the standard deviate

$$u = \frac{X-\mu}{\sigma_X} \qquad du = \frac{dX}{\sigma_X} \qquad \text{(b)}$$

whence

$$X = \mu + \sigma u, \qquad \text{(c)}$$

having removed the subscript X of σ in order to simplify the notation.

In the new variable $u(X)$ we have

$$M_X(t) = \frac{1}{\sqrt{2\pi}} \int_{-\infty}^{+\infty} \exp\left\{ -\frac{1}{2}u^2 + \mu t + \sigma u t \right\} du. \qquad \text{(d)}$$

The argument of the exponential can be written by collecting the terms in u:

$$-\tfrac{1}{2}(u^2 - 2\sigma u t) + \mu t.$$

After adding and subtracting $\frac{1}{2}\sigma^2 u^2$ to obtain a perfect square in the brackets,

$$(u-\sigma t)^2 = u^2 - 2\sigma t u + \sigma^2 t^2.$$

Having removed from within the sign $\int$

$$\exp\{\mu t + \tfrac{1}{2}\sigma^2 t^2\}$$

as a constant factor in u, within the $\int$ sign we now have

$$\exp\{-\tfrac{1}{2}(u^2 - 2\sigma t u)\} = \exp\{-\tfrac{1}{2}[u-\sigma t]^2 + \tfrac{1}{2}\sigma^2 t^2\}.$$

One can therefore write

$$M_X(t) = \exp\left\{\mu t + \frac{1}{2}\sigma^2 t^2\right\} \underline{\frac{1}{\sqrt{2\pi}} \int_{-\infty}^{+\infty} \exp\left\{-\frac{1}{2}(u-\sigma t)^2\right\} du}. \qquad \text{(e)}$$

Since σ and t are constants in regard to u, therefore

$$d(u-\sigma t) = du,$$

and the underlined integral above is equal to unity. Hence, explicitly restating the subscript X of σ,

$$M_X(t) = \exp\{\mu t+\tfrac{1}{2}\sigma_X^2 t^2\}. \qquad \text{(f)}$$

Such is the generating function for the initial moments μ_i' of a Gaussian variable (μ, σ_X^2).

As the expected value of $E[e^{tX}]$, the generating function of the initial moments is written as

$$M_X(t) = E[e^{tX}] = \int_{-\infty}^{+\infty} \left(1+\frac{tX}{1}+\frac{t^2X^2}{2!}+\ \dots\right) g(X)\, dX$$

$$= 1+\mu_1'\frac{t}{1}+\mu_2'\frac{t^2}{2!}+\ \dots\ . \qquad \text{(g)}$$

When the explicit value of $M_X(t)$ is known, the successive moments μ_i' are calculated without difficulty by simple operations of differentiation.

We thus write

$$\begin{aligned} \mu_0' &= 1 = M_X(0) \\ \mu_1' &= M_X'(0) \\ \mu_2' &= M_X''(0) \text{ etc.} \\ &\dots \end{aligned} \qquad \text{(h)}$$

Differentiating $M_X(t)$ with respect to t

$$M_X'(t) = (\mu+\sigma_X^2 t)\exp\{\mu t+\tfrac{1}{2}\sigma_X^2 t^2\}. \qquad \text{(i)}$$

The first initial moment is obtained by calculating this derivative at $t =$ Thus

$$\mu_1' = M_X'0) = \mu. \qquad \text{(j)}$$

In order to obtain the second moment μ_2', we take the second derivative

$$M_X''(t) = \sigma_X^2 \exp\{\mu t+\tfrac{1}{2}\sigma_X^2 t^2\}+(\mu+\sigma_X^2 t)^2 \exp\{t\mu+\tfrac{1}{2}\sigma_X^2 t^2\}, \qquad \text{(k)}$$

whence, at the point 0,

$$\mu_2' = M_X''(0) = \sigma_X^2+\mu^2. \qquad \text{(l)}$$

The variance is written as

$$V(X) = \mu_2'-\mu_1'^2 = M_X''(0)-M_X'^2(0) = (\sigma_X^2+\mu^2)-\mu^2 = \sigma_X^2 \qquad \text{(m)}$$

The generating functions for the initial moments thus yield the parameters μ and σ_X which define the particular Gauss law concerned.

One would also calculate μ_3' and μ_4'.

6.322.32. Generating function for the central moments of X

The central Gaussian deviate

$$x = X - \mu$$

of mean 0 and variance $\sigma_x^2 = \sigma_X^2$ possesses a generating function for the central moments $\mu_1 = 0$, $\mu_2 = \sigma^2$, μ_3, μ_4 and so on.

The central moment generating function is the expected value of $e^{tx} = e^{t(X-\mu)}$

$$M_x(t) = E[e^{tx}] = E[e^{t(X-\mu)}] = e^{-\mu t}E[e^{tX}] = e^{-\mu t}M_X(t). \quad \text{(a)}$$

Substituting into (a) the value of $M_X(t)$ given in 6.322.31 (f),

$$M_x(t) = e^{-\mu t} \cdot \exp\left\{\mu t + \frac{1}{2}\sigma_X^2 t^2\right\} = \exp\left\{\frac{1}{2}\sigma_X^2 t^2\right\}$$

$$= 1 + 0 + \sigma_x^2 \frac{t^2}{2!} + 0 + \frac{1}{2^2}\sigma_x^4 \frac{t^4}{2!} \cdots . \quad \text{(b)}$$

The factorial of the corresponding power of t appears in the denominator of the terms. It will also be borne in mind that $\sigma_x^2 = \sigma_X^2$.

Likewise, analogously to 6.322.31 (g), we have

$$M_x(t) = 1 + \mu_1 \frac{t}{1} + \mu_2 \frac{t^2}{2!} + \mu_3 \frac{t^3}{3!} + \mu_4 \frac{t^4}{4!} + \cdots . \quad \text{(c)}$$

Equating the coefficients of $t^i/i!$, for the Gaussian deviate x it is found that

$$
\begin{aligned}
1 &= 1 \\
\mu_1 &= 0 \text{ (expected result)} \\
\mu_2 &= \sigma_x^2 \\
\mu_3 &= 0 \\
\mu_4 &= \tfrac{1}{8}\sigma_x^4 \cdot 1 \cdot 2 \cdot 3 \cdot 4 = 3\sigma_x^4.
\end{aligned} \quad \text{(d)}
$$

6.322.33. Moment generating function of the standard Gaussian deviate

The standard Gaussian deviate

$$u = \frac{x}{\sigma_x} = \frac{X - \mu}{\sigma_X} \quad \text{(a)}$$

is of mean 0 and variance 1.

The generating function for the central moments of u is written without difficulty from 6.322.32 (b) by replacing σ^2 by unity

$$M_u(t) = e^{t^2/2}. \qquad \text{(b)}$$

Proceeding as in the foregoing sub-section, for the standard Gaussian deviate u we have

$$\begin{aligned} \mu_1 &= 0 \\ \mu_2 &= 1 \\ \mu_3 &= 0 \\ \mu_4 &= 3 \end{aligned} \qquad \text{(c)}$$

Formula (b) shows that the moment generating function of u, except for a change of sign in the exponent, is equal to the standard Gaussian probability density $\varphi(u)$, ignoring $1/\sqrt{2}M$.

6.322.34. Characteristic function

As we have seen (6.202.4), the characteristic function of a random variable is the battle steed of the calculus of probabilities and of mathematical statistics. It is defined as the expected value of itX, of itx or of itu as the case may be.

The characteristic function

$$\varphi_u(t) = e^{-t^2/2} \qquad \text{(a)}$$

in the standard Gaussian deviate is proportional to the standard Gaussian probability density

$$\varphi(u) = \frac{1}{\sqrt{2\pi}} e^{-u^2/2}. \qquad \text{(b)}$$

This reproducibility of form in the Fourier transform is interesting (see 6.302.4b).

6.322.35. Cumulant generating function

Following 6.302.6(b), this function is written as

$$K_X(t) = \ln M_X(t) = \ln \exp \{tu + \tfrac{1}{2}\sigma_X^2 t^2\} \qquad \text{(a)}$$

or

$$K_X(t) = \mu t + \tfrac{1}{2}\sigma_X^2 t^2. \qquad \text{(a')}$$

Formally and assuming the convergence of the series to t, the function $K_X(t)$ is in general written as

$$K_X(t) = \varkappa_1' t + \varkappa_2 \frac{t^2}{2!} + \varkappa_3 \frac{t^3}{3!} + \varkappa_4 \frac{t^4}{4!} + \ \dots \tag{b}$$

Upon identifying (a′) and (b) or differentiating (b) with respect to $t = 0$, we have immediately

$$K_X^{\text{I}}(0) = \varkappa_1' = \mu = \mu_1',$$

$$K_X^{\text{II}}(0) = \varkappa_2 = \sigma_X^2 = \mu_2,$$

and, necessarily, (c)

$$K_X^{\text{III}}(0) = \varkappa_3 = 0 \quad \text{(symmetry)},$$

$$K_X^{\text{IV}}(0) = \varkappa_4 = 0 \quad \text{(mesokurtosis)},$$

and so on.

On the other hand, in accordance with 6.322.32 (b) the cumulant generating function of $x = X - \mu$ is written as

$$K_x(t) = \ln M_x(t) = \tfrac{1}{2}\sigma_x^2 t^2. \tag{d}$$

Hence for the absolute deviation x, the second cumulant is

$$\varkappa_2(x) = \sigma_x^2 = \varkappa_2(X) = \sigma_X^2, \tag{d′}$$

in indicating the relevant variable for $\varkappa_2$ and σ^2.

Finally, the cumulant-generating-function of the standard deviate

$$u = \frac{x}{\sigma_x} = \frac{X-\mu}{\sigma_X}, \tag{e}$$

in accordance with 6.322.33 (b), is

$$K_u(t) = \ln M_u(t) = \tfrac{1}{2}t^2. \tag{f}$$

Hence

$$\varkappa_2(u) = 1. \tag{f′}$$

These results are summarized in Table 6.322.35A.

TABLE 6.322.35A. CUMULANTS OF THE THREE TYPES OF GAUSSIAN VARIABLES

	$\varkappa_1'$	$\varkappa_2$	$\varkappa_3$	$\varkappa_4$
X	μ	$\sigma^2 X$	0	0
$x = X - \mu$	0	$\sigma^2 x$	0	0
$u = \dfrac{X-\mu}{\sigma_X}$	0	1	0	0

6.322.4. *Random Gaussian (normal) deviates*

Starting from the M. G. Kendall–Babington Smith (1939) tables of random numbers and reading them by columns, Wold (1948) has produced satisfactory tables of 25,000 random normal deviates. There are 500 deviates per page, 1000 deviates (two opposite pages) per first, second, . . ., twentieth, fiftieth, thousandth. In one page, there are 10 columns of 50 deviates with the sign and at the bottom of each column is indicated Σu and Σu^2. These tables are useful in problems such as:

1. Constructing of samples of n Gaussian values X_i with known mean μ and standard deviation σ_X

$$X_i = \mu + u_i\sigma_X. \qquad (i = 1, 2, \ldots, n) \tag{a}$$

2. Checking the stochastic independence of $\bar{X}$ and s_x^2 on such samples.
3. As starting point for *simulation* of statistical models with Gaussian (normal) data.
4. Construction of random observations for other standard distribution (Quenouille, 1959); see also sections 6.311.6, 6.331.5, 6.335.18.

6.322.5. *Useful approximation to the Gaussian distribution. Application to simulation*

Let us consider a random variable of distribution $p = F(X)$. Simulation is a mathematical experiment in which random values of X ($X_1, \ldots, X_n$) are generated from equiprobable random values of p ($p_1, \ldots, p_n$); these follow a rectangular or uniform distribution on the range (0–1). Equiprobable p_i may be practically obtained either simply by hand, or by using computers.

(a) The set of the Japanese *icosahedric dice* of colours red, blue, yellow (considered in 6.111.22) may be (and is) used to generate random numbers with three independent random digits according to a prefixed order of reading the coloured dice. Then, from a *random p*, the inverted distribution

$$X = F^{-1}(p) \tag{a}$$

gives the mathematically (not random) corresponding value of X; the result is the random X corresponding to the operational value of p. This process is highly appreciated by small groups of students but is evidently time consuming. Although we have not verified the fact, the procedure is thought to be very near pure randomness.

(b) A quicker procedure is to generate random (or better pseudo-random) numbers with a *computer*. For the standard Gaussian distribution $\varphi(u)$ (see 6.322.22) there is the difficulty raised in 6.322.221, because $\Phi(u)$ is not expressible in terms of elementary functions whose combination should be invertible. Amongst other empirical solutions to this problem, Burr (1967) has presented an approximation to the standard Gaussian distribution by an algebraic relation between u and p with

$$u = \frac{X-\mu}{\sigma_X} \tag{b}$$

as introduced in 6.322.211.

In the crude variable X which is *essentially positive*, a distribution having a wide region of combinations of shape parameters $\sqrt{\beta_1}$ and β_2 (or α_3 and α_4) is given by

$$F(X) = 1 - \frac{1}{(1+X^c)^k} \tag{c}$$

$$X \geqslant 0 \qquad c, k > 0$$

From the relations between β_1, β_2 and the two parameters c, k, it is found that even with simple values such as $c = 5$, $k = 6$, one gets $\sqrt{\beta_1} = -0.013$ and $\beta_2 = 3.010$ suggesting an honest approximation to the Gaussian distribution for which $\beta_1 = 0$ and $\beta_2 = 3.0$. The mean and the standard deviation are respectively $\mu = 0.655$ and $\sigma_X = 0.161$, leading to a coefficient of variation of about 25%. A closer approximation to the standard Gaussian distribution $\varphi(u)$ occurs with the values

$$c = 4.874 \qquad k = 6.158$$

leading to

$$\mu = 0.645 \qquad \sigma_X = 0.192$$

$$\beta_1 = 0.000 \qquad \beta_2 = 3.000$$

In this case, the normal deviate is

$$u = \frac{X-0.645}{0.162} \tag{d}$$

and a better approximation is written in replacing (d) in (c) after solving for X

$$X = 0.645+0.162u. \tag{e}$$

One still better approximation $H(u)$ gives the corresponding values in the range $-3.6 \leqslant u \leqslant 3.6$. We give here an extract of Burr, 1967; table 1, p. 649.

TABLE 6.322.5A

u	$\varphi(u)$	Burr's $H(u)$
−3.6	0.00016	0.00016
−2.4	0.00820	0.00819
−1.2	0.11507	0.11509
0	0.50000	0.50000
+1.2	0.88493	0.88491
+2.4	0.99180	0.99181
+3.6	0.99984	0.99984

Burr's contribution is interesting, not only for simulation purposes but also for the *practical description* of the so-called Gaussian variability of essentially positive variates.

6.323. LOGISTIC LAW

6.323.1. Logistic law in three parameters

6.323.11. Logistic law in three parameters as the law of population growth

Historically the logistic law in three parameters was grafted on the exponential law of growth in two parameters, expounded in the following subsection. An interesting commentary is given by d'Arcy Wentworth Thompson (1942, pp. 142 *et seq.*).

6.323.111. Positive exponential law in two parameters. The Reverend Malthus (1798) proposed a model of exponential growth for a population in free expansion according to which the *absolute rate of growth* at time t is proportional to the value $N(t)$ already reached at this instant:

$$\frac{dN}{dt} = \varepsilon N(t) \qquad (t \geqslant 0), \tag{a}$$

where $N(t)$ represents an integer number of individuals, but is regarded as a continuous variable.

The constant ε of dimension T^{-1} can be isolated by writing the *relative rate of increase:*

$$\frac{1}{N}\,\frac{dN}{dt} = \frac{d\ln N}{dt} = \varepsilon, \tag{b}$$

or

$$d \ln N = \varepsilon\, dt. \tag{b'}$$

We integrate between $t = 0$ and t by adjusting the constant of integration to the initial population N_0. Thus

$$\Big[\ln N\Big]_0^t = \varepsilon \Big[dt\Big]_0^t,$$

i.e.

$$N(t) = N_0 e^{\varepsilon t}, \tag{c}$$

where $N(t) \to \infty$ as $t \to \infty$.

Such is the law of (positive) exponential growth, otherwise known as *Malthus' law.*

We give two examples.

(a) Kostitzin (1937, p. 51) has shown that a rodent couple generates a population which increases exponentially at the relative rate $\varepsilon = 0.4$ months^{-1}.

(b) In cellular physiology, Lison and Pasteels (1951), on the basis of histophotometric measurements, have shown that the synthesis of DNA during the segmentation of the sea urchin egg *(Paracentrotus lividus)* is exponential up to the stage of XXXII blastomeres.

For both types of mitotic figures observed, Martin (1951) shows that the relative growth rate ε from one stage to the next is practically the same:

mitotic divisions in metaphase: $\varepsilon_M = 0.0223$ min^{-1};
mitotic divisions in interkinesis: $\varepsilon_I = 0.0246$ min^{-1}.

From these results, the t test shows that there is no difference between the relative growth rates for the two types of division. An interesting result.

6.323.112. Introduction of a limitative term. Logistic law with three parameters. The means of subsistence do not however develop at the exponential rate, but at the slower pace of a linear increase, hence the certainty of disequilibrium according to Malthus. Doubtless impressed by the conclusions of Malthus, Quetelet (1835 and 1869, p. 433) sought for a model at saturation point by introducing a damping term inspired from Newton's law in mechanics:

$$\frac{dN}{dt} = \varepsilon N(t) - \varkappa\left(\frac{dN}{dt}\right)^2. \tag{a}$$

It was left to his compatriot Verhulst (1838, 1845 and 1847), then, independently, to Pearl and Reed (1920), to introduce the correct damping term, proportional to the square of the actual effective value of the population

$$\frac{dN}{dt} = \varepsilon N - \eta N^2$$

$$= \varepsilon N\left(1 - \frac{\eta}{\varepsilon} N\right), \tag{b}$$

and not to the square of the rate (velocity). This relation shows that $\frac{dN}{dt} \to 0$ as $\frac{\eta}{\varepsilon} N \to 1$, and so the introduction of the term in $-\eta N^2$ certainly realized the required finite limit:

$$N_\infty = \frac{\varepsilon}{\eta} = \omega. \tag{c}$$

Hence the equation of Verhulst–Pearl–Reed is written as

$$\frac{dN}{dt} = \varepsilon N\left(1 - \frac{N}{\omega}\right). \tag{d}$$

We integrate this differential equation by writing successively

$$\frac{\omega\, dN}{N(\omega - N)} = \varepsilon\, dt \tag{e}$$

$$\frac{[N + (\omega - N)]\, dN}{N(\omega - N)} = \varepsilon\, dt$$

$$\frac{dN}{\omega - N} + \frac{dN}{N} = \varepsilon\, dt \tag{f}$$

$$-d \ln (\omega - N) + d \ln N = \varepsilon\, dt.$$

We integrate the two members between 0 and t

$$\left[\ln \frac{N}{\omega - N}\right]_0^t = \varepsilon t,$$

and, inserting the initial condition $N(0) = N_0$, we get

$$\ln \frac{N(\omega - N)}{N_0(\omega - N_0)} = \varepsilon t. \tag{g}$$

Since, in the phenomenon of growth of a population, the variable t is essentially non-negative, and considering that N is a discrete variable (treated as continuous), we have $N_0 \geqslant 1$.

Taking the anti-logarithm, this becomes

$$\frac{N}{\omega - N} = \frac{N_0}{\omega - N_0} e^{\varepsilon t}; \tag{h}$$

thus the quotient of the population at instant t to the remaining population increases exponentially with time. We expand (h) in order to isolate N:

$$N = (\omega - N) \frac{N_0}{\omega - N_0} e^{\varepsilon t}$$

$$N\left(1 + \frac{N_0}{\omega - N_0} e^{\varepsilon t}\right) = \omega \frac{N_0}{\omega - N_0} e^{\varepsilon t}. \tag{i}$$

Hence, explicitly setting the argument t of N is $N(t)$,

$$N(t) = \frac{\omega}{1 + \frac{\omega - N_0}{N_0} e^{-\varepsilon t}}. \tag{j}$$

Such is the equation of the *logistic law* in the following two parameters present in the differential relation (b):

ε—purely exponential relative growth rate (in the absence of damping);
$\omega = \varepsilon/\eta = N_\infty$—asymptotic value of the population as $t \to \infty$.

As the damping coefficient η tends to zero, ω tends to ∞, and then *Malthus*' law is re-obtained (6.323.111c).

In fact a *third parameter* is added to ε and ω, involving adjustment to the initial value N_0. The (median) time $t_{\omega/2}$ at which half the asymptotic value ω is reached, is given by the relation

$$\frac{\omega}{2} = \frac{\omega}{1 + \frac{\omega - N_0}{N_0} e^{-\varepsilon t}}. \tag{k}$$

Hence

$$\frac{\omega - N_0}{N_0} e^{-\varepsilon t} = 1$$

and, finally,

$$t_{\omega/2} = \frac{1}{\varepsilon} \ln \frac{\omega - N_0}{N_0} = \frac{1}{\varepsilon} \ln \left(\frac{\omega}{N_0} - 1\right). \tag{k'}$$

Hence the three properties of the non-negative variable $t_{\omega/2}$:

(a) $t_{\omega/2}$ varies inversely as the relative growth rate ε in the absence of the damping factor;

(b) For fixed ε, $t_{\omega/2}$ is correspondingly longer for a greater distance between the initial and final effective population;

(c) $t_{\omega/2}$ exists if $N_0 < \omega/2$, $t_{\omega/2} = 0$ if $N_0 = \omega/2$, and $t_{\omega/2}$ does not exist if $N_0 > \omega/2$.

It is seen that N_0 contributes to fixing on a third parameter in the logistic describing the growth of populations or biological or economic phenomena. In order to be explicit in regard to the third parameter, we again take the relation (d) and show that $t_{\omega/2}$ corresponds to the point of inflection designated as t_i.

We re-write (d) and differentiate $\left(N' = \dfrac{dN}{dt}\right)$:

$$\frac{d^2N}{dt^2} = \varepsilon N'\left(1-\frac{N}{\omega}\right)-\varepsilon N\frac{N'}{\omega} = \varepsilon N'-\frac{2\varepsilon NN'}{\omega} = \varepsilon N'\left(1-\frac{2N}{\omega}\right). \tag{l}$$

The second derivative vanishes at point t, being a root of the equation

$$1-2\frac{N}{\omega} = 0, \tag{l'}$$

i.e. at the point $t_{\omega/2}$ where the population $N(t_{\omega/2})$ is seen to reach the level $\omega/2$.

The point of inflection t_i, or $t_{\omega/2}$, is written as a function of the initial value (N_0) and of the final value (ω):

$$t_i = t_{\omega/2} = \frac{1}{\varepsilon}\ln\left(\frac{\omega}{N_0}-1\right) = \frac{1}{\varepsilon}\ln\frac{\omega-N_0}{N_0}. \tag{m}$$

We isolate the argument of the logarithm

$$\frac{\omega-N_0}{N_0} = e^{\varepsilon t_i} \tag{m'}$$

and then set it in (j), thus obtaining

$$N(t) = \frac{\omega}{1+e^{-\varepsilon(t-t_i)}}. \tag{n}$$

This is another form of the *logistic law in three parameters* ε, ω and $t_i = t_{\omega/2}$. One immediately verifies that for $t = t_i$, we have

$$N(t_i) = \frac{\omega}{2}. \tag{o}$$

The parameter t_i has a biological significance for it marks the moment at which the damping factor enters manifestly into play.

The relation (d) may also be written so:

$$\frac{dN}{dt} = \frac{\varepsilon}{\omega}N(\omega-N) = \eta N(\omega-N). \tag{p}$$

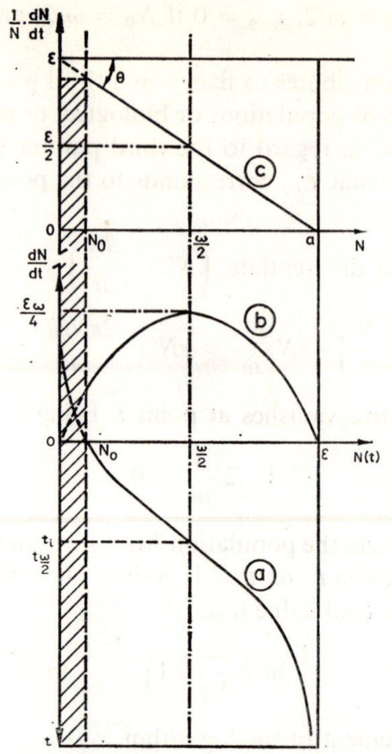

FIG. 6.323.112 I. The logistic law as a growth function and related rates.
(a) The logistic law:

$$N(t) = \frac{\omega}{1+e^{-\varepsilon(t-t_i)}} \qquad t \geqslant 0$$

$N(0) = N_0$
$t_i = t_{\omega/2}$ (inflectional time)
(b) Absolute growth rate in function N

$$\frac{dN(t)}{dt} = \eta N(\omega - N)$$

$\eta = \varepsilon/\omega$.
(c) Relative growth rate in function of N

$$\frac{1}{N}\frac{dN}{dt} = \eta(\omega - N) = \varepsilon - \eta N$$

$$\tan\theta = -\eta.$$

For each of these functions, the taboo zone is indicated by that within which the corresponding line is shown as interrupted.

In this form one recognises the rate of an autocatalytic monomolecular reaction which is proportional to the product of the concentration N already attained with the concentration $\omega - N$ still remaining to be attained. If the law is definitely logistic, this relation is symmetrical about the axis $\omega/2$ (see part b of Fig. 6.323.112 I).

Likewise, the logistic $N(t)$ which theoretically extends from zero to ∞ as t passes from $-\infty$ to $+\infty$, is theoretically symmetrical about the point of inflection $(t_i, \omega/2)$ (part a).

It should be borne in mind that the useful part of the logistic goes from N_0 to ω, i.e. in the range $0 \leqslant t < \infty$.

Finally, the relative growth rate $1/N\, dN/dt$ progressively decreases linearly

$$\frac{1}{N}\frac{dN}{dt} = \varepsilon - \eta N \tag{q}$$

from a theoretical value

$$\eta\omega = \varepsilon \quad \text{for} \quad t = -\infty \quad \text{and} \quad N = 0.$$

The *initial practical value* is written as

$$\left(\frac{1}{N}\frac{dN}{dt}\right)_{N=N_0}; \qquad t = 0; \qquad \eta(\omega - N_0) = \varepsilon - \eta N_0. \tag{r}$$

The slope $\tan\theta = -\eta$ of the line segment (q) measures the coefficient of growth inhibition, whilst the ordinate to the origin $N = 0$ measures the relative rate of increase.

Finally, from the relation (g), one can write

$$\ln\left(\frac{\omega}{N} - 1\right) = \ln\left(\frac{\omega}{N_0} - 1\right) - \varepsilon t, \tag{s}$$

where $\omega/N \geqslant 1$.

If, therefore, *the asymptotic term* ω *is known*, the slope of the line (s) furnishes a graphical estimate of ε. If ω is not known, the value of ω is guessed or estimated roughly on a plot. If the ω value is not correct, the points $\ln\left(\frac{\omega}{N_i} - 1\right)$, plotted as a function of t_i, do not lie disposed along a straight line.

Continuing by trial and rectification of error, it is possible to arrive at a linear graph to which the "correct" value of ω corresponds and in which the slope measures ε.

6.323.113. Examples. Verhulst gave as an example the case of the French and Belgian populations. Pearl and Reed (1920) applied the logistic law in three parameters to the population growth of the U.S.A. from 1790 to 1910:

$$\omega = 197{,}273{,}000, \quad \varepsilon = 0.33134, \quad \eta = 1.6\times10^{-10}.$$

This particular example, taken up by Lotka (1925 and 1956, p. 66), is discussed by Kostitzin (1937, pp. 53–56). Pearl (1930, p. 420) studied in detail the adjustment of a logistic to the population growth of Sweden 1750–1920.

In the works of the authors cited, other examples will be found, such as growth of *B, dendroides* and yeast cultures; experimental verification of the validity of applying the logistic to the growth of *Paramecia* by Gause (1935); growth of a population of *Drosophiles*, and growth of sunflower seedlings. In laboratory animals, the growth of the rat does not follow the classical logistic law after the age of 100 days, as noted also by Lotka (1956, p. 73) according to the data of Donaldson and Robertson.

On the other hand, Brody (1945, p. 550) produced a graph showing a good adjustment of rat growth with a cubic logistic (see section 6.323.4) between the age of 10 days and 1 year. For the sake of pharmacologists, however, we would mention that this author does not use the logistic law, but (p. 543) describes the pre- and post-inflectional growth by means of two arcs of different curvature, a method which he applies with success to numerous examples. It would be interesting to have a schedule of growth curves for laboratory animals, and to set them in relation to different reference curves, such as the logistic curve and the curves of Brody.

In tests on sub-acute and chronic toxicity, experimenters would appreciate being able to show any toxic effect on growth in terms of estimations of parameters for the control and treated groups.

6.323.114. Some references on the adjustment of a logistic law in three parameters. The method of adjustment is not under consideration here. In this regard the reader is referred to Pearl (1930) and Lotka (1925, 1956). The methods of adjustment of a logistic law to the data on the population growth of the U.S.A. from 1800 to 1910 have been the subject of a remarkable critical study by Nair (1954).

The cited authors consider the adjustment from the analytical point of view.

A statistical technique is indicated by Finney (1952b, pp. 183 and 253) (see section 6.323.3) and finally, a computer method of adjustment is proposed by Nelder (1962).

Bliss (II, 1970, p. 178) refers to other authors "for a definitive analysis of the three parameter logistic curve". He presents a method of successive approximations with the introduction of an auxiliary variate as has been used by Stevens (1951) in the fitting of an asymptotic regression.

6.323.115. Significance of the logistic law. According to the biologist and biometrician Teissier (1928), the "mathematician Verhulst" proposed a formal theory of growth based on considerations of simple logic.

From 6.323.112 (d), it was thus considered that

$$\frac{1}{N}\frac{dN}{dt} = \varepsilon\left(1-\frac{N}{\omega}\right). \tag{a}$$

Therefore the relative rate of growth is a decreasing linear function of the value already reached by the population or the mass of living matter. Adopting the same standpoint as Verhulst (see Pearl, 1930, p. 417), Pearl and Reed (1920) "independently derived the logistic curve, as an empirical curve to meet certain postulates for a curve to describe the growth of a population". With reference to a *chemical theory* leading formally to the same result as Verhulst and Pearl and Reed, (see Monod, 1942, p. 120), Robertson proceeded further to consider that "the processes of growth are dependent upon an autocatalytic reaction which governs them entirely. Hence the reason why growth is expressed by a formula which is also that for autocatalytic monomolecular reactions":

$$\frac{dx}{dt} = Kx(a-x). \tag{b}$$

This point of view is criticized by Teissier (1936) and Monod (1942, pp. 122–3) who put forward a *physiological theory* involving the concentration of available food. This reasoning leads to a relation of type (a) where the bracket contains an exponential function of N instead of a linear function.

In addition to these important contributions of the French School, a reference should be made also to the contribution of Reed and Berkson (1929) who report the application of the method of logits (see 6.323.22) to biological tests by Berkson. Berkson's point of view (1951) is codified in an article with the provocative title "Why I prefer logits to probits".

Kostitzin (1940) studied how the form of the solution $N(t)$ is influenced by the vital coefficients ε and η which vary with time or, in one special case, are periodic variables. The author also considers the influence of the population density.

Finally, we cite the important critical contribution of Feller (1940) concerning the empirical verification of the validity of the logistic law in three parameters for describing the phenomenon of growth.

This author took two examples from the published literature, i.e. surface growth of *B. dendroides* and height growth of sunflower seedlings, mentioned in 6.323.113.

He considered the adjustment of the classical logistic law to these data and compared it with two other curves in three parameters:

(1) $y(t) = a\frac{2}{\pi}\arctan 10^{b(t-c)}$, (c)

(2) a probabilistic form with slightly different notations from Feller:

$$z(t) = \frac{a}{2}\{1+\Theta[\sqrt{2}b(t-c)]\}, \qquad \text{(d)}$$

where

$$\Theta(u) = \frac{2}{\sqrt{2\pi}}\int_0^u e^{-u^2/2}\,du. \qquad \text{(d')}$$

These two curves produce results which in the main are equivalent to those furnished for the logistic curve, but with an advantage for the probabilistic form. Since the same data can reasonably be reconciled with the three types of curves, the biological implications of the logistic model are seriously weakened. This point of view is taken up with more vigor by Feller (1966, II, p. 52).

6.323.2. Logistic law in two parameters in the analysis of tests in all-or-nothing (Berkson)

6.323.21. Logistic law in two parameters

In the case of all-or-none responses the logistic law in two parameters is written as

$$p = 1-q = \frac{1}{1+e^{-(\alpha+\beta X)}}, \qquad \text{(a)}$$

α is a pure number and β has the dimension $[X]^{-1}$.

From the observations $(X_i, r_i/n_i)$, i.e. proportions r_i over n_i subjects exposed to the dose X_i, it is required to estimate the parameters α and β of this functional relation which relates p to X (J. Berkson, 1944, 1953).

In expression (a) the term p corresponds to the distribution $F(X)$ in the domain X extending from $-\infty$ to $+\infty$, with the relations

$$F(-\infty) = 0 \qquad F(+\infty) = 1.$$

Furthermore,

$$X_p = F^{-1}(p) \qquad \text{(b)}$$

is the quantile X corresponding to the total probability p of the all-or-none response.

In particular, the median point $X_{0.5}$ corresponds to the relation

$$\frac{1}{2} = \frac{1}{1+e^{-(\alpha+\beta X)}}. \tag{c}$$

Hence, at this point,

$$\alpha+\beta X_{0.5} = 0 \tag{d}$$

and

$$X_{0.5} = -\alpha/\beta. \tag{d'}$$

The argument of the exponential in (a) becomes

$$\alpha+\beta X = \beta\left(\frac{\alpha}{\beta}+X\right) = \beta(X-X_{0.5}) \tag{e}$$

and we have

$$p = \frac{1}{1+\exp\{-\beta(X-X_{0.5})\}}. \tag{f}$$

We introduce the *absolute deviate from the median*

$$x = X-X_{0.5}, \tag{g}$$

then the *reduced logistic deviate*

$$l = \beta x = \beta(X-X_{0.5}), \tag{h}$$

which is a pure, dimensionless number and will be considered in section 6.323.221.

Upon considering the definition of the *standard Gaussian deviate*

$$u = \frac{X-u}{\sigma}, \tag{i}$$

it is seen that in the expression (h) the parameter β plays the rôle of $1/\sigma$ in the Gauss curve; hence $1/\beta$ can be interpreted as a measure of dispersion for the logistic curve, a comment already made by Yule (1925).

The logistic probability density in X is written as

$$f(X) = \frac{dF(X)}{dX} = \frac{\beta e^{-(\alpha+\beta X)}}{[1+e^{-(\alpha+\beta X)}]^2}. \tag{j}$$

Hence the slope at the point of inflection of the logistic curve, which is also the median point (see section 6.323.112 (l') and (o)), is given by

$$F'(X_{0.5}) = f(X_{0.5}) = \beta/4, \tag{k}$$

since at this point $\alpha+\beta X_{0.5} = 0$.

As for the Gaussian (section 6.322.121) and Laplacian (6.321.12) sigmoid, the slope at the point of inflection of the logistic sigmoid is proportional to the inverse of the parameter of dispersion.

The logistic probability density is symmetrical about the median, as shown for the reduced density in 6.323.221 (c′).

Therefore the mean of X is equal to the median

$$E[X] = X_{0.5} = -\alpha/\beta. \tag{l}$$

This result is confirmed from consideration of the moment generating function (section 6.323.24).

In addition, one calculates the variance

$$V(X) = \frac{\pi^2}{3\beta^2}, \tag{m}$$

which depends only on β.

The logistic standard deviation is inversely proportional to

$$\sigma_X = \frac{\pi}{\beta\sqrt{3}}. \tag{m'}$$

Inversion of the logistic law in two parameters. We explicitly give relation (b) from (a). One has successively in the quantile X_p

$$\frac{1}{p} - 1 = e^{-(\alpha+\beta X_p)}$$

$$\ln\frac{q}{p} = -(\alpha+\beta X)$$

$$X_p = \frac{\ln(p/q) - \alpha}{\beta}. \tag{n}$$

In accordance with the relation 6.323.222 (c) in the foregoing sub-section, the quantile X_p as a linear function of the logit l_p and of the parameters α and β which fix the logistic law, is

$$X_p = \frac{l_p - \alpha}{\beta}. \tag{o}$$

6.323.22. The reduced logistic law in l

6.323.221. Logistic law in reduced deviate l. Into the logistic distribution in the original variable X

$$p = \frac{1}{1+e^{-(\alpha+\beta X)}},$$

we introduce the reduced deviate

$$l = \alpha+\beta X = \beta(X-X_{0.5}) \tag{a}$$

which is defined in 6.323.21 (h).

One thus obtains the reduced logistic distribution

$$p = F(l) = \frac{1}{1+e^{-l}} \tag{b}$$

where no parameter appears and whose range of l is from $-\infty$ to $-\infty$.

The median of l is equal to zero.

The reduced logistic probability density is

$$f(l) = \frac{dF(l)}{dl} = \frac{e^{-l}}{(1+e^{-l})^2} = F(l)[1-F(l)]. \tag{c}$$

The ordinate to the origin

$$f(0) = 0.25$$

is less than that for the standard Gaussian curve (0.4).

The reduced logistic density is symmetrical about the origin. It is verified without difficulty that

$$f(-l) = f(l). \tag{c'}$$

The moment generating function (section 6.323.23) shows that the expected value of l is zero, i.e.

$$E[l] = 0 \tag{d}$$

and the variance is

$$V(l) = \frac{\pi^2}{3}. \tag{e}$$

Therefore

$$\sigma_l = \frac{\pi}{\sqrt{3}}. \tag{e'}$$

The shape parameters of the reduced logistic density are (Moore, 1951):

$$\begin{aligned} \gamma_1 &= 0 \quad \text{(symmetry)}, \\ \gamma_2 &= 1.2 \quad \text{(lepto-kurtosis)}. \end{aligned} \tag{f}$$

6.323.222. *Concept of logit.* The logistic distribution in the reduced variable, written in the quantile l_p

$$p = \frac{1}{1+e^{-l_p}}, \tag{a}$$

can be inverted without difficulty; one thus obtains l_p as a simple explicit function of p, which is one of the advantages of the logistic law.

Successively we have

$$1+e^{-l_p} = \frac{1}{p} \tag{b}$$

and

$$e^{-l_p} = \frac{1-p}{p} = \frac{p}{q}. \tag{c}$$

Hence

$$l_p = -\ln\frac{q}{p} = \ln\frac{p}{q}, \tag{d}$$

or, as a function of p alone,

$$l_p = \ln\frac{p}{1-p}. \tag{e}$$

According to Berkson (1944 and 1949), the function of p thus defined is the *logit of p* or, the reduced logistic deviate corresponding to the total probability of an all-or-nothing response of biological material up to the dose X_p.

Rectification (Emmens, 1940, p. 221) by the *methods of logits* (Berkson, 1944). If the response to the dose X is definitely logistic, the consideration of formula 6.323.21 (o) shows that one must have a linear relation between l_p and X_p, i.e.

$$l_p = \alpha+\beta X_p. \tag{f}$$

Logit ruling. Such is the basis of the arithmetic logit ruling† of Berkson (1951), viz. abscissae: X on an arithmetical scale, and ordinates: p corresponding to X in logit scale $p = \frac{100}{1+e^{-l}}\%$.

Logit Tables. Tables of Berkson (1953) reproduced in *Geigy Tables* (1963, p. 56) (French edition). It should be noted that in the latter tables the numbers printed in italics in the body of the logit table are negative in sign.

Examples: *left-hand part*, entry with p ($p = 0.5$, $1-p = 0.5$); the logit l in the body of the table is read as

$$l_{0.5} = \ln\frac{0.5}{0.5} = \ln 1 = 0.0000;$$

right-hand part, entry with l; reading of p or of the antilogit of l_p.

† No. 32.450, Codex Book Co. Inc., Norwood, Mass., U.S.A.

We look for p corresponding to $l_p = -1$:

$$\ln \frac{p}{1-p} = -1; \qquad \text{therefore} \qquad \frac{p}{1-p} = e^{-1} = 0.368,$$

$$p = 0.368(1-p) \qquad 1.368p = 0.368$$

$$p = \frac{0.368}{1.368} = 0.2690.$$

The table gives 0.26894.

In general the antilogit p is obtained from l_p using the inverted formula (e):

$$\frac{p}{1-p} = e^{l_p}, \tag{g}$$

which is no other than the original logistic written in another form

$$p = \frac{e^{l_p}}{1+e^{l_p}}. \tag{h}$$

Table XI of Fisher–Yates (1957, p. 72). This table bears the title "The Logit or r, z Transformation".

The Fisher–Yates logit transformation

$$z = \frac{1}{2} \ln \frac{p}{q} \tag{i}$$

is equivalent to the transformation of the observed coefficient of correlation r into z with $r = 2p-1$.

For $p < 0.5$, the logits are negative, but equal in absolute magnitude to the tabular value for $1-p$:

$$z_p = -z_{1-p} \qquad (p < 0.5).$$

The logits tabulated in Fisher–Yates are exactly half the original logits of Berkson. This derives from the fact that for Fisher–Yates the point of departure is the classical transformation $r \rightarrow z$ defined by

$$z = \frac{1}{2} \ln \frac{1+r}{1-r}, \tag{j}$$

and tabulated in Fisher–Yates (1957, p. 59).

We make $r = 2p-1$; we get

$$z = \frac{1}{2}\ln\frac{1+2p-1}{1-2p+1}$$

$$= \frac{1}{2}\ln\frac{2p}{2(1-p)}$$

$$= \frac{1}{2}\ln\frac{p}{1-p}. \qquad \text{(k)}$$

The relation (i) is thus obtained again.

In the notation of Fisher–Yates, the antilogit is written from the relation (i) as

$$\frac{p}{1-p} = e^{2z}, \qquad \text{(l)}$$

or

$$p = \frac{e^{2z}}{1+e^{2z}}, \qquad \text{(m)}$$

or even

$$p = \frac{1}{1+e^{-2z}}. \qquad \text{(n)}$$

The function e^{2z} may in the name of $F_{\nu 1,\ \nu 2}$ play an important rôle in the comparison of two independent estimates of the same variance (tables V on the right-hand side of Fisher–Yates, 1957, where $n_1 \rightarrow \nu_1$ and $n_2 \rightarrow \nu_2$).

If one fixes on the notation, there is no risk of confusion between the logit of Berkson (1944)

$$l_p = \ln\frac{p}{1-p}. \qquad \text{(o)}$$

derived from the reduced logistic law, and the logit (*h*) of Fisher–Yates

$$z_p = \frac{1}{2}\ln\frac{p}{1-p}$$

associated with the distribution z, or the transformation r into z.

In any case we have the clear relationship

$$l_p = 2z_p. \qquad \text{(o')}$$

It would be a convenience if authors specified the table used by them in calculating the logits from percentages of observed reactions.

Re-definition of logit (Anscombe). If r reactors out of n exposed are observed, the logit of Berkson in (o) is estimated by

$$l = \ln \frac{r}{n-r}. \tag{p}$$

In order to remove the bias implied by this formula in the estimation of $\alpha+\beta X$ (see (f)), Anscombe (1956) re-defined the logit as follows:

$$l_A = \ln \frac{r+\frac{1}{2}}{n-r+\frac{1}{2}}, \tag{q}$$

which we shall designate as l_A in order to distinguish it from the logit defined in (p).

Anscombe shows also that

$$E[l_A] = \alpha+\beta X+0(n^{-2}),$$

$$V(l_A) = \frac{1}{npq} \tag{r}$$

$$\gamma_1(l_A) \approx \frac{2(p-q)}{\sqrt{npq}}.$$

The skewness of the distribution of l_A is opposite in sign to that of the binomial $(q+p)^n$ and is double in value. It is of interest to note that this consideration concerns the classical adjustment by maximum likelihood (Finney 1947, 1952b and 1964) or the minimum logit X^2 of Berkson (1953).

6.323.223. *Another form of the reduced logistic density*. In formula (c) above for the logistic probability density in its classical form

$$f(l) = \frac{e^{-l}}{(1+e^{-l})^2}, \tag{a}$$

we reduce the numerator to obtain

$$f(l) = \frac{1}{[e^{l/2}(1+e^{-l})]^2} \tag{b}$$

and

$$f(l) = \frac{1}{(e^{l/2}+e^{-l/2})^2}. \tag{c}$$

This form is reminiscent of the structure of the hyperbolic cosine and of the hyperbolic secant (Defares and Sneddon, 1960, p. 234).

For an argument m, we have in fact

$$\operatorname{sech} m = \frac{1}{\cosh m} = \frac{2}{e^m+e^{-m}}. \tag{d}$$

Hence the logistic probability density is thus written:

$$f(l) = \tfrac{1}{4}\,\text{sech}^2\, l/2. \tag{e}$$

Vessereau (1965) uses this form in the study of organoleptic characteristics.

Continuing in the hyperbolic state, one integrates $f(l)$ to re-find $F(l)$:

$$F(l) = \int_{-\infty}^{l} f(\text{l})\, d\text{l} = \frac{1}{2}\int_{-\infty}^{l} \text{sech}^2 \frac{1}{2}\, d\frac{1}{2} = \frac{1}{2}\left[\tanh \frac{l}{2}\right]_{-\infty}^{l}$$

$$= \frac{1}{2}\left[\tanh \frac{l}{2} - (-1)\right] = \frac{1}{2}\left(1 + \tanh \frac{l}{2}\right). \tag{f}$$

This form will be used in section 6.323.3.

6.323.224. Logistic law in the standard logistic deviate. The variable

$$l = \alpha + \beta X = \beta(X - X_{0.5}) \tag{a}$$

introduced in 6.323.21 (e) and (h), is a reduced deviate without dimension, but, since β is not the inverse of σ_X, l is not a standard deviate. In reality the standard deviate of the logistic variable is expressed as a function of β in relation to 6.323.21 (m'):

$$\sigma_X = \frac{\pi}{\beta\sqrt{3}}, \tag{b}$$

hence (see also Cochran and Davis, 1963, p. 187)

$$\beta = \frac{\pi}{\sigma_X \sqrt{3}}. \tag{b'}$$

The reduced deviate l becomes

$$l = \frac{\pi}{\sqrt{3}}\,\frac{X - X_{0.5}}{\sigma_X}, \tag{c}$$

or also

$$l = \frac{\pi}{\sqrt{3}}\lambda \tag{c'}$$

by introducing the standard logistic deviate

$$\lambda = \frac{X - X_{0.5}}{\sigma_X}. \tag{d}$$

We set ϕ_V (where V is the first letter of Verhulst) for the logistic probability density in the logistic standard deviate λ.

The law of conservation of the local probabilities in the transformation (c′) is written as

$$f(l)\,dl = \phi_V(\lambda)\,d\lambda. \qquad \text{(e)}$$

Hence

$$\phi_V(\lambda) = f(l)\cdot\frac{dl}{d\lambda} = \frac{\pi}{\sqrt{3}}f\left(\frac{\pi}{\sqrt{3}}\lambda\right). \qquad \text{(f)}$$

In particular, the ordinate to the origin is written as

$$\phi_V(0) = \frac{\pi}{\sqrt{3}}f(0) = \frac{3.1416}{1.732}\times 0.25 = 0.4535. \qquad \text{(g)}$$

One thus obtains a value ten percent greater than $\varphi(0) = 0.3989$ for the Gauss curve in the standard deviate u.

The logistic standard deviate λ corresponds to the N.E.D. of Gaddum (1933) for the Gaussian or normal law. Similarly one might refer to λ as the logistic equivalent deviate L.E.D. and

$$Y_{LO,\,P} = 5+\lambda \qquad \text{(h)}$$

as the logistic probability unit. We insist on the fact that the logit in common use is not the logistic equivalent deviate.

6.323.225. Use of the logit transformation. From its very definition, the logit transformation of observed proportions of r_i reactors on n_i exposed is used to get a linearization of the response on the variable dose or log dose in the *univariate case*. But this is far from being the unique situation. The analysis of the dependence of all-or-none responses (such as coronary heart disease coded 0 if absent and 1 if present) on *several independent variables* is of utmost importance in upper grade epidemiology. After Cox (1966), Walker and Duncan (1967), working on the results of the Framingham Study (Dawber, Meadors and Moore, 1951), have used the logit transformation of observed percentages and get a workable linearization of the all-or-none data. From this it is possible to use the techniques of multiple regression, but also extensions such as analysis of covariance.

6.323.23. Logistic imminence rate or hazard function

The conditional probability of meeting with a random logistic reduced deviate $\boldsymbol{l}$ within the interval $(l, l+dl)$, given that $\boldsymbol{l} \geqslant l$ (see 6.122) is written as

$$dJ(l) = I(l)\,dl, \qquad \text{(a)}$$

where

$$dJ(l) = \frac{f(l)\,dl}{1-F(l)}. \tag{b}$$

Following the relations 6.323.221 (b) and (c), the imminence rate corresponds to the formula

$$I(l) = \frac{e^{-l}}{(1+e^{-l})^2} \cdot \frac{(1+e^{-l})}{e^{-l}} = \frac{1}{1+e^{-l}} = F(l). \tag{c}$$

The logistic hazard function therefore reproduces the distribution of l, i.e. a monotonously increasing function of l.

Clearly, the conditional probability density of seeing l in a (prospective) interval $(l, l+dl)$ is no other than the unconditional probability of l being incident in the (retrospective) interval $(-\infty, l)$.

Gumbel (1960, p. 20) adopted actuarial notation and put $\mu(l)$ for the hazard or intensity function which we write as $I(l)$ and refer to as the imminence rate (Martin, 1961a).

In Gumbel (1960, p. 22) this function is represented in coordinates $\log_{10} I(l)$ as a function of l.

6.323.24. *Moment generating function associated with the logistic law*

Irwin (1950), Armitage and Irene Allen (1950), Gumbel (1960, pp. 126–7) aremarked that the logistic probability density $f(l)$ is expressible simply as function of the distribution $F(l)$.

In the reduced deviate l, we have

$$F(l) = \frac{1}{1+e^{-l}}; \qquad 1-F(l) = \frac{e^{-l}}{1+e^{-l}}, \tag{a}$$

$$f(l) = \frac{e^{-l}}{(1+e^{-l})^2} = \frac{1}{1+e^{-l}} \cdot \frac{e^{-l}}{1+e^{-l}} = F(l)\,[1-F(l)]. \tag{b}$$

Furthermore, after (a) and 6.323.222 (g) where $F \equiv p$,

$$e^{l} = \frac{F}{1-F}. \tag{c}$$

The moment-generating-function of the reduced variable l is written as

$$M_l(t) = \int_{-\infty}^{+\infty} e^{lt} f(l)\,dl = \int_0^1 e^{lt}\,dF = \int_0^1 F^t(1-F)^{-t}\,dF. \tag{d}$$

In accordance with the definition of the beta function (6.315.11 c and d), one writes

$$M_l(t) = \beta(1+t,\,1-t) = \frac{\Gamma(1+t)\Gamma(1-t)}{\Gamma(t-t+1)} = \Gamma(1+t)\cdot\Gamma(1-t). \qquad \text{(d')}$$

The derivative of $M_l(t)$

$$M_l'(t) = \Gamma'(1+t)\Gamma(1-t) - \Gamma(1+t)\Gamma'(1-t),$$

taken at the point $t = 0$, will vanish and the result is again found to be

$$E[l] = 0.$$

The second derivative of $M_l(t)$ at $t = 0$ is

$$M_l''(0) = 2\Gamma''(1) - 2\Gamma'(1)^2.$$

The variance of l is written as (see Gumbel, 1958, p. 16)

$$V(l) = \frac{\pi^2}{3}. \qquad \text{(e)}$$

For the logistic in X, we have

$$l = \alpha + \beta X,$$
$$V(l) = \beta^2 V(X). \qquad \text{(f)}$$

Hence

$$V(X) = \frac{\pi^2}{3\beta^2}, \qquad \text{(g)}$$

$$\sigma_X = \frac{\pi}{\beta\sqrt{3}}, \qquad \text{(g')}$$

and the result is found to be the same as in 6.323.21 (m'); in the logistic law in two parameters the parameter of dispersion is inversely proportional to the slope β of the density in coordinates logit–X.

6.323.25. Comparison of symmetrical probability laws

Biologists are interested in the comparison of symmetrical probability laws, one or the other of which could adequately and simply describe the variability of the phenomenon which they utilize. These two *criteria of adequacy and simplicity* dominate the operational application of the laws of probability to biological phenomena.

The comparison is made by Martin (1971) between the probability densities and between the distributions. In Fig. 6.323.25 I we give the comparison between the standard Gaussian (normal) law in u and the standard logistic law in λ, both being considered from three aspects: probability density, distribution, imminence rate. The parameters of each law are as follows:

	Mean	σ	γ_1	γ_2	Representation	
Gaussian (normal)	0	1	0	0	------	(a)
Logistic	0	1	0	1.2	———	

If the standard Gaussian is mesokurtic, the standard logistic is leptokurtic, which character can be seen in Fig. 6.323.25 I (a).

The relevant equations are

$$------ \quad \varphi(u) = \frac{1}{\sqrt{2\pi}} e^{-1/2u^2}, \quad -\infty < u < +\infty \tag{b}$$

$$——— \quad \varphi_V(\lambda) = \frac{\pi}{\sqrt{3}} \frac{e^{-\frac{\pi}{\sqrt{3}}\lambda}}{\left(1+e^{-\frac{\pi}{\sqrt{3}}\lambda}\right)^2} \quad -\infty < \lambda < +\infty. \tag{b'}$$

In Fig. 6.323.25 I (b) are shown the standard distributions

$$------ \quad \Phi(u) = \frac{1}{\sqrt{2\pi}} \int_{-\infty}^{u} e^{-1/2u^2} du, \tag{c}$$

$$——— \quad \Phi_V(\lambda) = \frac{1}{1+e^{-\frac{\pi}{\sqrt{3}}\lambda}}. \tag{c'}$$

The inflexional tangents of respective slopes $\varphi(0)$ and $\varphi_V(0)$ are shown.

In Fig. 6.323.25 I (c) are given the two imminence rates in standard deviates.

$$------ \quad I(u) = \frac{\varphi(u)}{1-\Phi(u)}, \tag{d}$$

$$——— \quad I(\lambda) = \frac{\pi}{\sqrt{3}} \Phi_V(\lambda) = 1.8138\, \Phi_V(\lambda). \tag{d'}$$

For the standard logistic, the imminence rate of λ is equal to the distribution of λ multiplied by $\pi/\sqrt{3}$, i.e. the standard deviation of the logistic reduced deviate (6.323.221 (e')). Roughly speaking, the densities and probabilities show about the same *general* (not local) trend in the range of the standard deviates. But the imminence rates are fundamentally different for positive values of u and λ; they diverge definitely for values greater than 1. This difference in behavior may serve in discriminating between two important and widely used laws for describing the variability of a biological response.

6.323.3. Logistic law with three parameters in biological tests on graded response

Let $U(X)$ be the continuous response of a biological material to the dose X. H is the *asymptote of the response*. Formally and if H is known, we have the following homology between $N(t)/\omega$ (6.323.21 a) and the quotient

$$p(X) = \frac{U(X)}{H}, \qquad \text{(a)}$$

where $p(X)$ is the fraction of the maximal response achieved for the dose X.

From formula 6.323.21 (a) and (a) above, the continuous response $U(X)$ obeys the logistic law in three parameters:

$$U(X) = \frac{H}{1+e^{-(\alpha+\beta X)}}. \qquad \text{(a')}$$

This function was introduced by Emmens (1940) in the study of the response–dose relation in biological tests on hypophyseal extracts of serum, or urine of pregnant women and gravid mares. The quantitative response is the weight increase of the ovaries of immature female rats.

The asymptote H (or the L of Emmens) is determined "from points which clearly mark the ceiling of the curve" (p. 200).

The method has been applied to thyrotropine tests based on the weight of the thyroid and the response to prolactin of the gullet of pigeons. Emmens (1940, p. 213) used an approximate test of quality of the adjustment with k doses, comparing the residue of adjustment of the logistic with $(k-2)$ degrees of freedom in the sampling variance; the latter is obtained from similar experiments and varies according to the preparation being studied.

We will raise three further interesting points in regard to this pioneer work:

(1) the concept of a unique "general logistic", the abscissae of which are in the standardized variable and the ordinates in $p(X)$ (see section 6.323.221); (see also Jacquet's method in 6.322.225.5 and Daniel's general purpose half normal grid in 6.322.227).

(2) the comparison of logistic adjustment with adjustment by a linear function of the log-dose;

(3) the concept of logit (although not by name) and of logistic weight in adjustment by least squares.

The residue decreases in magnitude with improving estimation of the asymptote.

We can conclude with Emmens "It seems probable, therefore, that a great number of dose–response curves are in fact logistic in type, in particular when the response is the weight of an organ or tissue, or a simple function of it" (1940, p. 216).

In their study of the protective action of the antihistamines in anaphylactic micro-shock in guinea-pigs, Armitage *et al.* (1952) used a estimate of the upper limit of the percentage of protection afforded by the various drugs, and then adjusted the logistic curve by relating the % protection to the logarithm of the dose of antihistamine.

In the present discourse we place the accent on the response as a function of the metameric variable X; the response in respect of the original variable D will be considered below in treating the log-logistic.

According to Hotelling (1927), the logistic of response as a function of X may be written in a different form (see 6.323.223 f) as

$$U(X) = \tfrac{1}{2}H[1+\tanh \tfrac{1}{2}(\alpha+\beta X)], \qquad \text{(b)}$$

where $X = \log D$,

$H =$ asymptote,

and

$$\tanh \phi = \frac{\sin \phi}{\cosh \phi} = \frac{e^{\phi}-e^{-\phi}}{e^{\phi}+e^{-\phi}} = \frac{e^{2\phi}-1}{e^{2\phi}+1}. \qquad \text{(b')}$$

For the purpose of pharmacologists, we will use the notations of Finney (1964, p. 69) introducing, however, the factor $\frac{1}{2}$ in the argument of the hyperbolic tangent (see 6.323.222 o').

The relation (b) is certainly a logistic in three parameters, namely, α, β and H. In effect: we expand the second term of (b) so

$$1+\tanh \phi = 1+\frac{e^{2\phi}-1}{e^{2\phi}+1} = \frac{2}{1+e^{-2\phi}}. \qquad \text{(c)}$$

Taking (c) into (b), we have

$$U = \frac{H}{2}\cdot\frac{2}{1+e^{-2\phi}} = \frac{H}{1+e^{-2\phi}}. \qquad \text{(d)}$$

The response U follows a logistic of asymptote H, where ϕ is associated with X by the linear relation

$$\phi = \tfrac{1}{2}(\alpha+\beta X). \qquad \text{(e)}$$

If the asymptote H is known, then (b) and (c) are written as

$$\frac{U}{H} = \frac{1}{2}\left[1+\tanh \frac{1}{2}(\alpha+\beta X)\right], \qquad \text{(f)}$$

We thus return to the case of a logistic in two parameters where the response U/H is expressed as a fraction of the asymptote.

In this case, we may write

$$2\frac{U}{H}-1 = \tanh\frac{1}{2}(\alpha+\beta X) \tag{g}$$

and

$$\alpha+\beta X = 2\tanh^{-1}\left(\frac{2U}{H}-1\right). \tag{h}$$

If therefore the response U/H is replaced by the transform (metamer)

$$Y = 2\tanh^{-1}\left(\frac{2U}{H}-1\right), \tag{i}$$

the experimental points at the (X_i, Y_i) will lie along a straight line, the parameters α and β of which may be estimated by the method of maximum likelihood already used for the probits.

If, however, the asymptote H is not known, the same method can still be used by calculating the correstions δa, δb and δH in successive cycles of iteration (Finney, 1952b, pp. 183 and 253). After having *in extenso* treated in probits an example of repellent on bees (p. 188), this author points out that the results would not be very different if the analysis were made in logits.

Such calculations are slow, but now one may be happy that electronic computers will facilitate the task of the experimenter.

A method which does not seem to be as prevalent amongst biologists as it deserves to be, is that due to Stevens (1951) which derives from the exponential in three parameters of the Mitscherlich type (asymptotic regression):

$$U = \alpha+\beta\varrho^{x} \qquad (0<\varrho<1), \tag{j}$$

or, further,

$$U = \alpha+\beta e^{-\gamma x}, \tag{j'}$$

where $\varrho = e^{-\gamma}$.

Stevens comments (p. 249) that the logistic in three parameters is no other than the inverse of (j),

$$Z = \frac{1}{\alpha+\beta\varrho^{x}} = \frac{1}{\alpha+\beta e^{-\gamma x}}. \tag{k}$$

If, therefore, one plots the points $(x_i, 1/z_i = y_i)$ graphically, the estimation of the parameters α, β, ϱ or γ of the asymptotic regression is again obtained.

This law is playing an ever increasing role in biological analysis. The present exposition shows only the methods of adjustment; the reader may appreciate a reference to other relevant work (Hartley, 1948); Pimentel-Gomez, 1953; Patterson, 1953 and 1956; Schneider, 1963).

Remark: Turner, Monroe and Lucas (1951) have introduced *generalized asymptotic regressions* and considered (pp. 126–7) a *family of "extinction curves"* including the following cases: straight line, parabola, negative exponential, inverse square law, rectangular hyperbola and also the inverse square root law.

6.323.4. Generalized logistics

The linear regression of the logit on $\alpha+\beta X$, i.e.

$$l = \alpha+\beta X, \tag{a}$$

can evidently be extended to the parabolic case

$$l = \alpha+\beta X+\gamma X^2, \tag{b}$$

to which the quadratic logistic of unit (or exactly known) asymptote corresponds:

$$p = \frac{1}{1+e^{-(\alpha+\beta X+\gamma X^2)}}\,. \tag{c}$$

Some cubic logistics with fixed asymptote have been applied by Greenwood (1925) to the population growth in England and Wales.

Gallo (1950, pp. 130–1) used a quadratic logistic in the adjustment of a hemolytic test. The research worker will find a guide *fil d'Ariane* in the plot (X, logit p). If the trend is linear, one is content with two parameters. Otherwise, three parameters are necessary if the trend is parabolic, and four parameters would be necessary to take into account any tendency of the cubic type.

In the study of vegetal growth, Nelder (1961 and 1962) defined a generalized logistic in four parameters by the differential equation

$$\frac{dW}{dt} = \varkappa W\left[1-\left(\frac{W}{A}\right)^{1/\theta}\right] \qquad (\theta > 0), \tag{d}$$

where W represents a weight, and A is the asymptote.

After introduction of the constant of integration λ, the integral contains indeed four parameters

$$\frac{W}{A} = \frac{1}{\left[1+\exp\left\{-\frac{\lambda+\varkappa t}{\theta}\right\}\right]^{\theta}}. \quad \text{(d')}$$

Here the following relation in $W^{1/\theta}$ and $A^{1/\theta}$ corresponds to the relation (6.323.112 h) which holds good for the classical logistic in $N(t)$ with asymptote ω:

$$\frac{(W/A)^{1/\theta}}{1-(W/A)^{1/\theta}} = \exp\left\{\frac{\lambda+\varkappa t}{\theta}\right\}, \quad \text{(e)}$$

where $1/\theta$ is the power of the weight attained as a fraction of the asymptote A. If $\theta = 1$, the relation which holds for the ordinary logistic is re-found.

Having presented a good example of adjustment in his work of 1961, Nelder (1962, p. 616) devised a Mercury Autocode program to make the adjustment.

Finally, to complete the information, we would specially mention that von Bertalanffy (1951, pp. 278, 282–3; 1957) gave a differential equation

$$\frac{dW}{dt} = AW^{a}-BW^{b} \quad \text{(f)}$$

in four parameters, the integral of which contains five parameters when it is realizable explicitly.

6.323.5. Generalized logit-normal laws

Kapteyn and van Uven (1916), in reference to Ch. Henry (1906), showed the value of performing a simple logarithmic transformation in order to obtain a reduced Gaussian deviate. Then Johnson (1949) defined a more flexible system of logarithmic transformations involving several parameters. The system of Johnson curves is exposed in M.G. Kendall and Stuart (1963, I, pp. 171–2).

Mead (1965) took up one of the Johnson transformations with finite range ($\xi \leqslant x \leqslant \xi+\lambda$) in the form

$$u = \gamma+\delta \ln \frac{x-\xi}{\xi+\lambda-x} \quad \text{(a)}$$

where u is the standard normal deviate.

In particular, for $\xi = 0$ and $\xi+\lambda = A$, one obtains

$$u = \gamma+\delta \ln \frac{x}{A-x}, \qquad \text{(b)}$$

or, further,

$$u = \gamma+\delta \ln \frac{x/A}{1-x/A} = \gamma+\delta \operatorname{logit} \frac{x}{A}. \qquad \text{(c)}$$

The variable x obeys an ordinary logit-normal law. It is at this point that Mead uses the idea of Nelder (1961), cited above in section 6.323.4. The transformation is rendered more flexible by writing

$$u = \gamma+\delta \ln \frac{(x/A)^\theta}{1-(x/A)^\theta} = \gamma+\delta \operatorname{logit} (x/A)^\theta. \qquad \text{(d)}$$

If u is really a standard Gaussian deviate, the variable x is said by Mead to obey a *generalized logit-normal law*, the probability density of which is written as

$$g_1(x) = \frac{\theta}{\sigma\sqrt{2\pi}x[1-(x/A^\theta)]} \exp\left\{-\frac{[\operatorname{logit} (x/A)^\theta-u]^2}{2\sigma^2}\right\}. \qquad \text{(e)}$$

The author discusses the different types of curves with finite range A that arise for the values of σ^2, μ and θ. He also proposes a two-stage method of adjustment, i.e. begin from A and θ to adjust μ and σ for minimum x^2, then find the combination (A, θ) which minimizes this family of x^2 minima. These calculations are performed on a computer using a program in Mercury Autocode. Finally, four examples of adjustment with a constant A are given, one density here being negatively skew, and one being positively skew. We have seen that formula (e) represents a generalized *logit-normal law*, i.e. a Gaussian density whose argument is

$$u = \frac{\operatorname{logit} (x/A)^\theta-\mu}{\sigma}. \qquad \text{(f)}$$

Following a method analogous to that used by Johnson (1949) in the normal case, Gart (1963) has introduced a density we might call "*generalized Gaussian-logistic law*"

$$F(t) = \frac{1}{2}\left\{1+\tanh\frac{1}{2}\cdot\frac{\pi}{\sqrt{3}}u\right\} \qquad \text{(g)}$$

with

$$u = \frac{\Phi(t)-u}{\sigma} \qquad \text{(h)}$$

and where $\Phi(t)$ is a monotonic function of t such that

$$-\infty < \Phi(t) < +\infty, \quad \text{when} \quad 0 < t < \infty.$$

Guide lines for getting expression (b) are given in 6.323.3 formula (f) for the logistic distribution and in 6.323.224 formulae (a), (c) and (d) for the relations between the reduced (l) and the standard (λ) logistic deviate.

6.324. Cauchy's Law

6.324.1. Reduced Cauchy law

6.324.11. Cauchy probability density

The Cauchy probability density in the reduced variable x over the domain $(-\infty, +\infty)$ is written as

$$f(x) = \frac{1}{\pi} \cdot \frac{1}{1+x^2}. \tag{a}$$

This curve is symmetrical about $x = 0$ and is represented in Fig. 6.324.12 I, lower portion.

M. G. Kendall and Stuart (1963, I, p. 59) note that it is a question of *convention* to consider whether the mean exists or not. The mean, regarded as the *principal value of the integral*

$$\lim_{n \to \infty} \frac{1}{\pi} \int_{-n}^{+n} x \frac{dx}{1+x^2}, \tag{b}$$

exists and is equal to zero.

In this restricted sense, one may say that zero is a parameter of centering for the Cauchy density.

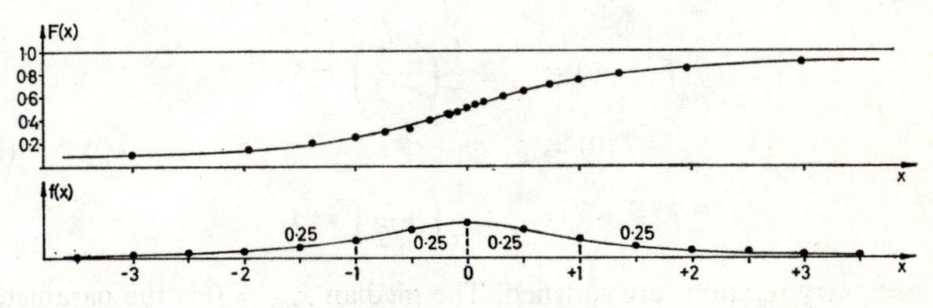

Fig. 6.324.12 I. Reduced Cauchy law. Probability density $f(x)$. Distribution $F(x)$.

In all rigor, only the moment of order less than 1 exists, i.e. μ'_0; the moments μ'_2, μ'_3 and μ'_4 do not exist at all. In particular, it is impossible to define the the variance of the Cauchy variable; whence there is no Cauchy standard deviate.

6.324.12. Cauchy reduced distribution

The probability of observing a value x equal at most to x fixed is written as

$$F(x) = \frac{1}{\pi} \int_{-\infty}^{x} \frac{dx}{1+x^2}$$

$$= \frac{1}{\pi} \int_{-\infty}^{0} \frac{dx}{1+x^2} + \frac{1}{\pi} \int_{0}^{x} \frac{dx}{1+x^2}$$

$$= \frac{1}{\pi}\Big[\arctan x\Big]_{-\infty}^{0} + \frac{1}{\pi}\Big[\arctan x\Big]_{0}^{x}$$

$$= \frac{1}{\pi}\left[0-\left(-\frac{\pi}{2}\right)\right] + \frac{1}{\pi}[\arctan x - 0].$$

Hence the Cauchy reduced distribution is

$$F(x) = \frac{1}{2} + \frac{1}{\pi} \arctan x$$

$$= \frac{1}{2} + \frac{1}{\pi} \tan^{-1} x \qquad \text{(a)}$$

$$(-\infty < x < +\infty),$$

as represented in Fig. 6.324.12 I, upper portion.

$$F(-\infty) = \frac{1}{2} + \frac{1}{\pi}\left(-\frac{\pi}{2}\right) = 0,$$

$$F(0) = \tfrac{1}{2}, \qquad \text{(b)}$$

$$F(+\infty) = \frac{1}{2} + \frac{1}{\pi}\left(+\frac{\pi}{2}\right) = 1.$$

The necessary relations are satisfied. The median $x_{0.5} = 0$ is the parameter of centering for the distribution.

6.324.13. Inversion of the Cauchy reduced distribution

We rewrite the relation (a) of the foregoing sub-section in the form

$$p = \frac{1}{2} + \frac{1}{\pi} \arctan x_p, \tag{a}$$

where x_p is the *reduced Cauchy variable* corresponding to the probability p:

$$\arctan x_p = \pi(p - \tfrac{1}{2}). \tag{b}$$

Hence

$$x_p = \tan \pi(p - \tfrac{1}{2}). \tag{c}$$

Thus, for x_p explicitly as a function of p, we have in particular

$$x_{0.5} = \tan 0 = 0,$$

$$x_{0.25} = \tan \pi\left(\frac{1}{4} - \frac{1}{2}\right) = \tan\left(-\frac{\pi}{4}\right) = -1,$$

$$x_{0.75} = \tan \pi\left(\frac{3}{4} - \frac{1}{2}\right) = \tan\left(+\frac{\pi}{4}\right) = +1.$$

Although σ does exist for the Cauchy variable, a parameter of dispersion can nevertheless be defined as the Cauchy *semi-inter-quartile deviation:*

$$Q = \frac{x_{0.75} - x_{0.25}}{2} = \frac{2}{2} = 1.$$

6.324.14. Relation between the Cauchy reduced law and the law of t with v degrees of freedom

For an arbitrary ν, the density of t with ν degrees of freedom is written as

$$f(t, \nu) = \frac{1}{\sqrt{\pi\nu}} \frac{\Gamma[(\nu+1)/2]}{\Gamma(\nu/2)} \left[\left(1 + \frac{t^2}{\nu}\right)^{\nu+1}\right]^{-1/2}. \tag{a}$$

For $\nu = 1$, we obtain the Cauchy law

$$f(t, 1) = \frac{1}{\sqrt{\pi}\sqrt{\nu}} \frac{\Gamma(1)}{\Gamma(\frac{1}{2})} [(1+t^2)^2]^{-1/2}$$

$$= \frac{1}{\sqrt{\pi}} \cdot \frac{1}{\sqrt{\pi}} \cdot \frac{1}{1+t^2} = \frac{1}{\pi} \cdot \frac{1}{1+t^2}.$$

The probability densities of t for v passing from 1 to infinity, vary continuously between two fixed curves, i.e.

for $v = 1$, Cauchy law:

$$f(t, 1) = \frac{1}{\pi} \cdot \frac{1}{1+t^2}, \quad f(0, 1) = \frac{1}{\pi} = 0.318;$$

for $v = \infty$, standard Gaussian law:

$$f(t, \infty) = \frac{1}{\sqrt{2\pi}} e^{-t^2/2}, \quad f(0, \infty) = \frac{1}{\sqrt{2\pi}} = 0.399.$$

6.324.2. *Cauchy law in the original variable*

6.324.21. *Cauchy probability density in X*

With the qualification made in section 6.324.11, we put μ for the parameter of centering of the original variable X, and α for a scale-parameter having the same dimension as X. The reduced variable x, which here is a reduced but not a standard deviate,† is written as

$$x = \frac{X-\mu}{\alpha}, \qquad dx = \frac{dX}{\alpha}. \tag{a}$$

Hence the Cauchy law in X is

$$h(X) = \frac{1}{\pi\alpha} \frac{1}{1+\left(\frac{X-\mu}{\alpha}\right)^2} \quad (-\infty < X < +\infty). \tag{b}$$

$$= \frac{1}{\pi} \frac{\alpha}{\alpha^2+(X-\mu)^2}. \tag{b'}$$

This probability density as obtained in the classical problem of the distribution of the intersection of a roulette pointer with an infinitely long ruler is discussed *inter alia* in Sverdrup (1967, I, pp. 106–9).

† Since σ_X does not exist!

6.324.22. Cauchy distribution in X

The Cauchy distribution in X is written as

$$F(X) = \int_{-\infty}^{X} h(X)\, dX = \frac{1}{\pi} \int_{-\infty}^{X} \frac{1}{1+[(X-\mu)/\alpha]^2}\, dX$$

$$= \frac{1}{\pi}\left[\arctan\frac{X-\mu}{\alpha} - \underbrace{\arctan(-\infty)}_{-(-\pi/2)}\right].$$

Hence

$$F(X) = \frac{1}{2} + \frac{1}{\pi}\arctan\frac{X-\mu}{\alpha}. \qquad \text{(a)}$$

6.324.23. Inversion of F(X)

The inversion of $p = F(X_p)$ starts from

$$p = \frac{1}{2} + \frac{1}{\pi}\arctan\frac{X_p-\mu}{\alpha}. \qquad \text{(a)}$$

Hence

$$\arctan\frac{X_p-\mu}{\alpha} = \pi\left(p-\frac{1}{2}\right) \qquad \text{(b)}$$

and

$$\frac{X_p-\mu}{\alpha} = \tan\pi\left(p-\frac{1}{2}\right). \qquad \text{(c)}$$

Therefore the p quantile of the Cauchy distribution is written

$$X_p = \mu+\alpha\tan\pi(p-\tfrac{1}{2}). \qquad \text{(d)}$$

6.324.3. Applications

(α) In *mathematical statistics* the Cauchy law corresponds to the density of t for one degree of freedom (section 6.324.14). The changeover from a rectangular or uniform law to a Cauchy law is treated in Hald (1952a, pp. 100–3). A *deceiving* property is that the mean of n *reduced* Cauchy variables follows the same law as each original variable (Gumbel, 1960).

(β) *Psychometric applications* in the analysis of all-or-none responses (Urban, 1909, 1910).

32*

(γ) *Pharmacological application.* When a variable X tends to infinity in both directions, Dufrenoy (1961) proposed the arctan transformation so as to obtain a transform y which ranges between the finite limits $(-M, +M)$:

$$y = \frac{2M}{\pi} \arctan \frac{X}{K\pi}, \tag{a}$$

where M and K are positive constants.

The modulus $K\pi$ on the scale of X corresponds to $y = M/2$.

One verifies the following relations:

$$\begin{aligned} &\text{For} \quad X = -\infty, \quad y = \frac{2}{\pi} M\left(-\frac{\pi}{2}\right) = -M; \\ &\text{For} \quad X = 0, \quad y = 0; \\ &\text{For} \quad X = +\infty, \quad y = \frac{2}{\pi} M\left(+\frac{\pi}{2}\right) = +M. \end{aligned} \tag{b}$$

(δ) *Biometric application.* Transformation of slopes b into arctan b. Døssing (1952) studied the slope (b) for the small weight increase per unit of increase in height amongst growing children in Denmark. The histogram of the individual slopes (2308 boys and 2433 girls) is positively skew. Døssing showed that the transformation of slopes (b)

$$y = \arctan b = \tan^{-1} b \tag{c}$$

leads to a practically Gaussian representation in y (see Døssing, fig. 5, p. 52).

6.325. Feller's Probability Law in $\frac{1}{\cosh x}$

This law will be presented in the variable x of reduced type.

6.325.1. *Probability density*

Feller (1940 and 1966, II, pp. 476–7) introduced the following ingenious generalization of the Cauchy law:

$$f(x) = \frac{1}{\pi \cos x} \qquad (-\infty \leqslant x \leqslant +\infty), \tag{a}$$

or, explicitly,

$$f(x) = \frac{2}{\pi} \frac{1}{e^x + e^{-x}} = \frac{2}{\pi} \frac{e^{-x}}{1+(e^{-x})^2}. \tag{b}$$

The formal analogy with the Cauchy law is evident if e^{-x} is substituted for x in the relation 6.324.11 (a).

6.325.2. Feller distribution

In accordance with the relation 6.325.1 (b) above, the Feller distribution is written as

$$F(x) = \frac{2}{\pi} \int_{-\infty}^{x} \frac{e^{-x}}{1+(e^{-x})^2}\, dx$$

$$= \frac{2}{\pi}\Big[-\arctan e^{-x}\Big]_{-\infty}^{x} = \frac{2}{\pi}\left(-\arctan e^{-x}+\frac{\pi}{2}\right), \qquad \text{(b)}$$

or, finally,

$$F(x) = 1-\frac{2}{\pi} \arctan e^{-x}. \qquad \text{(b)}$$

The following relations hold good:

$$F(-\infty) = 1-\frac{2}{\pi} \arctan \infty = 1-\frac{2}{\pi}\cdot\frac{\pi}{2} = 0,$$

$$F(0) = 1-\frac{2}{\pi} \arctan 1 = 1-\frac{2}{\pi}\cdot\frac{\pi}{4} = \frac{1}{2}, \qquad \text{(c)}$$

$$F(+\infty) = 1-\frac{2}{\pi} \arctan 0 = 1-\frac{2}{\pi}\cdot 0 = 1.$$

The inflectional tangent has the same slope as in the Cauchy law,

$$F'(0) = f(0) = \frac{1}{\pi} = 0.318$$

and is less steep than in the standard Laplacian distribution (6.321.22) and in the standard Gaussian distribution (6.322.221).

6.325.3. Characteristic function

The characteristic function (6.302.4) of the law in $1/\cosh x$ is

$$\phi_x(t) = \frac{1}{\cosh (\pi/2)t}. \qquad \text{(a)}$$

It is analogous in form to the density of x (6.325.1 a) and it shares this interesting property with the Gaussian law (6.322.34).

6.326. First Upper Extreme Value Probability Law

6.326.1. *First extreme value distribution*

From a *stability postulate* first produced by Fréchet (1927), R. A. Fisher and Tippett (1928) derived a distribution such that the largest value of samples of size n taken from this distribution has the same asymptotic distribution. This important work is discussed by Gumbel (1960, pp. 157–162). The first distribution of initial or crude data which is stable with respect to the largest value is equal to the first asymptotic distribution

$$\text{Prob}\,(x \leqslant x) = \exp\{-e^{-\alpha(x-v)}\} \qquad -\infty < x < +\infty. \tag{a}$$

The parameters v (localization) and α (shape) are positive and independent of n.

We write the reduced form in the variate

$$y = \alpha(X - v) \tag{b}$$

$$F(y) = \exp\{-e^{-y}\} \qquad -\infty < y < +\infty \tag{c}$$

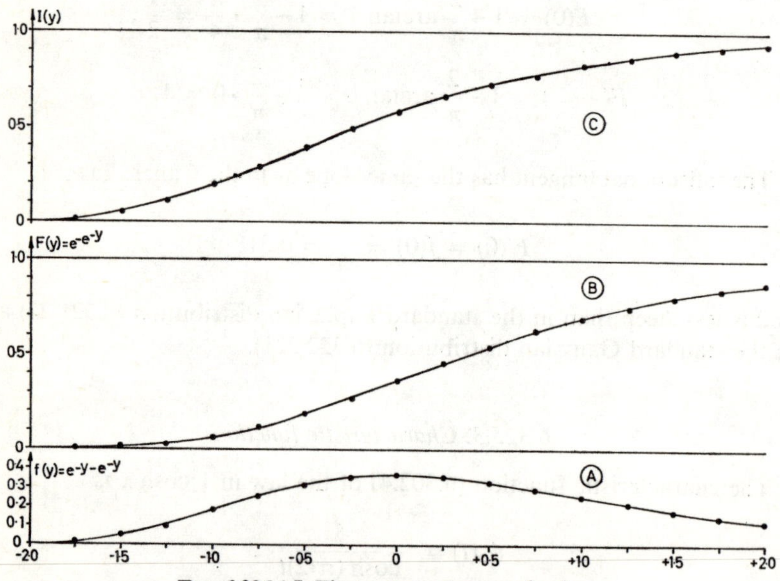

Fig. 6.326.1 I. First upper extreme value law.
(A) Probability density $\exp\{-y-e^{-y}\}$.
(B) Distribution.
(C) Imminence rate or hazard function.

with

$$F(-\infty) = 0 \qquad F(0) = e^{-1} = 0.368 \qquad F(+\infty) = 1. \tag{c'}$$

The relation (c) is the double exponential distribution of the largest reduced value and is represented in Fig. 6.326.1 I (B).

6.326.2. First upper extreme value reduced density

Taking the derivative of (c) in the preceding section, we have the density

$$f(y) = e^{-y} \cdot \exp\{-e^{-y}\} = \exp\{-y-e^{-y}\} \tag{a}$$

represented in Fig. 6.326.1 I (A).

The modal value of y, root of $f'(y) = 0$ is zero. Formula (a) is also written

$$f(y) = e^{-y}F(y). \tag{b}$$

The moment generating function (R. A. Fisher and Tippett, 1928) is

$$M_y(t) = \Gamma(1-t) \tag{c}$$

with $M_y(0) = \Gamma(1) = 0$.

We may then calculate (Gumbel, 1960, pp. 15–16 and 174)

$$\mu_1' = \Gamma'(1-t)_{t=0} = \gamma, \quad \text{(Euler's constant} = 0.5772\ldots) \tag{d}$$

$$\mu_2' = \Gamma''(1-t)_{t=0} = \gamma^2 + \frac{\pi^2}{6} = 1.978. \tag{e}$$

Whence the variance

$$\mu_2 = \sigma_y^2 = \frac{\pi^2}{6} = 1.645 \tag{e'}$$

and the standard deviation

$$\sigma_y = 1.282. \tag{e''}$$

The density is positively skew $\left(\sqrt{\beta_1} = \gamma_1 = 1.140\right)$ and the positive kurtosis is defined by

$$\beta_2 - 3 = \gamma_2 = 2.4$$

6.326.3. Inversion of the first extreme value distribution. (First) extremal probability paper

We write 6.326.1 (c) in the form

$$p = F(y_p) = \exp\{-e^{-y_p}\}. \tag{a}$$

Whence successively

$$-\ln p = e^{-y_p}$$

and

$$y_p = -\ln [-\ln p] \tag{b}$$

Just for exercise, we calculate the probabilities p responding to three cardinal values of the reduced deviate, i.e. $y_p = -1, 0, 1$:

(1) $y_p = 0$

From (b), we write successively

$$0 = -\ln [-\ln p]$$

$$0 = -\ln 1; \quad \text{whence} \quad -\ln p = +1 \quad \text{and} \quad p = e^{-1} \approx 0.37.$$

(2) $y_p = -1$ (c)

$$-1 = -\ln [-\ln p]$$

$$-1 = -\ln e; \quad \text{whence} \quad -\ln p = +e \quad \text{and} \quad p = e^{-e} = 0.07.$$

(3) $y_p = +1$ $\qquad p = e^{0.369} \approx 0.76.$

Continuing for other values of y_p, we find the two horizontal scales of the (*first*) *extremal probability paper* shown in Gumbel (1960, p. 177). Great advantage is taken by using table XII of Fisher–Yates (1957, p. 74) giving the *complementary log–log transformation*

$$v = \ln [-\ln (1-p)] \tag{d}$$

and entering with $1-p$ instead of p, since apart from the sign minus, formula (b) is the *direct log–log transformation*

$$v = \ln [-\ln p]. \tag{e}$$

The Fisher–Yates notation $\log_e p$ is our $\ln p$.

These authors remark that for small values of p, we have approximately

$$v \approx \ln p + \tfrac{1}{2} p. \tag{f}$$

6.326.4. *Imminence rate (hazard function)*

From 6.122 (c), 6.326.1 (d) and 6.326.2 (b) we write the imminence rate (hazard function) of the first reduced extremal value.

$$I(y) = \frac{f(y)}{1-F(y)} = \frac{e^{-y} \cdot F(y)}{1-F(y)} \cdot \tag{a}$$

But, since

$$F(y) = \exp \{-e^{-y}\} \tag{b}$$

we have

$$-\ln F(y) = e^{-y} \tag{b'}$$

and relation (a) becomes after division by $F(y)$

$$I(y) = \frac{-\ln F(y)}{[1/F(y)]-1} = \frac{\ln [1/F(y)]}{[1/F(y)]-1}. \qquad \text{(c)}$$

The reduced imminence rate is represented in Fig. 6.326.1 I (c) and shows a sigmoid trend between 0 and 1.

6.326.5. Use of the first extreme value probability law

We refer to Gumbel (1960, pp. 236–254) for applications of the first *upper* extreme value distribution to floods of rivers, meteorology (extreme value of pressure, precipitations, snowfalls, rainfalls (see also Sneyers, 1961) and temperature), aeronautics (instead of the Pearson III law), breaking strength for metals, textiles, electrotechnics (breakdown voltage of high voltage transformers), naval engineering, geology. Finally, in *Biology of Ageing and Dying*, the Gompertz distribution (6.337.11) of the ages at death about the modal age is the first asymptotic distribution of the *smallest* value. Thus says Gumbel (1960, p. 247) "at each age death takes away the weakest members".

6.326.6. Transition from the first to the second type of extreme value distribution

Starting again from the first upper extreme value distribution, we write

$$F(y) = \exp\{-e^{-y}\} \qquad -\infty < y < +\infty. \qquad \text{(a)}$$

We make the following transformation of variable

$$y = \ln z \qquad 0 \leqslant z < +\infty \qquad \text{(b)}$$

so that

$$e^{-y} = 1/z. \qquad \text{(b')}$$

Then (a) becomes the second extreme value distribution

$$F(z) = e^{-1/z} \qquad z \geqslant 0 \qquad \text{(c)}$$

which was derived by Frechet (1927) and is briefly considered for its own sake in section 6.338.

6.33. Continuous Probability Laws with Infinite Domain in One Direction

Such distributions of essentially positive random variables have a great importance in the calculus of probabilities (negative exponential, gamma function, χ^2, F) and in the biological and technical sciences (negative exponential, Weibull, log-normal, Erlang, Gompertz and other laws).

6.331. Negative Exponential Law

In view of the methodological interest of the negative exponential law, it will be treated in detail.

6.331.1. Negative exponential law in the original variable

6.331.11. Negative exponential density

6.331.111. Definition. The negative exponential law associates a continuous random variable X moving in the positive infinite domain $(0, \infty)$ with the density:

$$f(X) = \frac{1}{a} e^{-X/a} \quad (a > 0, X \geq 0). \tag{a}$$

This function is the immediate extension of a triangular distribution with finite range and of the type

$$\frac{1}{a}\left(1 - \frac{X}{a}\right);$$

the bracketed term is none other than the expansion of the exponential limited to the first two terms:

$$e^{-X/a} = 1 - \frac{X}{a} + \frac{1}{2} \cdot \frac{X^2}{a^2} \cdots . \tag{b}$$

One verifies without difficulty the necessary relation

$$\int_0^\infty f(X)\, dX = 1, \tag{c}$$

which expresses the fact that the total probability is equal to unity over the range of definition $(0, \infty)$ of the negative exponential.

By using the function X/a, we in effect have

$$\int_0^\infty e^{-X/a}\, d\frac{X}{a} = -\left[e^{-X/a}\right]_0^\infty = -(0-1) = 1. \tag{d}$$

Figure 6.331.111 I represents the exponential with parameter a which tends asymptotically towards 0 as $X \to \infty$.

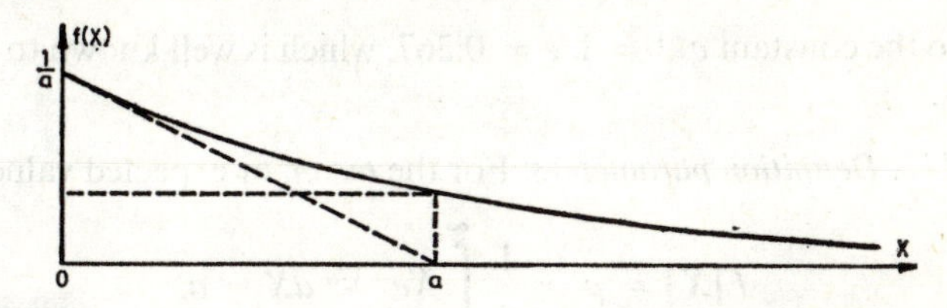

FIG. 6.331.111 I. Negative exponential probability density with mean a.

One calculates successively:

(a) *the ordinate at the origin*

$$f(0) = \frac{1}{a} \quad (a > 0); \tag{e}$$

(b) *the derivative in X*

$$f'(X) = -\frac{1}{a^2} e^{-X/a}, \tag{f}$$

the negative exponential here being a monotonically decreasing function for any X in $(0, \infty)$;

(c) *the derivative at the origin*

$$f'(0) = -\frac{1}{a^2}; \tag{g}$$

(d) *the equation of the tangent at the origin*

$$Y = f(0) - f'(0)X$$
$$Y = \frac{1}{a} - \frac{1}{a^2} X = \frac{1}{a}\left(1 - \frac{X}{a}\right). \tag{h}$$

The tangent at the origin, the slope of which is negative, cuts the OX axis at the point

$$X = +a.$$

This point is important because the mean μ and the standard deviation σ are to be found at this value (6.331.112 a and e).

For $X = a$, the ordinate of the negative exponential density is

$$f(a) = \frac{1}{a} e^{-1}.$$

Therefore the ratio of $f(a)$ to the ordinate $f(0)$ at the origin

$$\frac{f(a)}{f(0)} = e^{-1} \tag{i}$$

is equal to the constant $e^{-1} = 1/e = 0.367$, which is well known to electronic engineers.

6.331.112. Definition parameters. For the *mean,* or expected value, we have

$$E[X] = \mu = \frac{1}{a} \int_0^\infty X e^{-X/a} \, dX = a. \tag{a}$$

The negative exponential density can therefore be written as

$$f(X) = \frac{1}{\mu} e^{-X/\mu}. \tag{a'}$$

Variance: in general the variance is the expected value of the square of the deviation of variable X from its mean. It is therefore a weighted average of $(X-\mu)^2$, the local weight here being $dF = f(X)\, dX$ for a continuous random variable X:

$$\begin{aligned} V(X) = \sigma_X^2 &= E[(X-\mu)^2] \\ &= \int_D (X-\mu)^2 f(X)\, dX \qquad (D = \text{domain}). \end{aligned}$$

For the negative exponential density with parameter a, considering that $\mu = a$, we have

$$\begin{aligned} \sigma_X^2 &= \frac{1}{a} \int_0^\infty (X-a)^2 e^{-X/a}\, dX \\ &= \frac{1}{a} \int_0^\infty (X^2 - 2aX + a^2) e^{-X/a}\, dX. \end{aligned} \tag{b}$$

The integral of a sum being the sum of the integrals, we get

$$\sigma_X^2 = \frac{1}{a} \int_0^\infty X^2 e^{-X/a}\, dX - \frac{2a}{a} \int_0^\infty X e^{-X/a}\, dX + \frac{a^2}{a} \int_0^\infty e^{-X/a}\, dX.$$

In accordance with the definition of the original moments and the relation (6.331.111 d), one may write

$$\sigma_X^2 = \mu_2 = \mu_2' - 2a\mu_1' + a^2 .$$

Since $\mu_1' = \mu = a$,

$$\sigma_X^2 = \mu_2' - \mu_1'^2 . \tag{c}$$

One then calculates μ_2' applying integration by parts to obtain

$$\mu_2' = 2a^2 .$$

Therefore,

$$\sigma_X^2 = 2a^2 - a^2 = a^2 . \tag{d}$$

In the negative exponential density, with parameter a, the variance is equal to the square of the parameter and the standard deviation is equal to the parameter itself:

$$\sigma_X = a. \tag{e}$$

The negative exponential density can also be written as

$$f(X) = \frac{1}{\sigma} e^{-X/\sigma} . \tag{f}$$

In the negative exponential one has the important relation:

$$\mu = \sigma. \tag{g}$$

Therefore the coefficient of variation of X is equal to unity:

$$c_X = \frac{\sigma_X}{\mu} = 1, \tag{h}$$

Wherever the relation $\mu = \sigma$ holds good, the negative exponential applies to phenomena having a high degree of variability, such as the distribution of the intervals of quiescence between Poissonian events (e.g. radioactive disintegrations). If the relative variability of the phenomenon is greater than unity, it is well to envisage distributions of the hyper-exponential type (see Morse, 1958, p. 52).

Shape parameters. The parameters of skewness and kurtosis are calculated from the central moments μ_3 and μ_4; the latter are obtained by the formulae of section 6.302.1 for the initial moments μ_3' and μ_4' respectively, these being calculated by applying integration by parts as for μ_2'.

The following results are obtained:

(a) *Coefficient of asymmetry:*

$$\gamma_1 = \frac{\mu_3}{\sigma^3} = \frac{2a^3}{a^3} = 2, \tag{i}$$

hence strong and positive asymmetry for the negative exponential density, as expected from another quarter.

(b) *Coefficient of kurtosis:*

$$\gamma_2 = \frac{\mu_2}{\sigma^4} - 3 = \frac{9a^4}{a^4} - 3 = 6, \quad \text{(j)}$$

i.e. highly positive kurtosis which would best be interpreted as a very marked spread.

6.331.12. *Negative exponential distribution in X*

The negative exponential distribution in the original variable X corresponds to the probability of meeting with a random X at most equal to X fixed:

$$\text{Prob}\,(X \leqslant X) = F(X) = \frac{1}{a}\int_0^X e^{-X/a}\,dX = -\left[e^{-X/a}\right]_0^X = 1 - e^{-X/a}. \quad \text{(a)}$$

In this respect $F(0) = 0$ and $F(\infty) = 1$.

The median is given by:

$$F(X_{0.5}) = 1 - e^{-X_{0.5}/a} = \tfrac{1}{2}. \quad \text{(b)}$$

Therefore the quantile 0.5

$$X_{0.5} = F^{-1}(\tfrac{1}{2})$$

is obtained from the relations

$$e^{-X_{0.5}/a} = \tfrac{1}{2},$$
$$X_{0.5} = a \ln 2 = 0.693a. \quad \text{(c)}$$

For example, if X marks the age of radioactive atoms, the median $X_{0.5}$ is the period or half-life T.

6.331.13. *Complementary distribution of X*

The probability of meeting with a random X greater than a fixed X corresponds to the complementary distribution of X:

$$\text{Prob}\,(X > X) = 1 - F(X) = e^{-X/a}. \quad \text{(a)}$$

This total probability plays a fundamental rôle in the probabilistic study of the intervals X between Poissonian random events. In effect, the probability $P_0(X)$ of meeting with no event 0 in the interval X as from $X = 0$, is no more than the probability of seeing the event occur after X fixed; this will be so if the interval X between the event which occurred at $X = 0$, and the next event, is greater than X.

In general, the probability that two consecutive events may be separated by an interval greater than X, is written as

$$P_0(X) = \text{Prob}\,(\mathbf{X} > X). \quad \text{(b)}$$

In the special case of the negative exponential, the formulae (a) and (b) yield

$$P_0(X) = e^{-X/a}. \quad \text{(c)}$$

This is also the first term of a Poisson law in X and of mean $\mu = X/a$ (6.231.1).

6.331.14. *Inversion of the complementary distribution. Complementary log–log transformation*

As a corollary of the complementary distribution, we can introduce the *complementary log–log transformation* (Fisher–Yates, 1957, p. 15 and p. 74).

In fact, if a phenomenon obeys the law

$$p = 1-e^{-\beta t}, \quad \text{(a)}$$

where β has to be estimated, the complementary distribution is

$$1-p = e^{-\beta t}. \quad \text{(b)}$$

We take the natural logarithm of the two members:

$$\ln(1-p) = -\beta t. \quad \text{(c)}$$

One can estimate β by the negative slope of the straight-line segment in coordinates (t, $\ln(1-p)$), that is to say, on the standard semi-logarithmic ruling.

However, on again taking the ln of (c), we get:

$$\ln[-\ln(1-p)] = \ln\beta + 1\ln t. \quad \text{(d)}$$

If the process is truly exponential and in accordance with (a), one obtains a straight line of fixed angular coefficient 1 in the coordinates ($\ln t$, $\ln[-\ln(1-p)]$).

6.331.15. *Imminence rate or hazard function*

From the formulae 6.122 (c) and (d), the conditional probability of observing an event in the interval (X, $X+dX$), given that the event has not as yet happened in the interval (0, X), is in general written as

$$dJ(X) = \frac{f(X)\,dx}{1-F(X)} = I(X)\,dX. \quad \text{(a)}$$

For the exponential law, the conditional probability of the event occurring between X and dX is

$$I(X)\,dX = \frac{1}{a} \cdot \frac{e^{-X/a}}{e^{-X/a}} \cdot dX = \frac{dX}{a}. \tag{b}$$

Hence, for a negative exponential random variable, the *imminence rate* is constant within the range $(0, \infty)$ and equal to the inverse of the mean of X:

$$I(X) = \frac{1}{a}. \tag{c}$$

By excluding $X = 0$, it is seen that the general relation $I(X) > f(X)$ is well verified.

In the case of the phenomenon of typically exponential radioactive degradation, the relation (c) is read as the radioactive atom being spent without ageing.

6.331.16. Example: radioactive disintegration

Deterministic and probabilistic aspects. We will illustrate the negative exponential distribution by considering the disintegration of artificial radioisotopes expressed as the number of counts per 1 min after correction of the background activity of a preparation so as to retain only those impacts which are really due to the preparation.

From the deterministic standpoint, we proceed at the zero instant from a solution containing N_0 radioactive atoms. At real instant t, or after a lapse of time $(0, t)$, there are $N(t)$ atoms which are still active, i.e. a fraction $N(t)/N(0)$ of elements that are not disintegrated.

The determinist law for the activity still remaining at time t is written as

$$N(t) = N(0)e^{-\lambda t}. \tag{a}$$

The period T is the length of time after which the initial activity is reduced by half; the period or half-life is defined by the relation:

$$\frac{N(T)}{N(0)} = e^{-\lambda T} = \frac{1}{2}. \tag{b}$$

Therefore

$$T = \frac{0.693}{\lambda}. \tag{c}$$

The deterministic constant λ, which is the reciprocal of a period of time, formally corresponds to the reciprocal of the parameter of variability σ; if time is given in days, we have

$$\lambda_{\text{days}^{-1}} = \frac{1}{\sigma_{\text{days}}}. \tag{d}$$

In probabilistic terms, the life X of a radioactive atom is a random variable such that the probability of finding an arbitrary atom *still active at the age* X (i.e. the life of which exceeds X) is equal to the quotient $N(X) : N(0)$, where $N(0)$ is the initial number of atoms of the assembly subjected to the risk of disintegration, i.e. of passing from the active state (1) to the inactive state (0):

$$\text{Prob}\,(\boldsymbol{X} > X) = e^{-X/\sigma}. \tag{e}$$

The probability of the complementary event, i.e. to be disintegrated or in the state (0), is

$$\text{Prob}\,(\boldsymbol{X} \leqslant X) = 1 - \text{Prob}\,(\boldsymbol{X} > X) \tag{f}$$

$$F(X) = 1 - \exp\{-X/\sigma\}. \tag{g}$$

The total probability of observing a disintegrated atom in the interval $(0, X)$ is none other than the negative exponential distribution with real age X. Although both are expressed in units of time (e.g. in days), the *chronological time t* present in (a) is a "fixed", or at least fixable, variable. On the other hand, the *age or life* $\boldsymbol{X}$ of the radioisotope is a random variable of mean μ and standard deviation σ which follows the negative exponential law.

6.331.2. *Negative exponential law in the reduced variable*

6.331.21. *Negative exponential density in the reduced variable x*

In the density we introduce the reduced variable x

$$x = \frac{X}{\sigma}, \tag{a}$$

which is a pure number. The original variable is

$$X = \sigma x, \quad \text{or} \quad X(x) = \sigma x. \tag{b}$$

The reduced exponential density $\varepsilon(x)$ satisfies the relation

$$f(X)\,dX = \varepsilon(x)\,dx, \tag{c}$$

which expresses the conservation of the elementary probability when X is transformed into x in accordance with (a).

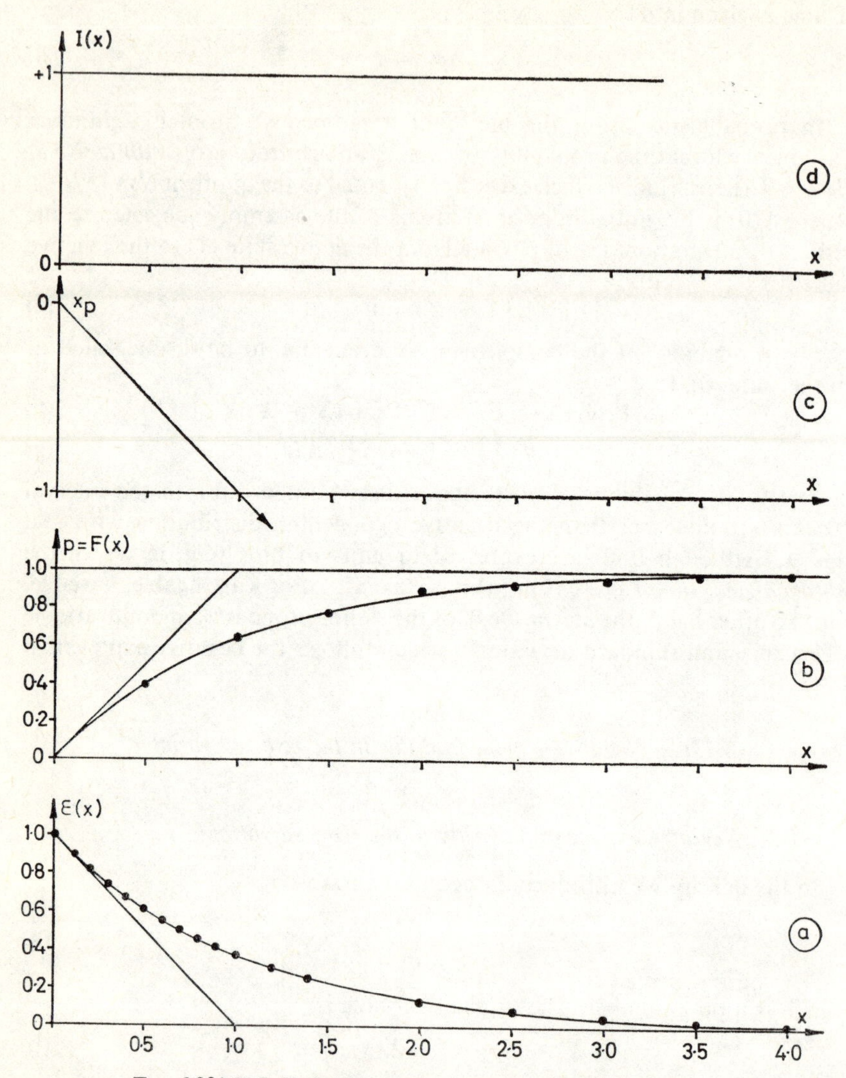

FIG. 6.331.21 I. Reduced negative exponential probability law.
(a) Probability density.
(b) Distribution.
(c) x_p = reduced variable corresponding to p.
(d) Imminence rate.

Therefore

$$\varepsilon(x) = f[X(x)]\frac{dX}{dx} = \frac{\sigma}{\sigma}e^{-x\sigma/\sigma} = e^{-x}. \quad \text{(d)}$$

Such is the reduced negative exponential in the parameter 1 which is represented in Fig. 6.331.21I (curve a).

By virtue of the general theorems relating to random variables, the parameters of (d) are readily written as

$$E[x] = \frac{1}{\sigma}E[X] = \frac{\mu}{\sigma} = \frac{\sigma}{\sigma} = 1,$$

$$V(x) = \frac{1}{\sigma^2}V(X) = \frac{\sigma^2}{\sigma^2} = 1, \quad \text{(e)}$$

$$\sigma_x = 1.$$

Coefficient of variation:

$$\gamma_x = 1. \quad \text{(f)}$$

This value C.V. % = 100% of the coefficient of variation for the exponential law is taken as a reference value in the definition of functions having greater kurtosis.

Relative excess of variance (see 6.231.2 d):

$$\delta_x = \frac{\sigma^2 - \mu}{\mu} = \frac{1-1}{1} = 0. \quad \text{(g)}$$

Shape parameters:

$$\gamma_1 = 2,$$
$$\gamma_2 = 6. \quad \text{(h)}$$

The coefficients (f) and (h) are invariant under the transformation (a), while (g) has not that property.

N.B. The argument of the reduced negative exponential is a *reduced* variable x which is the quotient of X days to σ days and not a *reduced deviate*, i.e. a deviation about the mean expressed in units σ.

6.331.22. *Reduced exponential distribution*

The reduced exponential distribution

$$\text{Prob}\,(x \leqslant x) = F(x) = 1 - e^{-x} \quad \text{(a)}$$

is represented by curve b in Fig. 6.331.21I.

For phenomena evolving in time, the reduced exponential distribution represents the probability of observing some *reduced* age taken at random

$$x = \frac{X}{\sigma},$$

which is at most equal to the fixed *reduced* age

$$x = \frac{X}{\sigma}.$$

A reduced age x thus divides the population exposed to risk into two mutually exclusive categories:

(a) the proportion p

$$p = 1-e^{-x} \tag{b}$$

of individuals who have undergone the transition from state 1 to the state 0, and

(b) the complementary proportion

$$q = e^{-x} \tag{c}$$

of individuals *remaining* in the state 1.

We thus have the *complementary distribution*

$$q = 1-F(x) = e^{-x}. \tag{d}$$

Explicitly giving the reduced age x as the argument, for each x we have the relation

$$p(x)+q(x) = 1. \tag{e}$$

If one compares the reduced exponential density (6.331.21):

$$\varepsilon(x) = e^{-x}$$

with the reduced complementary distribution (d), it is seen that they are formally identical, although having fundamentally different significances.

The reduced exponential distribution $F(x)$ thus corresponds to a simple differential equation where, one recalls, the reduced variable x is a pure number without dimension:

$$\frac{dF}{dx} = 1-F. \tag{f}$$

We have successively

$$\frac{d(1-F)}{dx} = -(1-F),$$

$$\frac{d(1-F)}{1-F} = -dx.$$

Without difficulty we get

$$d \ln (1-F) = -dx = -1 \cdot dx. \tag{g}$$

Integrating between the limits 0 and x,

$$\Big[\ln (1-F)\Big]_0^x = -\Big[x\Big]_0^x.$$

Therefore,

$$\ln [1-F(x)] = -x$$

and

$$1-F(x) = e^{-x}. \tag{h}$$

Definition (d) is thus again obtained for the complementary distribution. From 6.122 (c), 6.331.21 (d) and 6.311.22 (c), the imminence rate (hazard function) of the reduced negative exponential variable is the simplest one (see Fig. 6.331.21 (d))

$$I(x) = \frac{\varepsilon(x)}{1-F(x)} = \frac{e^{-x}}{e^{-x}} = 1. \tag{i}$$

In the radioactive field, the constancy of the imminence rate of disintegration is the statistical expression for the fact that any atom dies at a random moment without ageing.

6.331.23. Inversion of the reduced exponential distribution

Semi-logarithmic ruling. We write the distribution being considered as

$$p = F(x_p), \tag{a}$$

or

$$p = 1-e^{-x_p}. \tag{b}$$

The inversion of (a) is written symbolically as

$$x_p = F^{-1}(p). \tag{c}$$

The solution is obtained from (b) in explicit form:

$$e^{-x_p} = 1-p,$$

$$x_p = -\ln(1-p) = \ln\frac{1}{1-p} = \ln\frac{1}{q} \tag{d}$$

and is represented in Fig. 6.331.21I (c).

In terms of radioactivity, the reduced variable corresponding to the fraction p of cases that are disintegrated, is equal to minus the Napierian logarithm of the fraction $1-p = q$ of the atoms which remain active.

The reduced variable x_p develops in the essentially positive range $(0, \infty)$; hence the minus sign before the logarithm of a number less than 1. Table 6.331.23A shows the deciles of the reduced random variable which are obtained in accordance with (d).

TABLE 6.331.23A. REDUCED VARIABLE x_p CORRESPONDING TO THE FRACTION p OF CASES ACCORDING TO THE NEGATIVE EXPONENTIAL LAW

p	0	0.10	0.20	0.30	0.40	0.50	0.60	0.70	0.80	0.90	1
x_p	0	0.105	0.223	0.357	0.511	0.693	0.916	1.204	1.609	2.303	∞

In practice, the relation (d) is transformed as follows:

(1) one expresses x_p as a common logarithm in the following way:

$$\log_{10} A = 0.4343 \ln A = \frac{1}{2.30103} \ln A, \tag{e}$$

where

$$\log_{10} e = 0.4343 = \frac{1}{\ln 10} = \frac{1}{2.30103};$$

(2) the term p no longer denotes a fraction of unity $(p<1)$, but a percentage

$$100p\% < 100\%.$$

The practical formula is written so:

$$\begin{aligned} \eta_p &= \log_{10} 100(1-p) \\ &= \log_{10} 100+\log_{10}(1-p) \qquad \text{(f)} \\ &= 2+0.4343 \ln(1-p); \end{aligned}$$

$$\eta_p = 2-0.4343 x_p. \tag{g}$$

Since x_p varies between 0 and infinity (see Fig. 6.331.21I c), η_p varies between 2 and $-\infty$.

The quantity η_p could be referred to as the "expit" after the humorous appellation of J. O. Irwin (1963), the name being perfectly consistent with the terms: probit, normit, logit, angit and even "trigit".

On returning to the original variable X, or better still $X_p = \sigma x_p$, the relation (g) becomes

$$\eta_p = 2 - \frac{0.4343}{\sigma} X_p. \tag{h}$$

To base 10, the equation of the survival curve is written as

$$p = \frac{N(t)}{N(0)} = 10^{-0.4343\lambda t}.$$

In a semi-logarithmic ruling where the line (h) departs from the fixed point ($t = 0$, $\eta_p = 2$), the slope is

$$\underset{\text{deterministic}}{0.4343\lambda} = \underset{\text{probabilistic}}{\frac{0.4343}{\sigma}}$$

In this ruling the abscissae are marked off on an arithmetic (or metric) scale, whilst the ordinates are on a logarithmic scale of base 10 where the limits of the decades are, for instance, 1, 10 and 100, or perhaps 0.1, 1 and 10, according to the phenomenon being studied.

6.331.3. Forming an exponential density of given mean value

For $\mu = 3$, for example, the density is

$$f(X) = \tfrac{1}{3} e^{-X/3}, \tag{a}$$

with $\mu = \sigma = 3$.

The reduced variable $x = X/\sigma$ obeys the law

$$\varepsilon(x) = e^{-x}, \tag{b}$$

with

$$x = \frac{X}{3}. \tag{c}$$

Therefore

$$f(X) = \frac{1}{\sigma}\, \varepsilon(x). \tag{d}$$

Following Table 6.331.3A, the ordinate $f(X_p)$ corresponding to a given quartile X_p is calculated using the table for e^{-x} (e.g. Geigy Tables, 1963, pp. 16–17).

TABLE 6.331.3A. CONSTRUCTING NEGATIVE EXPONENTIAL LAW WITH μ (= σ) = 3

p	$1-p = e^{-x_p} = \varepsilon(x_p)$	x_p	$X_p = 3x_p$	$f(X_p) = \frac{1}{3}\varepsilon(x_p)$
0.0	0.1	0	0	$\frac{1}{3} = 0.333$
0.5	0.5	0.693	2.08	$\frac{1}{3}\times 0.5 = \frac{1}{6} = 0.1567$
0.9	0.1	2.3	6.9	$\frac{1}{3}\times 0.1 = \frac{1}{30} = 0.033$
0.95	0.05	3.0	9.0	$\frac{1}{3}\times 0.05 = \frac{5}{300} = 0.017$
0.99	0.01	4.6	13.8	$\frac{1}{3}\times 0.01 = 0.0033$
0.999	0.0001	7.9	20.7	$\frac{1}{3}\times 0.001 = 0.0003$

6.331.4. Generating functions associated with the negative exponential law

Firstly we consider the *generating function for the non-central moments.* Given the negative exponential density of mean μ,

$$f(X) = \frac{1}{\mu} e^{-X/\mu}, \tag{a}$$

the generating function of the central moments is written as

$$M_X(t) = E[e^{tX}] = \frac{1}{\mu}\int_0^\infty \exp\left\{tX - \frac{X}{\mu}\right\} dX$$

$$= +\frac{1}{\mu}\,\frac{1}{t-\frac{1}{\mu}}\int_0^\infty \exp\left\{X\left(t-\frac{1}{\mu}\right)\right\} dX\left(t-\frac{1}{\mu}\right) \tag{b}$$

$$= +\frac{1}{\mu}\,\frac{1}{t-\frac{1}{\mu}}\cdot\left[\exp\left\{X\left(t-\frac{1}{\mu}\right)\right\}\right]_0^\infty.$$

For values of the dummy variable t less than $1/\mu$ ($t < 1/\mu$), the exponent is negative and tends to zero as $X \to \infty$.

Therefore

$$M_X(t) = \frac{1}{\mu} \frac{1}{\frac{1}{\mu} - t} = \frac{1}{1-\mu t}. \tag{c}$$

From this generating function, no difficulty arises in locating the mean. We differentiate (c)

$$M_X'(t) = +(1-\mu t)^{-2}\mu \tag{d}$$

and, for $t = 0$, we obtain

$$\mu_1' = M_X'(0) = \mu. \tag{d'}$$

The *cumulant generating function* is written as

$$K_X(t) = \ln M_X(t) = -\ln(1-\mu t). \tag{e}$$

We differentiate twice, and then

$$K_X'(t) = \frac{\mu}{1-\mu t}, \tag{f}$$

$$K_X''(t) = \mu(1-\mu t)^{-2} \cdot \mu = \mu^2(1-\mu t)^{-2}, \tag{g}$$

so that for $t = 0$, we again obtain the variance written in 6.331.112 (d), which is equal to the square of the mean value:

$$\varkappa_2 = K_X''(0) = \mu^2. \tag{g'}$$

Going on, we get $\varkappa_3$ and $\varkappa_4$ for obtaining the shape parameters given in 6.331.112 (i) and (j).

6.331.5. *Random negative exponential standard deviates*

The standard negative exponential density (mean 0 and standard deviation 1) is written as

$$f(x) = e^{-(x+1)} \qquad x \geqslant -1. \tag{a}$$

The standard distribution is

$$F(X) = 1 - e^{-(x+1)} \tag{b}$$

with

$$F(-1) = 0 \qquad F(\infty) = 1$$

To get the negative exponential deviate x_p (Quenouille, 1959, x_4) corresponding to the standard normal deviate u_p, we write the relation

$$p = 1 - e^{-(x_p+1)} = \frac{1}{\sqrt{2\pi}} \int_{-\infty}^{u_p} e^{-u^2/2}\, du = \Phi(u_p). \tag{c}$$

Solving for x_p and using the distribution $\Phi(u_p)$, we get

$$e^{-(x_p+1)} = 1-\Phi(u_p) = \frac{1}{\sqrt{2\pi}} \int_{u_p}^{\infty} e^{-u^2/2}\, du, \tag{d}$$

whence

$$x_p = -\ln \frac{1}{\sqrt{2\pi}} \int_{u_p}^{\infty} e^{-u^2/2}\, du - 1 \tag{e}$$

is the standard negative exponential deviate corresponding to the standard normal deviate u_p.

6.331.6. *Asymptotic regression*

The asymptotic regression (Stevens, 1951)

$$Y = \alpha+\beta\varrho^X \qquad 0 < \varrho < 1 \tag{a}$$

is a *deterministic* relation between X and Y and has an important role in biometry. Putting

$$\varrho = e^{-\gamma} \tag{b}$$

the relation

$$Y = \alpha+\beta e^{-\gamma X} \tag{c}$$

is *formally* related to the negative exponential distribution of a random variable of mean $1/\gamma$.

In this case, Y plays the role of $p\,(0 \leqq p \leqq 1)$, $\alpha = 1$ and $\beta = -1$. Wilfrid Stevens, who died too young to expand its full capabilities, had realized the importance of the fitting of a three parameter asymptotic regression to many kinds of biological, agricultural (Mitscherlich law) and biometrical data. Bliss (1970, II, chap. 16) should be consulted for the different methods of fitting for cases such as growth curves of rats, prothrombin times of chick blood at different concentrations of vitamin K_1 (Bliss and Grimminger, 1969, and others). Stevens (1951) has also indicated the transformations of variables leading to the logistic and the Gompertz curve.

6.331.7. *Variance test for the exponential distribution*

Shapiro and Wilk (1965–8 have introduced a W test for testing for normality (p. 434). An analog procedure is appropriate for evaluation of the composite hypothesis of exponentiality, i.e. origin and scale unspecified (Shapiro and Wilk, 1972).

6.332. Rayleigh's Law

6.332.1. *Rayleigh's law in real time*

The law of Rayleigh (1880) has been employed in acoustics and in naval engineering (see Gumbel, 1960, p. 251).

6.332.11. *Distribution in real time*

We will here introduce Rayleigh's law as in fact presented to us in the study of the distribution of the time between two consecutive systoles of the heart (interval *RR* of the electrocardiogram) in patients suffering from auricular fibrillation (see 6.332.3).

In one other case, i.e. in the interpretation of an acaricide test (Martin and Christiane Cotteleer, 1958), a law of the Rayleigh or Weibull type (see 6.333) was envisaged in order to describe the variation in time of the mortality curve of mites subjected at time 0 to the action of a toxic agent. Martin (1961a, 1962b) considered the phenomena by advancing the simple hypothesis that the imminence rate of death ("mortality hazard function") is proportional to the lapse of time as from the toxic application, i.e.

$$I(t) = \frac{t}{\tau^2}\,. \tag{a}$$

The parameter τ has the dimension T of real time and is an *index of the susceptibility of the organisms to the toxic agent.* The instantaneous imminence of death ("instantaneous rate of mortality") at the time t of contact is written as

$$\frac{dF(t)}{1-F(t)} = \frac{t}{\tau^2}\,dt. \tag{b}$$

The mortality curve is therefore obtained by direct integration

$$F(t) = 1-e^{-t^2/2\tau^2}, \tag{c}$$

with $F(0) = 0$ and $F(\infty) = 1$.

The survival curve or complementary distribution is

$$S(t) = 1-F(t) = e^{-t^2/2\tau^2} \tag{d}$$

and yields the absolute probability of survival after the period of time t

which has already elapsed and also the "reliability at mission time" for which $S(t)$ is denoted as $R(t)$ (Basu, 1964, p. 216).

Putting $s(t)$ for the observed proportion which corresponds to the probability $S(t)$, the relation (d) enables one to verify whether or not an *ensemble* of data $(t, s(t))$ satisfies a Rayleigh law.

In fact, from (d) we obtain

$$\ln S(t) = -\tfrac{1}{2}t^2/2\tau^2, \tag{e}$$

i.e. a parabolic relation between time t and the Napierian logarithm of the observed proportion of survivors at time t.

Further, one may write

$$\ln[-\ln S(t)] = \ln\frac{1}{2\tau^2}+2\ln t. \tag{f}$$

In coordinates $\{\ln t, \ln[-\ln S(t)]\}$, providing the data are of Rayleigh type, one thus obtains a straight line of slope 2 and of ordinate at the origin $\ln 1/2\tau^2$.

Finally, Gumbel (1960, p. 252) obtained an estimate of τ by the method of maximum likelihood in the form

$$\hat{\tau} = \frac{1}{2}\sqrt{\frac{\Sigma t_i^2}{n}}. \tag{g}$$

We thus have just seen a simple example of a useful distribution being obtained from the notion of *imminence rate* (or *intensity*, or *hazard function*) of death, using a concept which is employed by actuaries (6.337.1).

6.332.12. *Rayleigh probability density in real time*

The probability density

$$f(t) = \frac{dF(t)}{dt} = \frac{t}{\tau^2}\cdot e^{-t^2/2\tau^2} \tag{a}$$

is known under the name of Rayleigh density (Parzen, 1960, p. 181; Gumbel, 1960, p. 251).

With reference to the expressions 6.332.21 (m) to (n) below in regard to the *reduced* Rayleigh law, we will now give the definitive parameters of the Rayleigh law in real time and in the parameter τ.

Parameters of localization:

$$E[t] = \tau\sqrt{\frac{\pi}{2}} = 1.253\tau, \tag{b}$$

$$Me = \tau\sqrt{2\ln 2} = 1.117\tau, \tag{b'}$$

$$Mo = \tau. \tag{b''}$$

Variance:

$$V(t) = \tau^2\left(2-\frac{\pi}{2}\right) = \frac{\tau^2}{2}(4-\pi) = 0.4292\tau^2. \quad \text{(c)}$$

Standard deviation:

$$\sigma_t = \tau\sqrt{\frac{4-\pi}{2}} = 0.655\tau. \quad \text{(c')}$$

The parameters γ_t, γ_1 and γ_2, which are pure numbers, are shown in 6.332.21 (o), (r) and (s) respectively.

6.332.2. Reduced Rayleigh law

6.332.21. Reduced Rayleigh density

The structure of relation (a) of the foregoing sub-section requires the introduction of a dimensionless *Rayleigh reduced variable:*

$$y = \frac{t}{\tau}, \quad \text{(a)}$$

where τ is a scale factor proportional to σ_t (see 6.332.12 c').

By virtue of the invariance of the elementary probability in this change of variable, we have

$$f(t)\,dt = \varphi(y)\,dy, \quad \text{(b)}$$

with

$$\frac{dy}{dt} = \frac{1}{\tau}. \quad \text{(b')}$$

Therefore

$$\varphi(y) = ye^{-y^2/2} \quad \text{(c)}$$

is the (unique) Rayleigh law in reduced form which is represented in Fig. 6.332.21I.

The derivative $\varphi'(y)$ vanishes at the modal value of y equal to unity:

$$Mo_y = 1; \quad \text{(d)}$$

The reduced parameters are calculated from the initial moments $\mu'(y)$. The calculation of $\mu'_r(y)$ is carried out by integrating by parts:

$$\mu'_r(y) = \int_0^\infty \underbrace{y^r}_{u} \cdot \underbrace{ye^{-y^2/2}\,dy}_{dv}, \quad \text{(e)}$$

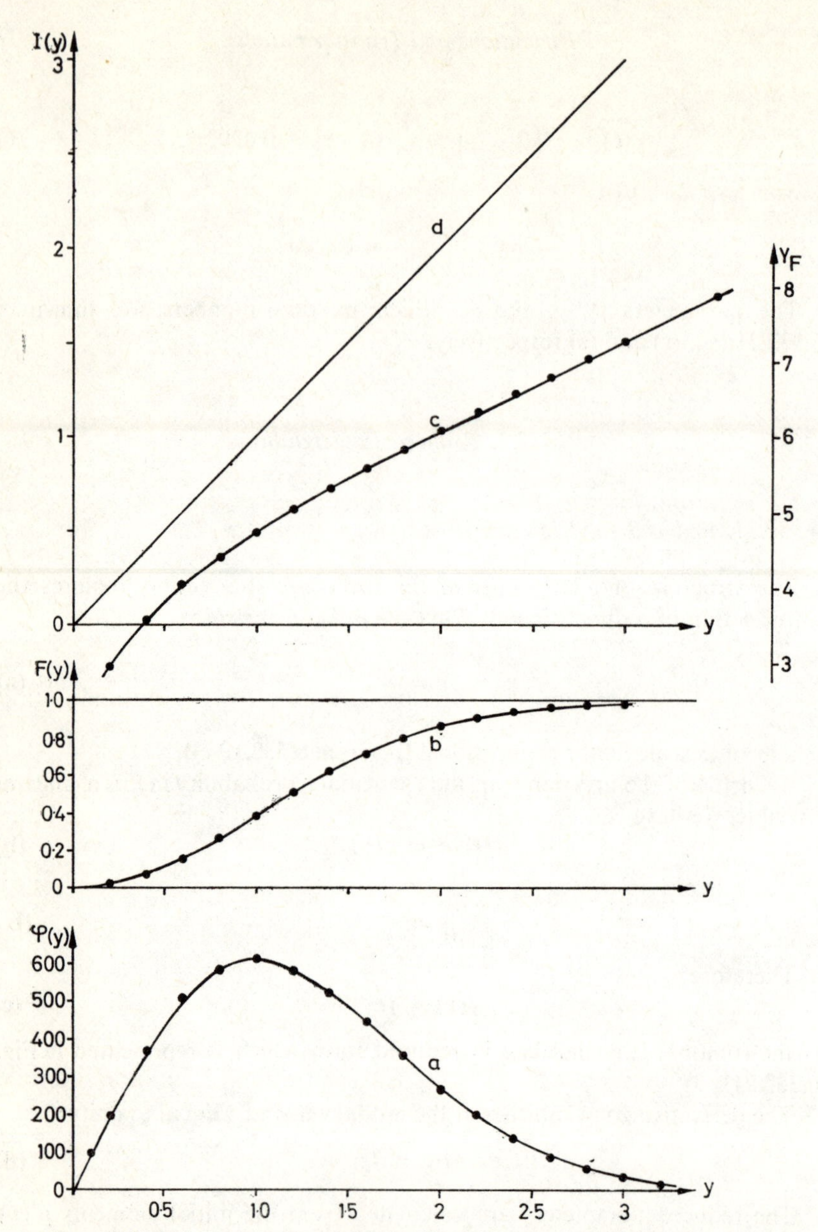

FIG. 6.332.21 I. Reduced Rayleigh law in the reduced Rayleigh variable y.
(a) Probability density.
(b) Distribution.
(c) Probit $-y$ line.
(d) Imminence rate.

where

$$u = y^r \rightarrow du = ry^{r-1}\,dy,$$

$$dv = -de^{-y^2/2} \rightarrow v = -e^{-y^2/2}.$$

Therefore,

$$\mu_r'(y) = -\left[y^r \cdot e^{-y^2/2}\right]_0^\infty + r\int_0^\infty e^{-y^2/2}y^{r-1}\,dy. \tag{f}$$

The term within the brackets vanishes and, dissociating y^{r-1} into $y \cdot y^{r-2}$, we obtain the recurrence relation

$$\mu_r'(y) = r\mu_{r-2}'(y). \tag{g}$$

If, therefore, $\mu_0'(y)$ and $\mu_1'(y)$ are known, no difficulty arises in writing the series of the $\mu_r'(y)$.

The relation

$$\mu_0'(y) = 1 \tag{h}$$

should be verified since (c) is the Rayleigh density in *reduced form*.

Then

$$\mu_1'(y) = \int_0^\infty y \cdot ye^{-y^2/2}\,dy = -\int_0^\infty y \cdot de^{-y^2/2}$$

$$= -\left[ye^{-y^2/2}\right]_0^\infty + \int_0^\infty e^{-u^2/2}\,du \tag{i}$$

$$= \sqrt{\frac{\pi}{2}} \quad \text{(see 6.322.224.1 a'').}$$

In accordance with (g), we then have in succession:

$$\mu_2'(y) = 2\mu_0' = 2, \tag{j}$$

$$\mu_3'(y) = 3 \cdot \mu_1' = 3\sqrt{\frac{\pi}{2}}, \tag{k}$$

$$\mu_4'(y) = 4\mu_2' = 4 \times 2 = 8. \tag{l}$$

The *definition parameters* of the reduced Rayleigh law are:

Mean:

$$E[y] = \mu_1'(y) = \sqrt{\frac{\pi}{2}} = 1.253. \tag{m}$$

Two other parameters of localization are considered as well:

Mode Mo = 1 (see formula (d) of the present sub-section); (m′)

Median Me = $\sqrt{2 \ln 2}$ = 1.117 (see 6.332.22 c). (m″)

Variance:

$$V(y) = \mu_2' - \mu_1'^2 = 2 - \frac{\pi}{2} = 2 - 1.5708 = 0.4292. \tag{n}$$

Standard deviation:

$$\sigma_y = \sqrt{0.4292} = 0.655. \tag{n'}$$

Coefficient of variation:

$$\gamma_y = \frac{\sigma_y}{E[y]} = 0.523; \tag{o}$$

the Rayleigh law implies a coefficient of variation of the order 50%.

The *shape parameters* are obtained from μ_3 and μ_4, the values of which are calculated from the initial moments: Thus

$$\begin{aligned} \mu_3 &= (\pi - 3)\sqrt{\frac{\pi}{2}} \\ &= 0.1416 \times 1.2533 = 0.177; \end{aligned} \tag{p}$$

$$\begin{aligned} \mu_4 &= 8 - 3\frac{\pi^2}{4} = 8 - 3 \cdot \frac{9.8696}{4} = 8 - 2.4674 \times 3 \\ &= 8.0000 - 7.4022 = 0.5978. \end{aligned} \tag{q}$$

Hence

$$\begin{aligned} \gamma_1 &= \frac{\mu_3}{\sigma_3} = \frac{(\pi-3)\sqrt{\pi/2}}{\frac{4-\pi}{2}\sqrt{\frac{4-\pi}{2}}} = \frac{2(\pi-3)\sqrt{\pi}}{(4-\pi)^{3/2}} \\ &= \frac{0.177}{0.4292 \times 0.655} = \frac{0.177}{0.281} \approx 0.63; \end{aligned} \tag{r}$$

$$\begin{aligned} \gamma_2 &= \frac{\mu_4}{\sigma_4} - 3 = \frac{32 - 3\pi^2}{4} \cdot \frac{4}{(4-\pi)^2} - 3 = \frac{32 - 3\pi^2}{(4-\pi)^2} - 3 \\ &= \frac{0.5978}{0.1842} - 3 = 3.245 - 3 \approx 0.25. \end{aligned} \tag{s}$$

6.332.22. Reduced Rayleigh distribution

From 6.332.11 (c) and 6.332.21 (a), the *reduced Rayleigh distribution* is written

$$F(y) = 1-e^{-y^2/2}, \tag{a}$$

where y varies in the range $(0, \infty)$, and

$$F(0) = 0, \quad F(\infty) = 1.$$

It is represented in Fig. 6.332.21I b.

The median $y_{0.5}$ corresponds to the relation

$$\tfrac{1}{2} = 1-e^{-y^2/2}. \tag{b}$$

Therefore

$$y_{0.5} = \sqrt{2 \ln 2} = 1.117. \tag{c}$$

We thus have the sequence:

Mode : 1
Median : 1.117
Mean : 1.253,

as expected in a density which is positively skew.

The complementary distribution

$$S(t) = 1-F(t) \tag{d}$$

yields the probability of survival after time t has elapsed, or alternatively, the "reliability at mission time", in which case $S(t)$ is written as $R(t)$ (Basu, 1964, p. 216).

6.332.23. Inversion of the reduced Rayleigh distribution

The relation (a) of the foregoing sub-section can be written as

$$p = 1-e^{-y_p/2}. \tag{a}$$

Therefore

$$e^{-y_p/2} = 1-p, \tag{b}$$

$$y_p^2 = -2 \ln (1-p), \tag{c}$$

$$y_p = \sqrt{-2 \ln (1-p)} = \sqrt{-\ln (1-p)^2}. \tag{d}$$

Such is the quantile y_p corresponding to the *reduced Rayleigh* law. The radicand is zero if $p = 0$, and positive if $p < 1$.

6.332.24. Imminence rate (hazard function)

According to the definition of section 6.332.22 and rewriting for $\tau = 1$ expression 6.332.11 (a), we have for the imminence rate or hazard function

$$I(y) = \frac{-dF(y)}{1-F(y)} = \frac{ye^{-y^2/2}}{e^{-y^2/2}} = y. \tag{a}$$

This rectilinear imminence rate is represented in Fig. 6.332.21I (d) above. The imminence rate of the reduced Rayleigh variable, as counted from $y = 0$, is precisely equal to the reduced random variable.

6.332.3. Application of Rayleigh's law in cardiac physiology

In the electrocardiographic field, Macrez (1960), Duboucher (1960) and Martin (1961a) have all analysed the frequency curve for the interval RR occurring between two ventricular systoles in a man suffering from complete arythmia due to pure auricular fibrillation. This interval is a random variable, the unimodal and positively skewed probability density of which corresponds to a Rayleigh law of parameter τ. Macrez refers to the parameter $r = 1:\tau$ as the ventricular "index of receptivity" which corresponds to the index of sensitivity to a toxic agent (see section 6.332.11).

We set t_0 for the absolute refractory period of the ventricle, hence only the excess $u = t-t_0$ is random in nature.

In accordance with Figs. 6.332.3 I and II, we write the imminence rate for the occurrence of the systole which is next in succession to that at time 0 (the systole which cannot occur before the supposedly fixed absolute refractory period t_0) in the form

$$I(t) = \frac{t-t_0}{\tau^2}. \tag{a}$$

This linear imminence rate has the dimension of T^{-1} and leads to the Rayleigh distribution of the interval RR (Macrez, 1960; Martin, 1961a)

$$F(t) = 1-\exp\left\{-\frac{1}{2}\,\frac{(t-t_0)^2}{\tau^2}\right\}. \tag{b}$$

The corresponding frequency function

$$f(t) = \frac{t-t_0}{\tau^2}\exp\left\{-\frac{1}{2}\,\frac{(t-t_0)^2}{\tau^2}\right\} \tag{c}$$

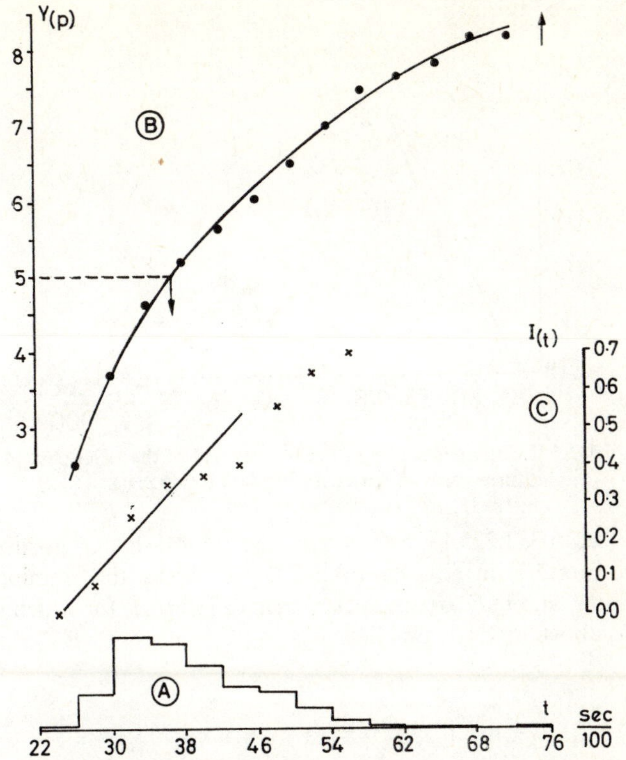

FIG. 6.332.3 I. Rayleigh law applied to the interval RR of ECG for a case of auricular fibrillation.
(A) Histogram of the RR interval in sec/100.
(B) Probit – RR graph.
(C) Imminence rate plot (from Martin 1961a).

yields the probability density of the variable $u = t - t_0$, which is the excess over the absolute refractory period t_0 of the interval t between two successive RR responses of the electrocardiogram. The period t_0 is homologous to the threshold of Claude Bernard in renal physiology.

If $I(t)$ is plotted as a function of t, the points are found to lie along a straight-line segment, at least for values of t that are not too high.

This notable observation does not apply to every case of auricular fibrillation, but only to cases belonging to the class defined for the property (a) above.

Figure 6.332.3I shows the histogram of 1000 RR intervals in the case of auricular fibrillation (A), the distribution curve of $t = t_0 + u$ in probit ordi-

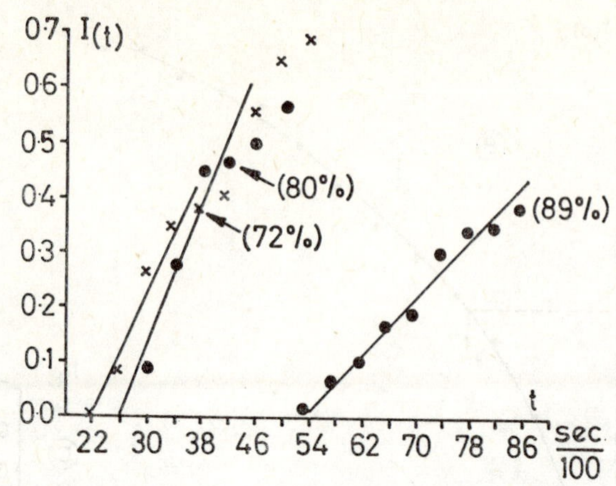

FIG. 6.332.3 II. Imminence rate of the RR interval of the ECG plotted for three cases of auricular fibrillation (see text).

nates (B) (see 6.332.225.1), and finally the almost linear portion of the imminence rate $I(t)$ in (C). Figure 6.332.3II shows the imminence rate for three cases, and indicates the percentage of intervals for which the linear relation (a) above holds in practice.

6.333. WEIBULL'S LAW

6.333.1. Reduced Weibull's law with one parameter

6.333.10. Introduction

The immediate extension of the relation 6.332.11 (a) in the form

$$I(t; \beta) = \frac{\beta}{\tau}\left(\frac{t}{\tau}\right)^{\beta-1} \qquad (\beta > 0), \tag{a}$$

which implies a note of duration in the wear of materials, and leads quite naturally to the type of law which was introduced by Weibull (1939, 1951) in the study of resistance to wear.

Rayleigh's law in this respect is a special case for $\beta = 2$; it is to be noted that in this case the factor $\frac{1}{2}$ of the argument of the exponential is replaced by unity.

Explicitly, in relation (a) the instantaneous imminence is

$$I(t)\,dt = \frac{dF}{1-F} = \frac{\beta}{\tau}\left(\frac{t}{\tau}\right)^{\beta-1} dt = \beta\left(\frac{t}{\tau}\right)^{\beta-1} d\left(\frac{t}{\tau}\right). \tag{b}$$

We introduce the dimensionless *reduced Weibull variable*

$$y = \frac{t}{\tau}. \tag{c}$$

In succession we then have

$$\frac{dF}{1-F} = \beta y^{\beta-1}\,dy, \tag{d}$$

$$-d\ln(1-F) = \beta y^{\beta-1}\,dy,$$

$$\Big[\ln(1-F)\Big]_0^y = -\beta\int_0^y y^{\beta-1}\,dy,$$

$$\ln[1-F(y)] = -\frac{\beta}{\beta}y^{\beta} = -y^{\beta},$$

and

$$F(y) = 1-e^{-y^{\beta}}, \tag{e}$$

for which

$$F(0) = 0, \quad F(\infty) = 1.$$

The pure number β is a shape parameter as will be seen from Fig. 6.333.10I, curves (a).

6.333.11. Reduced Weibull density

The reduced Weibull density with parameter β is written so:

$$f(y;\beta) = \frac{dF(y)}{dy} = \beta y^{\beta-1}e^{-y^{\beta}} \qquad (\beta > 0). \tag{a}$$

For $\beta = 1$, we again find the negative exponential law (section 6.331); for $\beta = 2$, Rayleigh's law is again obtained in section 6.332.21 (c), after changing y to $\frac{1}{\sqrt{2}}y$.

The family of density curves for $\beta = 3, 5, 7$ is shown in Fig. 6.333.10I (b). Let us now differentiate the relation (a): we get

$$f'(y;\beta) = \beta y^{\beta-2}e^{-y^{\beta}}[(\beta-1)-\beta y^{\beta}]. \tag{b}$$

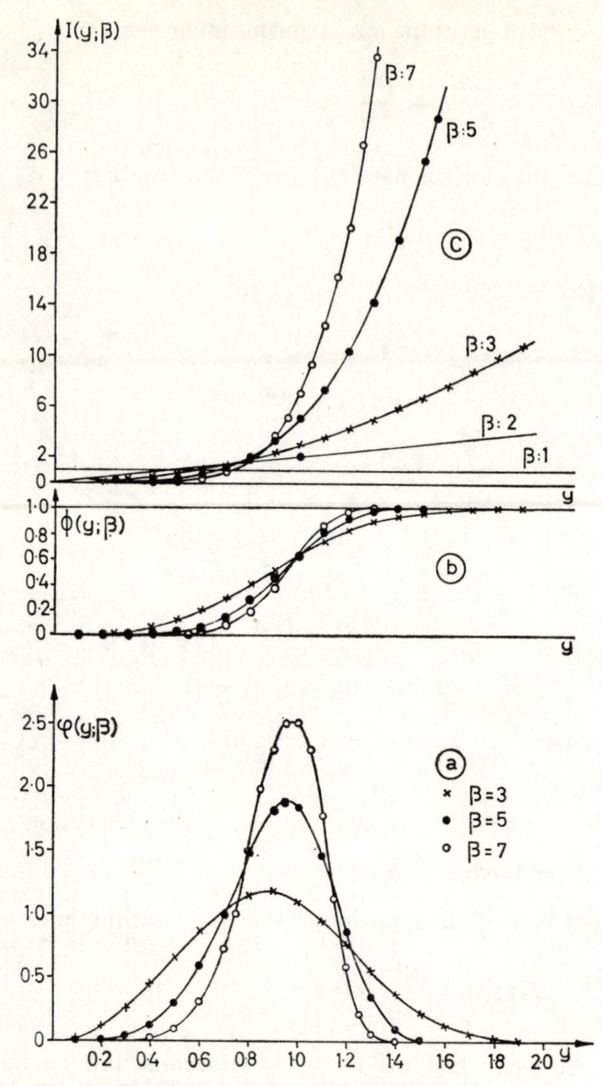

FIG. 6.333.10 I. Reduced Weibull laws with parameter β.
(a) Probability density with $\beta = 3, 5, 7$. $f(y; \beta)$ in the text.
(b) Distributions $\beta = 3, 5, 7$. $F(y; \beta)$ in the text.
(c) Imminence rate $\beta = 1$ (negative exponential), 2 (Rayleigh), 3, 5, 7.

Thus

$$f'(y;\beta) = 0 \quad \text{for} \quad y = 0 \quad \text{if} \quad \beta > 2,$$
$$y = \infty, \quad \text{and}$$
$$(\beta-1)-\beta y^{\beta} = 0.$$

Therefore the mode is

$$Mo_y = \sqrt[\beta]{\frac{\beta-1}{\beta}} = \sqrt[\beta]{1-\frac{1}{\beta}}. \tag{b'}$$

One may also write

$$Mo_y = \left(1-\frac{1}{\beta}\right)^{1/\beta}. \tag{c}$$

Consider now the *definition parameters* of the reduced Weibull law. In order to evaluate the initial kth order moments, Kao (1959) obtained the following result which we shall reproduce for the reduced case over the interval $(0, \infty)$:

$$\mu'_k = \int_0^\infty y^k f(y;\beta)\, dy \qquad Df$$

$$\mu'_k = k\int_0^\infty y^{k-1} S(y;\beta)\, dy \quad \text{(corollary 2, Kao 1959, p. 406)},$$

where (section 6.333.12 b) for the complementary distribution we have

$$S(y;\beta) = e^{-y^{\beta}},$$

whence

$$\mu'_k = k\int_0^\infty y^{k-1} e^{-y^{\beta}}\, dy = \Gamma\left(\frac{k}{\beta}+1\right). \tag{d}$$

Mean: By way of example, we will evaluate μ'_1.

$$\mu'_1 = \int_0^\infty e^{-y^{\beta}}\, dy. \tag{e}$$

We set

$$y^{\beta} = u, \quad du = \beta y^{\beta-1}\, dy,$$
$$dy = \frac{du}{\beta y^{\beta-1}};$$
$$y = u^{1/\beta} = u^b, \quad y^{\beta-1} = u^{1-(1/\beta)} = u^{1-b},$$

where $b = 1/\beta$.

Returning to (e), we have

$$\mu_1' = b\int_0^\infty e^{-u}u^{b-1}\,du$$

and

$$E[y] = \mu_1' = b\Gamma(b) = \Gamma(b+1), \tag{e'}$$

using the gamma function introduced later in 6.334.10.

Variance: as calculated by Kao (1959, p. 395), the variance in the reduced case is

$$V(y) = \sigma_y^2 = \Gamma(2b+1) - \Gamma^2(b+1). \tag{f}$$

From the same author, the *parameter of skewness* (written as α_3) is

$$\gamma_1 = \frac{\Gamma(3b+1) - 3\Gamma(2b+1)\Gamma(b+1) + 2\Gamma^3(b+1)}{[\Gamma(2b+1) - \Gamma^2(b+1)]^{3/2}}. \tag{g}$$

From 6.302.1 (e′) and relation (d) above, γ_2 is obtained.

By setting $b = 1$ in the relations (e), (f) and (g), the reader will again find the values given for the reduced negative exponential law (see section 6.331.112).

6.333.12. *Reduced Weibull distribution*

From 6.333.10 (e), this distribution is written as

$$F(y;\beta) = 1 - e^{-y^\beta}, \tag{a}$$

and is represented in Fig. 6.333.10I (b).

For the complementary distribution we have

$$S(y;\beta) = e^{-y^\beta}. \tag{b}$$

The median corresponds to the relation

$$\tfrac{1}{2} = 1 - e^{-y^\beta},$$

or

$$e^{-y^\beta} = \tfrac{1}{2}.$$

We then have, in succession,

$$y_{0.5}^\beta = +\ln 2;$$

$$y_{0.5} = (\ln 2)^{1/\beta}. \tag{c}$$

For $\beta > 1$, we always have

$$1 - \frac{1}{\beta} < 2.$$

Therefore, in Weibull's law with the shape parameter β greater than unity, the mode is less than the median.

6.333.13. *Inversion of the reduced Weibull complementary distribution*

We consider the *reduced variable corresponding to the probability p*. From the reduced complementary distribution, or survival curve $S(y)$ introduced in the relation 6.332.22 (d),

$$S(y;\beta) = 1-p = \exp\{-y_p^\beta\}, \tag{a}$$

we take the natural logarithm of the two expressions

$$y_p^\beta = \ln[-(1-p)]. \tag{b}$$

Therefore

$$y_p = \sqrt[\beta]{\ln[-(1-p)]}. \tag{c}$$

Repeating the operation, we obtain

$$\ln y_p = \frac{1}{\beta} \ln \ln [-(1-p)], \tag{d}$$

or

$$\ln \ln [-(1-p)] = \beta \ln y_p, \tag{e}$$

or also

$$\ln \ln \frac{1}{1-p} = \beta \ln y_p. \tag{f}$$

Such is the basis of the coordinates of the Weibull probability ruling (see 6.333.23).

6.333.14. *Imminence rate or hazard function of the reduced Weibull law*

The imminence rate (hazard function) of the reduced Weibull law with parameter $\beta > 0$ is written (6.122 c)

$$I(y;\beta) = \frac{f(y;\beta)}{1-F(y;\beta)} = \frac{\beta y^{\beta-1} e^{-y^\beta}}{e^{-y^\beta}} = \beta y^{\beta-1}. \tag{a}$$

This is a set of monotonically increasing continuous functions whose three curves are represented in Fig. 6.333.10I (c) for the following values of β:

$\beta = 1$	negative exponential:	I constant	
$\beta = 2$	Rayleigh law:	I increases linearly	(b)
$\beta = 3, 5, 7$	Weibull laws:	I increases as a parabola of degree 2, 4 and 6 respectively.	

$\beta < 1$

In this case, the hazard function decreases monotonically from infinity to zero; example: $\beta = \frac{1}{2}$.

$$I\left(y; \frac{1}{2}\right) = \frac{1}{2} y^{-1/2} = \frac{1}{2\sqrt{y}}. \tag{c}$$

$1 < \beta < 2$; the imminence rate increases with downwards concavity.

6.333.2. Weibull's law in the original variable with 3 parameters

6.333.20. Introduction

Introduced in 1939, there is as yet no standardized notation for Weibull's law.

We will consider the distribution in its form assumed in Kao (1959, p. 391)

$$F(t) = 1 - \exp\left\{-\frac{(t-\gamma)^\beta}{\alpha}\right\} \qquad (t \geqslant \gamma) \tag{a}$$

where γ is a parameter of threshold rather than of localization (x_n);
β is a shape parameter (m);
α is a scale parameter (x_0),

which plays the part of τ^β in the relations 6.333.10 (c) and (e) and has the dimension of T^β.

The symbols used by Weibull (1951) are shown here in brackets.

The Weibull distribution is also written:

$$F(t) = 1 - \exp\left\{-\left(\frac{t-\gamma}{\tau}\right)^\beta\right\}, \tag{b}$$

where τ has the dimension T.

When considered with the reduced variable (without dimension)

$$y = \frac{t-\gamma}{\tau}, \tag{c}$$

τ seems better indicated as parameter of scale than α. A notation equivalent to τ is also used by Menon (1963) for the scale parameter.

6.333.21. Weibull probability density in real time

The probability density is written so:

$$f(t; \beta, \gamma, \alpha) = \frac{\beta(t-\gamma)^{\beta-1}}{\alpha} \cdot \exp\left\{-\frac{(t-\gamma)^{\beta}}{\alpha}\right\} \tag{a}$$

$$(t \geqslant \gamma).$$

The mode is obtained by equating to zero the derivative of (a), which is written as

$$f'(t; \beta, \gamma, \alpha) = \frac{\beta}{\alpha}(t-\gamma)^{\beta-2} \exp\left\{-\frac{(t-\gamma)^{\beta}}{\alpha}\right\}\left[(\beta-1)-(t-\gamma)^{\beta}\frac{\beta}{\alpha}\right]. \tag{b}$$

The mode Mo is given by the root of the bracketed quantity which is set equal to zero:

$$(t-\gamma)^{\beta} = \frac{\alpha}{\beta}(\beta-1).$$

Thus

$$Mo = \gamma + \sqrt[\beta]{\frac{\alpha}{\beta}(\beta-1)}. \tag{c}$$

We set $\alpha = \tau^{\beta}$ in order to have the dimension T of γ, τ and Mo explicitly. The mode then becomes

$$Mo = \gamma + \tau\sqrt[\beta]{\frac{\beta-1}{\beta}}. \tag{c'}$$

Parameters of reduction. One passes from the reduced Weibull law to the law in real t by means of the transformation

$$y = \frac{t-\gamma}{\tau}$$

$$= \frac{t-\gamma}{\alpha^{1/\beta}} = \frac{t-\gamma}{\alpha^{b}}, \tag{d}$$

where $\tau = \alpha^{1/\beta} = \alpha^b$, and $b = 1/\beta$. (d')

By virtue of the general relations of 6.333.10 and of the parameter values of the reduced variable y (6.333.11 e, f and g) we have (Kao, 1959, p. 395):

$$E[t] = \gamma + \tau E[y] = \gamma + \tau\Gamma(b+1), \quad \text{(e)}$$

$$V(t) = \tau^2 V(y) = \tau^2[\Gamma(2b+1) - \Gamma^2(b+1)], \quad \text{(f)}$$

$$\gamma_1(t) = \gamma_1(y). \quad \text{(g)}$$

The estimation of the shape (β) and the scale (τ) parameters is considered by Menon (1963).

N.B. From the point of view of the third asymptotic distribution of the lower extremes (Gumbel, 1960, p. 278), the Weibull density is written as

$$f(t) = \frac{\beta}{v-\gamma}\left\{\frac{t-\gamma}{v-\gamma}\right\}^{\beta-1} \cdot \exp\left\{-\left(\frac{t-\gamma}{v-\gamma}\right)^{\beta}\right\} \quad \text{(h)}$$

$$(t > \gamma;\ v > \gamma;\ \beta \geqslant 1).$$

6.333.22. Weibull distribution in real time

This distribution has appeared already in (a) and (b) in section 6.333.20; the median is obtained by putting

$$\frac{1}{2} = 1 - \exp\left\{-\left(\frac{t-\gamma}{\tau}\right)^{\beta}\right\}.$$

Therefore

$$\left(\frac{t-\gamma}{\tau}\right)^{\beta} = \ln 2,$$

and the median $t_{0.5}$ is

$$t_{0.5} = \gamma + \tau(\ln 2)^{1/\beta}$$

or, according to 333.21 (d') (a)

$$t_{0.5} = \gamma + (\alpha \ln 2)^{1/\beta}.$$

6.333.23. Inversion of the Weibull complementary distribution in real time. Weibull rulings. Weibull plots

We write the Weibull distribution in the form

$$p = 1 - \exp\left\{-\left(\frac{t-\gamma}{\tau}\right)^{\beta}\right\}, \quad \text{(a)}$$

and then seek the value t_p corresponding to the total probability p.

We have in succession:

$$\exp\left\{-\left(\frac{t-\gamma}{\tau}\right)^{\beta}\right\} = 1-p, \tag{b}$$

$$\left(\frac{t-\gamma}{\tau}\right)^{\beta} = -\ln(1-p), \tag{c}$$

$$\beta \ln\left(\frac{t-\gamma}{\tau}\right) = \ln[-\ln(1-p)],$$

or

$$\begin{aligned} \ln[-\ln(1-p)] &= \beta \ln(t-\gamma) - \beta \ln \tau \\ &= \beta \ln(t-\gamma) - \ln \alpha. \end{aligned} \tag{d}$$

Thus a straight line is obtained in the following frame of reference:

Abscissae: $\ln t$ or $\ln(t-\gamma)$ if γ is known,

Ordinates: $\ln[-\ln(1-p)] = \ln \ln \dfrac{1}{1-p}$,

corresponding to the complementary log–log transformation tabulated by Fisher–Yates (1957, p. 74).

In conjunction with a table of natural logarithms, this complementary log–log table permits verification of the linear nature of data t, $p(t)$ which are transformed into $\ln t$ on the one hand, and on the other into

$$\ln\left[\ln \frac{1}{1-p(t)}\right].$$

From a straight-line segment, one can estimate the slope β and $\ln \alpha$ from the ordinate at the origin.

In order to avoid recourse to these tables, Kao (1959) has produced a Weibull probability ruling in which the Weibull plot can be made.

6.333.24. Estimation of the parameters of a Weibull law

Owing to the increasing use of the Weibull law in biology, we give a set of references on the estimation of parameters from actual data. Although biased, this set may be helpful for pharmacologists and other research workers in applied biology (Menon, 1963; Harter and Moore, 1965; Cohen, Jr., 1965; Dubey, 1967). The use of cumulative control charts for the Weibull distribution is considered by Johnson (1966). A simple method for obtaining exact lower confidence bounds for reliabilities (tail probabilities) for items whose life times follow a Weibull distribution is given by Johns and Lieberman (1966). Kao also gives a dissection of a composite phenomenon of failure due to a primary catastrophic mode of wear verifiable at the autopsy, and a secondary mode of destruction due to wear and tear in service (wear-out failure). We would refer the reader to this interesting publication. See also Gumbel (1960, p. 289), Bain and Antle (1970).

6.333.25. Application to microbiology and virology.

More recently, in the field of experimental microbiology, Shortley and Wilkins (1965) have shown that Weibull plots can reveal the heterogeneity of the hosts after inoculation with a dose of microorganisms or of viral particles.

It is generally known that an exponential response

$$p = 1 - e^{-\alpha d} \tag{a}$$

testifies to homogeneity of the doses and is represented by a straight line in a semi-logarithmic or "expit"–dose diagram (see 6.331.23).

In a p-log dose frame of reference the actual exponential response takes a sigmoidal trend which could suggest the possibility of rectification by the method of probits (6.322.225) or of logits (6.323.222). But, and this is the value of the present analysis, a notable rectification is obtained in Weibull coordinates $\log_{10}[-\ln(1-p)]$ plotted as a function of the $\log_{10}$ dose.

In this way Shortley and Wilkins produced their plots 2 and 3 representing the relative frequency of mortality of chick embryos injected with the virus of Newcastle disease. These authors (figs. 6, 7 and others) show how the composite responses due to two types of hosts are disposed and conclude: "*the Weibull analysis of biological-response data is much more meaningful than the more common probit analysis, and will in many cases serve to divide the hosts into susceptible and resistant groups. It is recommended that the Weibull type of analysis of response probabilities be more widely used*" (contributor's italics). See also Pike (1966) and Lee (1970) for analysis of painting experiments in cancer research.

6.334. Law of Gamma-l Probability

Before considering the law in the original variable X, the gamma probability law in the parameter l will firstly be studied in its reduced form (variable x), having regard to the important mathematical properties of the gamma function.

The original variable X is associated with the *reduced variable* x by a relation of the type $X = \alpha x$. In the asymmetrical densities where the mean loses part of its authority, a *reduced variable* is the original variable X expressed as a unit of the same dimension as X and referred either to the origin of the X's, or to an origin which is appropriate to each density (Gumbel, 1960, p. 7; Feller, 1966, II, p. 44). The scale factor is denoted as $\boldsymbol{a}$ or α by these authors, whilst β or b is used as the parameter of centering In the present chapter, the *reduced* gamma-l variable is $x = X/\alpha$, the "parameter of centering" here being zero. However, in section 6.334.5, we use a threshold parameter X_0.

It should be noted that here x is not an absolute deviate, but the reduced variable $x = X/\alpha$.

6.334.1. Reduced gamma-l law

6.334.10. The Euler function $\Gamma(l)$

The reduced gamma-l law, which depends only on a single parameter l, is written as

$$\gamma(x; l) = \frac{1}{\Gamma(l)} e^{-x} x^{l-1}, \qquad \text{(a)}$$

where the reduced variable x varies over the range $(0, \infty)$.

The integral of relation (a) over the domain $(0, \infty)$

$$\Gamma(l) = \int_0^\infty e^{-x} x^{l-1}\, dx \qquad (l > 0) \qquad \text{(b)}$$

plays a fundamental rôle in mathematics.

If one integrates (b) by parts and if $(l-1)$ is > 0, we have the recurrence formula:

$$\Gamma(l) = (l-1)\,\Gamma(l-1). \qquad \text{(c)}$$

Therefore, if l is a positive integer,

$$\Gamma(l) = (l-1)! \qquad \text{(d)}$$

In particular,

$$\Gamma(1) = 0!$$

Let us show that in fact

$$\Gamma(1) = 0! = 1.$$

By setting $l = 1$ in (b), we get

$$\Gamma(1) = \int_0^\infty e^{-x}\,dx = -\left[e^{-x}\right]_0^\infty = -(0-1) = 1.$$

Let us slightly change the formula (c) by writing

$$\Gamma(m+1) = m\Gamma(m). \tag{e}$$

For m integer, we may write (de la Vallée Poussin, 1928, II, p. 73)

$$\Gamma(m+a) = a(a+1)\ldots(a+m-1)\,\Gamma(a). \tag{f}$$

In particular, for the important case $a = \frac{1}{2}$, we have

$$\Gamma\left(m+\frac{1}{2}\right) = \frac{1\cdot 3\cdot 5\ldots(2m-1)}{2^m}\Gamma\left(\frac{1}{2}\right). \tag{g}$$

To calculate $\Gamma(\frac{1}{2})$, we start from Euler's complementary relationship (*ibid.*, p. 75)

$$\Gamma(a)\cdot\Gamma(1-a) = \frac{\pi}{\sin a\pi}. \tag{h}$$

For the denominator, we have $\sin \pi/2 = 1$ and

$$\Gamma(\tfrac{1}{2}) = \sqrt{\pi}. \tag{i}$$

In factorial notation, sometimes used whether a is integer or not (Fry, 1928, pp. 21 *et seq.*; Weatherburn, 1949, p. 146), we write

$$\Gamma(\tfrac{1}{2}) = (-\tfrac{1}{2})! = \sqrt{\pi}. \tag{i'}$$

Let us now put $m = 1$ in (g) in order to calculate

$$\Gamma(\tfrac{3}{2}) = (\tfrac{1}{2})!$$
$$\Gamma(1+\tfrac{1}{2}) = \tfrac{1}{2}\Gamma(\tfrac{1}{2}).$$

So

$$\Gamma\left(\frac{3}{2}\right) = \left(\frac{1}{2}\right)! = \frac{\sqrt{\pi}}{2}. \tag{j}$$

Finally, putting $a = 0$ in (h)

$$\Gamma(0) = (-1)! = \infty.$$

The evaluation of $\Gamma(m+1) = m!$ in the range $(-1, +\infty)$ is sketched in Fig. 6.334.10I. A graph extended to the negative values of m is given in Fry (1928, p. 24).

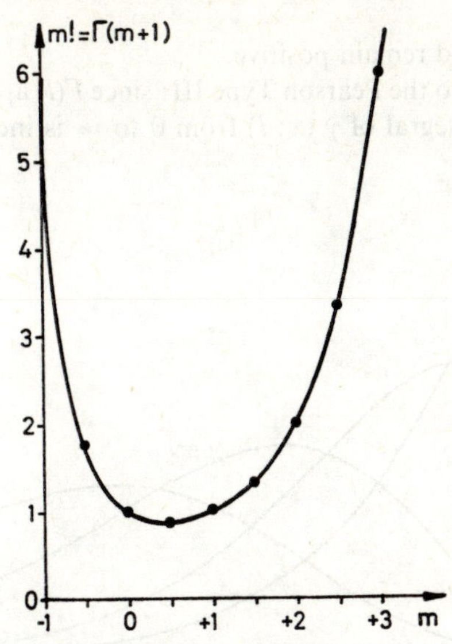

FIG. 6.334.10 I. Euler function $\Gamma(m+1)$.

6.334.11. Reduced gamma-l probability density

We re-write the reduced gamma-l density from the preceding section (formula a)

$$\gamma(x; l) = \frac{1}{\Gamma(l)} e^{-x} x^{l-1}. \tag{a}$$

The family of such curves for $l = 2, 3, 5, 7$ is represented in Fig. 6.334.11I. In order to outline the form of the gamma-l law, we differentiate (a) with respect to x:

$$\begin{aligned} \gamma'(x; l) &= \frac{1}{\Gamma(l)} [-e^{-x} x^{l-1} + (l-1) e^{-x} x^{l-2}] \\ &= \frac{e^{-x} x^{l-2}}{\Gamma(l)} [(l-1) - x]. \end{aligned} \tag{a'}$$

For $x = 0$, the derivative vanishes if $l > 2$; for $x = \infty$, the function is asymptotic to Ox.

There exists a mode

$$Mo = l-1 \tag{b}$$

if $l > 1$, for x should remain positive.

This law belongs to the Pearson Type III: since $\Gamma(l)$ appears in the denominator of (a), the integral of $\gamma(x; l)$ from 0 to ∞ is indeed equal to unity.

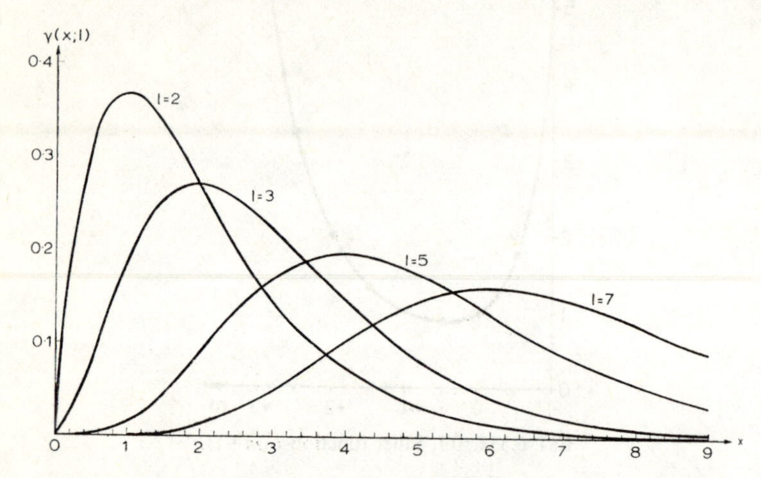

FIG. 6.334.11 I. Reduced gamma-l probability density for $l = 2, 3, 5, 7$.

For $l = 1$, the negative exponential law is again found, but the origin is not a mode, even though corresponding to the absolute maximum of the density.

The ordinate at the mode $l-1$ is

$$\gamma(l-1; l) = \frac{1}{\Gamma(l)} e^{-(l-1)}(l-1)^{l-1}. \tag{c}$$

Taking advantage of the Stirling formula ((q) infra) we may write

$$\gamma(l-1; l) \approx \frac{1}{\sqrt{2\pi}\sqrt{(l-1)}} \approx \frac{0.4}{\sqrt{l-1}} \tag{c'}$$

for l positive and large. However, for $l = 5$, the value of (c′) is approximately 0.2, in reasonably good correspondence with the ordinate at $x = 4$ for $\gamma(x; 5)$ as shown in Fig. 6.334.11 I.

Parameters. The initial moments of order $r = 1, 2 \ldots$ are found without difficulty from (a) and (b) of 6.334.10.

$$\mu'_r = \int_0^\infty x^r \gamma(x; l)\, dx = \frac{1}{\Gamma(l)} \int_0^\infty e^{-x} \cdot x^{r+l-1}\, dx, \tag{d}$$

$$\mu'_r = \frac{\Gamma(r+l)}{\Gamma(l)} = l(l+1) \ldots (l+r-1) = l^{(-r)},$$

using the symbolic notation of the exponents within brackets. Therefore

$$\begin{aligned} \mu'_1 &= l \\ \mu'_2 &= l(l+1) \\ \mu'_3 &= l(l+1)(l+2) \\ \mu'_4 &= l(l+1)(l+2)(l+3). \end{aligned} \tag{d'}$$

In accordance with the formulae of 6.302.1, the *central moments* of order 2, 3, 4 are

$$\begin{aligned} \mu_2 &= \mu'_2 - \mu'^2_1 = l^2 + l - l^2 = l \\ \mu_3 &= 2l \\ \mu_4 &= 3l^2 + 6l. \end{aligned} \tag{e}$$

The parameters are thus found to be:
Mean

$$E[x] = \mu'_1 = l. \tag{f}$$

Variance

$$V(x) = \sigma^2 = \mu'_2 - \mu'^2_1 = l(l+1) - l^2 = l. \tag{g}$$

Standard deviation

$$\sigma_x = \sqrt{l}. \tag{h}$$

Coefficient of variation

$$\gamma_x = \frac{\sqrt{l}}{l} = \frac{1}{\sqrt{l}}. \tag{i}$$

Relative excess of variance

$$\delta_x = \frac{\sigma^2 - \mu}{\mu} \equiv 0. \tag{j}$$

The relative excess of variance of the reduced gamma-l variable is identically equal to zero as in the negative exponential law (6.331.21 g) as well as in the Poisson law (6.231.2 d) for the discrete case.

There is an interesting relationship between the *discontinuous law* of Poisson which governs the *number* of *discrete* manifestations per cell of observation and, on the other hand, the two continuous laws governing the length of interval between successive events (negative exponential law), or the length of the l next successive interval (gamma-l law), but only in the reduced case.

One also calculates the central moments.

The shape *parameters* are calculated from μ_3 and μ_4:

$$\gamma_1 = \frac{2l}{l\sqrt{l}} = \frac{2}{\sqrt{l}}, \tag{k}$$

$$\gamma_2 = \frac{3l^2+6l}{l^2} - 3 = \frac{6}{l}. \tag{l}$$

It is generally known that $\gamma_1 = \beta_1^{1/2}$ and $\gamma_2 = \beta_2+3$; the locus of the points (β_1, β_2) calibrated in l for the reduced gamma law is represented by the straight line

$$\beta_2 = 3\left(1+\frac{\beta_1}{2}\right) \tag{m}$$

of the diagram of K. Pearson (6.302.8).

Stirling formula

In formulae 6.334.10 (b) and (d), we set $l = n+1$ so that

$$n! = \Gamma(n+1) = \int_0^\infty e^{-x}x^n\,dx \qquad x \geqslant 0. \tag{n}$$

In this context $n!$ is the area bounded by the curve

$$y(x) = e^{-x}x^n \qquad x \geqslant 0 \tag{o}$$

whose maximum occurs at $x = n$.

The corresponding ordinate is

$$y(n) = e^{-n}n^n. \tag{p}$$

The Stirling formula

$$n! = n^n e^{-n}\cdot\sqrt{2\pi n}\left(1+\frac{1}{12n}+\dots\right) \tag{q}$$

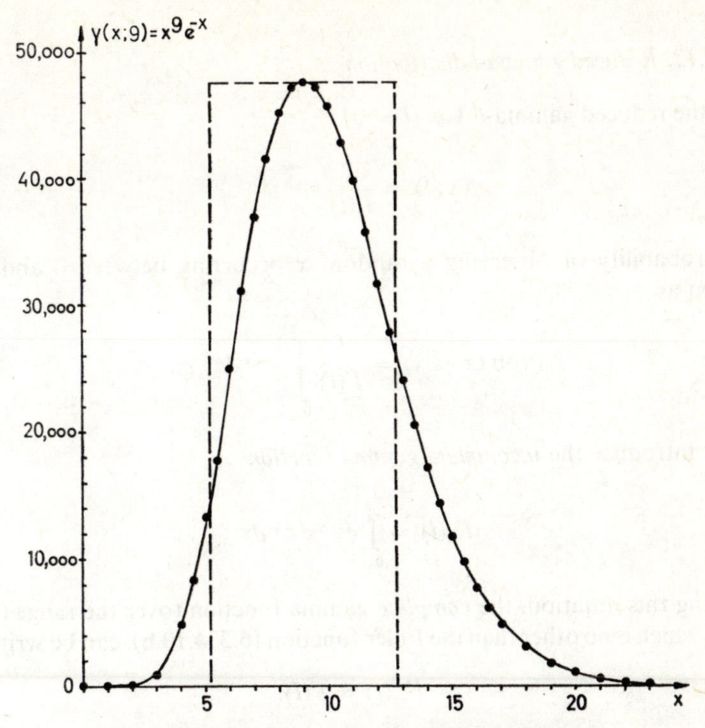

FIG. 6.334.11 II. Two representations of 9! (see text).

states that $n!$ is asymptotically equal to the product of the modal ordinate $y(n)$ by

$$\sqrt{2\pi n}\left(1+\frac{1}{12n}+\ldots\right).$$

(see Fig. 6.334.11 II) in the sense that for $n \to \infty$, the limiting value of the ratio of the two members of (q) is unity.

Even for n as small as 9, formula (g) without the term $1/12n$ gives the result

$$9^9 e^{-9} \cdot \sqrt{2\pi 9} = 359{,}533$$

as approximation for $9! = 362{,}880$ (see Fisher–Yates, 1957, p. 128). In other terms, we reconstruct 99.08% of 9!. Taking into account the factor $(1+1/108) = 109/108$, formula (q) gives 362,862, i.e. 99.995% of 9!. The rectangular area of base $3 \cdot \sqrt{2\pi} \cdot 109/108$ and height $y(9)$ is practically equal to the area (9!) underlying the curve (p).

6.334.12. Reduced gamma-l distribution

In the reduced gamma-l law ($l > 0$),

$$\gamma(x; l) = \frac{1}{\Gamma(l)} e^{-x} x^{l-1}, \qquad \text{(a)}$$

the probability of observing a random x occurring between 0 and x, is written as

$$\text{Prob}\,(x \leqslant x) = \frac{1}{\Gamma(l)} \int_0^x e^{-x} x^{l-1}\, dx. \qquad \text{(b)}$$

We introduce the *incomplete gamma function*

$$\Gamma_x(l) = \int_0^x e^{-x} x^{l-1}\, dx. \qquad \text{(c)}$$

Using this notation, the *complete* gamma function (over the range $(0, \infty)$ of x), which is no other than the Euler function (6.334.10 b), can be written as

$$\Gamma_\infty(l) = \Gamma(l). \qquad \text{(c')}$$

The reduced gamma-l distribution is finally written as

$$F(x) = \int_0^x \gamma(x; l)\, dx = \frac{\Gamma_x(l)}{\Gamma(l)}. \qquad \text{(d)}$$

By analogy with the incomplete beta function $I_x(l, m)$ which depends on the two constants l and m, it is preferable to write the distribution of the reduced gamma variable x in the parameter l as $I_x(l)$ rather than $I(x, l)$. We thus have

$$I_x(l) = \frac{\Gamma_x(l)}{\Gamma(l)}. \qquad \text{(e)}$$

K. Pearson (1922) has edited tables for the incomplete gamma function with entries

$$u = \frac{x}{\sqrt{l}} \quad \text{and} \quad p = l - 1.$$

6.334.13. *Inversion of the reduced gamma-l distribution*

We write the relation 6.334.12 (b) introducing the quantile x_p

$$p = F(x_p; l) = \frac{1}{\Gamma(l)} \int_0^{x_p} e^{-x} x^{l-1}\, dx. \tag{a}$$

It is not possible to invert explicitly this relation to the form

$$x_p = F^{-1}(p; l). \tag{b}$$

This important problem is considered in section 6.334.5 which considers the gamma-l probability plots.

6.334.14. *Imminence rate, hazard or intensity function*

The reciprocal of this function is written by Gumbel (1960, p. 144)

6.334.2. *Gamma probability law for the original variable X*

6.334.21. *Gamma probability density for X*

Let there be a dimensional continuous variable X. Up to the present we have used the reduced gamma-l variable, a pure number

$$x = \frac{X}{\alpha} \tag{a}$$

where α is a scale factor. We have

$$dx = \frac{dX}{\alpha}. \tag{a'}$$

Hence, performing the substitution (a) in $\gamma(x; l)$ which becomes $\gamma(X; l, \alpha)$, we have

$$dF(X) = \gamma(X; l, \alpha)\, dX = \frac{1}{\Gamma(l)} e^{-X/\alpha} \left(\frac{X}{\alpha}\right)^{l-1} \frac{dX}{\alpha}. \tag{b}$$

The gamma-l probability density in the original variable X measured in units α is therefore written so:

$$\gamma(X; l, \alpha) = \frac{1}{\Gamma(l)} \cdot \frac{1}{\alpha} \cdot e^{-X/\alpha} \left(\frac{X}{\alpha}\right)^{l-1}. \tag{c}$$

Since l/α remains constant in the derivation, the mode of X is that of $x = X/\alpha$ (see 6.334.11 b) multiplied by α

$$Mo_X = \alpha(l-1). \tag{d}$$

Parameters. Since $X = \alpha x$, the reduction parameters of $\gamma(X, l, \alpha)$ are deduced without difficulty from those for $\gamma(x; l)$:

$$E[X] = \alpha E[x] = \alpha l, \tag{e}$$

$$V(X) = \alpha^2 E[x] = \alpha^2 l, \tag{f}$$

$$\sigma_X = \alpha \sqrt{l}. \tag{g}$$

The standard deviation of the gamma-(l, α) variable X is proportional to the scale factor α and to the standard deviation $\sqrt{l}$ of the reduced variable.

$$\gamma_X = \frac{\sigma_X}{E[X]} = \frac{\alpha\sqrt{l}}{\alpha l} = \frac{1}{\sqrt{l}}. \tag{h}$$

The coefficient of variation of a gamma-(l, α) variable is independent of the scale factor, but depends only on the shape factor.

We also have

$$\begin{aligned} \mu_3(X) &= \alpha^3\mu_3(x) = 2\alpha^3 l, \\ \mu_4(X) &= \alpha^4\mu_4(x) = 3\alpha^4 l^2 + 6\alpha^4 l. \end{aligned} \tag{i}$$

Therefore the shape *parameters* are

$$\begin{aligned} \gamma_1(X) &= \frac{2\alpha^3 l}{\alpha^3 l\sqrt{l}} = \frac{2}{\sqrt{l}}, \\ \gamma_2(X) &= \frac{3\alpha^4 l^2 + 6a^4 l}{\alpha^4 l^2} - 3 = \frac{6}{l}. \end{aligned} \tag{j}$$

Here, too, the shape coefficients of the gamma-(l, α) law for X are independent of the scale factor.

6.334.22. Estimation of the parameters l and α of the gamma law

We consider a random sample $X_1, \ldots, X_n$ extracted from a population governed by the gamma density

$$\gamma(X; l, \alpha) = \frac{1}{\Gamma(l)} \frac{1}{\alpha} e^{-X/\alpha} \left(\frac{X}{\alpha}\right)^{l-1}. \tag{a}$$

The likelihood of this sample is written as

$$L = \prod_{i=1}^{n} \frac{1}{\Gamma(l)} \frac{1}{\alpha^l} \cdot X_i^{l-1} e^{-X_i/\alpha}. \tag{b}$$

Changing over to natural logarithms, we have

$$\ln L = \sum_{i=1}^{i=n} \left[-\ln \Gamma(l) - l \ln \alpha + (l-1) \ln X_i - \frac{X_i}{\alpha} \right]. \tag{b'}$$

The estimates $\hat{l}$ and $\hat{\alpha}$ of the parameters satisfy the system:

$$\frac{\partial \ln L}{\partial l} = 0 \qquad \frac{\partial \ln L}{\partial \alpha} = 0. \tag{c}$$

We condense the relation (b′). Since $\Gamma(l)$ and $l \ln \alpha$ are constants in the operation of summation, we have

$$\ln L = -n \ln \Gamma(l) - nl \ln \alpha + (l-1) \sum \ln X_i - \frac{\Sigma X_i}{\alpha}. \tag{d}$$

In accordance with (c), we write

$$\left\{ \begin{aligned} \frac{\partial \ln L}{\partial l} &= -n \frac{\Gamma'(l)}{\Gamma(l)} - n \ln \alpha + \sum \ln X_i = 0, & \text{(d')} \\ \frac{\partial \ln L}{\partial \alpha} &= \qquad -\frac{nl}{\alpha} + \frac{1}{\alpha^2} \sum X_i \quad = 0. & \text{(d'')} \end{aligned} \right.$$

The relation (d″) yields

$$\frac{nl}{\alpha} = \frac{1}{\alpha^2} \Sigma X_i,$$

and finally

$$\alpha = \frac{\Sigma X_i}{nl}.$$

This system can be written as

$$\left\{ \begin{aligned} \frac{\Gamma'(l)}{\Gamma(l)} + \ln \alpha &= \frac{1}{n} \Sigma \ln X_i, & \text{(e)} \\ \alpha &= \frac{\Sigma X_i}{nl}. & \text{(e')} \end{aligned} \right.$$

Therefore

$$\ln \alpha = \ln \frac{\Sigma X_i}{n} - \ln l. \tag{e''}$$

Upon substituting (e″) into (e), the equation in l becomes

$$\frac{\Gamma'(l)}{\Gamma(l)} - \ln l = \frac{1}{n} \cdot \Sigma \ln X_i - \ln \frac{\Sigma X_i}{n}, \tag{f}$$

the solution of which as a function of the X_i and n will yield the estimate $\hat{l}$; then (e′) will yield $\hat{\alpha}$.

The relation (f) can also be written so:

$$\frac{d \ln \Gamma(l)}{dl} - \ln l = \ln \left(\prod_{i=1}^{n} X_i \right)^{1/n} - \ln \bar{X}$$

$$= \ln MgX - \ln \bar{X} = \ln \frac{MgX}{\bar{X}}, \tag{g}$$

where MgX is the geometric mean of X.

Here the function of the left-hand side is tabulated in Chapman (1956); the right-hand side is a known quantity which is none other than the natural logarithm of the ratio of the geometric mean to the arithmetic mean of the observed values X_i.

Masuyama (1964) gives an empirical method of estimating parameters in a gamma distribution from a plot of $\sqrt{\chi^2}$ (see 6.334.42) against $\sqrt{X}$. The reader should also see Harter and Moore (1965).

6.334.3. Generating functions associated with the reduced gamma-l law

The generating function for the initial moments of the reduced variable x is written as (6.302.2 to 6.302.9)

$$E[(e^{xt})] = M_x(t) = \frac{1}{\Gamma(l)} \int_0^\infty e^{+xt} \cdot e^{-x} x^{l-1}\, dx$$

$$= \frac{1}{\Gamma(l)} \int_0^\infty e^{-x(1-t)} x^{l-1}\, dx \tag{a}$$

$$= \frac{1}{\Gamma(l)} \cdot \frac{1}{(1-t)^l} \int_0^\infty e^{-x(1-t)} [x(1-t)]^{l-1}\, d[x(1-t)]$$

$$= \frac{1}{\Gamma(l)} \cdot \frac{1}{(1-t)^l} \cdot \Gamma(l) = \frac{1}{(1-t)^l}.$$

Therefore

$$M_x(t) = (1-t)^{-l} \tag{b}$$
$$|t| < 1.$$

The value of the mean is again found by differentiating (b)

$$M_x'(t) = -l(1-t)^{-l-1}(-1). \tag{c}$$

Putting $t = 0$, we have as in 6.334.11 (d′)

$$\mu_1' = M'(0) = l(1-0)^{-l-1} = l. \tag{c′}$$

The mean of the reduced gamma-l variable is equal to $+l$.

In order to obtain μ_2', we differentiate (c)

$$M_x''(t) = \frac{d}{dt}\, l(1-t)^{-(l+1)} = l(l+1)\,(1-t)^{-l-2}. \tag{d}$$

Putting $t = 0$, there remains

$$\mu_2' = M_x''(0) = l(l+1) \tag{d′}$$

and so on.

In general, one writes the rth order derivative for $t = 0$

$$M^{(r)}(0) = \mu_r' = l(l+1)\ldots(l+r-1), \tag{e}$$
$$\mu_r' = l^{(-r)}$$

as a negative symbolic exponent of l (see p. 529).

The *cumulant generating function* is

$$K_x(t) = \ln M_x(t)$$
$$= \ln \frac{1}{(1-t)^l} = -l\cdot\ln(1-t); \tag{f}$$

$$K_x(t) = l\left(t+\frac{t^2}{2}+\frac{t^3}{3}+\ldots\right). \tag{g}$$

Therefore, by successively differentiating $K_x(t)$ and setting $t = 0$, the successive cumulants are generated without difficulty:

$$\varkappa_1 = \mu_1' = K_x^{\mathrm{I}}(t)\Big|_{t=0} = l(1+t+t^2+t^3+\ldots)\Big|_{t=0} = l \tag{h}$$
$$\varkappa_2 = \mu_2 = K_x^{\mathrm{II}}(t)\Big|_{t=0} = l(0+1+2t+3t^2+\ldots)\Big|_{t=0} = l \tag{i}$$
$$\varkappa_3 = \mu_3 = K_x^{\mathrm{III}}(t)\Big|_{t=0} = l(0+0+2!+3!\,t+\ldots)\Big|_{t=0} = 2!\,l \tag{j}$$
$$\varkappa_4 = \mu_4-3\mu_2^2 = K^{(\mathrm{IV})}(t)\Big|_{t=0} = l(0+0+0+3!+\ldots)\Big|_{t=0} = 3!\,l \tag{k}$$

From $\varkappa_2$, $\varkappa_3$ and $\varkappa_4$, we again find the shape parameters appearing in 6.334.11 (k) and (l), i.e.

$$\gamma_1 = \frac{\varkappa_3}{\varkappa_2^{3/2}} = \frac{2l}{l\sqrt{l}} = \frac{2}{\sqrt{l}},$$
$$\gamma_2 = \frac{\varkappa_4}{\varkappa_2^2} = \frac{6l}{l^2} = \frac{6}{l}. \quad \text{(l)}$$

6.334.4. Relations between the gamma and other probability laws

Like the Gaussian (normal) law, the gamma law plays a central role both in Theoretical and Applied Statistics.

6.334.41. Relation between the complementary distribution of Poisson and the incomplete gamma function

Given a Poisson law of mean μ

$$P_r = e^{-\mu}\frac{\mu^r}{r!} \qquad r = 0, 1, 2, \ldots. \quad \text{(a)}$$

The complementary distribution is

$$\text{Prob}\,(r \geqslant c \mid \mu) = \sum_{r=c}^{\infty} e^{-\mu}\frac{\mu^r}{r!}. \quad \text{(b)}$$

As in 6.232.1, we differentiate both sides of (b) with respect to the parameter μ

$$\frac{d\,\text{Prob}\,(r \geqslant c \mid \mu)}{d\mu} = \frac{e^{-\mu}\mu^{c-1}}{(c-1)!}. \quad \text{(c)}$$

We integrate the relation (c) termwise between 0 and μ, setting x for the current variable which is here a gamma-c reduced variable:

$$\text{Prob}\,(r \geqslant c) = \frac{1}{(c-1)!}\int_0^{\mu} e^{-x}x^{c-1}\,dx$$

$$= \frac{1}{\Gamma(c)}\int_0^{\mu} e^{-x}x^{c-1}\,dx. \quad \text{(d)}$$

Therefore, finally,

$$\text{Prob}\,(r \geqslant c) = \frac{\Gamma_\mu(c)}{\Gamma(c)} = I_\mu(c) \tag{e}$$

or

$$= I(\mu, c).$$

This incomplete gamma function was introduced earlier in 6.334.12 (d) and (e).

The calculations are facilitated by the tables of K. Pearson (1922), where one enters with $u = \mu/\sqrt{c}$ and $p = c-1$.

Clearly, the relation (e) becomes

$$P_c+P_{c+1}+\ \dots = \frac{1}{\Gamma(c)}\int_0^\mu e^{-x}x^{c-1}\,dx. \tag{f}$$

It is thus possible to express the sum of the Poisson probabilities of mean μ, i.e. $P_c+P_{c+1}+\ \dots$, in terms of a definite integral of the reduced gamma-c function, where the parameter c is the subscript of the first probability considered in (f) and where the upper limit is the mean of the Poisson law envisaged.

A complementary relation can also be written for a Poisson law of mean μ:

$$P_0+P_1+\ \dots\ P_{c-1} = \frac{1}{\Gamma(c)}\int_\mu^\infty e^{-y}y^{c-1}\,dy. \tag{g}$$

6.334.42. Relation between the gamma-(n/2) probability law and the χ_n^2 distribution

Let

$$u_i = \frac{x_i}{\sigma_i} \qquad i = 1, 2, \dots, n \tag{a}$$

be n independent standard normal deviates (each with zero mean). Then the sum of the squares of the u_i has chi-squared density with n degrees of freedom,

$$\chi_n^2 = \sum_{i=1}^{n} u_i^2, \tag{a'}$$

which is such that

$$x = \tfrac{1}{2}\chi^2 \tag{b}$$

is a gamma-l variate with parameter $l = n/2$.

Whence from 6.334.11 (a), the elementary probability is written

$$dF\left(\frac{1}{2}\chi^2\right) = \frac{1}{\Gamma(n/2)}\left(\frac{1}{2}\chi^2\right)^{(n/2)-1} e^{-\chi^2/2} d\left(\frac{1}{2}\chi^2\right). \quad \text{(c)}$$

From (b), χ_n^2 is twice a gamma-$(n/2)$ variate; its mean is twice the mean of χ (section 6.334.11); so

$$E[\chi_n^2] = 2E[x] = 2(n/2) = n. \quad \text{(d)}$$

Furthermore, the modal value of χ_n^2 is twice the mode of x, which is $l-1$:

$$Mo(\chi_n^2) = 2\cdot((n/2)-1) = n-2. \quad \text{(e)}$$

The variance is written as

$$V(\chi_n^2) = 4V(x) = 4(n/2) = 2n. \quad \text{(f)}$$

6.334.43. Relation between the gamma-1/2 reduced law and the standard Gaussian (normal) law

In 6.334.10 (j), we have seen that

$$\Gamma\left(\frac{3}{2}\right) = \left(\frac{1}{2}\right)! = \int_0^\infty e^{-x}x^{1/2}\,dx = \frac{\sqrt{\pi}}{2}. \quad \text{(a)}$$

We make the transformation

$$x = \frac{u^2}{2} \quad \text{(b)}$$

with $x^{1/2} = \dfrac{u}{\sqrt{2}}$ and $dx = u\,du$. (b′)

Putting (b) and (b′) in (a), we get

$$\frac{\sqrt{\pi}}{2} = \frac{1}{\sqrt{2}}\int_0^\infty e^{-u^2/2}u^2\,du. \quad \text{(c)}$$

This expression (c) is related to the second moment of u from the mean $E[u] = 0$ in the standard normal law. The latter is written, with respect to symmetry,

$$\mu_2 = \frac{2}{\sqrt{2\pi}}\int_0^\infty e^{-u^2/2}u^2\,du. \quad \text{(d)}$$

Combining (c) and (d) we get

$$\mu_2 = \frac{2}{\sqrt{2\pi}} \cdot \frac{\sqrt{2}}{\sqrt{2}} \int_0^\infty e^{-u^2/2} u^2 \, du$$

$$= \frac{2}{\sqrt{2\pi}} \cdot \sqrt{2} \cdot \frac{\sqrt{\pi}}{2} = 1 \quad \text{(e)}$$

which satisfies the definition of the standard normal deviate u

$$(E[u] = 0; \quad V(u) = \mu_2 = 1).$$

In fact, from the point of view of the distributions themselves this section is the particular case of that treated in 6.334.42 for $n = 1$, since χ_1^2 (one d.f.) is none other than the square of the standard normal deviate u

$$\chi_1^2 = u^2. \quad \text{(f)}$$

6.334.5. Probability plots for the gamma distribution

In the variable

$$x = \frac{X - X_0}{\alpha} \qquad X \geqslant X_0 \quad \text{(a)}$$

corresponding to the crude variable X with threshold X_0, the gamma-l density is written

$$\gamma(x; l) = \frac{1}{\Gamma(l)} x^{l-1} e^{-x} \qquad x \geqslant 0, \; l \geqslant 1 \quad \text{(b)}$$

and the distribution

$$p = F(x; l) = \frac{1}{\Gamma(l)} \int_0^{x_p} x^{l-1} e^{-x} \, dx. \quad \text{(c)}$$

By the definition of the quantiles-p of x and X respectively, we have the linear relation

$$x_p = \frac{X_p - X_0}{\alpha} \quad \text{(d)}$$

with abscissae X_p and ordinates x_p considered in section 6.334.13. Then the couples of cograduated quantiles (X_p, x_p) will fall along a straight line, of slope $1/\alpha$ and X_0 as intercept on the axis of abscissae.

If now we take a random sample from a gamma-l distribution in X and order it as $X_{(1)}, \ldots X_{(i)}, \ldots, X_{(n)}$, the $\mathcal{P}_i$ corresponding to $X_{(i)}$ may be defined as

$$\mathcal{P}_i = \frac{i-0.5}{n} \qquad i = 1, 2, \ldots, n. \tag{e}$$

This discrete and ordered suite of $\mathcal{P}_i$ is homologous to the continuous cumulative probability p of relations (c) and (d). Inverting formally the integral (c) we may write as in 6.334.13

$$x_p = F^{-1}(p; l) \tag{f}$$

for the convenient value of the parameter l.

In analogy with (c) for the particular realization considered we may write

$$\mathcal{P}_i = \frac{1}{\Gamma(l)} \int_0^{x_{\mathcal{P}_i}} x^{l-1} e^{-x} dx \tag{g}$$

or, with Wilk, Gnanadesikan and Miss Huyett (1962, p. 3)

$$\int_0^{x_{\mathcal{P}_i}} x^{l-1} e^{-x} dx = \Gamma(l) \mathcal{P}_i \tag{h}$$

$$i = 1, 2, \ldots, n.$$

Here l is known or estimated and the ordered sequence $x_{\mathcal{P}_i}$ has to be calculated.

After theoretical derivations, these authors give graphs of gamma-probability plots using a Monte Carlo method. Random samples from a gamma-l density with $l = r/2$ (see 6.334.42) are generated by taking the sum of squares of r random standard normal deviates, which follows a chi-square law with r degrees of freedom.

Inter alia, from variance data taken from Hald (1952a, p. 295) they produce a very nice gamma-probability plot with graphical evidence of a variance outlier.

6.334.6. Erlang law

In the particular case of l positive and integer the gamma-l law has been considered in its own right by an engineer, A. K. Erlang (1909, 1918, 1948), while Feller (1966, II, p. 47) considers this particular case as the gamma-l

law applied to queuing theory. The biologist is interested by the meaning of Erlang's law: let us consider a Poisson process developing in time t with intensity ξ, i.e. the mean number of events or impulses per unit time equals ξ. The mean number of impulses appearing in the interval (0, t) is

$$\mu(t) = \xi t. \tag{a}$$

Then one may show the following relations from the *evolutive* Poisson law

$$P_{[r(t)=r]} = e^{-\xi t}\frac{(\xi t)^r}{r!}: \tag{b}$$

Mean number of events in time interval (0, t):

$$E[r(t)] = \xi t; \tag{c}$$

Variance of the number of events in time interval (0, t):

$$V[r(t)] = \xi t.$$

Moreover, the time θ which elapses between two successive events follows a negative exponential density, which we may write as (6.331.111)

$$f_1(\theta) = \xi\, e^{-\xi\theta} \tag{d}$$

with

$$E[\theta] = \frac{1}{\xi}, \qquad V(\theta) = \frac{1}{\xi^2}. \tag{e}$$

The subscript 1 in (d) indicates that one interval between two successive Poissonian events is considered. For intervals made up of sums of a fixed number l of consecutive random intervals θ_i, the probability density of

$$\theta = \theta_1 + \theta_2 + \ldots + \theta_l \tag{f}$$

is *Erlang's density*

$$f_l(\theta) = \frac{1}{(l-1)!}\, e^{-\xi\theta}(\xi\theta)^{l-1}. \tag{g}$$

This formula is the gamma-l density for l integer greater than 1.

EXAMPLE: In some cases of auricular fibrillation (delirium cordis) in man, Macrez (1960, p. 607, note 1) considers the possibility of describing the variability of the RR interval of the ECG on the hypothesis that the left ventricle submitted to random impulses responds after having received a fixed number of impulses. This is another model that the Rayleigh model considered in section 6.332.3. The imminence rate or hazard function of the Erlang-l law is given by Gumbel (1960, p. 143).

6.334.7. Generalized gamma distributions

The reader may refer to Bain and Antle (1970).

6.335. LOG-GAUSSIAN LAW

6.335.1. Log-Gaussian law in two parameters

We consider a random variable such that the body weight Z, measured in kg weight/unit kg, is a pure number. The probability density of Z is positively skewed and is defined in the theoretical domain $(0, \infty)$.

The random variable Z is said to obey a log-Gaussian law if the logarithm of Z obeys a Gaussian law. Aitchison and Brown (1957) have produced an excellent book on this topic.

6.335.11. Log-Gaussian probability density

6.335.111. Definition. We set the logarithmic transformation

$$X = \ln Z, \tag{a}$$

the natural logarithm being used in order not to encumber the notation.

The Gaussian law in $\ln Z$ with the parameters $\mu_{\ln Z}$ and $\sigma_{\ln Z}$ is written as

$$g(X) = g(\ln Z) = \frac{1}{\sigma_{\ln Z}\sqrt{2\pi}} \exp\left\{-\frac{(\ln Z - \mu_{\ln Z})^2}{2\sigma^2_{\ln Z}}\right\}. \tag{b}$$

The probability element in $\ln Z$ is

$$g(\ln Z)\, d \ln Z = \frac{1}{\sigma_{\ln Z}\sqrt{2\pi}} \exp\left\{-\frac{(\ln Z - \mu_{\ln Z})^2}{2\sigma^2_{\ln Z}}\right\} d \ln Z \tag{c}$$

with $Z > 0$.

By explicitly giving the differential

$$d \ln Z = \frac{dZ}{Z} \tag{d}$$

and bearing in mind that Z is essentially positive, the equation of conservation of the probabilities for the corresponding ranges is

$$g(\ln Z)\, d \ln Z = h(Z)\, dZ \qquad (Z > 0). \tag{e}$$

Therefore the density of Z is

$$h(Z) = \frac{1}{Z\sigma_{\ln Z}\sqrt{2\pi}} \exp\left\{\frac{(\ln Z - \mu_{\ln Z})^2}{2\sigma^2_{\ln Z}}\right\}. \tag{f}$$

Such is the log-Gaussian probability density for the original variable Z. Many authors use the name log-normal (see 6.322.0).

The condition to be fulfilled in order that positive $h(Z)$ should be a probability density, is that the integral taken over the range $(0, \infty)$ of definition of Z should be equal to unity.

This is in fact so; let us therefore make change of variable (a) which changes the log-Gaussian Z over the range $(0, \infty)$ into the Gaussian X over the range $(-\infty, +\infty)$:

$$X = \ln Z, \quad dX = \frac{dZ}{Z}.$$

The relation (a) is written in X:

$$g(X)\, dX = \frac{1}{\sigma_X\sqrt{2\pi}} \exp\left\{-\frac{(X-\mu_X)^2}{2\sigma^2_X}\right\} dX, \tag{g}$$

and it is known that the integral of (g) extended over $(-\infty, +\infty)$ is equal to unity.

Between the range of Z and the range of X there exists a one-to-one correspondence.

It is now appropriate to express the parameters $\mu_{\ln Z}$ and $\sigma_{\ln Z}$ as a function of the parameters of the original asymmetrical distribution of Z such as the mean of $Z(\mu_Z)$, the median of $Z(Z_{0.5})$ and the variance of $Z(\sigma^2_Z)$.

6.335.112. Parameters of the log-Gaussian law

6.335.112.1. Parameters of localization. The study of the parameters of localization of the log-Gaussian variable is interesting from the theoretical point of view, for it leads to the quantification of the order relation

$$Mo < Me < \text{Mean},$$

which holds good for densities which are positively skew.

(a) *Median*. The median of Z, written as $Z_{0.5}$, is the point which corresponds to 50% of the cases which are less than or greater than $Z_{0.5}$.

We write

$$\mu_{\ln Z} = \ln Z_{0.5}. \tag{a}$$

Actually, the mean of ln Z, a Gaussian variable, is equally the median of ln Z which is no other than the quantile 0.5:

$$\mu_{\ln Z} = (\ln Z)_{0.5} = \ln Z_{0.5}. \tag{b}$$

Therefore

$$Z_{0.5} = e^{\mu_{\ln Z}}. \tag{c}$$

Such is the relation between the median of the original variable Z and the median of ln Z, which is also $\mu_{\ln Z}$.

On substituting the relation (c) into 6.335.111 (f) above, we obtain

$$h(Z) = \frac{1}{Z\sigma_{\ln Z}\sqrt{2\pi}} \exp\left\{-\frac{(\ln Z - \ln Z_{0.5})^2}{2\sigma_{\ln Z}^2}\right\}. \tag{d}$$

The argument of the exponential suggests the introduction of the relative variable or rather reduced variable (see 6.301)

$$z = \frac{Z}{Z_{0.5}}, \quad dz = \frac{dZ}{Z_{0.5}}, \tag{e}$$

i.e. the quotient of the random variable Z over its median $Z_{0.5}$.

The median of Z equals

$$z_{0.5} = \frac{Z_{0.5}}{Z_{0.5}} = 1.$$

However, since ln $Z_{0.5}$ is constant, we have

$$V(\ln z) = V(\ln Z - \ln Z_{0.5}) = V(\ln Z), \tag{f}$$

therefore

$$\sigma_{\ln z} = \sigma_{\ln Z}.$$

We now pass on from the density $h(Z)$ in the original variable Z, to the density $\lambda(z)$ in the reduced variable z; here the notation λ is the Greek version of the initial letter of the word logarithm.

Accordingly, we write the equation of conservation of the elementary probabilities:

$$h(Z)\, dZ = \lambda(z)\, dz.$$

Hence

$$\lambda(z) = h[Z(z)] \cdot \frac{dZ}{dz}.$$

Since $Z_{0.5}$ cancels out there remains

$$\lambda(z) = \frac{1}{z} \cdot \frac{1}{\sigma_{\ln z} \sqrt{2\pi}} \exp\left\{\frac{(\ln z)^2}{2\sigma^2_{\ln z}}\right\}. \tag{g}$$

Such is the *log-Gaussian density in the relative or reduced variable* z which has its axis (but is not centered) on the median $z_{0.5} = 1$.

In the special case where $z_{0.5}$ and $\sigma_{\ln z}$ are equal to unity, the log-Gaussian probability density $\lambda(z)$ is as shown in Fig. 6.335.112.1 I (a). The log-Gaussian density is positively skew. The median here plays the central role of the mean for the Gauss curve. Owing to the asymmetry, other parameters of localization are defined, namely, the mode *Mo* and the mean μ. This is not the case in the Gaussian curve, for the mean, median and mode all coincide, as also for any symmetrical density.

(b) *Mode of the log-Gaussian variable* Z. The probability density of Z is given by $h(Z)$ specified in the relation (f) above (see sub-section 6.335.111). By definition, the mode of Z, denoted as Mo_Z, is the point where $h(Z)$ is maximum, or the point at which the first derivative $h'(Z)$ vanishes and $h''(Z)$ is negative.

We will calculate the derivative of $h(Z)$ for $Z > 0$:

$$h(Z) = \frac{1}{Z\sigma_{\ln Z}\sqrt{2\pi}} \exp\left\{-\frac{(\ln Z - \mu_{\ln Z})^2}{2\sigma^2_{\ln Z}}\right\}. \tag{g'}$$

It is found that

$$h'(Z) = -\frac{1}{Z^2} \exp\left\{-\frac{(\ln Z - \mu_{\ln Z})^2}{2\sigma^2_{\ln Z}}\right\}\left[1 + \frac{\ln Z - \mu_{\ln Z}}{\sigma^2_{\ln Z}}\right].$$

The derivative $h'(Z)$ is zero at the modal point Mo_Z if we replace $\ln Z$ by $\ln Mo_Z$ in the bracket; whence

$$1 + \frac{\ln Mo_Z - \mu_{\ln Z}}{\sigma^2_{\ln Z}} = 0,$$

whence

$$\ln Mo_Z - \mu_{\ln Z} = -\sigma^2_{\ln Z},$$

$$\ln Mo_Z = \mu_{\ln Z} - \sigma^2_{\ln Z}.$$

Now

$$\mu_{\ln Z} = \ln Z_{0.5}.$$

Hence

$$\ln \frac{Mo_Z}{Z_{0.5}} = -\sigma^2_{\ln Z}.$$

Therefore

$$\frac{Mo_Z}{Z_{0.5}} = \exp\{-\sigma^2_{\ln Z}\},$$

or

$$Mo_Z = Z_{0.5} \exp\{-\sigma^2_{\ln Z}\}. \qquad (g'')$$

The mode of Z is thus less than the median; the mode/median ratio depends only on $\sigma_{\ln Z}$.

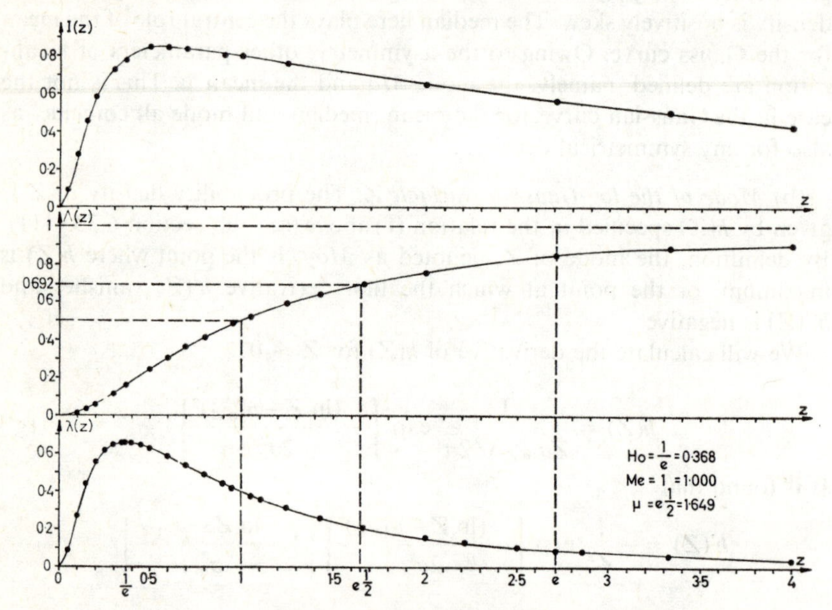

FIG. 6.335.112.1 I. Log-Gaussian (normal) law in the reduced variable z.
(a) Probability density $\lambda(z)$.
(b) Distribution $\Lambda(z)$.
(c) Imminence rate $I(z)$.

In Fig. 6.335.112.1 I, the mode of the reduced variable $z = Z : Z_{0.5}$ is equal to

$$e^{-1} = \frac{1}{2.718} = 0.368.$$

(c) *Expected value or (arithmetic) mean of the log-Gaussian variable Z.* The mean or expected value of Z is written as

$$\mu_Z = E[Z] = \int_0^\infty Zh(Z)\,dZ. \tag{h}$$

Upon changing to the relative variable z, we obtain

$$\mu_Z = \exp\left(\mu_{\ln Z} + \tfrac{1}{2}\sigma^2_{\ln Z}\right), \tag{i}$$

or, further,

$$\mu_Z = Z_{0.5} \cdot \exp \tfrac{1}{2}\sigma^2_{\ln Z}. \tag{j}$$

The principle underlying the operations is treated in Aitchison and Brown (1957, p. 8), whilst the details of the integration are to be found in Hald (1952a, p. 165).

The mean of Z is thus greater than the median; here, too, the mean/median ratio depends only on $\sigma^2_{\ln Z}$.

In Fig. 6.335.112.1 I (a) the mean of $z = Z : Z_{0.5}$ is

$$\mu_z = e^{1/2} = \sqrt{e} = 1.649.$$

(d) *Geometric mean of the log-Gaussian variable Z.* The arithmetic mean of $\ln Z$ is no other than the logarithm of the geometric mean of Z, written as Mg_Z.

In this respect, for a series of observed discrete values $Z_1, Z_2, \ldots, Z_n$, we have, on putting $Moy_{\ln Z}$ for the *observed* mean of the series Z_i:

$$Moy_{\ln Z} = \sum_{i=1}^{n} \frac{\ln Z_i}{n} = \ln\left(\prod_i Z_i\right)^{1/n} = \ln Mg_Z, \tag{k}$$

where the multiplication symbol $\prod$ plays a role which is analogous to the summation sign $\sum$. The symbol implies the product of n quantities $Z_1, \ldots, Z_n$:

$$\prod_{i=1}^{n} Z_i = Z_1 \cdot Z_2 \cdot \ldots \cdot Z_n.$$

The geometric mean of the Z_i is

$$Mg_Z = (Z_1 \cdot Z_2 \cdot \ldots \cdot Z_n)^{1/n} = \left(\prod_{i=1}^{n} Z_i\right)^{1/n}. \tag{l}$$

The geometric mean of n realizations of a log-Gaussian variable is no other than the median $Z_{0.5}$; in fact,

$$\mu_{\ln Z} = \ln Z_{0.5} = \ln Mg_Z. \tag{m}$$

Consider now some actual sample values of the log-Gaussian variable which are grouped in c classes that are equidistant in $\ln Z$ and of center $\ln Z_t$; hence the frequency table $(\ln Z_i, f_i)$ where

$$i = 1, 2, \ldots, c \quad \text{and} \quad \sum_1^c f_i = n.$$

We have

$$Moy_{\ln Z} = \frac{\sum_i^c f_i \ln Z_i}{\Sigma f_i} = \frac{\sum_i^c \ln Z_i^{f_i}}{n} = \ln \sqrt[n]{\prod_{i=1}^c Z_i^{f_i}}. \tag{n}$$

In this case the geometric mean of the sample values is

$$Mg_Z = \sqrt[n]{Z_1^{f_1} \cdot Z_2^{f_2} \cdot \ldots \cdot Z_n^{f_n}}. \tag{o}$$

Since the centers of classes $X_i = \ln Z_i$ are equidistant, the original Z_i are no longer so, but are arranged in a geometric progression. A useful system of coding the Z_i is given in Swaroop (1966, pp. 112–13).

(e) *Relation between the mode, mean and median of a log-Gaussian law.* We again take the relations (g″) and (j) in (p) and (q) respectively:

$$\frac{Mo_Z}{Z_{0.5}} = e^{-\sigma^2_{\ln Z}}, \tag{p}$$

$$\frac{\mu_Z}{Z_{0.5}} = e^{\frac{1}{2}\sigma^2_{\ln Z}}. \tag{q}$$

On squaring the two sides of (q), we get

$$\frac{\mu_Z^2}{Z_{0.5}^2} = e^{\sigma^2_{\ln Z}}, \tag{r}$$

the inverse of which is written

$$\frac{Z_{0.5}^2}{\mu_Z^2} = e^{-\sigma^2_{\ln Z}}. \tag{s}$$

Equating the left-hand sides of (p) and (s), we have

$$\frac{Mo_z}{Z_{0.5}} = \frac{Z_{0.5}^2}{\mu_Z^2}. \tag{t}$$

Since in proportion, the product of the mean terms is equal to the product of the extreme terms, we have

$$Z_{0.5}^3 = \mu_Z^2 \times Mo_Z. \tag{u}$$

In a log-Gaussian law, the cube of the median is equal to the product of the mode with the squared mean. This relation is verified without difficulty for the reduced variable $z = Z : Z_{0.5}$.

In effect,

$$1^3 = (e^{1:2})^2 \times e^{-1} = 1.$$

6.335.112.2. Parameters of variability

Relation between σ_Z^2 and $\sigma_{\ln Z}^2$:

With reference to 6.335.115 (m), one may write:

$$\sigma_Z^2 = \exp(2 \ln Z_{0.5} + \sigma_{\ln Z}^2) \cdot \left(e^{\sigma_{\ln Z}^2} - 1\right). \qquad \text{(a)}$$

By virtue of the relation (i) in the above sub-section, we introduce the squared mean μ_Z^2 into the second term of (a) and so obtain

$$\sigma_Z^2 = \mu_Z^2\left(e^{\sigma_{\ln Z}^2} - 1\right). \qquad \text{(b)}$$

Since the coefficient of variation is

$$\gamma_Z = \frac{\sigma_Z}{\mu_Z}, \qquad \text{(c)}$$

on setting, with Finney (1941),

$$\tau = e^{\sigma_{\ln Z}^2}, \qquad \text{(c')}$$

we get

$$\gamma_Z^2 = e^{\sigma_{\ln Z}^2} - 1 = \tau - 1, \qquad \text{(d)}$$

$$\tau = 1 + \gamma_Z^2, \qquad \text{(d')}$$

or

$$\sigma_{\ln Z}^2 = \ln(1 + \gamma_Z^2) = \ln \tau. \qquad \text{(e)}$$

The variance of $\ln Z$ is equal to the Napierian logarithm of $1+\gamma_Z^2$, where the coefficient of variation of Z is expressed as a fraction of unity. For the human body weight, $\gamma_Z = 0.13$ is the critical value which, according to Cochran (1938), divides the Gaussian adjustments from the log-Gaussian adjustments.

Special cases:

(a) Small γ_Z. Whenever the square of the coefficient of variation is a small quantity compared with unity ($\gamma_Z^2 \ll 1$), the higher powers γ_Z^4, ... are negligible in relation to 1 and to γ_Z^2. In this case the Maclaurin series expansion is limited to the first term:

$$\ln(1 + \gamma_Z^2) = \gamma_Z^2 + \ldots. \qquad \text{(f)}$$

Therefore

$$\sigma^2_{\ln Z} \approx \gamma^2_Z \tag{g}$$

and

$$\sigma_{\ln Z} \approx \gamma_Z \, . \tag{h}$$

The standard deviation of $\ln Z$ is approximately equal to the coefficient of variation of Z expressed as a fraction of unity. This relation is very useful for it enables *mechano-graphical mass calculations* to be verified without difficulty with reference to the original values Z, and to their transforms $\log_{10} Z$ as well.

The above method has been used on samples ranging from 1000 to 40,000 Belgian recruits (Martin, 1956), and for groups of schoolchildren of the Commune d'Uccle (Brussels) stratified according to age (Martin, 1962c).

(b) γ_Z of the order of unity. By virtue of the relation (e), whenever $\gamma_Z \approx 1$, i.e. when $\sigma^2_{\ln Z}$ or the coefficient of variation of Z is 100%, we have

$$e^{\sigma^2_{\ln Z}} = 1+1 = 2,$$

therefore

$$\sigma^2_{\ln Z} = \ln 2 = 0.693. \tag{i}$$

Values as high as 100% or more seldom occur in clinical data except for discrete (Poissonian or negative binomial) variables, or continuous (negative exponential) variables. However, when $\sigma^2_{\ln Z} = 1$, as is often assumed in graphical representations of the log-Gaussian law, the relation (f) yields:

$$e^1 = 1+\gamma^2_Z,$$

therefore

$$\gamma^2_Z = e-1 = 1.718,$$

or, further,

$$\gamma_Z = 1.31. \tag{j}$$

Thus, when the variance of $\ln Z$ is equal to unity, the coefficient of variation of the original variable is very high and equals 131%.

6.335.112.3. Shape parameters of the log-Gaussian law

Skewness: The calculations show that $\gamma_1(Z)$, the skewness parameter of the law $h(Z)$ which is positively skewed, depends only on the coefficient of variation of Z [(see d') above], or on τ (Aitchison and Brown, 1957, p. 8 and section 6.335.115).

$$\gamma_1(Z) = \gamma_Z(3+\gamma^2_Z) \tag{a}$$

$$= \sqrt{\tau-1}\,(\tau+2), \tag{b}$$

where γ_Z is the coefficient of variation of Z.

The skewness parameter is essentially positive since the coefficient of variation is always positive. Moreover, $\gamma_1(Z)$ increases with increasing γ_Z.

It is well to remember that $\gamma_1(\ln Z) = 0$, for the log-Gaussian law for $\ln Z$ is symmetrical and, in this case, the skewness is zero. This difference of behavior naturally involves specifying the variable, the shape parameter of which is under consideration; hence the notations

$$\gamma_1(Z) \quad \text{and} \quad \gamma_1(\ln Z).$$

Kurtosis: Calculation likewise yields the kurtosis parameter, which also depends only on the coefficient of variation of Z or on τ

$$\gamma_2(Z) = \gamma_Z^2(\gamma_Z^6+6\gamma_Z^4+15\gamma_Z^2+16) \tag{c}$$

$$= (\tau-1)(\tau^3+3\tau^2+6\tau+6). \tag{d}$$

The kurtosis is always positive, and it increases with increasing γ_Z.

Of course, $\gamma_2(\ln Z) = 0$ since $\ln Z$ is Gaussian.

From γ_1 and γ_2, the shape parameters of K. Pearson are

$$\beta_1 = \gamma_1^2 \quad \text{and} \quad \beta_2 = 3+\gamma_2. \tag{e}$$

These values, plotted in the classical (β_1, β_2) Pearson-diagram, lead to the S_L curve which inter-relates the shape parameters of the family of log-Gaussian densities (see below, 6.335.112.5 and K. Pearson, 1933, p. 215).

6.335.112.4. Theoretical and practical importance of the mean/median ratio of a log-Gaussian variable. In 6.335.112.1 (j) it was seen that

$$\frac{\mu_Z}{Z_{0.5}} = e^{\frac{1}{2}\sigma_{\ln Z}^2}. \tag{a}$$

Taking Napierian logarithms, this relation becomes

$$\ln \frac{\mu_Z}{Z_{0.5}} = \frac{1}{2}\sigma_{\ln Z}^2. \tag{b}$$

Hence the relation between the variance of $\ln Z$ and the mean/median ratio is

$$\sigma_{\ln Z}^2 = 2 \ln \frac{\mu_Z}{Z_{0.5}} = \ln \left(\frac{\mu_Z}{Z_{0.5}}\right)^2. \tag{c}$$

From the practical point of view, this relationship is of great importance. Thus, if the original distribution of Z is not even Gaussian, but simply symmetrical, we have $\mu_Z = Z_{0.5}$; hence, by virtue of (c), the theoretical variance of the logarithm of a random variable with a symmetrical probability density is rigorously zero. This surprising, but logical, result shows that one has to be very prudent in the transformation of variables and, in every case, to watch the symmetry of the law if it is real. Elsewhere, Gumbel (1960, p. 16) writes: "Some practical statisticians believe that every skewed distribution should be logarithmically normal and therefore use the logarithmic transformation indiscriminately."

We again take the relation (a). In accordance with 6.335.112.2 (c), we have

$$\frac{\mu_Z}{Z_{0.5}} = \exp\left\{\frac{1}{2}\ln(1+\gamma_Z^2)\right\} = \exp\{\ln\sqrt{(1+\gamma_Z^2)}\}$$
$$= \sqrt{1+\gamma_Z^2} = \sqrt{\tau}. \qquad \text{(a)}$$

Since γ_Z^2 is always positive, in a log-Gaussian law the mean μ_Z is always greater than the median $Z_{0.5}$.

6.335.112.5. The (β_1, β_2)*-diagram of K. Pearson. Locus of the log-normal points* (β_1, β_2). The parametric equations of the locus of points (β_1, β_2) corresponding to the log-normal densities have been presented by Wicksell (1917), K. Pearson (1933), Johnson (1949) and Martin (1972).

Owing to the intuitive meaning of the coefficient of variation of Z, i.e. γ_Z, and for the facility of the reader, we present here the parametric equations of the line (β_1; β_2) in terms of the parameter γ_Z (or of $\tau = 1+\gamma_Z^2$) as written in 6.335.112.3 (a), (b), (c) and (d):

$$\beta_1 = \gamma_Z^2(3+\gamma_Z^2)^2, \qquad \text{(a)}$$
$$\beta_2 = \gamma_Z^2(\gamma_Z^6+6\gamma_Z^4+15\gamma_Z^2+16)+3. \qquad \text{(b)}$$

It is thus easy to construct the log-Gaussian locus in the (β_1, β_2) diagram. If we put

$$\omega^2 = \tau \qquad \text{(c)}$$

we get the relations 6.64 and 6.65 written in Kendall and Stuart (1963, I, p. 169).

6.335.113. Ordinates of the log-Gaussian probability density. The log-Gaussian distribution

$$h(Z) = \frac{1}{Z}\,\frac{1}{\sqrt{2\pi}\sigma_{\ln Z}}\exp-\left\{\frac{\left(\ln\dfrac{Z}{Z_{0.5}}\right)^2}{2\sigma_{\ln Z}^2}\right\} \qquad \text{(a)}$$

has three points of particular importance for which we shall calculate the ordinates, namely, the mode, median and mean.

(a) *Mode*. By virtue of the relation 6.335.112.1 (p), we have

$$\frac{Mo_Z}{Z_{0.5}} = e^{-\sigma^2_{\ln Z}}.$$

Therefore

$$\ln \frac{Mo_Z}{Z_{0.5}} = -\sigma^2_{\ln Z}.$$

The ordinate at the mode is then written as

$$h(Mo_Z) = \frac{1}{Mo_Z} \cdot \frac{1}{\sqrt{2\pi}\sigma_{\ln Z}} \exp\left\{-\frac{\sigma^2_{\ln Z}}{2}\right\}. \tag{b}$$

Thus the ordinate at the mode of the log-Gaussian probability density can be calculated from the mode Mo_Z and the standard deviation $\sigma_{\ln Z}$.

(b) *Median:*

$$h(Z_{0.5}) = \frac{1}{Z_{0.5}} \frac{1}{\sqrt{2\pi}\sigma_{\ln Z}}.$$

Same remark as for the mode.

$$\frac{h(Z_{0.5})}{h(Mo_Z)} = \frac{\dfrac{1}{Z_{0.5}} \dfrac{1}{\sqrt{2\pi}\sigma_{\ln Z}}}{\dfrac{1}{Mo_Z} \dfrac{1}{\sqrt{2\pi}\sigma_{\ln Z}}} \cdot \frac{1}{\exp\left(-\dfrac{1}{2}\sigma^2_{\ln Z}\right)}$$

$$= \frac{Mo_Z}{Z_{0.5}} \cdot e^{\frac{1}{2}\sigma^2_{\ln Z}}.$$

By virtue of 6.335.112.1 (p), there remains

$$\frac{h(Z_{0.5})}{h(Mo_Z)} = e^{-\sigma^2_{\ln Z}} \cdot e^{+\frac{1}{2}\sigma^2_{\ln Z}} = e^{-\frac{1}{2}\sigma^2_{\ln Z}}. \tag{c}$$

As expected, the ordinate at the median is less than the ordinate to the mode.

(c) *Mean*. By virtue of the relation 6.335.112.1 (q), we have

$$\frac{\mu_Z}{Z_{0.5}} = e^{\frac{1}{2}\sigma^2_{\ln Z}}.$$

Therefore,

$$\ln\left(\frac{\mu_Z}{Z_{0.5}}\right) = \frac{1}{2}\sigma^2_{\ln Z}$$

and

$$\ln^2\left(\frac{\mu_Z}{Z_{0.5}}\right) = \frac{1}{4}\sigma^4_{\ln Z}.$$

Hence

$$h(\mu_Z) = \frac{1}{\mu_Z}\frac{1}{\sqrt{2\pi}\sigma_{\ln Z}}\exp -\frac{1}{4}\cdot\frac{\sigma^4_{\ln Z}}{2\sigma^2_{\ln Z}}$$

$$= \frac{1}{\mu_Z}\frac{1}{\sqrt{2\pi}\sigma_{\ln Z}}\exp\left\{-\frac{\sigma^2_{\ln Z}}{8}\right\}. \qquad \text{(d)}$$

The ratio of the ordinates to the mean and to the median is

$$\frac{h(\mu_Z)}{h(Z_{0.5})} = \frac{\dfrac{1}{\mu_Z}\dfrac{1}{\sqrt{2\pi}\sigma_{\ln Z}}\exp -\left\{\dfrac{\sigma^2_{\ln Z}}{8}\right\}}{\dfrac{1}{Z_{0.5}}\dfrac{1}{\sqrt{2\pi}\sigma_{\ln Z}}} = \frac{Z_{0.5}}{\mu_Z}\exp -\left\{\frac{\sigma^2_{\ln Z}}{8}\right\}. \qquad \text{(e)}$$

6.335.114. Frequency density of the transformed variable ln Z *from the frequencies of the original variable Z. Method of Sven Odén (1926).* Let $h(Z)$ be the frequency function of a log-Gaussian variable Z based on equidistant classes from the respective centers, and on $H(Z)$ which is the distribution of Z.

According to Herdan (1953, p. 126), Sven Odén (1926) gave a formula which enabled the frequency density of the transformed variable $X = \ln Z$, i.e. $g(\ln Z)$, to be calculated from $h(Z)$ without having to resort to a regrouping in classes for $\ln Z$.

By definition of $H(Z)$, we have

$$dH(Z) = h(Z)\,dZ. \qquad \text{(a)}$$

Multiplying the right-hand side top and bottom by Z, we obtain

$$h(Z)\,dZ = Zh(Z)\frac{dZ}{Z} = Zh(Z)\,d\ln Z. \qquad \text{(b)}$$

The probability density of $\ln Z$ is therefore written as

$$g(\ln Z) = Zh(Z). \qquad \text{(c)}$$

From the histogram (Z, f_i), we calculate the relative frequencies (pure numbers), knowing that $\Sigma f_i = n$ and that the class interval is equal to I,

$$g_i = \frac{Z_i f_i}{\Sigma f_i Z_i} = \frac{f_i Z_i}{n\bar{Z}}. \tag{d}$$

The g_i are centered on

$$X_i = \tfrac{1}{2}[\ln (Z_i + I/2) + \ln (Z_i - I/2)]. \tag{e}$$

This formula is valid in every case.

If, however, I is small in relation to the Z_i, one may write

$$\ln\left(Z_i + \frac{I}{2}\right) = \ln Z_i\left(1 + \frac{I}{2Z_i}\right)$$

$$= \ln Z_i + \ln\left(1 + \frac{I}{2Z_i}\right) \approx \ln Z_i + \frac{I}{2Z_i}$$

on limiting the expansion to the first term.

Likewise,

$$\ln\left(Z_i - \frac{1}{2}\right) \approx \ln Z_i - \frac{I}{2Z_i}.$$

Therefore the class center of the transformed Gaussian variable is approximately

$$X_i \approx \ln Z_i. \tag{f}$$

If the initial histogram is log-Gaussian, a new histogram (X_i, g_i) is thus obtained which is Gaussian.

In order to check this characteristic, the method of probits can be applied, using the following procedure:

(1) Accumulate the relative frequencies g_i defined in (d).
(2) Take the probits corresponding to these cumulative relative frequencies.
(3) Plot the probits as a function of the transformed ends of intervals $\ln (Z_i + I/2)$.
(4) Verify the correspondence between the co-graduated variables Z_p and $X_p = \ln Z_p$, and in particular for $p = 0.50$.

6.335.115. Moments of the log-Gaussian variable with two parameters. The log-Gaussian variable Z is such that the relation

$$X = \ln Z \tag{a}$$

obeys a Gaussian law:

$$dG(X) = g(X)\,dX = \frac{1}{\sigma_X\sqrt{2\pi}} \exp\left\{-\frac{(X-\mu_X)^2}{2\sigma_X^2}\right\} dX. \tag{b}$$

For the variable

$$Z = e^X, \tag{c}$$

we have

$$d\Lambda(Z) = \lambda(Z)\,dZ = \frac{1}{Z\sigma_{\ln Z}\sqrt{2\pi}} \exp\left\{-\frac{(\ln Z-\mu_{\ln Z})^2}{2\sigma_{\ln Z}^2}\right\} dZ. \tag{d}$$

The jth order moment of Z about the origin $Z = 0$ is

$$\mu_j' = E[Z^j] = \int_0^\infty Z^j\, d\Lambda(Z), \tag{e}$$

or, according to (c),

$$\begin{aligned} \mu_j' = E[e^{jX}] &= \int_{-\infty}^{+\infty} e^{jX}\, dG(X) \\ &= \frac{1}{\sqrt{2\pi}} \int_{-\infty}^{+\infty} \exp\left\{jX - \frac{(X-\mu)^2}{2\sigma_X^2}\right\} \frac{dX}{\sigma_X}. \end{aligned} \tag{f}$$

We introduce the standard normal deviate

$$u = \frac{X-\mu}{\sigma_X}, \quad du = \frac{dX}{\sigma_X} \tag{g}$$

and the value

$$jX = j\mu + j\sigma_X u. \tag{g'}$$

The relation (f) now becomes

$$\mu_j' = \frac{1}{\sqrt{2\pi}} \int_{-\infty}^{+\infty} \exp\left\{j\mu + j\sigma_X u - \frac{u^2}{2}\right\} du. \tag{h}$$

The sum of the terms of the exponential can be written by completing a square

$$\begin{aligned} j\mu + j\sigma_X u - \tfrac{1}{2}u^2 &= -\tfrac{1}{2}[u^2 - 2j\sigma_X u - 2j\mu] \\ &= -\tfrac{1}{2}(u - j\sigma_X)^2 + \tfrac{1}{2}(2j\mu + j^2\sigma_X^2). \end{aligned} \tag{i}$$

Taking the independent factor of u from the integral (h), we obtain

$$\mu_j' = \exp\left\{\frac{1}{2}(2j\mu+j^2\sigma_X^2)\right\}\cdot\underbrace{\frac{1}{\sqrt{2\pi}}\int_{-\infty}^{+\infty}\exp\left\{-\frac{1}{2}(u-j\sigma_X)^2\right\}du}_{1},$$

or

$$\mu_j' = \exp\{\tfrac{1}{2}(2j\mu+j^2\sigma_X^2)\}. \quad \text{(j)}$$

Hence the mean of the log-Gaussian variable $Z\,(X = \ln Z)$ is

$$\mu_1'(Z) = \exp\{\mu_{\ln z}+\tfrac{1}{2}\sigma_{\ln z}^2\} = \mu_Z, \quad \text{(k)}$$

this value being written from the relation 6.335.112.1 (i).

The second moment about the origin of Z is

$$\mu_2'(Z) = \exp\{2\mu+2\sigma_{\ln z}^2\}. \quad \text{(l)}$$

Hence the variance of Z

$$\begin{aligned}\sigma_Z^2 = \mu_2'-\mu_1'^2 &= \exp\{2\mu_{\ln z}+2\sigma_{\ln z}^2\}-\exp(2\mu_{\ln z}+\sigma_{\ln z}^2)\\ &= \exp\{2\mu_{\ln z}+\sigma_{\ln z}^2\}(\exp\sigma_{\ln z}^2-1),\end{aligned} \quad \text{(m)}$$

as indicated in 6.335.112.2 (a). The third and fourth moment lead to the shape parameters.

6.335.116. Invariance of the log-Gaussian law with regard to various transformations of variables. It is given that Z is a log-Gaussian variable, the elementary probability of which is

$$d\Lambda(Z) = \lambda(Z)\,dZ = \frac{1}{Z}\frac{1}{\sigma_{\ln z}\sqrt{2\pi}}\exp\left\{-\frac{(\ln Z-\mu_{\ln z})^2}{2\sigma_{\ln z}^2}\right\}dZ. \quad \text{(a)}$$

With Aitchison and Brown (1957, p. 11), the log-normal variable is written $\Lambda(Z;\mu,\sigma^2)$, implying a subscript $\ln Z$ for μ and σ.

The following properties can be claimed:

If Z is a log-Gaussian variable $\Lambda(Z;\mu,\sigma\ )$, then

cZ is distributed as $\Lambda(cZ;\mu+\ln c;\sigma^2)$, (b)

$1/Z$ is distributed as $\Lambda(1/Z;-\mu,\sigma^2)$, (c)

Z^2 is distributed as $\Lambda(Z^2;2\mu,4\sigma^2)$, (d)

Z^3 is distributed as $\Lambda(Z^3;3\mu,9\sigma^2)$. (d')

If a radius has a log-Gaussian distribution, the area of the circle or the volume of the sphere will also have a log-Gaussian distribution. This is one of the important properties of the log-Gaussian variable with two parameters.

In general, for Z being a $\Lambda(Z; \mu, \sigma^2)$ variable,

$$cZ^b \text{ is distributed as } \quad \Lambda(cZ^b; b\mu + \ln c; b^2\sigma^2). \tag{e}$$

Finally, the product Z_1Z_2 of two independent log-Gaussian variables with the means and variances (μ_1, σ_1^2) and (μ_2, σ_2^2) respectively, is distributed as

$$\Lambda(Z_1 Z_2; \mu_1+\mu_2, \sigma_1^2+\sigma_2^2),$$

where, it will be recalled, $\mu_1 = \mu_{\ln Z}$ and $\sigma_1^2 = \sigma_{\ln Z_1}^2$, the same applying for the subscript 2.

6.335.12. Log-Gaussian distribution. Co-graduation between ln z_p and u_p

The log-Gaussian distribution of Z, of median $Z_{0.5}$ and standard deviation $\sigma_{\ln Z}$, is written as

$$p = \Lambda(Z) = \text{Prob}\,(\mathbf{Z} \leqslant Z) = \int_0^Z h(Z)\,dZ. \tag{a}$$

This distribution, like the probability density, is positively skew and depends on two parameters.

The simplest case of log-Gaussian distribution of median and standard deviation 1 is given in Fig. 6.335.112.1I (b). Let us now refer to the *co-graduation* between the reduced log-Gaussian variable and the standard Gaussian deviate. Consider now Fig. 6.335.12I.

Abscissae: Reduced variable:

$$z = \frac{Z}{Z_{0.5}}. \tag{b}$$

Since $\sigma_{\ln Z} = 1$, the log-normal probability density of z is written

$$\lambda(z) = \frac{1}{z\sqrt{2\pi}} \exp\left\{-\frac{1}{2}(\ln z)^2\right\} \tag{c}$$

and is represented in part (A).

The standard normal density of u is represented in (B) on the upper right scale with $\varphi(u)$ on a perpendicular axis.

$$\varphi(u) = \frac{1}{\sqrt{2\pi}} e^{-u^2/2}.$$

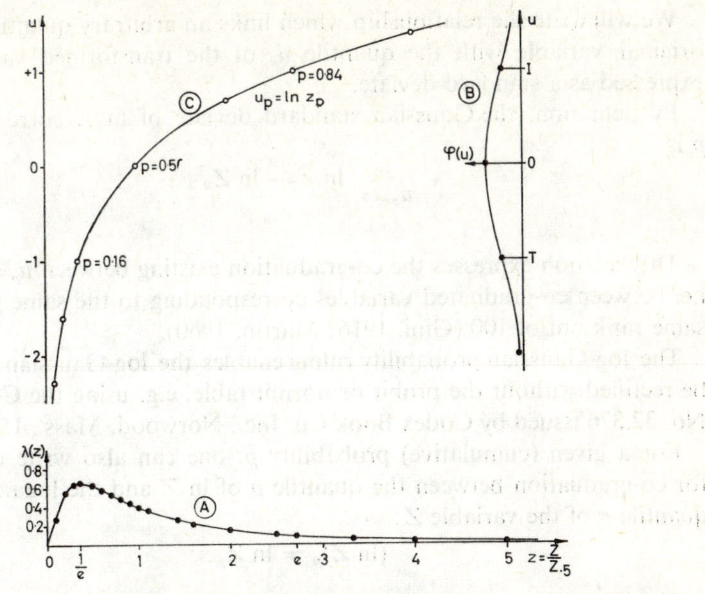

FIG. 6.335.12 I. Co-graduation between the reduced log-Gaussian (normal) variable z (A) and the standard Gaussian (normal) deviate u (B) via the co-graduation curve $u_p = \ln z_p$ (C)

The formula of co-graduation between u and z is written (Gini, 1916; Martin, 1960)

$$u_p = \ln z_p \tag{d}$$

and is represented in Fig. 6.335.12I (C). The three circled points in the curve (C) correspond to $p = 0.16$, $p = 0.5$ and $p = 0.84$ respectively. We are now looking for the general case.

Since, by definition, the transformed variable

$$X = \ln Z$$

is a Gaussian variable, we will write *the relation which associates the quantiles of Z with those of* $X = \ln Z$.

In fact, we require the functional relation which, using (b), links Z_p to X_p, or which associates the Z_p with the corresponding Gaussian standard deviate

$$u_p = \frac{X_p - X_{0.5}}{\sigma_X}. \tag{e}$$

We will write the relationship which links an arbitrary quantile Z_p of the original variable with the quantile u_p of the transformed variable ln Z expressed as a standard deviate.

By definition, the Gaussian standard deviate of ln Z corresponding to p is

$$u_p = \frac{\ln Z_p - \ln Z_{0.5}}{\sigma_{\ln Z}}. \qquad (e')$$

This relation expresses the co-graduation existing between u_p and $(\ln Z)_p$, i.e. between co-graduated variables corresponding to the same p, or to the same rank out of 100 (Gini, 1916; Martin, 1960).

The log-Gaussian probability ruling enables the log-Gaussian sigmoid to be rectified without the probit or normit table, e.g. using the Codex ruling No. 32.376 issued by Codex Book Co. Inc., Norwood, Mass., U.S.A.

For a given (cumulative) probability p, one can also write the relation for co-graduation between the quantile p of ln Z and the logarithm of the quantile p of the variable Z:

$$(\ln Z)_p = \ln Z_p.$$

The normal standard deviate (e′) then becomes

$$u_p = \frac{\ln (Z_p/Z_{0.5})}{\sigma_{\ln Z}}. \qquad (e'')$$

Therefore,

$$\ln \frac{Z_p}{Z_{0.5}} = u_p \cdot \sigma_{\ln Z},$$

$$\frac{Z_p}{Z_{0.5}} = \exp\{u_p \cdot \sigma_{\ln Z}\}, \qquad (f)$$

$$Z_p = Z_{0.5} \cdot \exp\{u_p \cdot \sigma_{\ln Z}\},$$

finally, or

$$Z_p = \exp\{(\mu_{\ln Z} + u_p \cdot \sigma_{\ln Z})\}. \qquad (f')$$

Thus one again finds, except for the notation, the formula (2.14) appearing in Aitchison and Brown (1957, p. 9).

Special cases. We have already seen the importance of the quantiles 0.025 0.16, 0.50, 0.84 and 0.975 in regard to the Gauss curve.

Table 6.335.12A, on the left-hand side, shows the relation (f) in detail as $Z_p/Z_{0.5}$.

The central column shows the values of the Gaussian standard deviate u_p which occurs in both series. The corresponding values for a Gaussian variable X appear on the right-hand side.

TABLE 6.335.12A. EQUI-p LOG-GAUSSIAN AND GAUSSIAN QUANTILES

Cumulative probability p	Log-Gaussian law in Z	u_p	Gaussian law in X
0.025	$\frac{Z_{0.025}}{Z_{0.50}} = \exp\{-1.96\sigma_{\ln Z}\}$	−1.96	$X_{0.025} - X_{0.50} = -1.96\sigma_X$
0.16	$\frac{Z_{0.16}}{Z_{0.50}} = \exp\{-1.\sigma_{\ln Z}$	−1	$X_{0.16} - X_{0.50} = -1.\sigma_X$
0.50		0	
0.84	$\frac{Z_{0.84}}{Z_{0.50}} = \exp\{+1.\sigma_{\ln Z}\}$	+1	$X_{0.84} - X_{0.50} = +1.\sigma_X$
0.975	$\frac{Z_{0.975}}{Z_{0.50}} = \exp\{+1.96\sigma_{\ln Z}\}$	+1.96	$Z_{0.975} - X_{0.50} = +1.96\sigma_X$

For the cumulative probabilities $p = 0.16$ and $p = 0.84$ respectively, one may write

$$\frac{Z_{0.84}}{Z_{0.50}} \times \frac{Z_{0.16}}{Z_{0.50}} = \exp\{(-1\sigma_{\ln Z} + 1 \cdot \sigma_{\ln Z})\} = e^0 = 1, \quad \text{(g)}$$

or, further,

$$Z_{0.16} \times Z_{0.84} = Z_{0.50}^2. \quad \text{(h)}$$

In a log-Gaussian distribution in Z, the median is the geometric mean of the quantile 0.16 and the quantile 0.84. This property will be used as the point of departure in measuring the degree of non-conformity of some variable V to the log-Gaussian model, and in order to calculate a correction V_0 which leads to a log-Gaussian variable $V + V_0$.

Generalization. We again take the relation (f) for two symmetrical fractions p and $(1-p)$ about 0.5:

$$\frac{Z_p}{Z_{0.50}} = \exp\{u_p \sigma_{\ln Z}\}, \quad \text{(i)}$$

$$\frac{Z_{(1-p)}}{Z_{0.50}} = \exp\{u_{(1-p)} \sigma_{\ln Z}\}. \quad \text{(j)}$$

Since u is a standard Gaussian deviate, and symmetrical in particular, therefore

$$u_{(1-p)} = -u_p\,. \tag{k}$$

Multiplying termwise the relations (h) and (j), and having regard to (k), we obtain the relation

$$Z_p \times Z_{(1-p)} = Z_{0.50}^2\,, \tag{l}$$

which extends to an arbitrary p $(0 < p < 1)$ the relation (h) written for $p = 0.16$.

In a log-Gaussian distribution the median is the geometric mean of the quantiles corresponding to probabilities p and $(1-p)$ which are symmetrical about 0.50.

To what quantile does the mean of the log-Gaussian law correspond?

We again take the relation (i)

$$\frac{Z_p}{Z_{0.5}} = \exp\{u_p \sigma_{\ln Z}\}.$$

Now, by virtue of 6.335.112.1 (q), we have

$$\frac{\mu_Z}{Z_{0.5}} = \exp\left\{\frac{1}{2}\sigma_{\ln Z}^2\right\}.$$

Therefore, if $Z_p = \mu_Z$, the arguments of the exponentials should be equal, i.e.

$$u_p \sigma_{\ln Z} = \tfrac{1}{2}\sigma_{\ln Z}^2\,, \tag{m}$$

or

$$u_p = \tfrac{1}{2}\sigma_{\ln Z}\,. \tag{m'}$$

Thus, in units of $\sigma_{\ln Z}$ we have

$$u_p = +0.5.$$

Hence the probability p which corresponds to the mean of a log-Gaussian law in Z is obtained by writing

$$p = \Phi(+\tfrac{1}{2}) = 0.692. \tag{n}$$

This relation is resolved by consulting Table 6.322.212A with $u = 0.5$ in column 1, and referring to the right-hand side column 6 with $p = 0.692$.

This relation can also be written in the original variable Z:

$$\mu_Z = Z_{0.692}\,, \tag{o}$$

as illustrated in Fig. 6.335.112I.

The three parameters of localization of the log-Gaussian variable Z are therefore written as functions of the median

$$\begin{aligned} &\text{Mode:} && Z_{0.158} \doteq Z_{0.5}\times e^{-1} \\ &\text{Median:} && Z_{0.500} = Z_{0.5}\times e^{0} \\ &\text{Mean:} && Z_{0.692} = Z_{0.5}\times e^{1/2}. \end{aligned}$$

6.335.13. Log-Gaussian law in common or decimal logarithms

Up to the present, the log-Gaussian law has been presented in terms of natural or Napierian logarithms. This treatment produces direct results because the Gauss law, usually written by means of the transcendental $e = 2.71\ldots$, opens the way to natural logarithms. The formulae are simplified by the relations of definition existing between the natural log of a number N and this number itself:

$$\begin{cases} e^{\ln N} = N & \ln e = 1, \\ \ln e^{N} = N & \ln 1 = 0. \end{cases} \qquad \text{(a)}$$

However, the majority of calculations are performed in common logarithms, as defined by the following relations:

$$\begin{cases} 10^{\log_{10} N} = N & \log_{10} 10 = 1, \\ \log_{10} 10^{N} = N & \log_{10} 1 = 0. \end{cases} \qquad \text{(b)}$$

It is therefore appropriate to give the main formulae of the log-Gaussian law, as written in common logarithms, explicitly introducing the basis 10 to avoid any hesitation of the reader between Log, log and ln.

The formulae for the changeover from the Napierian system to the decimal system (common logarithms) are:

$$\log_{10} U = \log_{10} e\times\ln U, \qquad \text{(c)}$$

The $\log_{10} e$, or modulus, is represented by M and equals

$$\log_{10} e = M = 0.4343. \qquad \text{(d)}$$

Table 6.335.13A shows the two versions of the log-Gaussian law in $\ln Z$ and $\log_{10} Z$, written log.

Table 6.335.13A

Log-Gaussian Law in Natural Logarithms and Common Logarithms
Original Log-Gaussian Variable Z

Localization:	$\mu_Z, Z_{0.5}, Mo_Z$
Dispersion:	σ_Z
Coefficient of variation:	$\gamma_Z = \dfrac{\sigma_Z}{\mu_Z}$
Coefficient of asymmetry:	$\gamma_1(Z) = \gamma_Z(3+\gamma_Z^2)$
Coefficient of kurtosis:	$\gamma_2(Z) = \gamma_Z^2(\gamma_Z^6+6\gamma_Z^4+15\gamma_Z^2+16)$

Logarithmic Transforms of Z

Natural logarithms (ln)	*Common logarithms* (log)
	$M = \log e = 0.4343$
$\ln Z$	$\log Z = M \ln Z$
$\sigma_{\ln Z}$	$\sigma_{\log Z} = M\sigma_{\ln Z}$
Median	
$Z_{0.5}$	$Z_{0.5}$
Mean	
$\mu_Z = Z_{0.5}e^{\frac{1}{2}\sigma^2_{\ln Z}}$	$\mu_Z = Z_{0.5}\, 10^{\frac{1}{2}\times(\sigma^2 \log Z)/M}$
Mode	
$Mo_Z = Z_{0.5}e^{-\sigma^2_{\ln Z}}$	$Mo_Z = Z_{0.5}\, 10^{-(\sigma^2 \log Z)/M}$
Variance	
$\sigma^2_{\ln Z} = \ln(1+\gamma_Z^2)$	$\sigma^2_{\log Z} = M \log(1+\gamma_Z^2)$
Standard Deviation	
$\sigma_{\ln Z} \approx \gamma_Z$	(γ_Z small) $\quad \sigma_{\log Z} \approx M\gamma_Z$

Probability density

$$h(Z) = \frac{1}{Z\sigma_{\ln Z}\sqrt{2\pi}} \exp\left\{-\frac{(\ln Z - \ln Z_{0.5})^2}{2\sigma^2_{\ln Z}}\right\}$$

$$h(Z) = \frac{M}{Z\sigma_{\log Z}\sqrt{2\pi}} \exp\left\{-\frac{(\log Z - \log Z_{0.5})^2}{2\sigma^2_{\log Z}}\right\},$$

exp meaning here $e = 10^M$.

6.335.14. Approximation of the log-Gaussian law by the Gaussian law

We again consider the relation 6.335.12 (i)

$$\frac{Z_p}{Z_{0.5}} = \exp\{u_p\sigma_{\ln Z}\}, \tag{a}$$

and expand the exponential of the right-hand side

$$\frac{Z_p}{Z_{0.5}} = 1 + u_p\sigma_{\ln Z} + \frac{1}{2!}u_p^2\sigma_{\ln Z}^2 + \dots . \tag{a'}$$

If $\sigma_{\ln Z}^2$ is negligible, we have the linear relation for first approximation

$$\frac{Z_p}{Z_{0.5}} = 1 + u_p\sigma_{\ln Z}, \tag{a''}$$

or,

$$u_p = \frac{1}{\sigma_{\ln Z}}\left(\frac{Z_p - Z_{0.5}}{Z_{0.5}}\right). \tag{b}$$

Now

$$\sigma_{\ln Z} = \frac{\sigma_Z}{\mu_Z}. \tag{c}$$

Therefore

$$\sigma_{\ln Z}\cdot Z_{0.5} = \sigma_Z\cdot\frac{Z_{0.5}}{\mu_Z}. \tag{d}$$

The denominator of relation (b) is equal to the standard deviation of Z multiplied by the ratio of the median of Z to the mean of Z, a ratio which is less than unity.

Under the conditions where $\sigma_{\ln Z}^2$ or the square of the coefficient of variation $(\sigma_z/\mu_Z)^2$ is negligible, one may write the standard Gaussian deviate (b) or N.E.D. in the form

$$u_p = \frac{Z_p - Z_{0.5}}{\sigma_Z}\cdot\frac{\mu_Z}{Z_{0.5}}, \tag{e}$$

or, according to 6.335.112.1 (q),

$$u_p = \frac{Z_p - Z_{0.5}}{\sigma_Z}\cdot\exp\left\{\frac{1}{2}\sigma_{\ln Z}^2\right\},$$

i.e. as the product of the reduced deviate of Z *with regard to the median of* Z multiplied by the ratio $\mu_Z : Z_{0.5}$, or the quantity

$$\exp\{\tfrac{1}{2}\sigma^2_{\ln z}\}$$

which is greater than unity.

If the coefficient of variation of Z is about 0.14, then its square $0.14^2 = 0.0196 \approx 0.02$ can be regarded as negligible compared with unity.

6.335.15. *Value of the logarithmic transformation*

Cochran (1938) and Hald (1952a, p. 164) have stated that even if the logarithmic transformation is useful, it still does not lead to results that differ appreciably to those of the analysis in non-transformed data if the coefficient of variation is less than about 12%. However, the 12–14% range for the coefficient of variation of the adult human body weight could not be closer to the critical value mentioned above. All the same, it seems to us that there is less risk of distortion if the known normal adult weight is regarded as log-Gaussian rather than Gaussian. More specifically, we refer the reader to Karpinos (1958, p. 410) who explains why he regards the weight as log-Gaussian (and the height as Gaussian), and to Martin (1956) for the representation of the probit-weight parabola of young adults, its rectification by the logarithmic transformation, and the linear correlation between the height (Gaussian) and the log-weight (Gaussian).

6.335.16. *Log-Gaussian imminence rate, hazard function*

For a law of median 1 and of unit variance $\sigma^2_{\ln z}$, the log-Gaussian hazard function or imminence rate is written as

$$I(z) = \frac{\lambda(z)}{1-\Lambda(z)}, \qquad \text{(a)}$$

or even

$$I(z) = \frac{\dfrac{1}{z}\varphi(\ln z)}{1-\Phi(\ln z)}. \qquad \text{(b)}$$

The curve for $I(z)$ is shown in the upper part of Fig. 6.335.112.1I (see above). Unlike the Gaussian hazard function which increases monotonically

(6.322.228), the log-Gaussian hazard function begins by increasing as z evolves from 0 to $z \approx \frac{1}{2}$, then decreases *slowly* when z tends towards ∞.

This behavior, which has escaped the attention of many biologists and clinicians, leads to conclusions which are not expected in this particular field. The position will be discussed in the following sub-section under the definition *use and abuse*.

Remark: related to the imminence rate (hazard function), the concept of elasticity introduced on a formal basis in 6.322.229 finds its full significance in logarithmic coordinates. For such a log-Gaussian elasticity, we refer the reader to the classical monograph of Aitchison and Brown (1957).

6.335.17. Utilization and limitation of the log-normal law

For multiplicative phenomena, the log-Gaussian law, introduced by Galton (1879) and McAlister (1879), plays a role analogous to that of the Gaussian law for additive phenomena.

Very many biological quantities are presented in the form of histograms which are positively skewed. Up to the present point, under the impetus of the excellent review of Gaddum (1945), the log-Gaussian law serves as prototype for these positive asymmetrical situations, whereas the Gaussian law is used as prototype for symmetrical situations.

In *biological tests on all-or-none response*, the distribution of the dose D associated with the response probability $p(D)$ is generally described as being log-Gaussian.

Examples: the *dose of a digitalic preparation* which is lethal for the cat (Bliss and Hanson, 1939), the *insecticide action* of rotenone on *Macrosiphonella sanborni* (Finney, 1947, p. 28), and the *fungicide action* of Bordeaux mixture (an agricultural spray) on *Alternaria tenuis* (Martin, 1952).

A comparison of the log-normal with the log-logistic law has been made by Aitchison and Brown (1957, p. 74), who cite several references in regard to this "heated controversy".

Clinical biology. In this field King and Wootton (1956, pp. 1–2) have described the following biochemical quantities having a positively skew distribution of the log-Gaussian type. They also have given the percentiles 1, 10, 90 and 99% which may be used by the clinician to define ranges of clinical normality.

Total blood

Urea (mg per 100 ml)
Non-proteic nitrogen (mg per 100 ml)

Creatinine (mg per 100 ml)
Cholesterol (mg per 100 ml)
The log-Gaussian character of urea has been confirmed by Healy (1968) on the basis of data collected from 1000 blood donors.

Serum and plasma

Potassium (Eq mol per l)
Cholesterol (mg per 100 ml)
Total acid-soluble phosphate (mg per 100 ml)
Alkaline phosphatase (K.A. units per 100 ml)
Acid phosphatase (units per 100 ml)
Formol phosphatase, stable (units per 100 ml)
Amylase (units per 100 ml)

Another type of logarithmic transformations of clinical variables

$$X = \frac{\log (U+C)}{K}$$

has been introduced with the aid of a computer. The constants C and K for each biological constituent have been determined by trial and repetitive evaluation of $\sqrt{\beta_1}$ and β_2 until a Gaussian density was approximated (Mason and Bulgren, 1964, p. 98). We may also note that Dixon (1961, p. 201) in one of BIMED computer programs uses for some clinical data the transform $\log_{10} (x+c)$ with c specified.

Serology. In this field Ipsen and Jerne (1944) have noted the log-normal nature of the determinations of serological activity expressed in titre or units per ml.

We glean a few other examples:

In *human biometry* the *body weight* of adults, 20-year-old army recruits, has an asymmetrical distribution which is rendered Gaussian by the logarithmic transformation (Yuan, 1933; Martin, 1956; Karpinos, 1958). Hence the linear correlation ($r \approx 0.65$) between the log-weight and height, these being quantities of the additive type for both of which the coefficient of variation is 3 to 4%. Moreover (Martin, 1956), the two regression lines associated with this correlation are straight lines.

In *human physiology* the *systolic* and *diastolic pressure* have histograms which are positively skewed, and these have been described as log-Gaussian (Winsor, 1950). The question of the definition (and of the distribution) of the normal pressure "has always been a great problem which is still partially unsolved" (Bechgaard, 1960, p. 199).

Other authors see in this asymmetry, however, the reflection of an interblending of Gaussian curves (Nagasaka *et al.*, 1967).

According to Linder and Grandjean (1950) the *threshold of excitation of the patellar reflex* is log-Gaussian.

Variability in "demand" consumption, which is the multiplicative phenomenon *par excellence*, is described as being log-Gaussian; e.g. the *consumption of alcohol* in France ("plus on boit, plus on a soif",[†] Lederman, 1956, p. 267); consumption of electricity (Hald, 1952a, p. 166).

In the study of *particle size of powders* and of dispersed particle systems, the diagram probit–log particle size is widely used. See, for example, Loveland and Trivelli (1927) and the important book of Herdan (1953).

In *clinical pharmacology* Schurmann and Schirduan (1948) have studied the threshold dose of penicillin active against gonorrhoea in soldiers; the distribution of these doses is of the log-Gaussian type with a coefficient of variation equal to 25%. More particularly, Wagner (1967, p. 209) has shown the log-Gaussian nature of the half-lives of lincomycin and of novobiocin in volunteers. All adjustments were made by electronic computer.

In studying the *follow-up* of cancer patients after having received radiotherapy treatment, Boag (1948 and 1949) used a model in three parameters, where c is the proportion of patients definitely cured, and the survival time of the patients who were not cured (as a proportion $1-c$), follows a log-Gaussian law of mean μ_t and of standard deviation $\sigma_{\log t}$. This log-Gaussian model has been considered by Tudway and Freundlich (1960) applying the maximum-likelihood method of estimation and using an electronic computer, and also by Haybittle (1962). In addition to this, Haybittle compares the results of this approach with those obtained by using the actuarial method of estimating the T-year survival rate (Littell, 1952).

The foregoing examples are of the continuous random variable type. In the discontinuous domain, Uemura (1964) has shown that the log-Gaussian law provides a sound adjustment to the egg emission rate of S. Mansoni at values between 1 and 300. The linguist uses the probit–log n diagram on a large scale, where n is a discrete variable (see e.g. Herdan, 1958). Other examples are given in Martin (1962a).

It is perhaps the moment at which to speak of the *use and abuse* of the log-Gaussian law. The law can be presented as an asymmetrical distribution or density, or as a straight line in probit–log X coordinates. But yet a fourth mode of presentation still remains, viz. the *imminence rate or hazard function* (see 6.122).

In the case of the log-Gaussian law, even the form of the imminence

[†] The more one drinks, the thirstier one becomes (*Tr.*).

rate provides for a conditional probability of decease which passes through a maximum, then decreases and tends asymptotically towards zero (Gumbel, 1960, p. 22; see also section 6.335.16).

In the studies of follow-up such a development is patently contrary to the facts, as clinicians and cancerologists have frequently been told by us. In a context of *life testing* the same objection is maintained by Buckland (1963, p. 25). It is therefore not because the probit–log survival times are sensibly aligned according to a linear tendency, that the log-Gaussian law can adequately describe the law which governs survival after treatment. Not only is it expedient, it is also a matter of urgency to find other types of laws for statistical studies of follow-up, as notably is the case with Irwin (1942), Kodlin (1961 and 1967) and Kaufmann (1967).

6.335.18. Random log-Gaussian standard deviates

From the results explained in 6.311.6 for the random standard rectangular deviates, we give according to Quenouille (1959, p. 179) the transform from $x_1 = u_p$ (standard normal deviate) to the log-normal standard deviate in Quenouille's notation

$$x_3 = \frac{e^{u_p} - e^{1/2}}{e^2 - e} \qquad \text{(a)}$$

with a lower limit for $x_3 = 0.763$ and a modal value of -0.593.

6.335.19. Log–log-Gaussian probability law

For data whose distribution is strongly positively skew, a log–log transformation may be convenient to normalize ("Gaussianize") the raw data.

Another solution may be also considered in the context of extreme value distributions (6.338).

6.335.2. Log-Gaussian law with three parameters

6.335.21. Densities in U positively skew

The distributions which are positively skewed, are not always rendered Gaussian by the logarithmic transformation. If after transformation of U into $\log U$, the probit–log U line is still parabolic with concavity towards the U

axis, it is expedient to find a transformation of U into log $(U-K)$, where K should be determined from data.

If then a linear diagram in probit–log $(U-K)$ coordinates is obtained, the variable $U-K$ can be considered to be log-Gaussian.

In such cases the constant K is the *threshold* of the *log-Gaussian variable*

$$Z = U-K, \tag{a}$$

which should be essentially positive.

The log-Gaussian probability density in three parameters K, μ and σ over the range $(0, \infty)$ is

$$f(U) = \frac{1}{(U-K)\sigma\sqrt{2\pi}} \exp\left\{-\frac{[\ln(U-K)-\mu]^2}{2\sigma^2}\right\}, \tag{b}$$

with

$$\begin{cases} \mu = \mu_{\ln(U-K)}, \\ \sigma^2 = \sigma^2_{\ln(U-K)}. \end{cases} \tag{c}$$

The constant K is the threshold of the variable U; the log-Gaussian variable Z is thus written in (a).

In distributions that are positively skewed, a distinct relationship is found between the three "cardinal" quantiles:

$$U_{84} - U_{50} > U_{50} - U_{16}; \tag{d}$$

this relation exactly shows the positive type of asymmetry.

In general, for any $p < 0.5$, we thus have

$$U_{1-p} - U_{0.50} > U_{50} - U_p. \tag{d'}$$

In order to estimate the third parameter K, an *approximate* method was described by Kapteyn and J. Van Uven (1916), and this was further considered by Gaddum (1945) as well as by Kenney and Keeping (1956, II, p. 123).

In reference to the Table 6.335.12A, it is desired to find a value of K such that the variable log $(U-K)$ is Gaussian. The percentiles 16, 50 and 84 of $(U-K)$ will then satisfy the *non-algebraic* relation

$$\frac{(U-K)_{16}}{(U-K)_{50}} \cdot \frac{(U-K)_{50}}{(U-K)_{84}} = 1, \tag{e}$$

where K is unknown.

Therefore

$$K \approx \frac{U_{16}U_{84}-U_{50}^2}{U_{16}+U_{84}-2U_{50}}. \tag{f}$$

This simple method is based on the three quantiles used. A more elaborate and up-to-date method is given by Mason and Bulgren (1964), already cited in section 6.335.17.

6.335.22. Densities in U negatively skew

Consider now the densities which are negatively skew and which are less frequent than the symmetrical densities and those positively skew. By way of example, we mention:

(1) the celebrated distribution of bean sizes which was first studied by Pretorius (1930), and then by others cited in Martin (1958). On the basis of the χ^2 test, it is found that none of the proposed solutions are satisfactory, considering the scale factor $N = 9440$ intervening as multiplier in χ^2;

(2) the serum globulins (King and Wooton, 1956, p. 2).

For such distributions, we have

$$U_{84}-U_{50} < U_{50}-U_{16},$$

or, further, (a)

$$U_{84}+U_{16} < 2U_{50}.$$

In this case the probit–random variable curve shows concavity towards the probit axis, and this concavity is accentuated by a simple logarithmic transformation of the type log U. The transformation to apply is of the type log (K_1-U), where K_1 can be estimated from the original distribution, and the quantiles U_{16}, U_{50} and U_{84} in the event of their occurrence.

The variable K_1-U must be log-Gaussian; therefore

$$(K_1-U_{16})(K_1-U_{84}) = (K_1-U_{50})^2.$$

By operating in the same way as in the above sub-section, the approximate solution is readily found to be

$$K_1 \approx \frac{U_{50}^2-U_{16}U_{84}}{2U_{50}-(U_{16}+U_{84})}. \tag{b}$$

Example: size of 9440 beans (Pretorius, 1930).

In Fig. 6.335.22I the quantiles are plotted graphically (%):

$$U_{16} = 13.90; \quad U_{50} = 14.52; \quad U_{84} = 15.02.$$

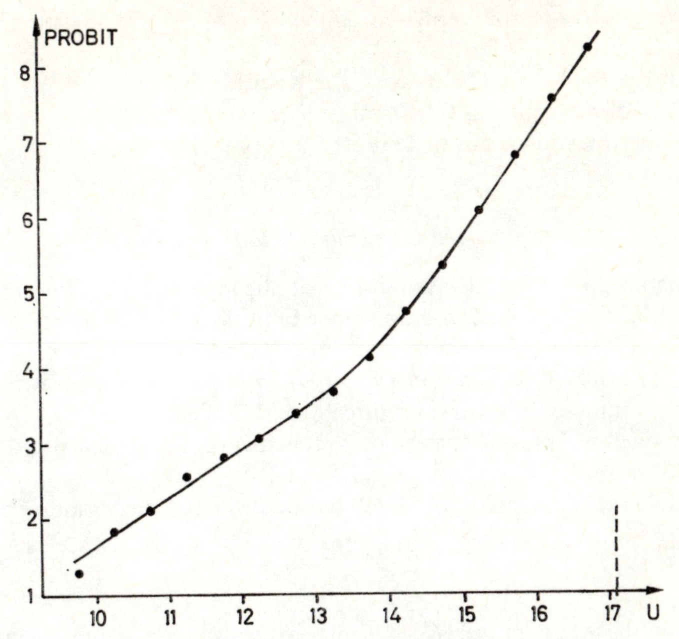

FIG. 6.335.22 I. Probit–size of beans diagram (see text).

Therefore

$$K_1 = \frac{14.52^2 - 13.90 \times 15.02}{2 \times 14.52 - (13.90 + 15.02)} = 17.10. \tag{c}$$

This approximate value plays the role of an asymptote for U; hence the differences $K_1 - U$ are positive and lead to a positively asymmetrical density of the type found in the above sub-section.

Another method proposed by Wicksell (1917) is used by Kendall and Stuart (1963, I, p. 170) giving an estimate $K_1 = 17.455$.

6.335.23. *Estimation of the three parameters*

Aitchison and Brown (1957, pp. 55–64) have expounded the methods available for estimating the three parameters appearing in 6.335.21 (b), namely, the method of maximum likelihood, Cohen's least sample method (Cohen, 1951), the method of moments, method of quantiles, the method of Kemsley which uses the quantiles and the moments at the same time, and finally a graphical method (p. 61, fig. 6.3); this procedure is illustrated by a diagram showing the correct value of the threshold K being approached by a process analogous to the approach of the asymptote of a logistic in three parameters (Brody, 1945).

Also on pp. 61–64 of the same vork, the Kemsley method is compared with the method of moments and the method of quantiles.

The reader may also consult Lambert (1964).

6.335.3. Log-Gaussian law in four parameters

In general, even if not always, biological phenomena have a lower limit K (a threshold, for instance) and an upper limit K_1 (e.g. the point at which the response is 100%). At the cost of introducing four parameters (μ, σ, K and K_1), one thus avoids the *conceptual error* (Aitchison and Brown, 1957, p. 16) of an infinite range in both directions.

Such a reference may also be made to this theoretical difficulty in the case of the Gauss curve.

Wicksell (1917) and Johnson (1949) have considered the variable

$$Z = \frac{U-K}{K_1-U} \qquad \text{(a)}$$

defined in the finite range (K, K_1) and which obeys a log-Gaussian law (μ, σ^2).

Readers should refer to Johnson (1949) as well as to Aitchison and Brown (1957, pp. 17–18) for the developments.

6.336. Log-logistic Law

6.336.1. Log-logistic law with three parameters in the original variable Z

We consider a reference curve for the dosage of antibiotics plotted by the method of circles of growth inhibition for sensitive germs. The graded response $U(X)$ is the diameter of the circle *visible* to the observer for an antibiotic dose Z (μg/ml) such that

$$X = \ln Z. \qquad \text{(a)}$$

Natural logarithms are used in order to avoid numerical coefficients.

The logistic with three parameters in X (6.323.3 a')

$$U(X) = \frac{H}{1+e^{-(\alpha+\beta X)}} \qquad (-\infty \leqslant X \leqslant +\infty), \qquad \text{(b)}$$

now becomes

$$U(X) = \frac{H}{1+e^{-\alpha-\beta \ln Z}} \qquad (Z \geqslant 0)$$

$$= \frac{H}{1+e^{-\alpha} \cdot e^{-\beta \ln Z}} = \frac{H}{1+aZ^{-\beta}}, \tag{c}$$

where

$$a = e^{-\alpha} > 0,$$

$$e^{-\beta \ln Z} = (e^{\ln Z})^{-\beta} = Z^{-\beta}.$$

Taking the inverse of (c), we obtain

$$\frac{1}{U(Z)} = \frac{1+aZ^{-\beta}}{H} = \frac{1}{H} + \frac{a}{H} Z^{-\beta}. \tag{d}$$

This relation expresses the *reciprocal of the response measured as a function of the dose Z administered.*

Theoretically, for $Z = 0$, we have $1/U = \infty$. For indefinitely increasing doses, $1/U$ tends to the reciprocal, $1/H$, of the asymptote of the response U. The metamer of the response $1/U$ monotonically decreases since the derivative of Z is negative if $0 \leqslant Z \leqslant \infty$.

In a recent paper concerning the microbiological dosage of antibiotics on large plates of primed agar Smets (1966) took advantage of this monotonic trend to trace a tangent at a point situated in a favorable zone, i.e. where the graphical accuracy is in the neighborhood of optimum.

This practical method enables graphical estimates $\hat{H}$, $\hat{a}$ and β_0 of first approximation to be obtained.

The value β_1 is retained which leads to a linear relation of type (d) in a $(Z^{-\hat{\beta}}; 1/U)$-diagram.

With some neighboring values of $\hat{\beta}_0$, one plots the points corresponding to the metamers $Z^{-\hat{\beta}_0}$ and $1/U$.

Finally, in accordance with (d), the classical formulae of linear regression permit the confirmation or correction of $\hat{\beta}_1$, in which case a $\hat{\beta}_2$ is retained which leads to a larger sum of the squares (S.S.) blocked by the linear regression. These latter calculations are performed from an 8×8 latin square (eight non-equidistant concentrations of Rifamycin S.V. ranging from 1.5 to 100 μg/ml). The computations were effected serially on the computer of the Centre de Calcul at the Université Libre of Brussels.

The author concludes that the set of test points seems to satisfy the log-logistic arc calculated in the wide range from 1.5 to 100 μg/ml of antibiotic.

Of course, within a *limited range* of concentrations between 1 and 8 μg/ml in a 8×8 Latin square, Smets (1966, p. 52) could verify the linearity of the log-dose response due to the easier handling of this linear approximation for the log-logistic arc.

The sequence of operations which has just been discussed, represents a hybrid sequence of *graphical plots* in a cycle of approximations which are followed by computer *calculations* of sums of squares depending on *one* parameter only.

The log-logistic is also considered in Bliss (II, 1970, p. 168).

A more elegant and more complete approach is given by Nelder (1961 and 1962).

6.336.2. Log-logistic law with fixed asymptote

6.336.21. Log-logistic distribution with two parameters

We again take the logistic law of 6.323.21 (a), which is rewritten as

$$p = \frac{1}{1+e^{-(\alpha+\beta X)}} \qquad \text{(a)}$$

and where X varies from $-\infty$ to $+\infty$ with $\beta > 0$.

One may also write [6.323.21 (f)]:

$$p = \frac{1}{1+\exp\{-\beta(X-X_{0.5})\}}. \qquad \text{(a')}$$

We set

$$X = \ln Z \qquad (Z \geqslant 0)$$
$$X_{0.5} = \ln Z_{0.5}. \qquad \text{(b)}$$

Therefore

$$X - X_{0.5} = \ln \frac{Z}{Z_{0.5}}. \qquad \text{(b')}$$

Thus the logistic law (a') in X becomes also the log-logistic law (c) in the essentially positive variable Z.

From the point of view of the log-logistic law, the variable Z is the operational variable which measures the dose of agent (active product) administered to specimens responding in all-or-nothing to this dose with a probability p.

By the transformation $X = \ln Z$, one transforms the real dose into a metamer X, the interest of which here is purely mathematical. The succession of transformations is as follows

$$p = \frac{1}{1+\exp\{-\beta \ln (Z/Z_{0.5})\}} = \frac{1}{1+\exp\{-\ln (Z/Z_{0.5})^{\beta}\}}. \tag{c}$$

Finally, the *log-logistic law in Z* with two parameters, $Z_{0.5}$ and β, is

$$p = \frac{1}{1+(Z/Z_{0.5})^{-\beta}}. \tag{d}$$

Wilson and Jane Worcester (1943, p. 80) write the log-logistic distribution in the real dose D as

$$p = \frac{D^{2\alpha}}{D^{2\alpha}+e^{2\alpha\gamma}}. \tag{e}$$

In order to avoid confusion with our notations, we will transform the relation (e) as follows:

$$D \to Z; \quad \alpha \to a; \quad \gamma \to c.$$

We proceed from the log-logistic (e) written in the real dose Z:

$$p = \frac{Z^{2a}}{Z^{2a}+e^{2ac}}, \tag{e'}$$

where c is one of the parameters of localization and a is a parameter of scale.

By introducing the transformation

$$x = \ln Z \quad \text{or} \quad Z = e^{x}, \tag{f}$$

we obtain

$$p = \frac{e^{2ax}}{e^{2ax}+e^{2ac}} = \frac{1}{1+e^{2a(c-x)}} = \frac{1}{1+\exp\{-2a(x-x_{0.5})\}}. \tag{g}$$

Having regard to the relation (f), the formula (g) reproduces the first relation (c) scaled by a factor of 2 in the exponent.

Considering now the formula of the hyperbolic tangent:

$$\tanh y = \frac{e^{y}-e^{-y}}{e^{y}+e^{-y}} = \frac{1-e^{-2y}}{1+e^{-2y}}, \tag{h}$$

we then write

$$e^{+2y} = \frac{1+\tanh y}{1-\tanh y}. \tag{h'}$$

From the formula (g),

$$e^{2a(c-x)} = \frac{1-p}{p}. \tag{i}$$

Combining (h′), (i) and (g), we get

$$e^{2a(c-x)} = \frac{1-p}{p} = \frac{1+\tanh a(c-x)}{1-\tanh a(c-x)}.$$

Therefore

$$p = \tfrac{1}{2}[1-\tanh a(c-x)], \tag{j}$$

as written by Wilson and Worcester (1943).

If the argument of the exponential in (g) is written explicitly with the sign −, we have

$$p = \tfrac{1}{2}[1+\tanh a(x-x_{0.5})], \tag{j′}$$

since $\tanh(-y) = -\tanh(y)$.

6.336.22. *Log-logistic distribution in the reduced variable z and with one parameter β*

Into the relation 6.336.21 (c), we introduce the *essentially positive reduced variable*

$$z = \frac{Z}{Z_{0.5}}. \tag{a}$$

The log-logistic distribution in this *reduced* variable and depending on the single parameter β is written as

$$F(z;\beta) = p = \frac{1}{1+z^{-\beta}}, \tag{b}$$

with

$$p = 0 \quad \text{for} \quad z = 0,$$
$$p = 1 \quad \text{for} \quad z = \infty,$$
$$p = \tfrac{1}{2} \quad \text{for} \quad z = 1.$$

Depending on the parameter of scale β, a bundle of curves is obtained, some of which are shown in Fig. 6.336.22I B for $\beta = 1, 2, 3, 4$.

In this representation the median point $z_{0.5} = 1$ is fixed.

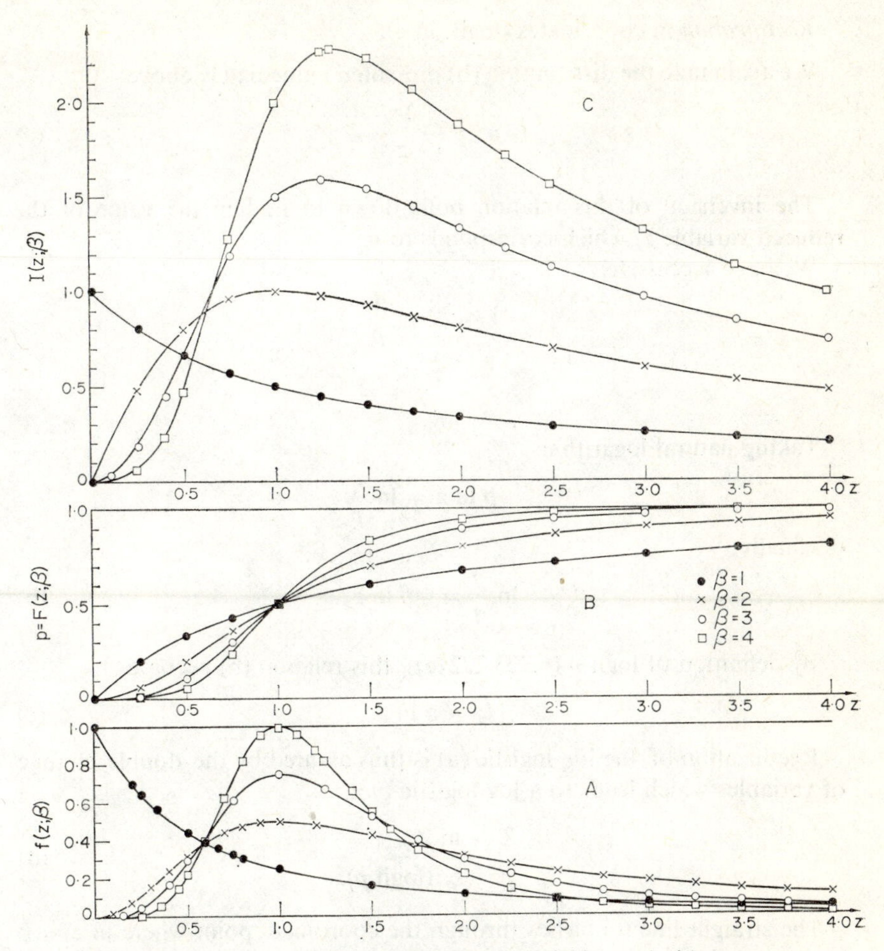

FIG. 6.336.22 I. Log–logistic law in the reduced variable z and with parameter $\beta = 1, 2, 3, 4$.

(A) Probability density.

(B) Distribution.

(C) Imminence rate.

6.336.23. Inversion of the reduced log-logistic law

Rectification in coordinates (logit, ln z).

We again take the distribution (b) presented immediately above:

$$p = \frac{1}{1+z^{-\beta}}. \qquad \text{(a)}$$

The inversion of this relation boils down to finding the value of the reduced variable z_p which corresponds to p.

We have successively

$$1+z^{-\beta} = \frac{1}{p},$$

$$z^{-\beta} = \frac{1}{p}-1 = \frac{q}{p}.$$

Taking natural logarithms,

$$-\beta \ln z = \ln \frac{q}{p},$$

or, finally,

$$\ln \frac{p}{q} = +\beta \ln z_p. \qquad \text{(b)}$$

By definition of logit p (6.323.222 (e)), this relation (b) becomes

$$l_p = \beta \ln z. \qquad \text{(c)}$$

Rectification of the log-logistic (a) is thus assured by the double change of variables which leads to a log-logistic plot

$$\begin{aligned} Z &\to \ln Z \\ p &\to l_p \text{ (logit } p). \end{aligned} \qquad \text{(d)}$$

The straight line (c) passes through the coordinate point where $\ln z = 0$ (corresponding to $z = 1$) and $l_p = 0$ (corresponding to $p = 0.5$).

6.336.24. Application to hemolysis by complement fixation

The log-logistic law in two parameters is applicable to responses of the type in all-or-nothing of probability p as a function of a dose Z of active product. The response in all-or-nothing can be of the type death versus survival for insecticides, germination versus non-germination for fungicides, or even hemolysis or non-hemolysis. The evolution of the *(total) probability of hemolysis* p of red corpuscles sensitized by a quantity Z of complement was described in this way by von Krogh (1916).

By convention, we put

$$K = Z_{0.5} \tag{a}$$

for the quantity of complement which assures 50% hemolysis (50% effective dose or ED_{50}).

The log-logistic *reduced variable* z measures the concentration of complement as a fraction of the median effective dose (ED_{50}).

Immunologists will recognize the log-logistic of hemolysis by complement in the following form (Thompson, 1948, p. 200):

$$\ln Z = \ln K + h \ln \frac{y}{1-y}, \tag{b}$$

or, alternatively,

$$\ln \frac{Z}{K} = h \ln \frac{y}{1-y}. \tag{c}$$

In this form, one well sees the homology with the formula 6.336.23 (c)

$$\beta = \frac{1}{h},$$

$$z = \frac{Z}{K},$$

$$p = y \quad \text{and} \quad q = 1-y.$$

Having introduced this immunological application in the general context of the log-logistic law, the reader is referred to Thompson for a critical study of the effect of inhibitors, carried out in plots with or without parallel lines.

It should also be noted that Waksman (1949) has used the log-Gaussian law to describe the course of hemolysis by complement fixation. In this sense the controversy of probit vs. logit remains open. We may add other examples.

For the definition of a hemolysis unit of streptolysin 0, Köhler and Grimm (1962) state that the only determination adequate for accurate measurement is that of the amount of streptolysin (on log scale) required for dissolving 50% of the erythrocytes present. The ordinate scale may be indifferently logit or probit.

Quoting Ipsen (1941, p. 67) we refer to a probit–log-dose graph for three reaction curves for α-staphylolysin acting on three concentrations of red blood corpuscles, 5, 10 and 20% respectively.

The probit straight-lines are parallel but displaced (by log 2) since a greater blood concentration requires more toxin to produce the same effect.

Recently, Bliss (II, 1970, pp. 169 and 170) has applied the log-logistic law to hemolytic time curves of diluted human blood following the addition of saponin. The percentage of hemolysis after t minutes is represented by a symmetric sigmoid in $\log t$. Moreover, the logits (6.323.222) of these percentages lay along a straight line with the same abscissae $\log t$.

6.336.25. Application to biological tests in all-or-none response

In this case the phenomenon is of the log-Gaussian or log-logistic type in the dose variable, Z.

Leaving aside here the log-Gaussian aspect, the transformation $Z = \ln X$ or $\log_{10} X$ leads to a logistic law with two parameters in the variable X which we have already discussed in section 6.323.2.

6.336.26. Imminence rate (hazard function) of the reduced log-logistic law

Starting from the distribution 6.336.22 (b), the complementary distribution of Z is written

$$1-F(z;\beta) = 1-p = \frac{1}{1+z^{\beta}} \tag{a}$$

Then differentiating 6.336.22 (b), we get the *reduced* log-logistic density

$$f(z;\beta) = \frac{\beta z^{\beta-1}}{(1+z^{\beta})^2} \tag{b}$$

drawn for $\beta = 1, 2, 3, 4$ in Fig. 6.336.221 (A). By definition of the imminence rate (hazard function) (6.122) we have

$$I(z;\beta) = \frac{\beta z^{\beta-1}}{1+z^{\beta}}\,. \tag{c}$$

This family of curves is shown in Fig. 6.336.221 (C). For $\beta = 1$, we have the particular case which is treated in 6.336.32 (a) and 6.336.33 (a) and (b).

The imminence rate of the reduced variable z for $\beta = 1$ is a monotonically decreasing function of z. On the other hand, for $\beta = 2, 3, 4$, the imminence rate curves starting from zero rise to a maximum for $z = 1, \sqrt[3]{2}, \sqrt[4]{3}$, respectively and then decrease monotonically with an asymptotic limit zero. This is shown in Fig. 6.336.22I (c).

6.336.3. *Log-logistic law for the reduced deviate*

The log-logistic distribution for the reduced deviate is remarkably simple and contains no parameter.

6.336.31. *Log-logistic distribution for the reduced deviate*

In the relation 6.336.21 (a′), the deviation from the median $X - X_{0.5}$, multiplied by β, is a pure number; the argument of the exponential, in the denominator, is the product $-\beta\,(X - X_{0.5})$.

The reciprocal of β serves as a parameter of scale.

The subsequent transformation 6.336.21 (b), where Z is expressed in terms of an adequate unit of measure

$$X = \ln Z \qquad X_{0.5} = \ln Z_{0.5}, \tag{a}$$

leads to the *absolute deviation* from the median

$$X - X_{0.5} = \ln Z - \ln Z_{0.5} = \ln Z/Z_{0.5}. \tag{a′}$$

Therefore, by multiplying by β, we have the *reduced logistic deviate:*

$$l = \beta(X - X_{0.5}) = \ln (Z/Z_{0.5})^{\beta}. \tag{b}$$

Introducing the *reduced variable*

$$z = \frac{Z}{Z_{0.5}}, \tag{b′}$$

the *reduced logistic deviate* is written simply as

$$l = \beta \ln z = \ln z^{\beta}. \tag{c}$$

The relations between this logistic variable l and the log-logistic variable a are written as

$$l = \ln z^{\beta} = \ln a, \quad a = e^{l}, \tag{d}$$

with

$$a = z^{\beta}. \tag{d'}$$

From the *logistic law in l*, one thus derives a *log-logistic law in a:*

$$p = \frac{1}{1+e^{-l}} = \frac{1}{1+e^{-\ln a}} = \frac{1}{1+1/a}. \tag{e}$$

The reduced log-logistic distribution in the *(absolute)* reduced *variable* contains no parameter at all and is simply written

$$p = F(a) = \frac{1}{1+1/a} = \frac{a}{1+a}. \tag{f}$$

This satisfies the two essential properties:

$$F(0) = 0, \quad F(\infty) = 1. \tag{f'}$$

This distribution is represented in Fig. 6.336.31I.

The median corresponds to the relation

$$\frac{1}{2} = \frac{a}{1+a}. \tag{g}$$

Hence

$$Me = a_{0.5} = 1. \tag{g'}$$

The relation (d) is verified for the median. Thus

$$a_{0.5} = e^{l}_{0.5} = 1. \tag{g''}$$

The complementary distribution is

$$R(a) = 1-F(a) = \frac{1}{1+a}, \tag{h}$$

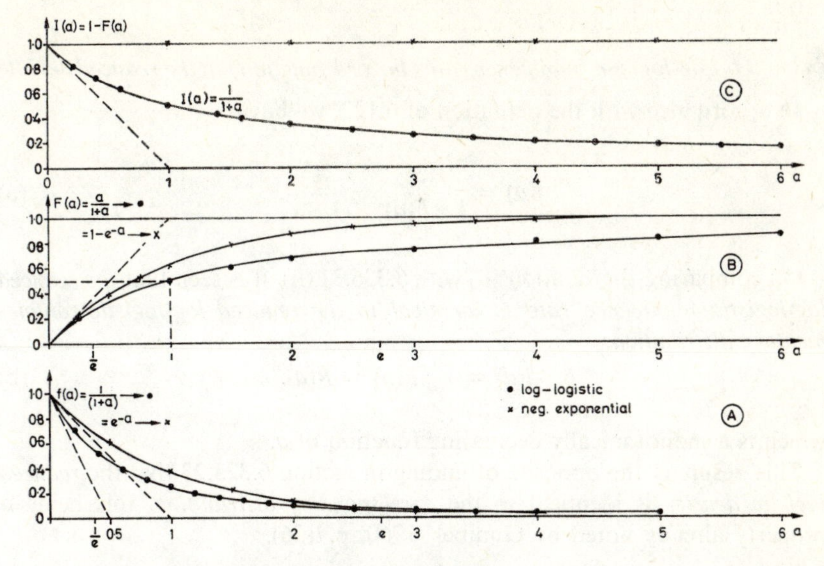

FIG. 6.336.31 I. Log-logistic law in the (absolute) reduced variable $a = z^{\beta}$.

(A) Probability density.
(B) Distribution.
(C) Imminence rate.

Comparison with the negative exponential.

with

$$R(0) = 1 \quad \text{and} \quad R(\infty) = 0. \tag{h'}$$

6.336.32. Log-logistic probability density for the reduced deviate

We differentiate the relation 6.336.31 (f)

$$f(a) = \frac{dF(a)}{da} = \frac{1}{(1+a)^2}. \tag{a}$$

This density function is shown in Fig. 6.336.31I (A). The tangent to the origin of $f(a)$ has the value 0.5 as abscissa at the origin. The comparison of $f(a)$ with the negative exponential is also given for the densities (A), the distributions (B) and the imminence rates or hazard functions (C).

6.336.33. Log-logistic imminence rate (hazard function) in the reduced deviate

In accordance with the definition of 6.122, we have

$$I(a) = \frac{f(a)}{1-F(a)} = \frac{1}{1+a}. \quad \text{(a)}$$

On comparing the relation (a) with 6.336.31 (h), it is seen that *the reduced log-logistic imminence rate is identical to the reduced log-logistic complementary distribution:*

$$I(a) = 1-F(a) = R(a). \quad \text{(b)}$$

which is a monotonically decreasing function of *a*.

This result is the opposite of finding in section 6.323.23 that the *reduced logistic density* is identical to the *corresponding distribution*, this being a property already noted by Gumbel (1960, p. 126).

6.336.4. *Generalized log-logistic law*

Recently, Richard and Pacault (1968) have considered the applicability of the log-logistic law to the DNA renaturation in the presence of Cu^{++} ions. They use the log-logistic plot technique (6.336.23 d) to reject the classical log-logistic in p and retain a log-logistic in p^2, it being a particular case of a generalized log-logistic

$$\alpha^m = \frac{1}{1+(A/Z)^\beta}$$

where A and m are positive constants.

In this case, α is the DNA *renaturation rate*, function of the time Z.

6.337. Laws of Cohort Survival

6.337.1. *Gompertz law*

6.337.11. *Gompertz distribution and complementary distribution*

In the study of the analytical representation of the law of survival $l(x)$ of a human cohort for the ages $x = 0, 1, 2, \ldots$, Gompertz (1825) proposed an expression which actuaries write l_x, while we also use the complementary distribution $S(x)$ and introduce explicitly the initial number N_0.

$$l(x) = kg^{c^x} = N_0 S(x), \tag{a}$$

$$l(0) = kg = N_0.$$

In this relation we write $l(x)$ with the argument x continuous, and l_x when x is not continuous, the x as a subscript indicating only the years of age reached. Indeed this law only applies to ages greater than about 10 years. (Here, the letter l is the initial of living.) The notation l_x represents the number of surviving (or still-living) at the beginning of the cohort age x.

The law governing the *mortality in the group concerned*, or the age-at-death distribution, is none other than

$$F(x) = 1 - S(x), \tag{b}$$

with

$$dF(x) = -dS(x). \tag{b'}$$

The relation between the distribution of the waiting time to death and a pure death process of intensity μ_x is clearly explained by Chiang (1968, pp. 60–61).

The conditional probability of death between x and $x+dx$ for someone having lived to the *instant* x is written without difficulty as the instantaneous imminence of death (see 6.122):

$$dJ(x) = I(x)\,dx = \frac{dF(x)}{1-F(x)}, \tag{c}$$

or, using the notation adopted in actuarial practice,

$$dJ(x) = \mu(x)\,dx. \tag{c'}$$

The function $\mu(x)$ is referred to as the "intensity" or "force" of mortality. But these terms, although sanctioned by usage, are not satisfactory for they no longer have their original significance in physics where they were first used. For this reason, we have introduced the general concept of the imminence rate of an all-or-nothing event occurring between the instants x and $x+dx$, providing that the event has not taken place before the instant x (Martin, 1961a and 1962a and b). This definition is consistent with the intuitive idea of imminence.

In this exposition on Gompertz law, and only here, we shall use in place of $I(x)$, the classical notation $\mu(x)$ used by the actuaries. In the body of the present text, $\mu(x)$ has not been used because μ is consistently employed throughout for the mean value of a population. Some authors, for example, Gumbel (1960, p. 20), use $\mu(x)$ for other phenomena in all-or-nothing besides human mortality, such as the failure of a lamp and so on.

The Gompertz law with three parameters given in (a) corresponds to a simple hypothesis regarding the form of $\mu(x)$, i.e. that the imminence rate $I(x)$ (hazard function), or "intensity" of mortality, increases exponentially with age:

$$\mu(x) = Ae^{Bx}. \tag{d}$$

The necessary condition 6.122 (e‴) is fulfilled.

Substituting into (c) this expression in two parameters, we get

$$dJ(x) = \mu(x)\,dx = \frac{dF(x)}{1-F(x)}, \tag{e}$$

whence

$$F(x) = 1-e^{-\int_0^x \mu(x)\,dx} = 1-e^{-J(x)}. \tag{e'}$$

Then, having regard to (b) and (b′), we have

$$\mu(x) = \frac{-dS(x)}{S(x)} = -\frac{d\ln S(x)}{dx}. \tag{f}$$

Substituting (d) into (f), we form a first-order linear differential equation which gives $S(x)$ from $\mu(x)$:

$$d\ln S(x) = -\mu(x)\,dx = -Ae^{Bx}\,dx. \tag{g}$$

Hence

$$\ln S(x) = -A\int e^{Bx}\,dx+C = -\frac{A}{B}e^{Bx}+C \tag{h}$$

and, immediately,

$$S(x) = \exp\left\{C - \frac{A}{B}e^{Bx}\right\} = e^C \exp\left\{-\frac{A}{B}e^{Bx}\right\} = \exp\left\{\frac{A}{B}(1-e^{Bx})\right\}, \quad \text{(i)}$$

$$C = \frac{A}{B} \quad \text{because} \quad S(0) = 1. \quad \text{(i')}$$

Comparing (i) with (a), via the relevant logarithms, homologies are found between the *set of actuarial parameters* (k, c, g) and the *set of biometrical parameters* (A, B, N_0).

$$\begin{aligned} \ln l(x) &= \ln k + c^x \cdot \ln g, \\ \ln N_0 S(x) &= \ln (N_0 e^{A/B}) + e^{Bx} \cdot \left(-\frac{A}{B}\right), \end{aligned} \quad \text{(j)}$$

whence

$$\begin{aligned} &\ln k = \ln (N_0 e^{A/B}) & & k = N_0 e^{A/B} \\ &\ln c = B \quad \text{since} \quad c^x = e^{x \ln c}, & \text{therefore} \quad & c = e^B \\ &\ln g = -\frac{A}{B}, & \text{therefore} \quad & g = e^{-A/B}, \end{aligned} \quad \text{(k)}$$

or, inversely,

$$\begin{aligned} A &= -(\ln g)\cdot(\ln c), \\ B &= \ln c, \\ N_0 &= kg. \end{aligned} \quad \text{(k')}$$

Modal age or "normal age". Again taking the relation (d)

$$\mu(x) = Ae^{Bx}, \quad \text{(l)}$$

the relation $F''(x) = 0$, which defines the mode, now becomes

$$\mu'(x) = \mu^2(x). \quad \text{(m)}$$

Explicitly for $\mu'(x)$ and $\mu(x)$, we have for the modal age t

$$ABe^{Bt} = A^2e^{2Bt}. \quad \text{(m')}$$

Hence the mode t is given by the root of the equation

$$Be^{Bt} = Ae^{2Bt},$$

or

$$e^{Bt}[B - Ae^{Bt}] = 0, \quad \text{(n)}$$

or

$$t_{\text{modal}} = \frac{1}{B} \ln \frac{B}{A}. \quad \text{(n')}$$

This *modal age* at death is called "normal age" by Gumbel (1937, p. 15; 1960, p. 246).

6.337.12. Gompertz plots

The Gompertzian imminence rate (or intensity) of dying (6.337.11 d) increases exponentially with age (or time) of a cohort. It is easy to linearize this function by writing

$$\ln \mu(x) = \ln A + Bx. \tag{a}$$

In such coordinates

$$\text{abcissae: age} = x,$$
$$\text{ordinates: } \ln \mu(x),$$

we get a *Gompertz plot*.

For given x_i, it is possible to estimate $\ln \mu(x_i)$ from formula 6.122 (e″) where $I(x) = \mu(x)$.

Sutter and Tabah (1952) use this technique for intercountry comparisons of B for several causes of death. The same is done for animal experiments by Kimball (1960, p. 507) and for radiation mortality by Fürth, Upton and Kimball (1959).

6.337.13. Pharmacological application to oral contraceptives

Garg, Rao and Carol Redmond (1970) give a maximum likelihood estimation of the parameters A and B (their k and α); they study the effect of a time translation on the Gompertz formula. They use the data provided by Prof. Cioccio in one investigation to determine the effects of prolonged oral contraception on mortality in mice. These were divided in five groups of 204 with the following assessments: no treatment at all, intubation with olive oil only, three groups with intubation with olive oil plus Enovid, Norlestrin or Clomiphene. The M-L estimations of B for the three first groups are about 0.0016 while $\hat{B} = 0.0033$ for the fourth and $\hat{B} = 0.0026$ for the last group. For the five groups, $\hat{A}$ varies from 0.525 (4th), 0.555 (5th) to about 0.600 for the three first groups. The standard errors are also given.

6.337.2. Makeham's law

When someone is still living at the age x, death may ensue between the instants x and $x+dx$ either owing to "old age", or to an accident (purely by chance). The imminence rate of death by physiological wear, as we have seen in 6.337.11, is written as

$$\mu_1(x) = Ae^{Bx}, \tag{a}$$

where A is a constant of dimension T^{-1}, as also B.

It is reasonable to suppose with Makeham (1860, 1867) that the imminence rate of death by accident (pure random event, see 6.331.15) is

$$\mu_2(x) = D, \tag{b}$$

where D is a constant of dimension T^{-1}.

Since the factors act independently, the resultant imminence rate of death, or resultant intensity, is

$$\mu(x) = \mu_1(x)+\mu_2(x) = Ae^{Bx}+D. \tag{c}$$

Returning to the notation of the actuaries (6.337.11 e and f) in $l(x)$

$$\mu(x) = -\frac{1}{l(x)}\frac{dl(x)}{dx}, \tag{d}$$

we have:

$$d \ln l(x) = -(Ae^{Bx}+D)\, dx$$

$$\ln l(x) = -\int (Ae^{Bx}+D)\, dx+\text{const.}$$

$$\ln l(x) = -\frac{B}{A}e^{Bx}-Dx+\ln k. \tag{d'}$$

The survival distribution on allowing for the alternative of death due to age or accident is thus

$$l(x) = ke^{-Dx} \exp\left\{\left(-\frac{B}{A}e^{Bx}\right)\right\}, \tag{e}$$

or, setting $e^{-D} = s$, one finds the well-known actuarial formula

$$l(x) = ks^x g^{c^x}, \tag{e'}$$

which corresponds to the composite intensity or hazard function (c).

6.337.3. *Laws of Perks and Beard*

After Makeham's law, Perks (1932) introduced an "intensity", no longer exponential, but *logistic* of the type

$$\mu(x) = a+\frac{b}{d+e^{-\lambda x}}. \tag{a}$$

The survival curve which is deduced from this describes well human adult mortality.

We refer biologists to an excellent critique of Beard (1959, vol. 5, p. 302) which concludes the five volumes of the *Ciba Foundation Colloquia on Ageing*. The respective merits of different expressions of $\mu(x)$ are compared. In addition, Beard himself proposes an interesting generalization by introducing a factor of longevity. For a long time, these considerations have been the privilege of actuaries, but they now begin to permeate within the field of Medical and Health Statistics.

6.338 Second Extreme Value Probability Law

According to Frechet (1927) and R. A. Fisher and Tippett (1928), the second probability law for original values which is stable with respect to the largest value is the second asymptote

$$F(z) = \exp - \{(v/z)^k\} \qquad \begin{array}{l} z \geqslant 0 \\ v > 0 \\ k > 0. \end{array} \tag{a}$$

Referring to Gumbel (1960) for more details, we are interested only with the particular case of (a) for $v = 1$ and $k = 1$, i.e. (see 6.326.6 c)

$$F(z) = e^{-1/z} \qquad z \geqslant 0 \tag{b}$$

for which we evidently have

$$F(0) = 0 \quad \text{and} \quad F(1) = 1.$$

This relation transforms a variable z in the one sided infinite range $(0, \infty)$ into a variable $F(z)$ within the finite range (0,1). This transformation is useful in the treatment of physiological data whose distribution is very positively skew. In a critical study of physiological norms, Cara and Martin (1964) insisted on the existence of several types of laws to describe the variability in human physiopathology.

Amongst these laws, in the Gaussian framework alone, we cite the Gaussian law (6.322) the log-Gaussian law (6.335) and the log–log-Gaussian law (6.335.19). The second extreme probability law is useful whenever it is a question of studying small ratios. But, for a relatively great ratio, such as the maximal expiratory volume per second/vital capacity the complementary distribution of the second extreme-value

$$G(z) = 1 - F(z) = 1 - e^{-1/z} \tag{c}$$

is convenient.

When the type of probability law governing the quantity Z being studied is known, an interval of physiological normality of measure 0.95 is defined by the quantiles $Z_{0.025}$ and $Z_{0.975}$. In particular, intervals of the (mean $\pm 2\sigma$) type used in the classical works *Handbook of Circulation* (Hamilton and Dow (ed.), 1962, 1963, 1965) and *Handbook of Respiration* (Fenn and Rahn (ed.), 1964, 1965) is evidently legitimate *only* in the Gaussian case.

In engineering a formula of type (a) with $k = 1$ and v changed into β is used to describe the relation between the duration of hardening z and the compression strength S of concrete (Sverdrup, I, 1967, p. 238)

$$S(z) = \text{const. } e^{-\beta/z} \qquad z > 0 \tag{d}$$

6.339. Beta-law of Second Kind

This is just a reminder of section 6.315.2 which for pedagogic reasons was developed between sections 6.315.1 and 6.315.3.

REFERENCES†

Abbot, W. S. (1925) A method of computing the effectiveness of an insecticide. *J. Economic Entomology*, **18:** 265–267.

Aitchison, J. (1968) *Statistics*, I, Oliver & Boyd, Edinburgh and London.

Aitchison, J. and Brown, J. A. C. (1957) *The Log-normal Distribution with Special Reference to Its Uses in Economics*. Cambridge University Press.

Aitken, A. C. (1957) *Statistical Mathematics*, 8th ed. Oliver & Boyd, Edinburgh.

Andersen, Ib. Cited by Ipsen and Jerne (1944), p. 359.

Anscombe, F. J. (1949) The statistical analysis of insect counts based on the negative binomial distribution. *Biometrics*, **5:** 165–175.

Anscombe, F. J. (1950) Sampling theory of the negative binomial and logarithmic series distributions. *Biometrika*, **37:** 358–382.

Anscombe, F. J. (1956) On estimating binomial response relations. *Biometrika*, **43:** 461–464.

Armitage, P. (1971) *Statistical Methods in Medical Research*. Blackwell Sc. Pub., Oxford, Edinburgh.

Armitage, P. and Allen, I. (1950) Methods of estimating the LD_{50} in quantal response data. *J. Hygiene*, **48:** 298–322.

Armitage, P., Herxheimer, H. and Rosa, L. (1952) The protective action of antihistamines in the anaphylactic microshock of the guinea-pig. *Brit. J. Pharmacology and Chemotherapy*, **7:** 625–636.

Bain, L. J. and Antle, Ch. E. (1970) *Inferential Procedures for the Weibull and Generalized Gamma Distributions*. U.S.A.F. Aerospace Research Lab. 70-0266.

Bartko, J. J. (1961) The negative binomial distribution: a review of properties and applications. *Virginia J. Sci.*, **12:** 18–37.

† See also Addendum, p. 609.

BARTKO, J. J. (1966) Approximating the negative binomial. *Technometrics*, **8:** 345–350.

BARTLETT, M. S. (1960) *Stochastic Population Models*. Wiley, N.Y.

BASS, J. (1967) *Eléments de calcul des probabilités théorique et appliqué*. Masson, Paris.

BASU, A. P. (1964) Estimates of reliability of some distributions useful in life testing. *Technometrics*, **6:** 215–219.

BEALL, G. (1940) The fit and significance of contagious distributions when applied to larval insects. *Ecology*, **21:** 460–474.

BEARD, R. E. (1959) Note on some mathematical mortality models, in *Ciba Foundation Colloquia on Ageing*, vol. 5, pp. 302–311. Churchill, London.

BECHGAARD, P. (1960) The natural history of benign hypertension, pp. 198–213 in K. D. Bock and P. T. Cottler (eds.), *Essential Hypertension*, Springer, Berlin.

BERKSON, J. (1944) Application of the logistic function to bio-assay. *J. Amer. Stat. Assoc.* **39:** 357–365.

BERKSON, J. (1949) Minimum χ^2 and maximum likelihood solution in terms of a linear transformation, with particular reference to bio-assay. *J. Amer. Stat. Assoc.* **44:** 273–278.

BERKSON, J. (1951) Why I prefer logits to probits. *Biometrics*, **7:** 327–339.

BERKSON, J. (1953) A statistically precise and relatively simple method of estimating the bio-assay with quantal response, based on the logistic function. *J. Amer. Stat. Assoc.*, **48:** 565–599.

BERKSON, J. (1955) Estimate of the integrated normal curve by minimum normit chi square with particular reference to bio-assay. *J. Amer. Stat. Assoc.*, **50:** 529–549.

BERTRAND, I. and GAYET-HALLION, T. (1952) Induction mitogenétique dans la thyroide du rat. Ajustement statistique de données par les distributions de Poisson et de Polya. *C.R. Acad. Sci. Paris*, **235:** 214–216.

BIRNBAUM, A. (1959) On the analysis of factorial experiments without replication. *Technometrics*, **1:** 343–357.

BLACKMAN, G. E. (1942) Statistical and ecological studies on the dispersion of species in plant communities. *Ann. Bot.*, **6:** 351–366.

BLISS, C. I. (1935) The calculation of dosage-mortality rate. *Ann. Applied Biol.*, **22:** 134–167.

BLISS, C. I. (1953) Fitting the negative binomial distribution to biological data. *Biometrics*, **9:** 176–200.

BLISS, C. I. (1960) Some statistical aspects of preference and related tests. *Applied Statistics*, **9:** 8–19.

BLISS, C. I. (1965) An analysis of some insect trap records in: *Classical and Contagious Discrete Distributions*, G. P. Patil (ed.), pp. 385–397. Calcutta Statistical Pub. Cy. Pergamon Press, Oxford.

BLISS, C. I. (1967) *Statistics in Biology*, Vol. I. McGraw-Hill, New York.

BLISS, C. I. (1970) *Statistics in Biology*, Vol. II. McGraw-Hill, New York.

BLISS, C. I. and GRIMMINGER (1969) Response criteria for the bio assay of Vitamin K. *Biometrics*, **95:** 735–745.

BLISS, C. I. and HARISON, J. C. (1939) Quantitative estimation of the potency of digitalis by the cat method in relation to secular variation. *J. Amer. Pharmaceut. Assoc.*, **28:** 521–530.

BLOM, G. (1954) *Statistical Estimates and Transformed Beta-variables*. Wiley, N.Y.; Almqvist & Wicksell, Stockholm.

BOAG, J. W. (1948) The presentation and analysis of the results of radiotherapy. *Brit. J. Radiol.*, **21:** 128–138, 189–203.

BOAG, J. W. (1949) Maximum likelihood estimation of the proportion of patients cured by cancer therapy. *J. Royal Stat. Soc.*, **B11:** 15–53.

Borges, R. (1970) Eine Approximation der Binomialverteilung durch die Normalverteilung der Ordnung $1/n$. *Zeit. Wahrscheinlichkeitsth. verwandte Geb.*, **14:** 189–199.

Box, G. E. P. (1953) Non-normality and tests on variances. *Biometrika*, **40:** 318–355.

Brass, W. (1958) Models of birth distribution in human populations. *Bull. de l'Inst. Int. de Stat.*, **36:** 165–178.

Brody, S. (1945) *Bioenergetics and Growth*. Rheinhold, N.Y.

Buckland, W. R. (1963) *Statistical Assessment of the Life Characteristics. A Bibliographic Guide*. Griffin, London.

Bulmer, M. G. (1965) *Principles of Statistics*. Oliver & Boyd, Edinburgh.

Burr, I. W. (1967) A useful approximation to the normal distribution, with application to simulation. *Technometrics*, **9:** 647–651.

Camp, B. H. (1951) Approximation to the point-binomial. *Ann. Math. Statist.*, **22:** 130–131.

Campbell, G. A. (1923) Probability curves showing Poisson's exponential summation. *Bell System Technical J.*, **2:** 95–113.

Cara, M. and Martin, L. (1964) Etablissement des normes physiologiques, pp. 97–112. in *L'Exploration fonctionnelle pulmonaire*, Flammarion, Paris.

Chapman, D. G. (1956) Estimating the parameters of a truncated gamma distribution. *Ann. Math. Stat.*, **27:** 498–506.

Chatfield, C. and Goodhardt, G. J. (1970) The beta-binomial model for consumer purchasing behaviour. *Applied Statistics*, **19:** 240–250.

Chernoff, H. and Lieberman, G. J. (1954) Use of normal probability paper. *J. Amer. Stat. Assoc.*, **49:** 778–785.

Chiang, C. L. (1968) *Introduction to Stochastic Processes in Biostatistics*. Wiley, N.Y., London, Sydney.

Clapham, A. R. (1936) Overdispersion in grassland communities and the use of statistical methods of plant ecology. *J. Ecol.* **24:** 232–251.

Cochran, W. G. (1938) Some difficulties in the statistical analysis of replicated experiments. *Empire J. Exp. Agriculture*, **6:** 157–175.

Cochran, W. G. and Davis, M. (1963) Sequential experiments for estimating the median lethal dose, in *Le Plan d'expériences*. Editions du Centre National de la Recherche Scientifique, Paris, pp. 181–194.

Cohen, A. C. Jr. (1951) Estimating parameters of logarithmic normal distributions by maximum likelihood. *J. Amer. Stat. Assoc.*, **46:** 206–212.

Cohen, A. C. Jr. (1965) Maximum likelihood estimation in the Weibull distribution based on complete and on censored samples. *Technometrics*, **7:** 579–588.

Cornfield, J. (1967) Bayes theorem. *Rev. Int. Stat. Institute*, **35:** 34–49.

Cornish, E. and Fisher, R. A. (1937) Moments and cumulants in the specification of distributions. *Inst. Int. de statistique*, **5:** 307–320.

Cox, D. R. (1966) Some procedures connected with the logistic qualitative response curve, *Research Papers in Statistics: Essays in honour of J. Neyman's 70th birthday*. Florence N. David (ed.). Wiley, New York.

Dacie, J. V. (1956) *Practical Haematology*, 2nd ed. Churchill, London.

Daniel, C. (1959) Use of half-normal plots in interpreting factorial two-level experiments. *Technometrics*, **1:** 311–341.

David, Florence N. (1962) *Games, Gods and Gambling*. Griffin, London.

David, Florence N. (ed.) (1966) *Research Papers in Statistics: Essays in honour of J. Neyman's 70th birthday*. Wiley, N.Y.

David, Florence N. and Barton, D. E. (1962) *Combinatorial Chance*. Griffin, London.

DAVIES, O. L. (ed.) (1956) *Design and Analysis of Industrial Experiments.* Oliver & Boyd, Edinburgh.

DBAWER, T. R., MEADORS, G. F. and MOORE, F. E. JR. (1951) Epidemiological approaches to heart disease: the Framingham study. *Amer. J. Publ. Health*, **41:** 279–286.

DEFARES, J. G. and SNEDDON, I. N. (1960) *The Mathematics of Medicine and Biology.* North-Holland, Amsterdam.

DE JONGE, H. (ed.) (1961) *Quantitative Methods in Pharmacology.* North-Holland, Amsterdam.

DE JONGE, H. (1963) *Medische Statistiek I en II.* Verh. van het Nederlands Instituut voor praeventive geneeskunde te Leiden XLI.

DE LA VALLÉE POUSSIN, CH. J. (1903) *Cours d'Analyse Infinitésimale*, I. Gauthier-Villars, Louvain.

DE LA VALLÉE POUSSIN, CH. J. (1928) *Cours d'Analyse Infinitésimale*, II. Gauthier-Villars, Louvain.

DIXON, W. J. (1961) Some statistical uses of large computers. *Proc. Fourth Berkeley Symposium on Mathematical Statistics and Probability*, vol. IV, pp. 197–209.

DODGE, H. F. and ROMIG, H. G. (1963) *Sampling Inspection Tables—Single and Double Sampling*, 2nd ed. Wiley, N.Y.

DONALDSON and ROBERTSON (1956) In Lotka, A. J. *Elements of Mathematical Biology*, p. 73. Dover, New York.

DØSSING, J. (1952) Determination of individual normal weights of school children. *Opera ex Domo biologiae hereditariae humaniae universitatis Hafniensis*, vol. 28. Munksgaard, København.

DUBEY, S. D. (1967) Some percentile estimations for Weibull parameters. *Technometrics*, **9:** 119–130.

DUBOUCHER, G. (1960) Arythmie complète et méthode statistique. *Arch. Maladies du coeur et des vaisseaux*, **53:** 1122–1136.

DUFRENOY, J. (1961) Using the arc tangent scale for quantitative methods, in de Jonge, H. (ed.) *Quantitative Methods in Pharmacology*, pp. 368–372. North-Holland, Amsterdam.

EGGENBERGER, F. and POLYA, G. (1923) Über die Statistik verketteler Vorgänge. *Z. f. angewandte Mathematik und Mechanik*, **4:** 279–289.

EISENHART, CH. (ed.) (1950) *Tables of the Binomial Probability Distribution*, see NATIONAL BUREAU OF STANDARDS (1950).

ELDERTON, W. P. (1953) *Frequency Curves and Correlation.* Harren Press, Washington, D.C.

ELSTON, R. C. and PICKREL, J. C. (1962) A statistical approach to ordering and usage policy for a hospital blood bank. *Transfusion*, pp. 41–47.

EMMENS, C. W. (1940) The dose–response relation for certain principles of the pituitary gland, and of the serum and urine of pregnancy. *J. Endocrinology*, **2:** 194–225.

ERLANG, A. K. (1909) The theory of probability and telephone traffic. *Nyt. Tidsskr. Math.*, **20B**: 33–39.

ERLANG, A. K. (1918) Solution of some problems in the the oryof probabilities of significance in automatic telephone exchanges. *Post Office Electrical Engineer's Journal*, **10:** 189–197 (see FRY (1928)).

ERLANG, A. K. (1948) see HALSTRØM and JENSEN (1948) The life and works of A. K. Erlang. *Trans. Danish Acad. Tech. Sci.* No. 3.

FELLER, W. (1940) On the logistic law of growth and its empirical verifications in biology. *Acta Biotheoretica*, **5:** 51–65.

FELLER, W. (1958) *An Introduction to Probability Theory and Its Application*, Vol. I, 2nd ed. Wiley N.Y.; Chapman-Hall, London.

Feller, W. (1966) *An Introduction to Probability Theory and Its Applications*, Vol. II. Wiley, N.Y.; Chapman-Hall, London.

Fenn, W. P. and Rahn, H. (1964/5) *Handbook of Physiology*, Sect. III, *Respiration* (American Physiological Society). Williams & Wilkins, Baltimore, Vols. I and II.

Finney, D. J. (1941) On the distribution of a variate whose logarithm is normally distributed. *J. Roy. Stat. Soc.*, Suppl. 7, 155–161.

Finney, D. J. (1947) *Probit Analysis*, 1st ed. Cambridge Univ. Press.

Finney, D. J. (1949) On a method of estimating frequencies. *Biometrika*, **36:** 233–234.

Finney, D. J. (1952a) *Statistical Method in Biological Assay*, 1st ed. Griffin, London.

Finney, D. J. (1952b) *Probit Analysis*, 2nd ed. Cambridge Univ. Press.

Finney, D. J. (1964) *Statistical Method in Biological Assay*, 2nd ed. Griffin, London.

Fisher, R. A. (1922) On the dominance ratio. *Proc. Roy. Soc. Edinburgh*, **42:** 321–341.

Fisher, R. A. (1925) *Statistical Methods for Research Workers*, 1st ed. Oliver & Boyd, Edinburgh. 1958, 13th ed.

Fisher, R. A. (1935) Appendix to Bliss, C. I., The case of zero survivors. *Ann. Appl. Biology*, **22:** 164–165.

Fisher, R. A. (1941) The negative binomial distribution. *Ann. Eugenics*, **11:** 182–187.

Fisher, R. A. (1943) A theoretical distribution for the apparent abundance of different species. *J. Animal Ecology*, **12:** 54–58.

Fisher, R. A. (1947) *Les Méthodes statistiques*, translated by I. Bertrand from the 10th ed. of Fisher, R. A. (1925). P.U.F., Paris.

Fisher, R. A. (1950) *Contributions to Mathematical Statistics*. W. A. Shewhart (ed.). N.Y.

Fisher, R. A. (1953) Note on the efficient fitting of the negative binomial. *Biometrics*, **9:** 197–199.

Fisher, R. A. (1959) Mathematical probability in the natural sciences. *Technometrics*, **1:** 21–29.

Fisher, R. A. and Tippett, L. H. C. (1928) Limiting forms of the frequency distribution of the largest or smallest member of a sample. *Proc. Cambridge Philos. Soc.*, **24:** 180–190.

Fisher, R. A. and Yates, F. (1957) *Statistical Tables*, 5th ed. Oliver & Boyd, Edinburgh.

Fisher, R. A., Corbett, A. S. and Williams, C. B. (1943) The relation between the number of individuals and the number of species on a random sample of an animal population. *J. Animal Ecology*, **12:** 42–58.

Fisz, M. (1958) *Wahrscheinlichkeitsrechnung und Mathematische Statistik*. V.E.B. Deutscher Verlag der Wissenschaften, Berlin.

Frechet, M. (1927) Sur la loi de probabilité de l'écart maximum. *Ann. de la Soc. polonaise de Mathématiques (Cracovie)*, **6:** 93–122.

Fry, Thornton C. (1928) *Probability and Its Engineering Use*, 1st ed. Van Nostrand, N.Y.

Fürth, J., Upton, A. C. and Kimball, A. W. (1959) Late pathologic effects of atomic detonation and their pathogenesis. *Radiation Research*, Suppl. 1, 243–264.

Gaddum, J. H. (1933) Reports on biological standards, III. Methods of biological assay depending on a quantal response. *Medical Research Council, Spec. Report* No. 183. H.M.S.O., London.

Gaddum, J. H. (1945) Log normal distributions. *Nature*, **156:** 463–466.

Gallo, V. (1950) *Statistica Ematologica*. Biblioteca "Haematologica" XI Pavia, Tipografia del libro.

Gallo, V. (1961) Elementi di statistica ematologica e di ematologia teorica, in P. Introzzi (ed.), pp. 1033–1044.

Galton, F. (1879) The geometric means in vital and social statistics. *Proc. Roy. Soc. (London)*, **29:** 365–366.

GARG, M. L., RAJA, B. RAO and REDMOND, C. K. (1970) Maximum likelihood estimation of the parameters of the Gompertz survival function. *Applied Statistics*, **19:** 152–159.
GART, J. J. (1963) A median test with sequential application. *Biometrika*, **50:** 55–62.
GART, J. J. (1966) Alternative analyses of contingency tables. *J. Roy. Stat. Soc.*, B**28:** 164–179.
GATEZ, P. (1966) Personal communication.
GAUSE, F. (1935) *Vérification expérimentale de la théorie mathématique de la lutte pour la vie*. Hermann, Paris.
GEARY, R. (1947) Testing of normality. *Biometrika*, **34:** 209–242.
GEIGY, A. G. (1963) *Tables scientifiques—Mathématiques et Statistique*. Documenta Geigy, Bale. (Also English ed.)
GINI, C. (1916/17) Delle relazioni tra le intensita cograduate di due caratteri. *Atti del Reale Istituto Veneto di Scienze, Littere, Arti* **76:** 1147–1185.
GLASSER, G. J. (1967) The age replacement problem. *Technometrics*, *9*: 83–91.
GLASSER, O. (ed.) (1950) *Medical Physics*, I and II. Year Book Publ., Chicago.
GOMPERTZ, B. (1825) On the nature of the function expressing the law of human mortality and on a new mode of determining value of life contingencies. *Phil. Trans. Roy. Soc. (London)* **36:** 513–585.
GREENWOOD, M. (1925) The growth of population in England and Wales. *Metron.* **5:** 1–20.
GREENWOOD, M. (1946) The statistical study of infectious diseases. *J. Roy. Stat. Soc.*, A **109:** 85–110.
GREENWOOD, M. (1950) Accident proneness. *Biometrika*, **37:** 24–29.
GREENWOOD, M. and UDNY YULE, G. (1920) An inquiry into the nature of frequency distributions representative of multiple happenings with particular reference to the occurrence of multiple attacks of disease or of repeated accidents. *J. Roy. Stat. Soc. (London)*, **83:** 255–279.
GRIMM, H. (1962) Tafeln der negativen Binomialverteilung. *Biometr. Zeits.*, **4:** 239–262.
GRIMM, H. and MALÝ, V. (1962) Anwendung der Sequenzanalyse auf die negative Binomialverteilung. *Biometr. Zeits.*, **4:** 182–192.
GUMBEL, E. J. (1937) *La Durée extrême de la vie humaine*. Hermann, Paris.
GUMBEL, E. J. (1958) *Statistics of Extremes*. Columbia University Press, N.Y. 2nd printing 1960.
HAIGHT, J. A. (1961) Index to the distributions of Mathematical Statistics. *J. Research of the Nat. Bureau of Standards (USA): B. Math. and Math. Physics*, **65**b: 23–60.
HAIGHT, J. A. (1967) *Handbook of the Poisson Distribution*. Wiley, N.Y., London, Sydney.
HALD, A. (1952a) *Statistical Theory with Engineering Applications*. Wiley, N.Y.; Chapman & Hall, London.
HALD, A. (1952b) *Statistical Tables and Formulas*. Wiley, N.Y.; Chapman & Hall, London.
HALDANE, J. B. S. (1945) A labour-saving method of sampling. *Nature*, **155:** 49.
HALSTRØM, H. L. and ARNE JENSEN (1948) The life and works of A. K. Erlang. *Trans. Danish. Acad. Tech. Sci.* No. 3.
HAMILTON, W. F. and DOW, P. (ed.) (1962) *Handbook of Physiology*, Sect. II, *Circulation*, Vol. 1. (Vol. 2 1963; Vol. 3 1965.) (American Physiological Society) distributed by Williams & Wilkins, Baltimore.
HARTER, LEON H. (1961) Expected values of normal order statistics. *Biometrika*, **48:** 151–165.
HARTER, H. L. and MOORE, A. H. (1965) Maximum likelihood estimation of the parameters of gamma and Weibull populations from complete and from censored samples. *Technometrics*, **7:** 639–643.
HARTLEY, H. O. (1948) The estimation of non-linear parameters by "internal least squares". *Biometrika*, **35:** 32–45.

HAYBITTLE, J. L. (1962) The estimation of T-year survival rate in patients treated for cancer. *J. Roy. Stat. Soc. (London)*, A**124**: 268–283.

HAZEN, A. (1914) Storage to be provided in impounding reservoirs for municipal water supply. *Trans. Amer. Soc. Chem. Engineers*, **77**: 1539–1569.

HEALY, M. J. R. (1952) A table of Abbott's correction for natural mortality. *Ann. Applied Biol.*, **39**: 211–212.

HEALY, M. J. R. (1968) The disciplining of medical data. *Brit. Med. Bull.*, **24**: 210–214.

HENRY, CH. (1906) Mesure des capacités intellectuelle et énergetique. *Mémoires Inst. Solvay*, Fasc. 6. Misch et Thorn, Bruxelles.

HENRY, P. (1926) *Cours de probabilité du tir*. Ecole d'application de l'artillerie et du génie, Fontainebleau (France), Fasc. 1, 1894. Publ. in *Mémorial de l'artillerie*, **5**: 295–447.

HERDAN, G. (1953) *Small Particle Statistics*. Elsevier, Amsterdam.

HERDAN, G. (1957) The mathematical relation between the number of diseases and the number of patients in a community. *J. Roy. Stat. Soc.* A**120**: 320–330.

HERDAN, G. (1958) The relation between the dictionary distribution and the occurrence distribution of word length and its importance for the study of linguistics. *Biometrika*, **45**: 222–228.

HOEL, P. G. (1948) *Introduction to Mathematical Statistics*. Wiley, N.Y.

HOTELLING, H. (1927) Differential equations subject to error, and population estimates. *J. Amer. Stat. Assoc.*, **22**: 283–314.

HUZURBAZAR, V. S. (1965) Some invariants of some discrete distributions admitting sufficient statistics for parameters. *Classical and Continuous Discrete Distributions*. pp. 231–240. G. P. Patil (ed.). Calcutta Statistical Pub. Cy.; Pergamon Press, Oxford.

INDULSKI, J., NOFER, J. and ROLEJKO, A. (1966) Epreuve d'analyse statistique de la fréquence des cas d'absentéisme. *Proc. Intern. Congress on Occupational Health, Vienna, 19–24 Sept. 1966*. Reprint AVII, pp. 51–54.

INTROZZI, P. (ed.) (1961) *Trattato italiano di medicina interna*. Abruzzini, Roma.

IPSEN, J. (1941) *Contribution to the Theory of Biological Standardization*. Nyt Nordisk Forlag, Arnold Busck, Copenhagen.

IPSEN, J. and JERNE, N. K. (1944) Graphical evaluation of the distribution of small experimental series. *Acta pathologica (Copenhagen)*, **81**: 343–366.

IRWIN, J. O. (1942) The distribution of the logarithm of survival times when the true law is exponential. *J. Hygiene (Cambridge)*, **42**, 328–333.

IRWIN, J. O. (1950) Personal communication to Armitage and Miss Allen (1950), p. 321.

IRWIN, J. O. (1960) Re-edition for private circulation of McKendrick (1925/6).

IRWIN, J. O. (1963) Personal communication, February 1963.

IRWIN, J. O. (1963) The place of mathematics in medical and biological statistics. *J. Roy. Stat. Soc.*, A**186**: 1–44.

JAQUET, R. (1962) Contribution à l'étude de: Plages de confiance de la droite de Henry. Séparation en ses composants d'un mélange de deux populations normales indépendantes. Thèse de 3e cycle. Faculté des Sciences, Nancy.

JOHNS, M. V., JR. and LIEBERMAN, G. J. (1966) An exact asymptotically efficient confidence bound for reliability in the case of the Weibull distribution. *Technometrics*, **8**: 135–175.

JOHNSON, N. L. (1949) Systems of frequency curves generated by method of translation. *Biometrika*, **36**: 149–176.

JOHNSON, N. L. (1966) Cumulative sum control charts and the Weibull distribution. *Technometrics*, **8**: 481–492.

JOHNSON, N. L. and KOTZ, S. (1969) *Discrete Distributions*. Houghton-Mifflin, Boston.

KAO, J. H. K. (1959) A graphical estimation of mixed Weibull parameters in life-testing of electron-tubes. *Technometrics*, **1**: 389–407.

KAPTEYN, J. C. and VAN UVEN, M. J. (1916) *Skew Frequency Curves in Biology and Statistics.* Hoitsema, Groningen.

KARPINOS, B. D. (1958) Weight–height standards based on World War II experience. *J. Amer. Stat. Assoc.*, **53:** 408–419.

KAUFMANN, H. (1967) Der Vergleich von Überlebungsquoten bei tödlichen Erkrankungen mittels eines modifizierten Boag-Verfahrens. *Methods of Information in Medicine*, **6:** 174–177.

KELLEY, T. L. (1924) *Statistical Method.* Macmillan, N.Y.

KENDALL, D. G. (1948) On some methods of population growth leading to R. A. Fisher's logarithmic series distribution. *Biometrika*, **35:** 6–15.

KENDALL, M. G. and BABINGTON SMITH (1948) *Tracts for Computers*, No. XXIV. C. E. S. Pearson (ed.). University Press, Cambridge.

KENDALL, M. G. and STUART, A. (1963) *The Advanced Theory of Statistics.* Vol. I, *Distribution Theory*, 2nd ed. Griffin, London.

KENNEY, J. F. and KEEPING, E. S. (1956) *Mathematics of Statistics*, I and II. Van Nostrand, Princeton (N.J.).

KIMBALL, A. W. (1960) Estimation of mortality intensities in animal experiments. *Biometrics*, **16:** 505–521.

KING, E. and WOOTTON, I. (1956) *Micro-analysis in Medical Biochemistry*, 3rd ed. Churchill, London.

KLECZKOWSKI, A. (1949) The transformation of local lesion counts for statistical analysis. *Ann. Applied Biol.*, **36,** 139–152.

KLECZKOWSKI, A. (1953) A method for testing results of infectivity tests with plant viruses for compatibility with hypotheses. *J. Gen. Microb.*, **8:** 295–301.

KLECZKOWSKI, A. (1955) The statistical analysis of plant virus assays: a transformation to include lesion numbers with small means. *J. Gen. Microb.*, **13:** 91–98.

KNUDSEN, L. (1949) Sample size of parental solutions for sterility testing. *J. Am. Pharmaceutical Ass.*, **38:** 332–337.

KNUDSEN, L. and CURTISS, J. M. (1947) The use of angular transformations in biological assays. *J. Amer. Stat. Assoc.*, **42:** 282–296.

KODLIN, D. (1961) Survival time analysis for treatment evaluation in cancer therapy. *Cancer Research*, **21:** 1103–1107.

KODLIN, D. (1967) A new response time distribution. *Biometrics*, **23:** 227–239.

KÖHLER, W. and GRIMM, H. (1962) Zur Definition einer Hämolyse-Einheit des Streptolysin O. *Z. f. Immunitätsforschung und experimentelle Therapie*, **123:** 155–169.

KOJIMA KENICHI and KELLEHER, T. (1962) Survival of mutant genes. *The American Naturalist*, **96:** 329–346.

KOSTITZIN, V. A. (1937) *Biologie mathématique.* Colin, Paris.

KOSTITZIN, V. A. (1940) Sur la loi logistique et ses généralisations. *Acta Biotheoretica*, **5:** 155–159.

LABOUREUR, M., CHOSSAT, M. and CARDOT, CL. (1963) *Cours de calcul mathématique moderne.* Béranger, Paris.

LAMBERT, J. A. (1964) Estimation of parameters in the three-parameter log-normal distribution. *Austr. J. Statistics*, **6:** 29–32.

LANCASTER, H. O. (1950) Statistical control in haematology. *J. Hyg.*, **48:** 402–417.

LANDAHL, H. (1939) A contribution to the mathematical biophysics of psychophysical discrimination, II. *Bull. Math. Biophysics*, **1:** 159–176.

LEDERMANN, S. (1956) *Alcool, alcoolisme, alcoolisation.* Presses Universitaires de France, Paris.

LEDLEY, R. S. and LUSTED, L. B. (1959) Reasoning foundations of medical diagnosis: symbolic logic, probability and value theory aid our understanding of new physician's reason. *Science*, **130:** 9–21.

LEDLEY, R. S. and LUSTED, L. B. (1962) Medical diagnosis and modern decision making. *Mathematical Problems in the Biological Sciences. Proc. Symp. in Applied Math.*, **14:** 117–158.

LEE, P. N. (1970) The simulated population of analysis of animal painting experiments in cancer research. *Biometrics*, **26:** 777–785.

LINCOLN, TH. S. and PARKER, R. D. (1967) *Medical Diagnosis using Bayes Theorems.* Rep. Operations Research Dpt. School of Hygiene and Public Health, Johns Hopkins University.

LINDER, A. and GRANDJEAN, E. (1950) Statistical analysis of some physiological experiments. *Sankhya*, **10:** 1–12.

LINDLEY, D. V. (1964) The Bayesian analysis of contingency tables. *Ann. Math. Stat.*, **35:** 1622–1643.

LISON, L. and PASTEELS, J. (1951) Etudes histophotométriques sur la teneur en acide désoxyribonucléique des noyaux au cours du développement embryonnaire chez l'oursin *Paracentrotus lividus. Arch. de Biologie*, **72:** 1–43.

LITTELL, A. S. (1952) Estimation of the *T*-year survival rate from follow-up studies over a limited period of time. *Human Biology*, **24:** 87–116.

LOTKA, A. J. (1925) *Elements of Physical Biology.* Williams & Wilkins, Baltimore. New augmented edition 1956, *Elements of Mathematical Biology*, Dover, N.Y.

LOVELAND, R. P. and TRIVELLI, A. P. H. (1927) Mathematical methods of frequency analysis of size of particles. *J. Franklin Institute*, **204:** 193–217, 377–389.

MACREZ, C. (1960) Essai d'analyse de l'irrégularité ventriculaire dans l'arythmie complète (Notion de réceptivité ventriculaire). *Presse Médicale (Paris)*, **68:** 607–608.

MAKEHAM, W. M. (1860) On the law of mortality and the construction of annuity tables. *J. Inst. Actuaries*, **8:** 301–10; (1867) *ibid.* **13:** 325.

MALTHUS, THE REVEREND TH. R. (1798) *An Assay on the Principle of Population.* 2nd ed., 1813. French transl. by P. Theil, Seghers, Paris, 1963.

MARTIN, D. C. and KATTI, S. K. (1965) Fitting of certain contagious distributions to some available data by the maximum likelihood method. *Biometrics*, **21:** 34–48.

MARTIN, J. (1967) *Notions de base en mathématiques et statistiques*, 2nd ed. Gauthier-Villars, Paris.

MARTIN, L. (1951) Etude biométrique des mesures histophotométriques de L. Lison et J. Pasteels au cours du développement de l'oeuf de l'oursin *Paracentrotus lividus* (Appendice à Lison–Pasteels, 1951). *Arch. de Biologie*, **72:** 45–64.

MARTIN, L. (1952) Probits normaux classiques et probits triangulaires. *Bull. Inst. Agron. et Stations de Recherche de Gembloux*, **19:** 346–386.

MARTIN, L. (1955) *Stencilized Course of Medical Statistics.* Faculty of Medicine, Brussels.

MARTIN, L. (1956) *Etude biométrique de grandeurs somatiques accueillies sur les conscrits et recrues belges et de leur évolution.* Institut National de Statistique, Bruxelles.

MARTIN, L. (1958) Analyse et ajustement des courbes de fréquences unimodales par la méthode des probits. *Bull. Inst. Int. de Statistique*, **36:** 43–59.

MARTIN, L. (1960) Homométrie, allométrie et cograduation en Biométrie générale. *Biometr. Zeits.* **2:** 73–97.

MARTIN, L. (1961a) Stochastic processes in biology and medicine, in J. Neyman (ed.), *Proc. Fourth Berkeley Symp. on Math. Stat. and Probability*, Vol. IV, *Biology and Problems of Health*, pp. 307–320.

MARTIN, L. (1961b) Etude biométrique de la natalité en Belgique sur la base du recensement de 1947. *Bull. Assoc. Lic. en Sc.-Actarielles de l'Univ. libre de Bruxelles*, No. 2, 25–53.

MARTIN, L. (1962a) Transformations of variables in clinical therapeutical research. *Methods of Information in Medicine (Methodik der Information in der Medizin)*, **1:** 38–50.

MARTIN, L. (1962b) Imminence et temps d'attente d'une réaction en tout ou rien. *Bull. Inst. Int. de Statistique*, **39:** 239–249.

MARTIN, L. (1962c) Etude socio-biométrique, in G. Jacquemyns (ed.), *Une Commune de l'agglomération bruxelloise*, *Uccle*, Tome II, pp. 371–474. Ed. de l'Institut de Sociologie, Bruxelles.

MARTIN, L. (1966) *Cours de méthodes statistiques avec applications aux problèmes de la Santé*. Faculté de Médecine et Ecole de Santé publique, Université Libre de Bruxelles.

MARTIN, L. (1969) Essai sur la description de phénomènes discontinus en statistique médicale. *Mémoires de l'Académie Royale de Médecine de Belgique*, in press.

MARTIN, L. (1971) Comparison of symmetric probability densities, distributions and probability units. *Biométrie–Praximétrie* (in press).

MARTIN, L. (1972) K. Pearson's (β_1, β_2) diagram for continuous probability laws used in Biology. *Biométrie–Praximétrie* (in press).

MARTIN, L. and COTTELEER, C. (1958) Etude quantitative de la courbe de mortalité d'acares soumis à l'action du Gammexane et du DMC. *C.R. Soc. Biol.*, **152:** 1062–1068.

MASON, E. E. and BULGREN, W. G. (1964) *Computer Applications in Medicine*. Thomas, Springfield, Ill., U.S.A.

MASUYAMA, M. (1951) An improved binomial probability paper and its use with tables. *Reports of Statistical Application Research*, Vol. 1. Union of Japanese Scientists and Engineers.

MASUYAMA, M. (1964) An empirical method of estimating parameters in a gamma distribution. *Rep. Stat. Appl. Res. J.U.S.E.*, **11:** 152–153.

MATHER, K. (1949) *Biometrical Genetics*. Methuen, London.

MCALISTER, D. (1879) The law of the geometric mean. *Proc. Roy. Soc. (London)*, **29:** 367–376.

MCDOWELL, A. J. (1967) *Mean Blood Hematocrit of Adults, Unites States 1960–1962*. National Center for Health Statistics, Washington, D.C., Series II, No. 24, April 1967.

MCKENDRICK, A. G. (1925/6) Application of mathematics to medical problems. *Proc. Edinburgh Math. Soc.* **44:** 1–34.

MEAD, R. (1965) A generalised logit-normal distribution. *Biometrics*, **21:** 721–732.

MENON, M. V. (1963) Estimation of the shape and scale parameters of the Weibull distribution. *Technometrics*, **5:** 175–182.

MOLINA, E. C. (1942) *Poisson's Exponential Binomial Limit*. Table I, Individual terms; Table II, Cumulated terms. Van Nostrand, N.Y.

MONOD, J. (1942) *La Croissance des cultures bactériennes*. Hermann, Paris.

MOORE, P. G. (1956) Normality in quality control charts. *Applied Statistics*, **5:** 171–179.

MORAN, P. A. P. (1968) *An Introduction to Probability Theory*. Clarendon Press, Oxford.

MORRISON, L. M. (1960) Diet in coronary atherosclerosis. *J. Amer. Med. Assoc.*, **173:** 884–888.

MORSE, PHILIPP M. (1958) *Queues, Inventories and Maintenance*. Wiley, New York; Chapman & Hall, London,

Mosteller, F. and Tukey, J. (1949) The use and usefulness of binomial probability paper. *J. Amer. Stat. Assoc.*, **44:** 174–212.

Müller, L. C. (1950) Biological assays involving quantal responses. *Ann. N.Y. Acad. Sci.*, **52:** 903–919.

Nagasaka, M., Hanaoka, W., Seki, K. and Yoshitoshi, Y. (1967) A mathematical model of blood pressure distribution in the aged. *Acta Cardiologica*, **22:** 1–11.

Nair, K. R. (1954) The fitting of growth curves. In O. Kempthorne, Th. Bancroft, J. Gowen, J. Lush (eds.), *Statistics and Mathematics in Biology*, pp. 119–132. Iowa State College Press, Ames, Iowa, U.S.A.

National Bureau of Standards (1950) *Tables of the Binomial Probability Distribution*, presented by Ch. Eisenhart. National Bureau of Standards, Applied Math. Series 6, U.S. Govt. Printing Office, Washington D.C.

Nelder, J. A. (1961) The fitting of a generalisation of the logistic curve. *Biometrics*, **17:** 89–110.

Nelder, J. A. (1962) An alternative form of a generalized logistic equation. *Biometrics*, **18:** 614–616.

Newell, D. J. (1965) Unusual frequency distributions. *Biometrics*, **21:** 159–168.

Neyman, J. (1939) On a new class of "contagious" distributions, applicable in entomology and bacteriology. *Ann. Math. Stat.*, **10:** 35–57.

Neyman, J. (1950) *First Course in Probability and Statistics*. H. Holt, New York.

Neyman. J. (ed.) (1961) *Proc. Fourth Berkeley Symposium on Mathematical Statistics and Probability, June 20–July 30, 1960*. Vols. I, II, III, IV. Berkeley and Los Angeles, Univ. of California Press.

Overall, J. E. and Williams, C. M. (1965) Conditional probability program for diagnosis of thyroid function. *J. Amer. Med. Assoc.* **183:** 307–311.

Parzen, Em. (1960) *Modern Probability Theory and Its Applications*. Wiley, N.Y., London

Patil, G. P. (1960) On the evaluation of the negative binomial distribution with examples *Technometrics*, **2:** 501–505.

Patil, G. P. (ed.) (1965) *Classical and Contagious Discrete Distributions*. Calcutta Statistical Pub. Cy.; Pergamon Press, Oxford.

Patil, G. P. and Joshi, S. W. (1968) *A Dictionary and Bibliography of Discrete Distributions*. Published for the International Statistical Institute. Oliver & Boyd, Edinburgh.

Patil, G. P., Kamat, A. R. and Wani, J. K. (1964) Certain studies on the structure and statistics on the logarithmic series distribution and related tables. ALL 64–197 Aerospace Research Laboratories, Office of Aerospace Research USAF, Wright–Patterson Air Force Base, Ohio.

Patterson, H. D. (1953) The use of autoregression in fitting an exponential curve. *Biometrika*, **45:** 389–400.

Patterson, H. D. (1956) A simple method for fitting an asymptotic regression curve. *Biometrics*, **12:** 323–329.

Paulson, E. (1942) An approximate normalization of the analysis of variance distribution. *Ann. Math. Stat.* **13:** 233–235.

Pearl, R. (1930) *Medical Biometry and Statistics*, 2nd ed. Saunders, Philadelphia.

Pearl, R. and Reed, L. J. (1920) On the rate of growth of the population of the United States since 1790 and its mathematical representation. *Proc. Nat. Acad. Sci.*, **6:** 275–288.

Pearson, E. S. (1938) The probability integral transformation for testing goodness of fit and combining independent tests of significance. *Biometrika*, **30:** 134–148.

Pearson, E. S. (1942) The probability integral of the range in samples of *n* from a normal population. Foreword and tables. *Biometrika*, **32:** 301–308.

PEARSON, E. S. (1968) Some comments on the assumption of normality involved in the ISO documents. ISO/TC 691WG (United Kingdom).

PEARSON, E. S. and HARTLEY, H. O. (eds.) (1954) *Biometrika Tables for Statisticians*, Vol. I. University Press, Cambridge.

PEARSON, K. (ed.) (1922) *Tables of the Incomplete Gamma Function*. H.M.S.O., London.

PEARSON, K. (1933) Note on the fitting of frequency curves. *Biometrika*, **25:** 213–216.

PEARSON, K. (ed.) (1934) *Tables of the Incomplete Beta Function Ratio*. 1st ed. Biometrika Office, University College, London. 2nd ed. 1948.

PEATMAN, J. (1947) *Descriptive and Sampling Statistics*. Harper Brothers, N.Y.

PERKS, W. (1932) On some experiments in the graduation of mortality statistics. *J. Inst. Actuaries*, **63:** 12–57.

PIKE, M. C. (1966) A suggested method of a certain class of experiments in carcinogenesis. *Biometrics*, **22:** 142–161.

PIMENTEL-GOMEZ, F. (1953) The use of Mitscherlich's regression law in the analysis of experiments with fertilisers. *Biometrics*, **9:** 498–516.

POINCARÉ, H. (1912) *Calcul des probabilités*. Gauthier-Villars, Paris.

POLYA, G. (1931) Sur quelques points de la théorie des probabilités. *Ann. Institut Henri Poincaré*, **1:** 117–162.

PRETORIUS, S. (1930) Skew bivariate frequency curves. *Biometrika*, **22:** 109–223.

PYKE, R. (1965) Spacings. *J. Roy. Stat. Soc.*, **B27:** 395–449.

QUENOUILLE, M. H. (1949) A relation between the logarithmic, Poisson and negative binomial series. *Biometrics*, **5:** 162–164.

QUENOUILLE, M. H. (1958) *Fundamentals of Statistical Reasoning*. Griffin, London.

QUENOUILLE, M. H. (1959) Tables of random observations from standard distributions. *Biometrika*, **46:** 178–204.

QUETELET, AD. (1835) *Physique Sociale*, I and II, 1st ed. Bachelier, Paris. 2nd ed. 1869 Muquardt, Bruxelles; Baillière et fils, Paris; Issakoff, St Petersbourg.

QUETELET, AD. (1871) *Anthropométrie*, Muquardt, Bruxelles; Gand, Leipzig.

RAPOPORT, A. (1952) "Ignition" phenomena in random nets. *Bull. Math. Biophysics*, **14:** 35–44.

RASHEVSKY, N. (1940) *Advances and Applications of Mathematical Biology*, University of Chicago Press, Chicago.

RAYLEIGH, J. W. STRUTT, 3RD BARON (1880) On the resultant of a large number of vibrations of the same pitch and of arbitrary phases. *Phil. Mag.* **10:** 73. (Cited from Gumbel (1958), p. 366.)

REED, L. J. and BERKSON, J. (1929) The application of the logistic function to experimental data. *J. Phys. Chem.*, **33:** 760–779.

RICHARD, H. and PACAULT, A. (1968) Etude cinétique de la transformation structurale de l'acide désoxyribonucléique. *Bull. Soc. Chimie biologique*, **50:** 417–426.

ROY, B. (1965) *Aléas numériques et distributions de probabilité usuelles*. Fasc. 1. *Généralites sur les aléas numériques*. Dunod, Paris.

SARHAN, A. and GREENBERG, B. (1967) *Contribution to Order Statistics*. Wiley, N.Y.

SCHLAIFER, R. (1959) *Probability and Statistics for Business Decisions*. McGraw-Hill, N.Y.

SCHNEIDER, B. (1963) Die Bestimmung des Parameter im Ertragsgesetz von E. A. Mitscherlich. *Biometr. Zeits.* **5:** 78–95.

SCHUERMAN, H. and SCHIRDUAN, M. (1948) Untersuchungen über Fragen der Penicillindosierung. *Klin. Wschr.* **26:** 526–528.

SHAPIRO, S. S. and WILK, M. B. (1965) An analysis of variance test for normality (complete samples). *Biometrika*, **52:** 591–611.

SHAPIRO, S. S. and WILK, M. B. (1968) A comparative study of various tests of normality. *J. Amer. Stat. Assoc.*, **63**: 1343–1362.

SHORTLEY, G. and JUDD, R. WILKINS (1965) Independent action and birth–death models in experimental microbiology. *Bacterial Reviews*, **29**: 102–141.

SIBUYA, M., YOSHIMURA, I. and SHIMIZU, R. (1964) Negative multinomial distribution. *Ann. Inst. Stat. Math.*, **16**: 409–426.

SIMONS, L. E. and GRUBBS, F. E. (1952) *Tables of the Cumulative Binomial Probabilities.* Ordnance Corps Pamphlet 20–1. Washington D.C., U.S.A.

SKELLAM, J. G. (1948) A probability distribution derived from the binomial distribution by regarding the probability of success as variable between the sets of trials. *J. Roy. Stat. Soc.*, **B10**: 257–261.

SMETS, PH. (1966) *Méthode statistique de dosage des antibiotiques. Application au dosage d'antibiotiques en présence de bile.* Mimeo. Presses Universitaires, Université Libre de Bruxelles.

SNEYERS, R. (1961) *On a Special Distribution of Maximum Values.* Institut Royal Météorologique de Belgique, Contribution No. 65.

SPARKS, D. N. (1970) Half normal plotting algorithm AS 30. *Applied Statistics*, **19**: 192–196.

STEVENS, W. L. (1951) Asymptotic regression. *Biometrics*, **7**: 247–267.

STÖRMER, H. (1970) *Mathematische Theorie der Zuverlässigkeit.* Oldenburg, München.

STUDENT (W. GOSSET) (1907) On the error of counting with a haemacytometer. *Biometrika*, **5**: 351. See also: *Student's Collected Papers*, eds. E. S. Pearson and J. Wishart, Biometrika Office, London, 1942.

SUTTER, J. and TABAH, L. (1952) La mortalité, phénomène biométrique. *Population (Paris)*, **7**: 69–94.

SVEN ODÉN (1926) In *Colloid Chemistry*, J. Alexander (ed.) N.Y. (Cited from Herdan (1953), p. 142.)

SVERDRUP, E. (1967) *Laws and Chance Variations*, Vol. I, *Elementary Introduction.* North-Holland, Amsterdam.

SWAROOP (1966) Revised and edited by A. B. Gilroy in collaboration with K. Uemura, *Statistical Methods in Malaria Eradication.* W.H.O. Monograph Series No. 51.

TEISSIER, G. (1928) Croissance des populations et croissance des organismes. *Ann. Physiologie*, **3**: 342–425.

TEISSIER, G. (1942) In *La Croissance des cultures bactériennes*, pp. 122–123, Monod (ed.), Hermann, Paris.

THIELE, T. N. (1903) *Theory of Observations.* London. Reprinted in *Ann. Math. Stat.*, **2**: 165–307, 1931.

THOMPSON, D'ARCY WENTWORTH (1942) *On Growth and Form.* University Press, Cambridge.

THOMPSON, W. R. (1948) On the use of parallel or nonparallel systems of transformed curves in bioassay: illustration in the quantitative complement-fixation test. *Biometrics*, **4**: 197–210.

THORNDIKE, FRANCES (1926) Applications of Poisson's probability summation. *Bell System Technical J.*, **5**: 604–624.

TIPPETT, T. N. (1925) On the extreme individuals and the range of samples taken from a normal populations. *Biometrika*, **17**: 364–387.

TSENG TUNG CHENG (1949) The normal approximation to the Poisson distribution and a proof of a conjecture of Ramajunan. *Bull. Amer. Math. Soc.*, **55**: 396–501.

TUDWAY, R. C. and FREUNDLICH, H. F. (1960) The use of a cobalt 60 beam unit for the treatment of carcinoma of the larynx. *Brit. J. Radiol.*, **23**: 91–104.

TUKEY, J. W. (1949) Personal communication, Daniel (1959).

TURNER, M. E. and EADIE, G. S. (1957) The distribution of red blood cells in the hemacytometer. *Biometrics*, **13:** 485–495.

TURNER, M. E., MONROE, R. J. and LUCAS, H. J. (1961) Generalized asymptotic regression and nonlinear path analysis. *Biometrics*, **17:** 120–143.

TWEEDIE, M. C. K. (1945) Inverse statistical variates. *Nature (London)* **155:** 453.

UEMURA, K. (1964) Statistical analysis of data in bilharziosis research. W.H.O., Geneva, BILH/Exp. Com. 3/INF/8.64.

URBAN, F. (1909) Die psychophysische Massmethoden als Grundlagen empirischer Messungen. *Arch. f. Ges. Psychologie*, **15:** 261–355; (1910) *ibid.* **16:** 168–227.

VAN DER WAERDEN, B. L. (1967) *Statistique mathématique*, trad. Mme C. Guinchat et A. Degenne. Dunod, Paris.

VAN DEUREN, P. (1934) *Calcul des probabilités*, I. *La théorie des probabilités*. Gauthier-Villars, Paris; Wesmael-Charlier, Namur (Belgique).

VERHULST, P. F. (1838) Notice sur la loi que la population suit dans son accroissement. *Correspondance Math. et Phys. (Ad. Quetelet ed.) Bruxelles*, **10:** 113–121.

VERHULST, P. F. (1845) Recherches mathématiques sur la loi d'accroissement de la population. *Nouveaux Mém. Acad. Roy. des Sciences et Belles Lettres de Bruxelles*, **18:** 1–38.

VERHULST, P. F. (1847) Deuxième mémoire sur la loi d'accroissement de la population. *Nouveau Mém. Acad. Roy. des Sciences et Belles Lettres de Bruxelles*, **20:** 1–32.

VESSEREAU, A. (1965) Les méthodes statistiques appliquées au test des caractères organoleptiques. *Ann. de la Nutrition et de l'Alimentation*, **19**A: 103–140.

VON BERTALANFFY, L. (1951) *Theoretische Biologie*, II. *Stoffwechsel, Wachstum*, 2nd ed. A. Francke Verlag, Bern.

VON BERTALANFFY, L. (1957) Quantitative laws in metabolism and growth. *Quart. Rev. Biology*, **32:** 218–231.

VON KROGH, M. (1916) Colloidal chemistry and immunology. *J. Infectious Diseases*, **19:** 452–471.

WAGNER, J. G. (1966) Design and data analysis of biopharmaceutical studies in man. *Canad. J. Pharmaceut Sciences*, **1:** 55–68.

WAGNER, J. G. (1967) Use of computers in pharmacokinetics. *Clinical Pharmacology and Therapeutics*, **8:** 201–218.

WAKSMAN, B. H. (1949) A comparison of the von Krogh formula (logistic function) and the method of probits as applied to hemolysis by complement. *J. Immunology*, **63:** 409–414.

WALD, A. (1959) *Sequential Analysis*. Wiley, N.Y.; Chapman & Hall, London.

WALKER, S. H. and DUNCAN, D. B. (1967) Estimation of the probability of an event as a function of several independent variables. *Biometrika*, **54:** 167–179.

WEATHERBURN, C. (1949) *A First Course in Mathematical Statistics*. University Press, Cambridge.

WEBER, A. A. (1960) Problèmes de statistique mathématique posés par les programmes de Santé publique (Thèse no. 1287). Editions Médecine et Hygiène, Genève.

WEBER, A. (1967) Personal communication.

WEBER, A. A. and LINDER, A. (1960) Auswertung der Ergebnisse eines Versuches zur Trachombekämpfung. *Biometr. Zeits.* **2:** 217–229.

WEBER, E. (1961) *Grundriss des Biologischen Statistik*, 4te Aufl. Gustav Fischer Verlag, Jena.

WEIBULL, W. (1939) *A Statistical Theory of Strength of Materials*, No. 151 Ing. Vet. Akad. Handlingen, Stockholm.

WEIBULL, W. (1951) A statistical distribution function of wide applicability. *J. Applied Mechanics*, **5:** 293–297.

WICKSELL, S. D. (1917) On logarithmic correlation with an application to the distribution of ages at first marriage. *Medd. Lunds Astron. Observ.* No. 84.

WILK, M. B., GNANADESIKAN, R. and MISS M. J. HUYETT (1962) Probability plots for the gamma distribution. *Technometrics*, **4:** 1–20.

WILLIAMS, C. B. (1947) The logarithmic series and its application to biological problems. *J. Ecol.*, **34:** 253–272.

WILLIAMSON, E. and BRETHERTON, M. H. (1964) Tables of the logarithmic series distribution. *Ann. Math. Stat.*, **35:** 284–297.

WILSON, E. B. and JANE WORCESTER (1943) The determination of LD_{50} and its sampling error in bio-assay. *Proc. Nat. Acad. Sci. (U.S.A.)*, **29:** 79–85.

WINSOR, CH. (1950) Biometry, in *Medical Physics*, I and II. O. Glasser (ed.), Vol. II, pp. 105–126. Year Book Publ., Chicago.

WOLD, H. (1948) Random normal derivates. *Tracts for Computers*, No. XXV. E. S. Pearson (ed.). University Press, Cambridge.

YUAN PAE-TSI (1933) On the logarithmic frequency distribution and the semi-logarithmic correlation surface. *Ann. Math. Stat.*, **4:** 30–74.

YULE, G. U. (1925) The growth of populations and the factors which control it. *J. Roy. Stat. Soc.* **88:** 1–62.

YULE, G. U. and KENDALL, M. G. (1950) *An Introduction to the Theory of Statistics*, 14th ed. Charles Griffin, London.

ZELEN, M. (1959) Factorial experiments in life testing. *Technometrics*, **1:** 269–288.

ADDENDUM

The present contribution is based on extensive lecture notes in French referred to as Martin (1966) and which have been translated into English. The subject of distributions, both theoretical and applied, in general and in pharmacology has received great impetus and we think it useful for the reader to give specific or related references, some of which have been placed in the references: Cochran (1964), Deming (1966), Finney (1966), Gnedenko *et al.* (1969), Katz (1965), Lebedev and Fedorova (1960), Janet McArthur and Colton (1970), Molenaar (1970), Mosteller and Tukey (1968).

More specifically, for the distribution theory, we may add Patil and Joshi (1968) for the discrete distributions and the recent series of contributions by Johnson and Kotz (1969, I Discrete distributions; 1970, II and III Univariate continuous distributions) as well as by Kotz and Johnson (1969, 1970).

COCHRAN, W. G. (1959) Newer statistical methods, in: *Quantitative Methods in Human Pharmacology and Therapeutics*, pp. 119–143. Pergamon Press, London.

COCHRAN, W. G. (1963) *Sampling Techniques*. 2nd ed., 1964. Wiley, N.Y.

DEMING, W. E. (1966) *Some Theory of Sampling*. Dover, N.Y.

FINNEY, D. J. (1968) *Statistics for Mathematicians*. Oliver & Boyd, Edinburgh and London.

GNEDENKO, B. V., BELYAEV, YU. K. and SOLOVIEV, A. D. (1969) *Mathematical Methods of Reliability Theory* (transl. from Russian). Academic Press, N.Y., London.

JOHNSON, N. L. and KOTZ, S. *Distributions in Statistics*. I Discrete distributions (1969); II and III Univariate continuous distributions (1970). Houghton-Mifflin, Boston.

KATZ, L. (1965) Unified treatment of a broad class of discrete distributions, in Patil, G. P. (ed.), pp. 175–182.

KOTZ, S. and JOHNSON, N. L. (1969) Distribution theory in statistical literature. *Proceedings 37th Session of the International Statistical Institute (London)*, pp. 303–305.

KOTZ, S. and JOHNSON, N. L. (1971) Statistical distributions: A survey of the literature, trends and prospects. *Proceedings 38th Session of the International Statistical Institute (Washington)*, pp. 227–231.

LEBEDEV, A. V. and FEDOROVA, R. M. (1960) *Guide to Mathematical Tables*. Pergamon Press, Oxford.

MCARTHUR, JANET and COLTON, T. (1970) *Statistics in Endocrinology*. The M.I.T. Press, Cambridge, Mass. (U.S.A.), and London.

MOLENAAR, W. (1970) *Approximation to the Poisson, Binomial and Hypergeometric Distribution Functions*. Amsterdam, Mathematiek Centrum, Tract n° 31.

MOSTELLER, F. and TUKEY, J. W. (1968) Data analysis, including statistics. Chapter 10 in Lindzey and Aronson (eds.), *Revised Handbook of Social Psychology*. Addison-Wesley, Reading, Mass., U.S.A.

SHAPIRO, S. S. and WILK, M. B. (1972). An analysis of variance test for the exponential distribution (complete samples). *Technometrics*, **14**: 355–370.

References to General Statistical Literature

In addition to these publications, the research worker and the consumer of statistics will find it very useful to consult the following compendia on statistical literature:

KENDALL, M. G. and DOIG, A. G. (1962, 1965, 1968) *Bibliography of Statistical Literature* (3 books), which covers the period pre-1940 to 1958 according to the scheme:

Period	*References*	*Volumes*	*Year of publication*
Pre-1900	approx. 2350	pre-1940	1968
1900–1939	approx. 7600		
1940–1949	approx. 6500	1940–1949	1965
	approx. 140	suppl. in pre-1940	
1950–1958	approx. 10,000	1950–1958	1962
	approx. 1200	suppl. in pre-1940	

These three volumes are published by Oliver & Boyd, Edinburgh and London.

LANCASTER, H. O. *Bibliography of Statistical Bibliographies* (1968), published for the International Statistical Institute by Oliver & Boyd, Edinburgh and London. The author considers the subject according to three topics:

(1) Personal bibliographies; e.g.: GOSSET, WILLIAM SEALEY, 1876–1937. Article by FISHER, R. A. (1938) Student, *Ann. Eugen. (London)* **9**: 1–9.

(2) Subject bibliographies: e.g.: COCHRAN, W. G. (1957) Analysis of covariance: its nature and uses, *Biometrics*, **13**: 261–281.

(3) National bibliographies: generally scanty and not up to date. Four additional lists are given by LANCASTER (1969, 1970, 1971–1972) in the *Review of the International Statistical Institute*.

PATIL, G. G. and JOSHI, S. W. (1968) *A Dictionary and Bibliography of Discrete Distributions*. Oliver & Boyd, Edinburgh.

An international encyclopaedia is supposed to be used by research workers of different countries. Therefore we give *(inter aliis)* two references:

KENDALL, M. G. and BUCKLAND, W. R. (1957), *Dictionary of Statistical Terms*. Oliver & Boyd, Edinburgh, London.

PAENSON, I. (1970) *Systematic Glossary of the Terminology of Statistical Methods*. Pergamon Press, Oxford.

AUTHOR INDEX

SUBJECT INDEX